"This lusty tome generated by Bloom's voracious reading habit and extraordinary talent for explanation proclaims that groups of individuals—from people to vervet monkeys to bacteria—organize themselves, create novelty, alter their surroundings, and triumph to leave more offspring than loner individuals. A stunning commitment to scientific evidence, this sequel to *The Lucifer Principle* ought to purge the academic world of 'selfish genes' and the neodarwinist dogma of 'individual selection.'"

—Lynn Margulis, Distinguished University Professor,
University of Massachusetts, Amherst, and
recipient of a 1999 National Medal of Science

"The Thales of the Internet, H. Bloom thinks what he wants, writes what he thinks, and performs his synthesis with a good heart, uncompromising truth, creative brain, and mountains of evidence. From the bacterial web of Eshel Ben-Jacob to the scientific sidelining of Professor Ling, we see the daunting power of groups that interact and sacrifice their members in order to thrive and evolve."

—Dorion Sagan, author of *Biospheres* and coauthor of
Into the Cool: The New Thermodynamics of Life

"Howard Bloom's *Global Brain* is filled with scientific firsts. It is the first book to make a strong, solidly backed, and theoretically original case that we do not live the lonely lives of selfish beings driven by selfish genes, but are parts of a larger whole. It is the first to take this idea out of the realm of mysticism and into the sphere of hard-nosed, data-derived reality. And it is one of the few books which carry off such grand visions with energy, excitement, and keen insight."

—Elizabeth Loftus, former President,
American Psychological Society, and author of
Witness for the Defense and *The Myth of Repressed Memory*

"In a superbly written and totally original argument, Howard Bloom continues his one-man tradition of tackling the taboo subjects. With a marvelously erudite survey of life and society from bacteria to the Internet, he demonstrates that group selection is for real and the group mind was there from the start. What we are entering now is but the latest phase in the evolution of the global brain. This is a must read for professionals and laymen alike."

—Robin Fox, University Professor of Social Theory,
Rutgers University, and coauthor of *The Imperial Animal*

"A modern-day prophet, Bloom compels us to admit that evolution is a team sport. This is a picture of the universe in which human emotions find their basis in the survival of matter, and the atoms themselves are held together with love. I am awestruck."

—Douglas Rushkoff, author of
Media Virus, Coercion, and *Ecstasy Club*

"God, this is *great* stuff!"

—Richard Brodie, author of
Virus of the Mind: The New Science of the Meme
and original author/programmer of Microsoft Word

"Stunning! Howard Bloom has done it again. He is certainly on to something."

—Peter Corning, Director, Institute for the
Study of Complex Systems; President, International Society
for the Systems Sciences; and author of
The Synergism Hypothesis: A Theory of Progressive Evolution

"Howard Bloom's work is simply brilliant and there is nothing else like it anywhere—we've looked, as have our colleagues. *Global Brain* is powerful, provocative, and mind-blowing."

—Don Edward Beck, Ph.D., author of *Spiral Dynamics*
and Codirector, National Values Center

"Howard Bloom has a fascinating vision of the interplay of life and a compelling style which I found captivating."

—Nils Daulaire, President and CEO,
Global Health Council

"My head is still spinning from so much eloquence and content."

—Valerius Geist, President, Wildlife Heritage and
author of *Life Strategies, Human Evolution, Environmental Design*
and *Toward a Biological Theory of Health*

"Bloom paints a spirited and wide-ranging picture of the importance of information-sharing and other forms of cooperation in organisms ranging from bacteria to humans. Arguments on group versus individual selection are normally conducted in dense prose, but Bloom's overview is high, swift, and enjoyable."

—Peter J. Richerson, coauthor (with Robert Boyd) of
Principles of Human Ecology and *The Pleistocene and the
Origins of Human Culture: Built for Speed*

GLOBAL BRAIN

GLOBAL BRAIN

The Evolution of Mass Mind from the Big Bang to the 21st Century

HOWARD BLOOM

John Wiley & Sons, Inc.

NEW YORK • CHICHESTER • WEINHEIM • BRISBANE • SINGAPORE • TORONTO

Published by John Wiley & Sons, Inc.
Published simultaneously in Canada

Library of Congress Cataloging-in-Publication Data:

Bloom, Howard K.
    The global brain : the evolution of mass mind from the big bang to the 21st century/
Howard Bloom.
        p.   cm.
    Includes bibliographical references.
    ISBN 0-471-29584-1 (cloth : acid-free paper)
      1. Brain—Evolution. 2. Human evolution. I. Title.

QP376.B627 2000
576.8—dc21

                                                                99-085976

Printed in the United States of America

10 9 8 7 6 5 4 3 2 1

To
Richard Metzger
Florian Roetzer
Bradley Fisk
and
E. Barton Chapin, Jr.

There are few observers who possess a clear and comprehensive view of the revolutions of society, and who are capable of discovering the nice and secret springs of action which impel, in the same uniform direction, the bland and capricious passions of a multitude of individuals.

Edward Gibbon

DON JUAN [out of all patience]. By Heaven, this is worse than your cant about love and beauty. . . . Granted that the great Life Force has hit on the device of the clockmaker's pendulum, and uses the earth for its bob; that the history of each oscillation, which seems so novel to us the actors, is but the history of the last oscillation repeated; nay more, that in the unthinkable infinitude of time the sun throws off the earth and catches it again a thousand times as a circus rider throws up a ball, and that our age-long epochs are but the moments between the toss and the catch, . . . I, my friend am as much a part of Nature as my own finger is a part of me. If my finger is the organ by which I grasp the sword and the mandoline, my brain is the organ by which Nature strives to understand itself. . . .

THE DEVIL. What is the use of knowing?

DON JUAN. Why, to be able to choose the line of greatest advantage instead of yielding in the direction of the least resistance. Does a ship sail to its destination no better than a log drifts nowhither? The philosopher is Nature's pilot. And there you have our difference: to be in hell is to drift: to be in heaven is to steer.

George Bernard Shaw

CONTENTS

Prologue: Biology, Evolution, and the Global Brain   1

1. Creative Nets in the Precambrian Era   14

2. Networking in Paleontology's "Dark Ages"   20

3. The Embryonic Meme   29

4. From Social Synapses to Social Ganglions:
   Complex Adaptive Systems in Jurassic Days   39

5. Mammals and the Further Rise of Mind   49

6. Threading a New Tapestry   57

7. A Trip through the Perception Factory   64

8. Reality Is a Shared Hallucination   71

9. The Conformity Police   81

10. Diversity Generators: The Huddle and the Squabble—
    Group Fission   91

11. The End of the Ice Age and the Rise of Urban Fire   100

12. The Weave of Conquest and the Genes of Trade   109

13. Greece, Miletus, and Thales: The Birth
    of the Boundary Breakers   121

14. Sparta and Baboonery: The Guesswork
    of Collective Mind   129

15. The Pluralism Hypothesis: Athens' Underside   141

16. Pythagoras, Subcultures, and Psycho-Bio-Circuitry   151

17. Swiveling Eyes and Pivoting Minds:
    The Pull of Influence Attractors   164

18. Outstretch, Upgrade, and Irrationality:
    Science and the Warps of Mass Psychology   178

19. The Kidnap of Mass Mind: Fundamentalism,
    Spartanism, and the Games Subcultures Play   191

20. Interspecies Global Mind   207

21. Conclusion: The Reality of the Mass Mind's Dreams:
    Terraforming the Cosmos   217

NOTES   225

BIBLIOGRAPHY   291

ACKNOWLEDGMENTS   353

INDEX   355

PROLOGUE:

Biology, Evolution, and the Global Brain

◩ ◩ ◩

Since the infancy of the personal computer in 1983, authors and scientists have been churning out works on the subject of a coming global brain strung together by computer networking. Today the Internet, the World Wide Web, and its successors allow a neuroscientist in Strasbourg to swap ideas instantly with a philosopher of history in Siberia and an algorithm juggler in Silicon Valley. But according to the visionaries who predict a world-spanning intelligence, this is just the beginning. They tell us that a radical human transformation has begun,[1] one that will hook "the billions of minds of humanity together into a single system . . . [like] Gaia growing herself a nervous system."[2] We will soon come together, predict the techno-prophets, on a post–World Wide Web computer net that will learn our ways of thought and fetch us the knowledge we need before we know we want it, a web that will turn the human race into a single "spiritual super-being," a massive "collective conscious" that will even incorporate the brains of computer-equipped whales in distant seas.[3] The result will be "one of the greatest leaps in the evolution of our species."[4]

My twenty years of interdisciplinary work indicate that beneath these visions lurks a strange surprise. Yes, the computerized linking of individual minds is likely to bring considerable change. But a worldwide neocortex—complete with whales—is not a gift of the silicon age. It is a phase in the ongoing evolution of a networked global brain which has existed for more than 3 billion years. This planetary mind is neither uniquely human nor a product of technology. Nor is it a result of reincarnation, or an outgrowth of telepathy. It is a product of evolution and biology. Nature has been far more clever at connectionism than have we. Her mechanisms for information swapping, data processing, and collective creation are more intricate and agile than anything the finest computer theoreticians have yet foreseen.

From the beginning, we've been yanked together by the tug of sociality. Three and a half billion years ago, our earliest cellular ancestors, bacteria, evolved in colonies. Each bacterium couldn't live without the comfort of rubbing against its neighbors. If it was separated from its companions, a healthy bacterium would rapidly divide to create a new society filled with fresh compatriots.[5] Each colony of these single-celled foremothers faced warfare, disaster, the hunt for food, and windfalls of plenty as a megateam. From the beginning, we living beings have been modules of something current evolutionary theory fails to see, a collective thinking and invention machine.

This book will show how without microchips or mystic intervention we evolved as components of subgroups, overgroups, and cross-group hyperlinks. It will show how our nature as nodes in a larger net has affected our emotions, our perceptions, and our ways of bonding with or tyrannizing friends and enemies. It will show how even when we battle using ideas or weaponry we are parts of a greater mind constantly testing fresh hypotheses.

In coming pages, we'll see how our bacterial progenitors rose from the muck shortly after this planet congealed and built intricately organized settlements whose inhabitants numbered in the trillions, settlements which sent and received information globally. We'll see how our more "advanced" forebears gave up the swiftness of bacterial data-swaps in order to achieve the size and skill which came with multicellularity. We'll spy on the 1.2 billion years in which our forebears were severely retarded in their networking, and see how we finally learned to interlink our data in a manner which began to approach the high-speed give-and-take between our cousins, the champion data linkers of all time, microbes to whom we are mere cattle on whose flesh and blood they dine.

We'll glimpse the appearance of new information cabling nearly 300 million years ago among the spiny lobsters of Paleozoic seas, and see how this wiring upgrade would someday put us on the road to broadband connectivity. We'll watch as knee-jerk squabbles splintered early prehuman tribes, in the process upping the options tumbling through a networked mind. We'll see how Stone Age cities and their web of interconnects made it possible to harness the powers of strange emotions and of peculiar intellects. We'll note how ancient bakers and pickle makers plugged into the database of strange creatures to achieve things far beyond our native capabilities. We'll witness the manner in which groupthink literally shapes the tissues of a baby's brain, and how crowd power conjures up the shared hallucinations that we adults call reality.

We'll see through the battles of Spartans and Athenians how our species-wide IQ continued to be upped or lowered by the tussle between cultural points of view. We'll uncover a battle between *two* global minds, a bizarre world war which could blacken the glitter of this fresh, new century, and see why 3 billion human lives could be snuffed out if we fail to maximize our freedoms and our interconnectivity. Finally, we'll peer one hundred years ahead and summon up one way, with luck, we'll snare the energy flaring from the centers of black holes and harness it to fuel new opportunities.

Along the way, *Global Brain* will offer a new scientific theory—one I've been refining for the last twenty years—which explains the inner workings of something to which conventional evolutionary thinkers have been blind: a planet pulsing with a more-than-massive data-sharing mind.

Few scientists have spotted the clues to the prehistoric rise of global data connection. The reason for this clouded vision is a theory known as individual selection. Individual selectionism has provided powerful new ways of piercing such human and animal mysteries as love, hatred, and jealousy since it was first proposed in 1964.[6] But during the years since then the individual selectionist legacy has fallen into premature senility, going from an eye-opener to a conceptual cataract. To the blind keepers of an outworn scientific creed, the assumptions underlying the recognition of a global brain in the near future and the distant past would seem at best naive—at worst, a pseudoscience, a punishable heresy.

Yet the scientific credentials of those who predict a computerized worldwide intelligence are impeccable. Peter Russell, who wrote the first book on the coming global brain[7] in 1983, studied mathematics and theoretical physics at Cambridge, worked with Stephen Hawking, earned a degree in experimental psychology, then obtained a postgraduate degree (once again at Cambridge) in computer science. Joel de Rosnay, author of the 1986 book *Le Cerveau Planétaire (The Planetary Brain),* has been director of research applications at the Pasteur Institute, a research associate in biology and computer graphics at MIT, and was instrumental in the creation of France's Center for the Study of Systems and Advanced Technologies. Valentin Turchin, a key thinker on the future linked computers will bring, holds three degrees in theoretical physics. Gottfried Mayer-Kress, author of "The Global Brain as an Emergent Structure from the Worldwide Computing Network," holds a doctorate in theoretical physics from the University of Stuttgart and has been associated with such prestige institutions as CERN, Los Alamos National Lab, and the Santa Fe Institute. Francis Heylighen, chief organizer of the Global Brain Study Group to which all these gentlemen belong, possesses a doctorate in physics from the University of Brussels and is, among other things, associate director of Brussels's multidisciplinary Center Leo Apostel.

Why, then, would an army of equally august specialists be likely to deride the Global Brain Study Group's notion of superorganismic intelligence? Why would hard-nosed experts in a host of disciplines resist anticipating the future of such collective data processing, and, far more important, how could they have failed to see its multibillion-year-old ancestry?

The individual selectionists who dominate today's neo-Darwinism believe that humans and animals are driven by the voracity of genes. A gene sufficiently greedy to guarantee that many copies of itself make it into the next generation will rapidly expand its family tree. Genes which program for self-denial and give up what they have to help out strangers may fail to breed entirely. Their number will shrink decade after decade until the unselfish utterly fade away. Those who survive will be cynics preprogrammed by natural selection to commit an act of generosity only if their donations pay off in hordes of progeny.

Meanwhile, another school of evolutionary thought has been driven underground. It is known as group selection. Those few willing to admit to their belief in group selection argue that individuals *will* sacrifice their genetic legacy in the interests of a larger collectivity. Such a need to cooperate would have been necessary long ago to make a global brain and a planetary nervous system possible. On the other hand, if the *individual* selectionists prove correct, humans and earlier life-forms would have been unwilling to share knowledge which might have given others a competitive edge. If selfishness is the force that drives us, there are future consequences, too. The cyber-ocean of the World Wide Web and its coming technological successors could be a barracuda pit rather than a meta-intellect.

Numerous academics who allegedly shun emotional bias have turned group selectionism into a scientific crime.*[8] Meanwhile, Robert Wright, a scientific chronicler with clout, calls *individual* selectionist psychology "the new paradigm." Yet the scientific view that all behavior is ultimately based on self-interest isn't new at all—it began its climb early in the twentieth century. A primal imperative to save one's self underlay the concept of "the fight or flight" syndrome hinted at by William McDougall in 1908[9] and popularized by Walter Cannon in 1915.[10] However, as research psychologist Robert E. Thayer says, "certain aspects of the fight or flight response were never supported by scientific evidence."[11] What's more, creatures confronted with an overwhelming threat are frequently paralyzed by anxiety, resignation, and fear. In other words, instead of fighting or fleeing for their lives, real-world inhabitants often leave themselves open to the jaws of death and let themselves be hauled away as prey. David Livingstone, of "Dr. Livingstone, I presume" fame, describes the actual sensation:

> I saw the lion just in the act of springing upon me. . . . He caught my
> shoulder as he sprang, and we both came to the ground below
> together. Growling horribly close to my ear, he shook me as a terrier

---

*Robert Wright, author of the influential book *The Moral Animal,* is more gentle in his condemnation. Group selectionism, he says, is simply a seductive "temptation," luring those of us who may be weak of mind.

does a rat. The shock produced a stupor similar to that which seems to be felt by a mouse after the first shake of the cat. It caused a sort of dreaminess in which there was no sense of pain, nor feeling of terror, though [I was] quite conscious of all that was happening. It was like what patients partially under the influence of chloroform describe, who see the operation but feel not the knife.[12]

Livingstone felt no urge to either put up his dukes or run away. Yet the fight-or-flight hypothesis is gospel to this day.

Some thirty years after fight-or-flight analyses became the rage, biologist William Hamilton and others had the courage to face at least one small fly in the self-interest ointment. If individual survival is the be-all and end-all of existence, how could one account for altruism?

During the early 1960s, Hamilton focused on the selfless manner in which female worker bees* sacrifice their reproductive rights and chastely serve their queen. His triumph was a mathematical demonstration that the workers carried essentially the same genes as their regal ruler. Hence, when an individual lived out her life on behalf of her monarch, she only *appeared* to be ignoring her own needs. By pampering the colony's egg layer, each worker was coddling replicas of her seed. Altruism, asserted Hamilton, was merely a slick disguise for selfishly promoting your own genes.[13]

W. D. Hamilton's ideas and those built upon them† have contributed mightily to our understanding of evolutionary mechanisms in fields from medicine and ecology to psychology and ethology (the study of animals in the wild). But roughly twenty-five years after the Hamiltonian epiphany, examination of real-world bee colonies demonstrated that the Oxford professor's math didn't fit the facts. There were far more genetic variations in societies of unselfish insects than Hamilton's equations would allow.[14]

What's more, in 1992 Hans Kummer—a Swiss primatologist whose twenty years of work with the Hamadryas baboons of Ethiopia and Saudi Arabia had made him one of the acknowledged greats in his field—wrote a book, *In Quest of the Sacred Baboon,* summarizing his research experiences. In it he looked over not only his own evidence, but the accumulated results

---

*All worker bees are female. Male bees are famous for their casual lifestyle. They're called drones.

†Here's a typical post-Hamiltonian escape clause. Remember that according to individual selectionists an animal, plant, or human being will only give up on a good thing if the profit to his genes is greater than what he tosses away. At worst, a generous being's self-denial must benefit his relatives, the carriers of genes much like his own. This is called "kin selection." A living thing can give up a bit of its welfare on behalf of a nonrelative . . . but only if it has reason to expect that its favor will be repaid. This theoretical loophole is known as "reciprocal altruism." One way or the other, the rules for a successful gene remain the same. A creature is simply a gene's way of making yet more genes. Anything a creature donates, its genes must get back in spades.

of others who'd studied monkeys and apes. Kummer's research had demonstrated conclusively that primates do *not* ally themselves primarily with those who share their own genes. In fact, Kummer pointed out, when primates fight, they go at it hammer and tongs more often with their relatives than with anyone else.[15] So in the case of bees and baboons, individuals were *not* tossing aside their interests simply to protect near-clones of their own chromosomes. Apparently something else was going on.

Nonetheless, Hamilton's formulae for individual selection hardened into catechism. Many of those whose scientific observations have tempted them to stray have been stopped by the threat of excommunication from professional respectability, of banishment from career advancement, and of being blackballed from a vital achievement in the academic community—tenure, the guarantor of status and of job security.

In the mid-1990s a growing group of scientists risked ridicule by arguing for the simultaneous validity of group *and* individual selection. State University of New York evolutionary biologist David Sloan Wilson, who has produced papers championing group selection for over twenty-five years, is this band's acknowledged pioneer. I was the organizer of its only formal guerrilla brigade—the Group Selection Squad—which shook things up considerably in the mid-1990s. David Sloan Wilson has pointed to over four hundred studies that support the group selectionist point of view.[16] He has concentrated his attention on research indicating that humans who pool their reasoning usually make far better decisions than those lone rangers who keep their calculations to themselves.[17] I've focused my efforts elsewhere, introducing to the debate a scientific discipline whose data individual selectionists refuse to contemplate. This obdurately overlooked field is psychoneuroimmunology—the study of how friendships, lovers, marriage, rejection, isolation, personal devastation, and self-loathing can rev or wreck your body and your brain.

As we've already seen, individual selectionists insist that a creature—be he man or woman or beast—will only sacrifice his comfort if the payback to his genes is greater than what he gives.[18] Reality has sinned against this one true word of theory. As long ago as the early 1940s, researchers like Rene Spitz[19] discovered that among humans the genetic survival instinct had a dark twin of an unexpected nature. It was a physiological doppelgänger of Freud's fanciful Thanatos, the death wish. Spitz and other trackers of hard evidence documented the many ways in which isolation, loss of control, and a downward plunge in status provoke depression, listlessness, ill health, and death, then came up with separate labels ("anaclitic shock," "learned helplessness")[20] for each instance they identified. In an earlier book, *The Lucifer Principle: A Scientific Expedition into the Forces of History*, I took the liberty of introducing a blanket designation. Each investigator from Rene Spitz and Harry Harlow to Lydia Temoshok, Martin

Seligman, Hans Kummer, and Robert Sapolsky has unearthed an example of a "self-destruct mechanism"—an inner-judge built into our biology that is able to sentence us with harsh severity.

Rene Spitz showed that up to 90 percent of foundling home babies raised with excellent food, bedding, and sanitation but without love and cuddling died. Harry Harlow revealed how baby monkeys brought up without mothers and playmates sat in their cages whimpering and picking at their skin and flesh until they bled. When given the freedom to be with others, these pictures of pathos were too frightened, inept, and emotionally damaged to socialize. Robert Sapolsky discovered how wild baboons who couldn't gain status in their tribe were flooded with hormonal poisons which killed off their brain cells, made their hair fall out, invited illness to come in and stay a while, and threatened their very lives.

Humans are apparently the same. Investigations have revealed that the hospital patients who need help the most—those submerged in depression—are the least likely to receive their doctors' and nurses' tender, loving attention. Careful scrutiny indicates that the sufferers are unwittingly triggering their own rejection. Depressed patients whine, snarl, or turn their faces to the wall[21] in ways that alienate their doctors and nurses. They upset their caregivers through every means from facial expression and verbal intonation to body language.[22] An individual selectionist would explain that such self-damaging behavior must be the result of an adaptive response—one with a hidden benefit. The patient's death might boost the genetic success of close relatives by relieving them of a burden or by enriching them with insurance and bequests. In addition, by subtly doing themselves in the patients might benefit friends who could return the favor to their family someday.

However, studies show the opposite. The patients with the greatest number of relatives and friends are the *least* likely to be depressed. Instead, they tend to be the cheerful souls who even in the face of death remain charming and bring doctors and nurses flocking sympathetically to their beds. These are the folks in the best position to satisfy individual selectionists by conferring their earthly remains on the children and grandchildren who carry replicas of their genes. Yet they are the least likely to die and prematurely open their last will and testament.

Both animal and human studies demonstrate that the depressed souls who flirt unwittingly with the grim reaper are those the individual selectionists would not expect—those whose death is *least* likely to benefit kinfolk carrying genes similar to their own.[23] These patients' family ties are either frayed or nonexistent.[24] Friends? They often have none. In fact, they tend to feel there's no place in this world where they belong.[25] These unfortunates are apparently seized by something akin to the suicide mechanism called apoptosis. Apoptosis is a firecracker string of self-destruct routines preprogrammed into nearly every living cell. Its fuse is lit when the cell receives signals that it is no longer useful to the larger community.[26] Between self-crippling immune systems and self-defeating conduct, isolated individuals vastly increase their odds of death.[27] The payoff to their gene-mates is likely to be zilch.

When caught in a bind, individual selectionists frequently claim that we are witnessing an instinct which was helpful during our days in hunter-gatherer tribes[28]—an instinct which, when we were wandering the savannas of Africa, genuinely *did* enhance our genes' chances to survive. These apologists often declare that what benefited us in the era of the first stone ax has been perverted in its purpose by modern civilization.[29] For example, some evolutionary psychotherapists have proposed that the uncontrollable restlessness of attention deficit disorder came in handy on the African plains, where the more you wandered the more likely you were to come across prey, but that the urge to roam is frowned on now that we make children and teens sit unnaturally still to learn their ABC's.

This argument is intriguing, but it's unlikely to hold water in the cases we're talking about. When you snatch chimps, dogs, laboratory mice, and a wide variety of other animals away from the group they know and love,* exhaustion overwhelms them, their immune system down-shifts, and they begin to waste away.[30] Like us, these creatures dramatically increase their odds of death when they are severed from their social bonds, not when their disappearance stands to benefit other carriers of their genes. Too much time cramped up in postindustrial classrooms is unlikely to be these animals' problem—especially those who've been observed plummeting into this wretched state in the wilds of the Serengeti or of Ethiopia.

This is where the new model of the evolutionary process I introduced in *The Lucifer Principle* and will take to virgin territory in this book may come in handy. Let us suppose for a moment that *group* selectionists are correct. One notable proponent of group selection, a naturalist named Charles Darwin, argued in 1871 that groups do battle, and in the face of such competitions, "a selfish and contentious people will not cohere, and without coherence nothing can be effected. A tribe rich in the above qualities ['reasoning . . . and foresight . . . the habit of aiding his fellows . . .

---

*I've taken a bit of a liberty here by using the word "love" in relationship to animals. In science, the hypothesis that animals can feel emotions in the same way humans do is widely considered unprovable, anthropocentric, and hence unacceptable. However, the concept that, in fact, animals do have feelings similar to ours is gaining acceptance. For eighty-nine scholarly references making the case for animal feelings of emotional attachment—specifically, love—see: Jeffrey Moussaieff and Susan McCarthy. *When Elephants Weep: The Emotional Lives of Animals.* New York: Delacorte, 1995: 64–90, 247–250.

the habit of performing benevolent actions . . . social virtues . . . and . . . social instincts'] would spread and be victorious over other tribes. . . . Thus the social and moral qualities would tend slowly to advance and be diffused throughout the world."[31] In other words, individuals *will* sacrifice themselves for the good of a larger whole. When groups struggle, the ones which boast the most effective organization, strategy, and weapons win. Individuals who contribute to their group's virtuosity will be part of the team which survives. Those too busy serving themselves to lend a hand in defense of their band are likely to have more than just those hands lopped off when their homes are plundered by invading foes.[32]

The bridge between group and individual selectionists may be hidden in another concept—that of the complex adaptive system. A complex adaptive system is a learning machine, one made up of semi-independent modules which work together to solve a problem. Some complex adaptive systems, like rain forests, are biological. Others, like human economies, are social. And the ones computer scientists work with are usually electronic. Neural networks and immune systems are particularly good examples. Both apply an algorithm—a working rule—best expressed by Jesus of Nazareth: "To he who hath it shall be given; from he who hath not even what he hath shall be taken away."

High-tech neural nets are hordes of individual electronic switch points wired in a complex mesh. The network linking the switches together has an unusual property. It can beef up or turn down the number of connections and amount of energy channeled to any switch points in the grid. An immune system is a team of free agents on a far, far grander scale. It contains between 10 million and 10 billion different antibody types. In addition it possesses a flood of entities known as "individual virus-specific T cells." Both the immune system and the neural net follow the biblical precept. Agents which contribute successfully to the solution of a problem are snowed with resources and influence. But woe be unto those unable to assist the group. In the immune system, T cells on patrol encounter the molecular signs of an invader. Each T cell is armed with a different arrangement of receptors—molecular grappling hooks. A few of the T cells discover that their weaponry allows them to snag and disable the attackers. These champions are allowed to reproduce with explosive speed, and are given the raw material they need to increase their clones dramatically.[33] T cells whose receptors can't get a grip on the invaders are robbed of food, of the ability to procreate, and often of life itself. Each is subject to destruction from within via apoptosis's preprogrammed cellular suicide.[34]

In the computerized neural net, nodes whose guesswork contributes to the solution of a problem are rewarded with more electrical energy and with connections to a far-flung skein of switch points they can draw to join their cause. The nodes whose efforts prove useless are fed less electrical juice, and their ability to recruit connections from others is dramatically reduced. In fact, their failure drives connections away.

Both T cells and neural network nodes compete for the right to commandeer the resources of the system in which they abide. And both show a seeming "willingness" to live by the rules which dictate self-denial. This combination of competition and selflessness turns an agglomeration of electronic or biological components into a learning machine with a quandary-solving power vastly beyond that of any individual module it contains.

The same modus operandi is built into the biological fabric of most social beings. Look, for example, at evidence from the phenomenon which its discoverers call "learned helplessness."[35] Animals and humans who can solve a problem remain vigorous. But mice, monkeys, dogs, and people who cannot get a grip on their dilemma become victims of self-destruct built-ins. Experiments on the physiological impact of mastering a problem began in the 1950s, when Joseph Brady and his colleagues devised a cruel but clever contraption.[36] They placed two small chairs side by side. The chairs were wired to an electrical circuit which slammed simultaneous shocks of identical voltage to a pair of involuntary loungers. The experimental couples destined to be strapped into these hot seats were monkeys. Only one thing made the monkey on the left different from that on the right. The right-hand monkey was given a button with which he could solve the duo's joint dilemma. With it, he could turn off shocks when they arrived. Investigators assumed that the monkey with the switch would develop severe health problems. He was the "executive monkey," the one of the pair weighed down with responsibility. The beast sitting next to him was relieved of pain at the same instant, but didn't have to lift a finger to gain his swift reprieve. Indeed, early analyses seemed to demonstrate that the experimenters' assumption had been correct. The monkey with the ordeal of decision making was declared to have a far greater tendency to develop ulcers—a major scourge of human corporate executives at the time.

But later inquiry showed that the executive monkey experiments had been poorly designed.[37] Their results were declared invalid. Ten years down the road, Rockefeller University's Jay Weiss tried a wicked variation on the experiment, and demonstrated something rather different. Weiss hooked 192 rats by their tails to a live electric wire. He gave some of the rodents a control switch, but left the others to simply grin and bear it. When Thor's (or Edison's) lightning struck, the unsuspecting rodents would at first scurry and jump to find a quick way out. The luckier of the beasts would soon discover their control buttons. When the current sizzled their sterns, they would lunge for the switch and turn it off, rescuing both themselves and their switchless fellow sufferers. The rats whose frantic searches resulted in no discovery of a means of control, on the other hand, would eventually give up their struggle, lie down on the cage floor, and accept their jolts with an air of resignation. Even worse, the rats without the control levers would end up physical wrecks—scrawny, unkempt, and ulcerated—while those who could slam the current off stayed reasonably plump and fit. All this despite the fact that each and every rat received exactly the same surge of current for the same amount of time and at the same instant.

As "learned helplessness" experiments continued, it was discovered that more than mere laziness crippled the beasts who couldn't find a way to abort their punishment. Their immune systems no longer protected them from disease. If given a way to escape their situation, their perceptions were too bleary to notice something as simple as an open doorway out. Their self-destruct mechanisms had taken control. All indications were that these self-maiming reflexes were physiologically preprogrammed. Most telling was the fact that the experimental animals able to cope with the slings and arrows of a researcher's outrageous fortunes retained a vigorous immune system, a relatively keen perception of the world around them, and remained energetic—despite the periodic spurts of abuse.[38]

The rats without control switches had been sabotaged by their own bodies, poisoned by their own stress hormones. How could this internally inflicted damage aid the victims in projecting their precious genes into the next generation? Apparently no one asked.

A naturalist named V. C. Wynne-Edwards, however, had already observed the effects of these phenomena au naturel.[39] Most species in the wild are not isolated by a cage, but live as part of a herd, a flock, a colony, or a pack. Edwards studied communities of wild grouse in the Scottish moors. Here, punishments and rewards were handed out not by scientists, but by wind, rain, other grouse, and by poultry-loving prowlers of all kinds. Male grouse who mastered their surroundings and were socially adept managed to corner the best food and the largest plots of real estate. In the process, they became strong and self-confident. Those less able to forage successfully or to grab a large plot of land became droopy, dispirited, and unkempt. Weakened, they entered the seasonal competition for females, attempting to outdo their problem-mastering flockmates in tournaments of ferocity and of fancy display. Each morning they erected the combs on their heads in a feeble manner which showed their lack of confidence, fluttered in the air with as much flash as they could muster, battled for control of land, and usually lost. Their failure to find a way to dominate their natural environment led to a corresponding failure to gain control in their social milieu. By winter's end, almost all of the losing red grouse were dead . . . victims, says Wynne-Edwards, of "the after-effects of social exclusion."[40] The triumphant birds, on the other hand, were rewarded with avian harems and patches of land not only rich in food, but heavily fortified by high heather plants against passing predators.

Wynne-Edwards theorized that he was watching group selection at work. The birds whose failure had led to a physical decline, he reasoned, were unwittingly sacrificing themselves to adjust the group size to the carrying capacity—the amount of food and other necessities—in their locale. The Scot announced his conclusions in 1962. By 1964 William Hamilton's equations had taken the evolutionary community by storm. Wynne-Edwards became the poster boy for group selection and was driven from scientific respectability.[41]

What Wynne-Edwards had seen at work was a complex adaptive system devilishly similar to a neural net. Those individuals within the group capable of finding solutions to the problems of the moment were rewarded with dominance, desirable food, luxury lodging, and sexual privileges. The weak links in the group's neural net, the individuals who had not found a means of solving the puzzles thrown their way, were isolated and impoverished by the social system and disabled internally.

In other words, a flock of feathered aviators had shown all the characteristics of a collective learning machine. Later, Israeli naturalist Amotz Zahavi would postulate that bird roosts function as communal information-processing centers (more about this later when we make friends with the raven).[42] Now try a little twist of thought. If you put Zahavi's conjecture together with the observations of Wynne-Edwards, add in the evidence from "learned helplessness" experiments, and toss in the discoveries of complex adaptive systems researchers, an interesting pattern emerges.

My work since 1981 has been to put that jigsaw puzzle together. Here's how the picture looks once the various parts are snapped in place. Social animals are linked in networks of information exchange. Meanwhile, self-destruct mechanisms turn a creature on and off depending on his or her ability to get a handle on the tricks and traps of circumstance. The result is a complex adaptive system—a web of semi-independent operatives linked to form a learning machine. How effective is this collective learning mesh? As David Sloan Wilson discovered, a group usually solves problems better than the individuals within it. Pit one socially networked problem-solving web against another—a constant occurrence in nature—and the one which most successfully takes advantage of complex adaptive system rules, that which is the most powerful cooperative learning contraption, will almost always win.[43]

It is time for evolutionists to open their minds and abandon individual selectionism as a rigid creed which cannot coexist with its supposed opposite, group selection. For when one joins the two, one can see that the networked intelligence forecast by computer scientists and physicists as a product of emerging technologies has been around a very long time. In fact, it has sculpted the perverse makeup which manifests itself in our depressive lethargy, in our paralyzing anxiety, in the irritability which drives others away when we need them most, in our depressive resignation when success repeatedly eludes us, and in the failure of our health when we lose the status, goals, or people who give us our sense of meaning and even our very sense of being. Our pleasures and our miseries wire us humans as modules, nodes, components, agents, and microprocessors in the most intriguing calculator ever to take shape on this earth. It's the form of social computer which gave not only us but all the living world around us its first birth.

1

CREATIVE NETS IN THE
PRECAMBRIAN ERA

4.55 Billion B.C. to 1 Billion B.C.

Networking has been a key to evolution since this universe first flared into existence. Roughly 12 billion years ago, a submicroscopic pinpoint of false vacuum arose in the nothingness and expanded at a rate beyond human comprehension, doubling every 10^{-34} seconds.[1] As it whooshed from insignificance to enormity it cooled, allowing quarks, neutrinos, photons, electrons, then the quark triumvirates known as protons and neutrons to precipitate from its energy.

The instant of creation marked the dawn of sociality. A neutron is a particle filled with need. It is unable to sustain itself for longer than ten minutes.* To survive, it must find at least one mate, then form a family. The initial three minutes of existence were spent in cosmological courting, as protons paired off with neutrons, then rapidly attracted another couple to wed within their embrace, forming the two-proton, two-neutron quartet of a helium nucleus. Those neutrons which managed this match gained relative immortality.† Those which stayed single simply ceased to be. The rule at the heart of a learning machine was already being obeyed: "To he who hath it shall be given. From he who hath not even what he hath shall be taken away."

Protons, on the other hand, seemed able to survive alone. But even they were endowed with inanimate longing. Flitting electrons were overwhelmed by an electrical charge they needed to share. Protons found these elemental sprites irresistible, and more marriages were made.‡ From the

---

*10.3 minutes, to be more precise.[2]

†The staying power of helium atoms is so great that roughly 14 billion years later, the universe remains 25 percent helium.

‡The precise fit in charge between protons and electrons is remarkable. Nobel Prize–winning biologist George Wald points out several totally improbable facts. A proton has 1,842 times as much mass as an electron. Each is a radically different kind of entity. The proton is hodge-podged together from two quarks with positive charge and one quark with negative charge. The

mutual needs of electrons and protons came atoms. Atoms with unfinished outer shells bounced around in need of consorts, and found them in equally bereft counterparts whose extra electrons fit their empty slots.

And so it continued. A physical analogue of unrequited desire was stirred by allures ranging from the strong nuclear force to gravity. These drew molecules into dust, dust into celestial shards, and knitted together asteroids, stars, solar systems, galaxies, and even the megatraceries of multigalactic matrices. Through the connective compulsion "a terrible beauty was born."*

One of the products of this inorganic copulation was life. Gravity pulled this earth together 4.5 billion years ago. The latest findings suggest that before the new sphere's crust could even stabilize, the powers of chemical attraction yanked together the first detectable life. While massive rains of planetesimals were still smacking this sphere like a boxer pummeling the face of an opponent, self-replicating molecules paved the path for deoxyribonucleic acid (DNA). Massive minuets of DNA generated the first primitive cells—the prokaryotes—by 3.85 billion B.C. A geological wink after that—in roughly 3.5 billion B.C.[3]—the first communal "brains" were already making indelible marks upon the face of the early seas. Those marks are called stromatolites—mineral deposits ranging from a mere centimeter across to the size of a man. Stromatolites were manufactured by cooperating cellular colonies with more microorganisms per megalopolis than all the humans who have ever been.[4] These primordial communities throve in the shallows of tropical lakes and of the oceans' intertidal pools. Their citizens were cyanobacteria, organisms so internally crude that they had not yet gathered their DNA into a nucleus. But in their first eons of existence, these primitive cells had already mastered one of the primary tricks of society: the division of labor.[5] Some colony members specialized in photosynthesis, storing the energy of sunlight in ornately complex molecules of adenosine triphosphate (ATP). The sun-powered assemblers took in nutrients from their surroundings and deposited the unusable residue in potentially poisonous wastes. Their vastly different bacterial sisters, on the other hand, feasted on the toxic garbage which could have turned their photosynthetic siblings into paste.

---

electron is a single minuscule lepton. Yet the positive charge of a proton *precisely* matches the negative charge of an electron. And it does so no matter which of the zillions of existing protons and electrons you choose to introduce to each other.

What's more, the correspondence is so exact that if the two had been off by a mere two-billionths, this universe's matter would have fled apart with a force completely beyond the connective powers of gravity. "Hence," concludes Wald, "we should have no galaxies, no stars or planets, and—worst of all—no physicists."

*To paraphrase Yeats.

The mass of these interdependent beings were housed in an overarching shelter of their own construction. As cyanobacterial founders multiplied, they formed a circular settlement. The waters within which the homestead was established washed a layer of clay and soil over the encampment. Some of the bacteria sent out filaments to bind these carbonate sediments in place. Tier by tier, the colony created its infrastructure, a six- or seven-foot-long, sausage-shaped edifice as large compared to the workers who had crafted it as Australia would be to a solitary child with pail and sand shovel.

Another capacity of the colony outshone even its architecture. Evidence indicates that each bacterial megalopolis possessed a staggeringly high collective IQ. The clue is in a pattern extremely familiar to those few scientists who study the intelligence of microbial societies. The fossil remains of stromatolites spread like ripples from a common center—the key to a game plan of exploration and feeding unique to certain bacterial species. Let's call it for convenience the strategy of probe and feast.

For generations bacteria have been thought of as lone cells, each making its own way through the treacherous world of microbeasts. This is a misimpression, to say the very least. The work of the University of Chicago's James A. Shapiro and the University of Tel Aviv's Eshel Ben-Jacob has shown that bacteria are social to the nth degree.*[6] A bacterial spore lands on an area rich in food. Using the nutrients into which it has fallen, it reproduces at a dizzying pace.[7] But eventually the initial food patch which gave it its start runs dry. Stricken by famine, the individual bacteria, which by now may number in the millions, do not, like the citizens of Athens during the plague of 430 B.C., merely lay down and die. Instead these prokaryotes embark on a joint effort aimed at keeping their colony alive.

The early progeny of a colony's spores had been shaped as ace digesters—creatures built to feed on the chow around them and stay steadfastly in place. Being rooted to the spot made sense when their microbit of turf was overflowing with treats. As food ran low, sitting still and swallowing fast was no longer a winning strategy. Out of hunger or sheer restlessness,[8] the bacterial cells switched gears. Rather than reproducing stay-at-homes like themselves, they marshaled their remaining resources to generate daughters of a different kind[9]—rambunctious rovers[10] built to spread out in a search for new frontiers.[11] Members of this restless generation sported an array of external whips[12] with which they could snake their way across a hard surface or twirl through water and slime. The cohort of

---

*For example, two myxobacteria must be in body contact before either is "motivated" to go about its business. In fact, making body contact with as many other bacteria as possible is more important to an individual bacteria than sidling up to a food source.

the bold left[13] to seek its fortune, expanding outward[14] from the base established by its ancestors. When their travels finally brought them to new lands of milk and honey, they issued the signal to change ways and put down solid roots again. The successful explorers produced whipless daughters equipped to mine the riches of one spot just as their grandparents had done. These stolid homesteaders dug into the banquet discovered by parental kin. But when the homestead's food ran low, wanderlust again set in. The rooted exploiters sent forth new daughters whose whips propelled them on another expedition into the great unknown.

Each generation of explorers left few traces in the wilderness it roamed. Only when it found a land of plenty did it hunker down, forming a visible ring of crowds which sucked the riches from its virgin home. Modern bacteria still shift from pioneers to colonizers and back to pioneers again, leaving ripples of concentric circles clear as a dartboard's rings. Ripples of exactly this type appear in the ancient stromatolites—an indication that bacteria 3.5 billion years ago used the intricate social system of alternating probe and feast.

All modern bacterial colonies do not use the concentric ripple strategy to explore and to exploit. Some, like aquatic myxobacteria—ganghunters which pursue their prey—will stretch and twist[15] until they catch a victim's chemical whiff.*[16] This may explain why some stromatolites snake seven feet horizontally over shallow bottoms at the margins of the sea.

There was no "each woman for herself" in those deep, dark, early days. To the contrary. Modern research hints that primordial communities of bacteria were elaborately interwoven by communication links.[17] Their signaling devices would have been many: chemical[18] outpourings with which one group transmitted its findings to all in its vicinity;[19] fragments of genetic material drifting from one end to the other of the community. And a variety of other devices for long-distance data broadcasting.[20] These turned a colony into a collective processor[21] for sensing danger, for feeling out the environment,[22] and for undergoing—if necessary—radical adaptations to survive and prosper. The resulting learning machine was so ingenious that Eshel Ben-Jacob has called its modern bacterial counterpart a "creative web."[23]

One key to bacterial creativity was the use of attraction and repulsion cues. When famine struck, some bands of bacterial outriders blazed a long trail which led to territory as barren as that from which they'd fled. But the

---

*Other modern bacteria also writhe and snake. *Eschericia coli* will wait for a moment of weakness, then advance long lines of troops from the intestines of a human newborn to her rectum or her throat. And the canny adventurers called *Salmonella abortusovis* will extend their colonies in three weeks from the eyes to the branching river system of the lymph nodes, then establish beachheads in the liver, spleen, lungs, and brain.

failing expeditions did not suffer their fate in silence. For they were the sensory tentacles with which the larger group felt out its terra incognita.[24] As such, they had to communicate their findings. To do so, they broadcast a chemical message[25]—"avoid me." Other exploring groups heeded the warning and shunned their sisters stranded in the desert. By releasing chemical repulsers, the failed scouts had sealed their fate. They would die in the Sahara into which they'd wandered—unaided and alone.[26] But their death had served its purpose—adding survey reports to an expanding knowledge base.

Other bacterial cells encountered turbulent conditions whose menace destroyed them utterly. But they, too, managed to ship back information about their fate. The fragments of their shredded genomes sent a message of danger to the colony back home. Then there were the voyagers whose trek took them to a new promised land. These sent out a chemical bulletin of an entirely different kind.[27] Loosely translated, it meant, "Eureka, we've found it. Join us quickly and let's thrive. Ain't it grand to be alive?"[28]

If the colony's strategy of spread out and seek proved useless, the messages sent back to the center did not unleash new waves of emigrants. The incoming communiqués provided raw data for genetic research and development. Informationally linked microorganisms*[29] possessed a skill exceeding the capacities of any supercomputer from Cray Research or Fujitsu. In a crisis, bacteria did not rely on deliverance via a random process like mutation, but instead unleashed their genius as genetic engineers.[30] For bacteria were the first to use the tools which now empower biotechnology's genetic tinkerers: plasmids, vectors, phages, and transposons—nature's gene snippers, duplicators, long-distance movers, welders, and reshufflers. Overcoming disaster sometimes involved plugging in prefabricated twists of DNA and reverting to ancestral strategies. When tricks like this didn't work and the stakes were life or death, the millions—and often trillions—of bacteria in a colony used their individual genomes, says Ben-Jacob, as individual computers, meshing them together, combining their data, and forming a group intelligence capable of literally reprogramming their species' shared genetic legacy in ways previously untried and previously unknown.

The microbial global brain—gifted with long-range transport, data trading, genetic variants from which to pluck fresh secrets, and the ability to reinvent genomes—began its operations some 91 trillion bacterial generations before the birth of the Internet. Ancient bacteria, if they func-

---

*As John Holland, a pioneer in the development of complex adaptive systems studies, puts it, the "behavior of a diverse array of agents" when merged results in "aggregate capabilities" far beyond those of any individual.[30]

tioned like those today, had mastered the art of worldwide information exchange.[32] They swapped snippets of genetic material like humans trading computer programs.[33] This system of molecular gossip allowed microorganisms to telegraph an improvement from the seas of today's Australia to the shallow waters covering the Midwest of today's North America. The nature and speed of communication was probably intense. The earliest microorganisms would have used planet-sweeping currents of wind and water to carry the scraps of genetic code with which they telegraphed the news of their latest how-to tips and overall breakthroughs. At least eleven different bacterial species[34] apparently exchanged trade secrets by 3 billion B.C.

Microbes have continued to upgrade the speed and sophistication of their worldwide web to this day. Meanwhile future *multicellular* forms would come to land and sea with a plethora of new capabilities. Though networked intelligence would remain a key to the survival of these more "advanced" species, it would take over 2 billion years before a true brain of planetary scope would rise among the higher animals . . . along with the early spread of tools of stone.

2

NETWORKING IN
PALEONTOLOGY'S "DARK AGES"

3.5 Billion B.C. to 520 Million B.C.

Vitalism is not the only alternative to Darwinism. I propose a new
option, that of cooperative evolution based on the formation of
creative webs. The emergence of the new picture involves a shift
from the pure reductionistic point of view to a rational holistic one,
in which creativity is well within the realm of the Natural Sciences.

Eshel Ben-Jacob
Head of the School of Physics and
Astronomy
Tel Aviv University

In the previous chapter we caught a glimpse of the manner in which evolution equipped each bacterium to act as a component in an interconnected
brain. The result: a social colony capable of networking data, solving problems, creatively retooling genomes, and of transmitting and receiving
genetic upgrades globally. But a lot of evolution has gone over the dam
since 3.5 billion B.C. What, if anything, has happened to the global brain
since then?

The story is a strange one. Paleontological dogma has it that virtually nothing of significance occurred again until the Cambrian explosion
roughly 535 million years ago. One popular science writer, summing
up the opinion of the experts, calls this interim "three billion years of
non-events."[1] Oh, there was the occasional evolutionary burp, say the
yawning authorities. But such moments of indigestion are hardly worth
mentioning.

The clues are many that there was little to yawn about. Since new
forms of behavior and interaction leave few fossil records, and since paleontologists have been virtually blind to Proterozoic social activity, the
record *seems* barren. But evidence indicates that intimate forms of organi-

zation were undergoing long and ever more intricate tests to see which ones would be among the chosen few to flourish next.

At level after level, purposeful assemblies meshed to form processor/ responders which, in turn, became the micromodules in a grander networked web. The first bacterial cells—the prokaryotes—had been highly coordinated confederations of what, for lack of a better term, we would have to call "smart molecules." Each was dedicated to a vital function. Some talented molecules pumped sugars and amino acids, responding to needs in the locations they served. Others filled the demand for power, disassembling fuel to liberate its energy. Still others tuned the chemical balance—assembling the proteins, nucleotides, vitamins, and fatty acids cells need to stay alive. Molecular groupings within the prokaryotic cell sensed food or danger and passed the message to molecular squadrons manning the oars and rotors* which created movement, allowing their host to pounce on a promising morsel or to turn the cheek to a stalker and swiftly race away.

However, fossilized biochemicals indicate that new forms of interaction kicked through evolution's door as early as 2.7 billion years ago.[2] One of the first of the resulting creatures to leave the indentation of its entire shape in the fossil record was the macroscopic organism *Grypania,* which dates back to 2.2 billion B.C.[3] This hoop-shaped relative of cyanobacteria, the size of a wire wrapped around a penny and then let loose, was apparently one of the earliest large eukaryotes—life-forms with nuclei in the centers of their cells.[4] *Grypania* represents not only a major leap in size, but a form of life which thrived on radical breakthroughs in the biological nets wiring a creature's critical insides.[5] While a single *pro*karyote—such as a bacterium—had been a network of clever molecules, a eukaryote was the grander web which emerged when masses of molecular networks fused.

Current theory says that eukaryotic cells began as bacteria[6] with a hospitable disposition. They took in fellow bacteria as boarders[7] and put them to work in their innards as permanent residents.[8] The eubacterial visitors which made up mitochondria were given the task of generating internal energy.[9] Cyanobacteria-like solar converters were turned to chloroplasts and employed to handle interior photosynthesis. And the wiry and multitalented bacteria known as spirochetes were drafted as struts for the eukaryotic cell's internal framework, as contractile fibers for internal transport, as whirling weed whackers for external movement, and as organizers for the reproductive splitting of the eukaryote's enormous store of genes. The busily toiling houseguests made themselves so much at home that they eventually were reproduced along with each of their cellular host's daughter clones.[10] According to biologist Lynn Margulis, it was this

*The cilia and flagella.

collaborative approach which allowed life to survive the first toxic pollutant holocaust—the spread in the atmosphere of a gas lethal to earth's horde of early inhabitants. The killer gas was oxygen. But mitochondria living in the new eukaryotic cells saved the day, gulping oxygen before it could do its harm and turning the murderous vapor into food for their protectress and for the other members of her cellular commune.[11]

The fossil record hints that a billion years ago, single-celled eukaryotes used their internal teamwork for more than merely riding out catastrophe. Skeletons made up of former spirochetes equipped many of the eukaryotes called protozoans with a chassis capable of supporting a host of newfangled accessories. Some developed external shells and the ability not only to whisk through water but to crawl along thanks to spirochetic microtubules which pulled one shell segment together with another, then relaxed the pair again. Helping these protozoans rachet up their size and range of aptitudes were breakthroughs like interior tubes and bladders which collected water and spat it out before an overload could bloat the cellular insides. This bilge pump anticipated the later evolution of the kidney. Another major advance was "development"—the ability to assume a succession of physical forms each fitted to achieve a different goal. A protozoan might begin life as a high-speed surveyor propelling itself with a whiplike flagellum as it sought new territory to mine, then settled down to the slow-moving but powerful blob of an amoeba*[12]—a supreme exploiter of environmental finds. This was the equivalent of being a U-2 spy plane early in life and a harvesting machine once a field of grain had been espied.

There are tantalizing hints of innumerable as-yet-undiscovered steps in another key networking technology—the advent of a nervous system. The cyanobacterial cells of 3.5 billion years ago were already capable of transmitting data from their sensory molecules to their molecular motion makers, allowing a bacterium to scoot from trouble and zip toward opportunity. But the eukaryote—an assembly of formerly independent beings which live and die in unison—was a far larger and more intricate beast. Its equipment for internal communication included its cytoskeleton's matrix of tubular crossbars, which shifted shape instantly to fit the cell to its surroundings.[13] The cytoskeleton was such an agile interior coordinator that some audacious theorists have called it a cellular "brain."[14] Interior data traffic was also aided by "second messengers" like cyclic AMP (short for adenosine monophosphate).[15] Cyclic AMP was an undersized molecule which collected bulletins arriving at the ports of the outer membrane and

---

*Bacteria had also possessed the ability to go from swimmer to homesteader, but prokaryotes managed to pull this off not only in new variations—like the amoebic form—but on a vastly grander scale.

rushed them to their targets deep inside the cell. CAMP—cyclic AMP— readjusted the operation of membrane channels, turned on energy producers, activated enzymes,*[16] and even changed the cell's speed and direction[17]—altering the cell's mission in mid-course. Cyclic AMP's travels were notable not only for the accuracy of their routing but for the cluttered distances they covered. The average eukaryotic cell was ten times the size of a prokaryote—and some eukaryotes are many thousands of times that of their cellular predecessors. In a cell so large, rapid detection by the membrane and the equally swift reactions made possible by second messengers would prove extremely necessary.

Protozoans were endangered by fast-moving cousins, the carnivores of their world. Some single-celled eukaryotic hunters were equipped with poison launchers (toxicysts)[18] on their exterior along with the flagella and cilia needed for brisk movement. These protozoans on the prowl needed to coordinate a host of spirochetic whips and propulsive whiskers (cilia) to produce precision movement. Their potential prey, provided with similar propulsion devices, had to be equally exact in mobilizing their organs for evasion.

Additional prototypes of nervous system components were in the making. Prokaryotes like bacteria had long been able to detect chemical gradients—flows whose growing weakness or strength allowed a single bacterium to determine whether it was swimming toward or away from a chemical beacon's source—then to use its "molecular computing networks," as microbiologist James Shapiro calls them, "to make appropriate decisions and adjust its activity to coordinate with other cells in the group."[19] The chemical codes connecting protozoans used roughly a hundred signal-transmission proteins which would prove vital to the neurons and synapses of today. Single-celled eukaryotes took information-gathering a giant step further, developing specialized sentinels. One example was the eyespot of the *Euglena*.[20] Some *Euglena* used this photoreceptor in tandem with another light-detecting speck on one of the flagella near their mouths, thus evolving dual-organ phototaxis—double-eyespot light recognition—an early forerunner of stereoscopic vision.

Meanwhile, one of the first assemblies of smart molecules had been the crew of that ingenious unit called the gene.† An even grander molecular fusion had been the chromosome—a knottily twisted rope of genes

---

*For example, CAMP drives literally all of a slime mold amoeba's many developmental changes by activating the enzyme PKA. According to research reported in the leading American journal *Science,* among higher organisms, the CAMP and PKA team "are key actors in everything from embryogenesis in fruit flies to memory in mice."

†Genes are strings of the molecules we know as nucleotides.

which not only worked together,* but fused so tightly that they formed a massive mega-molecule. A prokaryotic bacterium possessed a single free-floating, ring-shaped chromosome. This made subdividing into clones a relative breeze. But there was a revolution in the offing toward which this simple cell division was the merest tease. First came a eukaryotic invention—the bundling of numerous chromosomes into separate compartments walled off in the central library of a nucleus.[21] A genome is a cell's complete collection of blueprints and instruction manuals. Eukaryotes could amass genomes a thousand times larger in size and infinitely greater in complexity than any genome which had come before.

Eukaryotes topped that trick with yet another innovation—an elaborately orchestrated breakthrough in cell-division called meiosis.[22] Genes on a chromosome come in pairs. Each of these twin genes[†] serves the same purpose, but the one on the right may itch to go about its business differently than its near double on the left, cranking out different features for the creature it helps build and regulate. One gene may code for the color brown, while its seatmate on the chromosomal train holds instructions for making blue. One may code for wrinkles, while the other holds the blueprint for smooth. As long as they stay together, only the stronger of each pair rules.

Eukaryotes took advantage of this disagreement between twins. Prokaryotes had xeroxed chromosomes in their entirety. But eukaryotes unzipped their chromosomal ribbons lengthwise, ever-so gently separating each genetic pair. This yielded two skinny juliennes,[‡] each with slightly different properties. As the mother cell split in two, she centered one of these slit-down-the-middle chromosomal filaments at the command post of her first daughter cell, and the second ever-so-slightly different chromosomal half in daughter number two. Then she left her chromosomally incomplete heirs to search for mates who could contribute their own unzipped half of a chromosomal strand and merge to form a unity. Eukaryotes thus launched a great leap forward in data mix-and-matching, one which roils and churns within us to this day. We latter-age eukaryotes call the resulting DNA cut-and-shuffle sexuality.

Sex would lead eukaryotes to even stranger novelties. Lynn Margulis contends that the tamed spirochetes laboring in the eukaryote's single cell were now overloaded with responsibility. They received the indoor chore

---

*Contrary to the implication of the phrase "the selfish gene," all genes function in teams. Even the genes of a bacterium are welded in a circular chromosome.

†Each member of a gene-pair on a chromosome is called an allele.

‡I've boiled a four-step process down to its essentials here. There's an intermediate phase of division so difficult to describe in simple terms that even its most succinct depictions are enough to make a super-IQ nonbiologist dizzy.

of guiding unzipped chromosomes through the minuet of cell division. And they kept their outdoors job of cellular propulsion. But they couldn't work inside the cell and out at the same time.[23] Leaving a cell immobilized through its "pregnancy" was a dangerous business, to say the least. The dividing eukaryote could not aggressively seek food. Nor could it avoid predatory one-celled creatures whipping through the water in search of someone to eat. The solution: to concentrate spirochetic propellers on the outside of one cell, then to generate an *attached* cell whose spirochetes could stay inside directing the dance of twining and dividing chromosomes. This, according to Margulis, would start the run-up to another massive leap in the evolution of networks: multicellularity.

One of the early stages of that leap may have been the spherical cooperative of 65,536 semi-independent eukaryotic cells called a volvox.[24] Fused gelatinously together to form a pinpoint-sized hollow ball, the colony members divided into two different forms. One group concentrated on composing the cooperative's balloon-like body. The other, located prophetically in the posterior, focused on reproduction. This division of responsibility would be critical to the development of "higher" organisms, creatures whose existence would depend on a split between the cells which build the body* and cells like sperm and egg† which are cloistered until the day they're called on to produce a new generation of young. Volvox were apparently not content to pioneer one sexual invention. They also were among the first forms to generate male and female colonies.[25]

One advantage of eukaryotic clustering was sheer size. *Prokaryotic* myxobacteria form a treelike "fruiting body"[26] when they congregate—however, it's so tiny that it has to be magnified roughly two hundred times before its details become clear. The height its stalk can offer as a takeoff point for spores is minuscule. However, *eukaryotic* amoebas can unite in a giant cell roughly a foot across. That blob, called a plasmodium, holds within it literally billions of nuclei, and is able either to undergo sexual reproduction or to take another route and become a plethora of fruiting bodies immensely loftier than the ones bacteria create. Should the plasmodium "choose" sexuality, it is able to complete a host of radically new processes which, in more advanced beings, would allow for the creation of an embryo.[27] This has led some scientists to conclude that plasmodial slime molds—as these colonies of talented eukaryotic amoebas are called—may be another missing link between single-celled animals and such multicelled beasts as you and me.[28]

But first there were more jumps to be taken in the information exchange *between* such joined-at-the-hip eukaryotes. Several forms of

*Somatic cells.
†Germ cells.

cilia-powered protozoans* produced a second generation which, unlike their unicellular parents, did not totally wall themselves off at birth. Their direct connection to each other allowed one cell to sense an obstacle or an opening and to flash the data so fast that the multitude could react almost instantly and in total coordination. This "wiring" between cells prefigured neural components. It was composed of remodeled spirochetic micro-tubules[29]—the same construction materials from which nerve cells would evolve. The odds are good, then, that in the 2 billion years now blank to us, numerous further elements of primal nervous systems evolved through trial, error, and—if the University of Tel Aviv's Eshel Ben-Jacob's suspicions are correct—purposeful invention.

These evolutionary achievements were incremental steps toward multicellularity. Colonies of single-celled organisms could be sieved apart, then, if given freedom, were (and still are) able to reconstruct their shattered community.[30] The multicellular beings which emerged at the end of the paleoproterozoic era had lost that capability. In exchange, they had gained the wherewithal to achieve far grander things. Some of the earliest remains found so far of multicellular organisms, crudely called carbon films, were probably the leaves and strands of early seaweeds—1.6-billion-year-old amalgamations of prokaryotic algae.[31] These precocious prokaryotes were, according to some paleontologists, passive multicellular sheets which could only wave in the currents or settle on rocks at the bottom of the seas. However, as Wurzburg University biologist Helmut Sauer puts it, "once multicellularity is established, all kinds of fungi, plants, and animals can evolve."[32] A scholarly way of saying that all hell was ready to break loose.

True to Dr. Sauer's words, 1.4 billion years[33] after the new eukaryotic refinements had begun, the first *really* exotic multicellular beings made their debut beneath the sun.[34] One recently discovered fossil clam dates to over 720 million B.C.[35] That clam was a terra-flop ahead of anything seen before, dwarfing interlaced protozoans in size, complexity, and internal wiring. It possessed two hinged shells operated by a pair of powerful muscles able to open them with exquisite control and clamp them shut with massive power; a tongue-like foot of muscle able to dig an emergency escape hole in the ocean bottom; a tube to penetrate above the marine floor and suck oxygen-and-food-rich water down to its hideout when the clam dove underground; and a filter system of cilia through which the clam could pump the liquid it had sucked, and with which it could then sift out protozoans and other edibles, passing them via mucous fluids to the mouth. The early mollusk even possessed a heart with three chambers. All of this was wired to a host of sensors and a nervous system controlled by

Carchesium and *Zoothamnium*.

three processing clusters (ganglia). Without exquisite synergy, these separate components would have been useless. When networked, they constituted something truly unprecedented: an immense and purposeful union of interacting parts—a nearly infinite agglomeration of networks within networks and machines within machines.

The era of the first clam's birth brought a mob of strange and as yet little understood creatures—the Ediacarans—to this earth. These wildly varied gifts of multicellularity are currently thought to have been soft-bodied creatures living near the surface of the sea or crawling through the muck and mire of its floor.[36] The many legs of some Ediacaran remains demonstrate that trillions of eukaryotic cells could at last pull off the high-speed integration needed for a skittering beast.

Networks stretch and alter to achieve new capabilities. Then they mesh as modules in yet grander webs of being. Creatures of a trillion cells achieved new peaks of networking. Yet they wove still higher ties—the ties of sociality. Trilobites—the scuttling army helmets of the ancient seas—dominated the period from 600 million to 500 million B.C. Not only did these bottom-scourers have heads, eyes, sensory antennae, and all the indications of a nervous system centralized in a brain, but their fossils tend to be found in group-grope-style remains. Some paleontologists, extrapolating backward from the behavior of such trilobitic living relatives as horseshoe crabs, suspect that the armored ancients gathered for mating orgies in which they shed their shells for nearly naked body contact. Trilobite specialist Kevin Brett cites evidence that males may have been larger than females (or vice versa), and that many trilobites were, in his words, "quite ornate"—their carapaces studded with elaborate ridges, bumps, and lines. From that and the positioning of trilobites in fossil beds, Brett proposes the sexual festivities may not have been entirely promiscuous. Modern "toads," he points out, "will mate with just about anything—so they don't necessarily recognize members of even their own species." Brett suspects that trilobites were a bit more discerning.[37]

But this was not the only Cambrian sociality. Invertebrate zoologist Kerry B. Clark theorizes that the foot-and-a-half-long, torpedo-shaped *Anomalocaris canadensis* swam in feeding herds. "The largest animals in most ecosystems are typically herding herbivores," he notes, "and I see nothing about *Anomalocaris* that precludes this."[38] However, says Dr. Clark, he's reduced to educated guesswork, since few of his fellow scientists have attempted to study Cambrian social life.

One thing seems certain: a huge step forward was also an enormous step back. As Lynn Margulis and Dorion Sagan point out in their brilliant book *Microcosmos,* multicelled organisms lost the rapid-fire external information exchange, quick-paced inventiveness, and global data sharing of bacteria. With their newly developed nervous systems and brains, multicellular creatures made awesome contributions to the elaboration of cell-to-cell communication. And with the elaborate facilities in their nuclei, they giant-stepped the powers of genetic memory. But their data was now stuck inside the body. Most of it would take a billion years to get back out again.

Ilya Prigogine, the Nobel Prize–winning pioneer of the study of self-organizing systems, has observed that a breakdown of progress is frequently an illusion. Under the shattered fragments new structures and processes ferment. And from these innovations come fresh orders whose wonders appear numberless.[39] The new organisms had vastly increased their capacities as individual information processors. If these advanced modules could be linked worldwide, the result would change the nature of the very game of life.

3

THE EMBRYONIC MEME

720 Million B.C. to 65 Million B.C.

In roughly 2.1 billion B.C. eukaryotes had lost their worldwide mind. Could they now evolve another of a very different kind?

Among bacteria, updating genetic files had been a snap. But the DNA archives of eukaryotes were sealed in the castle keep of that new cellular chamber—the nucleus. Their lengthy chains of nucleic acids were churned out by micro "factories" anchored to the nucleus's skeleton,[1] then wrapped around protein spools so tightly they were reduced to one ten-thousandth of their original length, beaded on ropelike arrays,[2] and "further compacted . . . approximately 250-fold."[3] Finally the compressed bundles were walled off in a double-membrane envelope.[4] This ruled out the freewheeling trade and test drive of genetic scraps with which bacteria still sped through their global learning acts. So while the functions individual eukaryotes could handle expanded exponentially, eukaryotes were crippled by adaptive disability.

A glimmer of hope for escape from this trap appeared in one of the dramatis personae of the previous act, the clam, which bowed into the fossil record at 720 million B.C. That bivalve possessed an information-processing device I failed to mention—memory.[5] Memory exists in many of the life-forms whose ancestors appeared over 500 million years ago.[6] Even the lowly fruit fly, descendant of an ancient Cambrian line, has a storage system which uses the same stages as do ours—short-term memory leading to mid-term memory and finally long-term memory[7]—all made possible, as with humans, only if the fly does not cram its lessons but sips them slowly, taking periods of rest so the data can digest.[8]

Researchers have recently pinpointed the pre-Jurassic macramé of genes responsible for this memory sequence in insects, shellfish, chicks, and human beings. Some of memory's coils and threads were borrowed from antiquity. Recall another actor in our previous episode—the internal cellular messenger known as cyclic AMP (adenosine monophosphate). Cyclic AMP was a holdover from bacterial days,[9] one which became essential to multicellular beings, and which continues to carry out its roles in you and me. Researchers at Cold Spring Harbor Lab are convinced that sometime before 200 million B.C., a knowledge-accumulator gene called DCREB2 harnessed cyclic AMP for a new purpose—rapid data storage.[10] In the rise of eukaryotes' global mind, this would prove a necessity.

When memory appeared, the effect was dramatic. A multicelled creature could quickly store experience in a nervous system's circuitry. This opened the way for a swift reprogrammer zoologist Richard Dawkins calls the meme*—a habit, a technique, a twist of feeling, a sense of things, which easily flips from brain to brain.[11]

Genes had been limited to transmitting the data they contained via chains of adenine, cytosine, guanosine, and thymine corkscrewed in a microscopic clump. Memes could carry their message via the swift intangibles of scent, sight, and sound.[12] The result would trigger a knowledge explosion and the evolution of data webs with a whole new style. The key to this revolution lies in the early rise of memory's child—learning—the medium in which memes thrive. We've seen how internal networks wired trillions of cells into a single organism. But learning would cable organisms *externally,* networking 20 million or more multicellular creatures into a *super*organism of impressive size,[13] one endowed with 20 million brains, trillions of scent receptors, 40 million ears, and 40 million eyes.

The clam had shown up 200 million years before the action really began. Virtually all the phyla that have crawled, walked, flown, or swum during the modern era arose roughly 520 million years ago in a blink of geologic time so brief it's called the "Cambrian explosion." Cambrian parvenus included Porifera (sponges); onychophorans (wormlike beasts with fourteen to forty-three pairs of legs found mostly today in South America, the Caribbean, and Africa); mollusks (snails, squid, octopi, oysters, and clams); echinoderms (starfish, sea urchins, sea cucumbers, and sea lilies); and crustaceans (spiders, shrimp, crabs, and insects). Meanwhile, a gang of old-timers who had inched onto the evolutionary highway a billion years ago showed signs of moving into the fast lane. They were our ancestors: chordates—early vertebrates.

Among the Cambrian crustaceans were Eurypterids,[14] prototypes of the scorpions which may have been the first life-forms to walk on land. Eurypterids took internal neural wiring a quantum leap farther than the clam. They possessed:

---

*The rhyme between "gene" and "meme" is deliberate. Dawkins views the gene as a replicating molecule which busily made hard copies of itself in the earth's primordial soup and continues cranking out copies to this day. He sees the meme as an immaterial replicator which duplicates itself in the virtual soup of minds.

- a central nervous system complete with brain
- a focal ganglionic cable similar to our spinal chord
- and an extensive lace of wiring which controlled their legs, mouth, gut, and everything in between

In addition, these proto-scorpions of the Cambrian boasted sensors to detect internal movement, to keep track of orientation in space, and the sight, touch, and smell detectors necessary to pinpoint any scourge or temptation[15] gliding in the waters around them. Some of these sensory organs were astonishingly intricate. Eurypterid eyes, according to invertebrate zoologist Kerry B. Clark, could be six inches long. The size of these enormous peepers, Clark feels, indicates that there was "one hell of a lot of neural processing going on."[16]

Once you have visual detectors and a central nervous system, you are equipped to do elaborate versions of something individual bacteria could only master in a limited way. Take, for example, a descendant of the pre-Cambrian mollusks—the octopus. Put a modern octopus in a large glass jar. Give it lots of room to move. Dangle something harmless outside the walls of its receptacle. Don't worry, it can see. Try, for example, a teddy bear. Whenever the stuffed animal appears, electrically zap the octopus. After a bunch of tries, unplug your painful shock producers, pop the bear within the octopus's viewing range, and whomp—the beast will jet itself in the opposite direction. Learning!

But can this form of prudence be networked—can it be passed from one octopus to another? Most certainly. Bring in an equally transparent container housing a second octopus. Place it next to the octopus you've trained. Now show the prepunished tentacle-bearer the stuffed toy. As it whooshes back in panic, its naive neighbor will watch. Try the experiment a few more times, just to make sure the newcomer gets the message. No, it has never been stung by shock. But yes, it has seen its fellow water denizen indicate that when a cuddly bear appears there may be trouble in the air (or water, as the case may be). Now isolate octopus number two and show it the plaything. It will follow the lead of its more experienced teacher and recoil with a speed that will astonish you. What's more, it will catch on faster by following the cues of another octopus than if it had been forced to learn about shocks and teddy bears on its own.

Congratulations. You have just uncovered one synapse of the social brain—imitative learning.[17] You have also witnessed the operation of a primordial meme. No cellular material was exchanged. Only photons connected the two creatures. Yet the neural response of one octopus showed up in the brain of the other, neat, complete, and ready to go.

Alas, we have no Cambrian trilobites or proto-scorpions on which to run this experiment. However, the number of Cambrian creatures with a central nervous column and a brain was vast. The eyes and sensors of these creatures were intricate and varied. It is a distinct possibility that Cambrian denizens may have been among the first to practice imitative learning—learning to do as others do.[18]

The imitative compulsion is one of the critical connectors with which collective brains are made. Shortly after 500 million B.C., there arose the fish—an imitator par excellence. Schooling is one of a fish's most pivotal defenses. A mob of potential seafood morsels swims together in unison, each carefully heeding the cues it gets from others in its vicinity. As long as the frontal portion of a fish's brain is intact, it will slavishly follow its neighbors.[19] The advantage: a group of relative midgets can ripple like a giant sheet, a writhing sea monster with light glinting off its scales in such a way that a predator is dazzled and has difficulty zeroing in on a single fish of prey.

How much do fish rely on imitative learning? To what extent can their neural hardware be reprogrammed by primitive memes? Regard the guppy—one of evolution's early experiments in fin-and-scale morphology. Female guppies are instinctively biased to prefer males of a deep orange hue. But this does not mean they are immune to the imitative learning triggers we call fashion cues. Isolate a guppy from the crowd, pop her into an eye-tricking apparatus conceived by biologist Lee Dugatkin, and make it look to an observer as if she prefers a male who is paler than the normal sex-arousing shade. Let her sisters take a peek. They will watch her apparent amorous attraction to pallid suitors they had previously shunned. Calibrating their behavior to that of the tastemaker, others will soon begin a piscine swoon over the formerly repulsive sallow beaus.[20] Richard Dawkins, when trying to explain the nature of a meme, gives the example of a melody which infects one mind after another until its presence is outrageous. In guppies, movement cues and preferences in skin tone are equally contagious.[21]

Once a group of animals, no matter how primitive, possesses imitative learning, it's in position to regain an old networking knack which multicellular creatures had for some time lacked. Like bacteria, creatures watching for each other's lead can pool their information to arrive at mass decisions far beyond a single mind's capacities. In a few minutes we'll see how information pooling hiked the mass IQ of bees.

Extrapolating backward once again, we can deduce that another Cambrian descendant introduced a second twiner of connective mind into the sea: the social hierarchy. Among the first crustaceans were tiny Cambrian shrimp. Their later relatives, crayfish and lobsters, emerged sometime after 300 million B.C. These decapods were early masters of the art of imitation. Some spiny lobsters engage in seasonal migration, a parade of tens of thousands per cortege in which each marcher follows the path of the one before it single file across the weedy bottom of the sea. It has been hypothesized that spiny lobsters *(Panulirus argus)* evolved this slavish march as the earth's climate cooled and rich but shallow feeding grounds were swallowed by a nearby glacier's spread.[22] Nonconformists who stayed behind could well have paid the price by ending up preserved in blocks of ice, solidly frozen and solidly dead.

Dominance hierarchies are pecking orders in which each animal knows who's on top, who is not, and who is in between. In crustaceans the pecking order showed up in its crudest form—forcing underlings to give up food to a leader who proved by size and strength that his sexual organs were likely to produce the group's best genes.

Lobsters, for example, have long lived in clusters of cavelike dugouts beneath the sea. At night, the males grow restless and roam about, tapping on the door of each neighbor, looking for a showdown. The lobster inside comes to the entrance and faces off with the intruder. The goal is to see who's larger. If the visitor can tower over his rearing host, the apartment dweller vacates his home. The larger lobster knocks around the new abode for a bit, then goes off to the next cave for a visit. If the lobster making these nighttime rounds is large enough, by evening's end he's flushed all his male neighbors from their dens. Later, he lets them return to take the rest they need. But he's proven a point. He is the one to whom all others must accede.[23]

Periodically an upstart manages to topple the old boss and position himself securely at the hierarchy's top. Such social climbers can radically reprogram the mass mind. Hormones are critical to the swiftness of this change. After a pushing match in which crustacean combatants whip their antennae and lock claws, the winner struts regally on the tips of his toes. The loser slinks subserviently backward as if bowing to his foe. The victor's confidence comes from serotonin,* the loser's dejection from octopamine. Studies of clashes in crayfish reveal that serotonin alters neuron activity so significantly that Stanford University's Russ Fernald says "the animal in some sense has a different brain."[24] This "different brain" rapidly reconfigures the beast's role in the social machinery. If he finds a bit of food and the top lobster, anxious to commandeer the crumb, starts to head his way, the loser of last night's shoving match will give the morsel up and rapidly back away.[25]

The hormones of hierarchy would continue to retool individuals far down the evolutionary road, quickening shape-shifts among communities.

---

*Yes, the same serotonin which antidepressants like Prozac and Zoloft boost in the human brain.

Later we'll see their influence on scientific battles, vendettas between nations, and the civilizations of humanity.

In 350 million B.C. another Cambrian descendant appeared—the insect. With it came fresh harbingers of the new group brain's agility. At first, says legendary entomologist Edward O. Wilson, insects were probably solitary.[26] The fossil evidence supporting this conclusion is strong but by no means definitive. Invertebrate zoologist Kerry Clark points out that "the most primitive living insects are very similar, morphologically, to the oldest fossils. They're solitary. These are things like springtails. But social behavior has arisen convergently in Hemiptera, Hymenoptera, Lepidoptera, Isoptera, and maybe a couple of other orders, so might occur earlier than noted."[27] Clark adds that even springtails are not as individualistic as they are generally portrayed. Their fossilized remains are often found in herdlike clumps. In *The Insect Societies* and his much later book *The Ants*, E. O. Wilson groups together those contemporary insects which live on their own, those which have a rough-hewn sociality, and those which build colonies of up to 3 million.[28] Using words like "undoubtedly"—a term which usually means "I'm making an educated guess"[29]—Wilson makes it clear that he and his fellow entomologists have *assumed* since at least 1923 that the loners must have evolved first.[30] Frankly, this presupposition is open to a bit of inquiry.

The notion that individualism came first runs against the very grain of cosmic history. As we've seen, grouping has been inherent in evolution since the first quarks joined to form neutrons and protons. Similarly, replicators—RNA, DNA, and genes—have always worked in teams . . . often teams so huge as to defy imagining. The bacteria of 3.5 billion years ago were creatures of the crowd. So were the trilobites and the echinoderms (protostarfish) of the Cambrian age.

If entomologists have things backward, their error has spawned a host of others central to modern evolutionary science. For E. O. Wilson is not only an expert on insects, he is the founder of a rich and fruitful discipline—sociobiology. And sociobiology has, in turn, helped lay the groundwork for the dogma of the "selfish gene."

The fossil record shows that 300 million years ago, gangs of protocockroaches were tunneling *group* homes in dead tree ferns,[31] and that as early as 220 million B.C. another extremely social insect was building hives by the hundreds in Arizona's Petrified Forest—Apoidea: the bee.[32] Thomas Seeley, perhaps the leading contemporary expert on bee behavior, has been awed for quite some time by the extent to which colonies of bees pool their meager intellects to create a massive computing device. Seeley presented a sophisticated account of this observation in a 1987 article he called "A Colony of Mind: The Beehive as Thinking Machine." Seeley's 1995 *The Wisdom of the Hive* fleshed out the details of this theme.

Like guppies, bees are slaves to the contagion of the meme. In one experiment, researchers put two dishes of an identical sugar-water solution close to a pair of hives. Then the scientists trained a few bees from hive A to visit dish A. The bees of hive A obediently followed their pre-trained scouts. Despite the high caloric content of dish B, all ignored it and drank only from the "preapproved" container, carrying drops of its contents back to their home base. The bees in the second hive were tricked by the same technique into following the leader and visiting *only* dish B.[33] There was no significant number of rebels in either hive. In a very real sense, the swarm of bees had been transformed from a chaos of individuals to a single mind. Their transmuter: imitative learning.

The group mind of a hive is capable of remarkable mental feats. I described in my book *The Lucifer Principle* an experiment in which bees were given an inadvertent group IQ test. A dish of sweetened water was placed outside the hive. The buzzing fliers soon found it and, following the leader, concentrated their collective attention on mining every glucose molecule the dish contained. The next day, the dish was moved to a location twice as far from the hive. The bees used three of those tricks which make a group brain thrive—hierarchy, information pooling, and imitation—to pinpoint the new target area. While the mass of followers clung meekly to their honeycombs, a handful of "independent thinkers" flew about at will, testing one spot, then another, for food. The division of labor soon resulted in the discovery of the sugar dish's location. Now the herd instinct which results from imitative learning took over. The sheeplike multitude followed those who had made the find and combined their efforts to exploit the food source for all it was worth.

The following day, the experimenters once again set the dish twice as far from the hive as on the previous occasion. And once again the scouts fanned out, a myriad of antennae and eyes gathering input for a collective mind. Once again the trailblazers spotted the dish and the herd of follower bees swarmed to maximize the prize.

Then came the part that astonished the researchers. Each day they doubled the distance from dish to hive. The flight path's length followed an arithmetic progression of the sort which trips up many a human taking an aptitude test. After several days the swarm no longer waited for its scouts to return with bulletins on the latest coordinates. Instead, when experimenters arrived to set down the sugar water, they found the bees had preceded them.[34] Like multiple transistors crowded on the chip of a pocket calculator, the massed bees had computed the next step in a mathematical series.[35]

There are more secrets to a swarm's collective intelligence than hierarchy, information pooling, and efficient imitation. A bee on scouting duty fans an eccentric path in search of food. If she spies a promising cache, she does not operate on impulse. She double- and triple-checks her conclusions, reflying the route several times to memorize its bearings.[36] Then she returns to the hive and uses one of the first forms of symbolic representation known in evolution—the waggle dance. Cakewalking on an upright wall of the hive's lightless interior, she performs a figure eight. Its orientation indicates the direction of her find relative to the position of the sun. The speed of her movement, the number of times she repeats it, and the fervor of her noisy waggling indicate the richness of the food source and the difficulty of the flight (half a mile in a stiff wind consumes far more energy than the same distance cruised through placid air). Her audience follows her as she wriggles her news, sniffing her body for the scent of the food, feeling her movements, alert not only to the instructions each motion imparts, but to the judgment of the target's worth implied by the performer's energy.[37]

Attention begets attention—a rule whose oddities we'll later see in many a higher being. Despite the initial messenger's caution in verifying her conclusions, the masses are not easily swayed. Other scouts make the trip, reach their own judgments, then return to waggle-dance their verdicts. The more vigorous and numerous the corroborative performances, the more persuasive is the data. Several bees usually make separate discoveries. Some of the finds are richer and easier to reach than others. The greater the payoff, the more scouts are impelled to fly out and verify the reports for themselves. The more returning believers who shake and wriggle their confirmations, the more bees allocated to work the patch. The number of converts is affected by the fact that a bee who has discovered a jackpot will jitterbug far longer than one who has encountered a mediocre flower zone. The longer the shimmy, the greater the number of indecisive foragers who'll have a chance to catch the show.

This process consumes time, but it swivels the group's focus from a flower patch nearly stripped of riches to one whose nectar and pollen are now ripe. A hive has just a few short months in which to store a honey supply. If it fails to manufacture more than fifty pounds[38] it is likely to run out of provisions before the winter's done. This means certain death—not just for the frailer bees among the bunch, but for the entire community.[39] It means the extinction of the swarm's gene lines and of its collective identity. Each incoming scout's dance has contained small errors. By pooling and averaging inputs, onlookers are able to home in on their destinations with impressive accuracy. The mass mind has once again made calculations beyond any single bee's capacity.

Hierarchy has also played a role—nonconformist leaders performed the risky task of exploration.[40] And conformist followers ensured that a crush of crowd power would swiftly mine the nutrients from even the most abundant patch of blooms.

Statistics may give a sense of how critical cooperation and hierarchy are to this collaborative task. Gathering enough nectar and pollen to make it through just one winter requires the members of an average colony to make over 4 million foraging trips and to log a distance of over 12 million miles—the length of 482 flights around the world.[41] Without a hierarchy coordinating the mass, the achievements of the honeybee would never come to pass. Each worker is capable of bearing eggs, but only queens use their chemical signal of privilege—the pheromone 9-ODA—to turn their nestmates into sterile workers who cater to the regals' needs.[42] Thus queens start the specialization of castes which provides the resources from which all will feed. It takes only fifty bees and a queen before the workers feel impelled to build the combs of a new domicile. Without a queen, it takes five thousand[43]—that's the magic of hierarchy.

When a colony runs out of resources, it splits. A huge swarm rallies around the old queen and leaves the hive in search of fresh quarters. Hanging in a balled clump from a branch,[44] the homeless pioneers execute a technique like that which allowed them to zero in on flower patches. Scouts comb the landscape for a location which will be safe from predators, will provide protection from blustery winds, and will be near fresh food. Then the surveyors deliver their conclusions. Crowds gather around the several spots in which the advocates of each site dance. Hyperenergized acrobats promoting the same destination gradually entice bees away from smaller and less enthusiastic groups of publicists. The performers who draw the biggest crowd convince the others that their site's the best. Finally the swarm heads out en masse to build its nest.[45] Nonconformity, information pooling, enthusiasm, popularity, cohesion, and hierarchy—all will prove their value in the future of mass intellect.

Ants, whose signs of sociality appear after 80 million B.C.,[46] use their networked mind for yet another purpose—warfare. So vital are the coordinating mechanisms which wire a crowd of ants into a single thinking machine that the most effective strategy for attacking a rival colony is to strike without notice and cause a panic, breaking the communicative bonds which connect the victims. But often, two ant armies meet unexpectedly. The shock scatters each phalanxed legion in a frenzied rout. Victory belongs to the group which can reconstitute its links with the greatest speed.[47]

While octopi and fish use collaborative information processing, their networks remain remarkably local. Ants, on the other hand, show signs of developing something old among bacteria but new among eukaryotes—a cosmopolitan web. The most important tools of data transmission among ants are chemical. A maverick ant, nosing about in unexplored territory, will stumble across food, eat her fill, then head slowly back toward the nest with her abdomen close to the ground. This is not postmeal lethargy. The ant is laying a liquid attractant for her sisters, who cannot resist the compulsion to follow in its wake. If they, too, find that the pickings at trail's end are good, they will return in the same manner, sprinkling the chemical traces of their jubilation behind them. Thus a widening or waning scent trail encodes data on the richness of a food source, its ease of exploitation, and its gradual depletion.[48] A team of Belgian biologists has called this odor track, which summarizes the experience of hundreds or of thousands, a form of collective memory.[49]

Equally important to the ant colony are its alarm sprays—pheromones which alert the legions to danger. Ants are able to read the chemical warning signals sent by other species, thus picking up on the fact that there's trouble in the vicinity, and turning nearby colonies into sensory extensions.[50] In turn, they act as sensors for nearby populations of "foreigners." A patchwork of rival ant cities thus forms a primitive internet.

We have now reached a point 1.9 billion years after the emergence of the first eukaryotic cells and 1.4 billion years after the first multicellular film. Those bacteria which were able to absorb internal guest workers and turn eukaryotic have churned out beasts with nervous systems. And now, with learning and new kinds of information exchange, multicellular animals have begun the rise toward a new form of global brain.

4

FROM SOCIAL SYNAPSES TO SOCIAL GANGLIONS:

Complex Adaptive Systems in Jurassic Days

> How ya gonna keep 'em down on the farm,
> after they've seen Paree?

For most of human history, the need to eke out a living from the earth kept over 90 percent of the human population in the countryside. But once a small number could produce food for multitudes, a formerly repressed desire went hog-wild—our urge to cram together. Today, more than 75 percent of Europeans and North Americans have crowded into cities. In Belgium the figure tops 95 percent. This lust for company has hit the developing world even harder. In a measly two generations, Mexico's urban congregants have leaped from 25 percent to 70 percent of the population. And Tokyo is now jammed with 27 million human beings, roughly three times the worldwide number of hominids alive at even the lushest moment of the Paleolithic age. This may be very well and good among bacteria, ants, and bees, but we are noble vertebrates. Why do *we* have this passion for gathering?

If paleontologist Robert Bakker[1] is right, the dinosaurs of 120 million years ago showed a potent social appetite, but one nowhere near the size of ours. Bakker hypothesizes that vegetarian dinosaurs like iguanodons browsed in herds, pooling the input from their nostrils, eyes, and ears, then mounting a carefully phalanxed defense against would-be feasters on their meat. Bakker also suggests that carnivorous dinosaurs hunted in teams. In one Bakker scenario, a single Utahraptor acts as a decoy, distracting the attention of an iguanodon pack. Meanwhile, her hunt-mates surround the prey and strike it from the back.[2]

But the really big-time need of vertebrates to aggregate wouldn't show its beak until well after crow-like birds first made the scene roughly

130 million years ago.[3] The fossil remains* of these early birds were already clustered in what appears to have been flocks.[4] Since then, many feathered species have been as attracted to their equivalent of the big city as are we. Some bird roosts have outdone the largest human municipalities by two to one—reaching 50 million inhabitants or more.[5] This sociable over-crowding courts extraordinary risk. The larger the flock, the bigger the territory it must cover to feed, and the greater the chances of triggering a famine. So why have avians become, like us, hypnotized by the need to join a crowd?

One of the first guesses ornithologists came up with was fuel econ-omy. In winter's freeze, they reasoned, birds could huddle, combine their warmth, and cut down on the metabolic cost of heat. But a hard look at the arithmetic put an end to this idea. A thickly populated roost quickly con-sumes everything edible in its neighborhood. From that point on the daily commute from home to a decent meal and back is likely to be over fifty miles.[6] The calories burned in travel can swallow 27 percent of a starling's intake, a figure which far outweighs the pittance saved by snuggling. Overnighting alone in a sheltered hollow—despite the need to generate extra body heat—would exact nowhere near that price.[7]

Why then, do birds congregate in avian megalopolises? There is something far more critical than energy to be gained—information. Like the octopi of our last chapter, birds rely for their perception of the world on those around them. Experimenters put a young, inexperienced black-bird and an older, wiser flier in cages side by side. The savvy elder was shown an owl, and attacked the potential killer furiously. The youngster witnessed the emergency response, but couldn't see the predator it was directed at. Sly researchers had placed a partition in his line of sight. On the younger's side of the opaque divider appeared a stuffed honeyeater, a congenial creature which does not feast on blackbird meat. The setup was designed to convey the impression that the elder's pugnacity had been roused by the harmless sweet-snacker. Later the young bird was put next to an unseasoned fledgling like itself. Both were shown the honeyeater. The newcomer was indifferent. But the bird who'd seen his elder go into a rage flew at the beehive connoisseur, assaulting it with might and main. Soon the novice picked up the message and joined in. Then it, too, was paired with a naive bird who didn't have a clue. Like his teacher before him, the bird who'd learned his lesson demonstrated the importance of mobbing honeyeaters and passed the practice on. Wrongheaded as it was, the tradition was handed down through six blackbird generations before the experimenters called it quits.[8]

---

*Groups of the ancient bird *Confuciusornis* found on the shore of a lake in northeast China's Liaoning Province.

Okay, birds have imitative learning. What's so astonishing about that? We've already shown the spread of imitation in creatures as primitive as spiny lobsters nearly 300 million years ago. And we've explained how imitative learning acted like a synapse, allowing information to leap the gap from one creature to another. But a whole new kind of information processor arises when neurons or independent beings join a mass. Huddled like roosting birds in the agglomeration called a ganglion, nerve cells can swap and compare data by the batch, piecing together the big picture by combining information scraps. In 1973, Amotz Zahavi, the eminent Israeli naturalist, posited that the roost, too, was an "information center"[9] in which birds exchanged their day's experience to arrive at a larger view.

From 1988 to 1990, Bernd Heinrich of the University of Vermont and John and Colleen Marzluff of the Sustainable Ecosystems Institute in Meridian, Idaho, tested the information center hypothesis on ravens in western Maine's pine forests. Their technique was to capture wild ravens and to keep these carrion consumers caged until all their existing knowledge about food locations was thoroughly out of date. Then the experimenters put a fresh carcass—the ultimate raven cold-cut buffet—in a previously unused place, let the birds in on its coordinates by showing them the lump of meat as the sun was going down, and set the newly enlightened ravens free. The next day, one of the twenty-six birds let in on the secret showed up leading thirty ravens from a roost over a mile away. During the next few days, two more of the experimentally isolated ravens also came back to feast on the cadaverous prey. Each had a trail of roost mates in its wake. From this and a variety of other experiments, the three researchers concluded that "raven roosts are [indeed] mobile information centres" in which the birds pass around their data on where succulent cadavers can be found.[10] Heinrich and the Marzluffs also uncovered another intriguing fact. Ravens don't hoard their information on food selfishly. They broadcast the news long-distance by performing high-altitude acrobatics, "a social soaring display" which can attract hungry and clueless fellow ravens from up to thirty miles away.[11]

Zahavi had been right. Roosts, at least among ravens, are data collection centers. What's more, they are part of extended networks, pooling data between distant strangers for the sake of all. So when we ask why creatures itch to gather by the millions—as raven relatives like rooks, jackdaws, and carrion crows do—information processing may be the vital clue.

A JOURNEY TO THE HEART OF A LEARNING MACHINE

But what keeps mobs of bacteria, insects, birds, and Jurassic kings and queens from lapsing into anarchy? What so consistently turns groups into social learning machines? Before we see the mass mind grow in power, subtlety, and depth, let's pry into its workings with a little help from theory. The principles of social learning machines which follow do *not* come from the Mecca of complex adaptive system thinking, the Santa Fe Institute in New Mexico, where most developments in this field originate. And unlike other theories on the subject, they aren't based on computer simulations. They're the result of thirty-two years of fieldwork observing the real thing—the mass mind in action. The insights of Santa Fe systems modelers like John Holland have helped me greatly in this enterprise. But the principles I'm about to enunciate come from a technique closer to that used by Darwin and by Margaret Mead—becoming a participant when major decisions were made in journalism, popular music, film, television, education, advertising, and politics. Roaming freely from Germany to England to Argentina and throughout the United States with notepad in hand, among the stars, the backstage crews, and crowds at stadiums where up to ninety thousand gathered for concerts and for other mass events. Spending thousands of hours in the private homes of celebrities, media executives, and of the public they were trying to reach. I had the privilege of an inside view during the birth of many a cultural movement, and acted as a counselor to whom individuals of all kinds poured the details of their private lives. Conference rooms, arenas, studios, dressing rooms, power restaurants, and the bedrooms and living rooms where teenage groups hang out are the Samoas and Galápagos Islands of modern mass behavior. Meanwhile, I conducted surveys, made anthropological forays into inner cities and suburbia, accumulated reports from other empirical explorers, compared them with the findings of fellow scientists, and ran vast quantities of data through numerous conceptual sieves in an effort to discover just what makes a mass mind tick.

The result indicates that a collective learning machine achieves its feats by using five elements. This quintet of essentials includes: (1) conformity enforcers; (2) diversity generators; (3) inner-judges; (4) resource shifters; and (5) intergroup tournaments.[12]

- **Conformity enforcers** stamp enough cookie-cutter similarities into the members of a group to give it an identity, to unify it when it's pelted by adversity, to make sure its members speak a common language, and to pull the crowd together in efforts sometimes so vast that no single contributor can see the larger scheme in its entirety.

   In humans, upcoming chapters will show how conformity enforcers lead, among other things, to a myriad of cruelties and to a shared worldview which both shapes the wiring of a baby's brain and literally changes the way that adults see, a collective perception which makes one group's reality another's mass insanity.

- **Diversity generators** spawn variety. Each individual represents a hypothesis in the communal mind. You can see this in one of nature's most superb learning machines, the immune system. The immune system contains between 10 million and 10 billion different antibody types. Each one is a guess, preconfigured to snag the weak points of an enemy.[13] If one antibody isn't properly shaped to lock onto an invader, another will have to sink its specialized hooks into the raider. It's vital for defensive flexibility to have numerous fallback antibodies on the scene. So the immune system maintains a horde of seemingly useless types in its population, though it keeps these idlers in a state of deprivation. When a previously unknown disease storms the body's barricades, the immune system's pool of misfits often contains a few individuals with exactly what it takes to trounce the foe. Among human beings, different personality types also embody approaches which, while they may not be necessary today, could prove vital tomorrow. As we move past the Ice Age toward modernity, we'll see how odd folks out have often supercharged our history.
- Next come **inner-judges.** Inner-judges are biological built-ins[14] which continually take our measure, rewarding us when our contribution seems to be of value[15] and punishing us when our guesswork proves unwelcome or way off the mark. If we've solved a knotty problem and can hear the cheers of bosses, friends, family, and admirers echoing in our ears, our inner-judges flood us with hormones similar to amphetamine and cocaine. These chemicals swell our chests, give us energy, and set our minds ablaze. Zest and confidence help us see new ways to achieve impossibilities. On the other hand, if we can't get a grip on our problems and no one seems to want what we're offering, the inner-judges activate our self-destruct machinery.[16] Our bodies overdose on stress hormones, kill off brain cells,[17] dull our wits,[18] sabotage our immune systems, make us ill, depress us, steal our pep,[19] and often fill our minds with an urge to curl up and disappear or die. Our inner-judges are sometimes generous, but are often far from kind. Yet inner-judges are critical to complex adaptive systems from those of single-celled creatures to those made up of human minds.
- Fourth are **resource shifters.** Resource shifters range from social systems to mass emotions, but all have one thing in common—they shunt riches, admiration, and influence to learning-machine members who cruise through challenges and give folks what they want. Meanwhile, resource shifters cast individuals who can't get a handle on what's going on into some equivalent of pennilessness and unpopularity. Jesus captured the resource shifters' algorithm—its working

rule—when he said, "To he who hath it shall be given; from he who hath not, even what he hath shall be taken away."

- Bringing up the rear are **intergroup tournaments,** gang wars ranging from the Lilliputian to the gargantuan, from friendly baseball games and corporate competition to terrorist raids and nuclear confrontation, face-offs which force each collective intelligence, each group brain, to churn out innovations for the fun of winning or for sheer survival's sake.

To understand how these five forces make the mass mind click, it may be helpful to reexamine the clockwork of group learning in an organism normally thought to have no intelligence at all: our old friend the bacterium. In the late 1980s, two scientists we've met before, University of Tel Aviv physicist Eshel Ben-Jacob and the University of Chicago's James Shapiro, were perplexed. The supposed lone rangers known as bacteria actually live in colonies which etch elaborate designs as they expand. Some colonies grow in rippling ringlets. Others snake outward in symmetric tracery.

Ben-Jacob detoured from normal physics and spent five years studying a new strain isolated from the common bacteria *Bacillus subtilis.*[20] Meanwhile, Shapiro focused on such organisms as *E. coli* and salmonella. Unlike the traditional biologists who had preceded him, Ben-Jacob applied an unconventional tool to his data: the insights he had absorbed from the mathematics of materials science. New developments in this field suggested that the elaborate patterns formed by bacterial colonies might be merely the result of inorganic processes which produce patterning in water, crystals, soil, and rocks. The Israeli physicist set out to separate the products of inanimate self-organization from those of microbial hyperactivity.

Meanwhile, another biological mystery was gumming up the works. Standard neo-Darwinism said that bacteria stumble from one innovation to another by accident. A passing gamma ray, a destructive chemical, or a bout of scorching temperature can scramble a line of genetic code, garbling its message and creating a mutation. Most mutations are harmful. But according to the evolutionary doctrine which MIT's Experimental Study Group has labeled neo-Darwinism's "Central Dogma,"[21] every great once in a while a random jumble spells out instructions for an evolutionary advance. Since 1974,[22] however, a growing body of evidence had accumulated indicating that useful bacterial mutations are *not* completely random.[23] By 1999, over 880 studies suggested that some mutations might, in fact, be genetic alterations "custom-tailored" to overcome emergencies.

Ben-Jacob's studies suggested that far more than the self-organization of inanimate matter was at work within the petri dish. In fact, his results led to a surprising conclusion, one which could produce radical advances in evolutionary theory. Ben-Jacob contended that the package of genes carried by each individual bacterium is more than a mere carrier of construction plans. He wrote in the queen of physics journals, *Physica A,* that the genome can "recognize difficulties and formulate problems," "collect and process information, both about internal state[s] and external conditions (including the state of other bacteria)," and arrive at an evaluation of current needs.[24] What's more, the genetic bundle seemed to accomplish something even computers cannot achieve. Said Ben-Jacob, "The genome makes calculations and changes itself according to the outcome."[25] It activates genes which were in limbo, deactivates others which had been going full steam, copies some, moves them to new locations, lengthens or shortens old strings, and comes up with fresh combinations of genetic code. It accomplishes a feat analogous to what William Shakespeare did when he rearranged existing words and phrases to write a play. Concluded Ben-Jacob, in bacteria's case "evolutionary progress is not a result of successful accumulation of mistakes, but is rather the outcome of designed creative processes."*[26]

Reaching this conclusion left a puzzle. Several researchers testing bacterial adaptivity have tormented colonies with problems so overwhelming that they dwarf any individual bacterium's solo computational powers. For example, experimenters have taken a community of the intestine-dwelling bacteria *Escherichia coli* away from the cuisine it normally eats and offered it only salicin—a pain-reliever squeezed from the bark of willow trees which, to the *E. coli* bacterium, is inedible as pitch. An individual bacterium can crank nourishment out of this unpalatable medication only if it undergoes a step-by-step sequence of two genetic breakthroughs, one of which entails taking a giant step backward. The odds of pulling this off through random mutation are less than 1 in 10,000,000,000,000,000,000,000—or, to put it in English, more than 10 billion trillion against one.†[27] Yet *E. coli* consistently manage it. How? The answer, Ben-Jacob hypothesized, lies in networking. A "creative net" of

---

*Says Ben-Jacob, "Random mutations do exist and also affect creativity. However I propose that the designed changes have a more crucial role in evolution."

†For those who want to check the arithmetic, here are the facts. "*Escherichia coli* K12 strain chi 342LD requires two mutations in the bgl (beta-glucosidase) operon, bglR0—bglR+ and excision of IS103 from within bglF, in order to utilize salicin. In growing cells the two mutations occur at rates of $4 \times 10(-8)$ per cell division and less than $2 \times 10(-12)$ per cell division, respectively. . . . The two mutations occur sequentially. . . . It is shown that the excision mutants are not advantageous within colonies; thus, they must result from a burst of independent excisions late in the life of the colony. Excision of IS103 occurs only on medium containing salicin, despite the fact that the excision itself confers no detectable selective advantage." Using the standard formulae for permutations and combinations, I suspect you'll find that the odds against the spontaneous evolution of salicin munching are far larger than the conservative figure given above.

bacteria, unlike a man-made machine, can invent a new instruction set with which to beat an unfamiliar challenge. Some colony members feel out the new environment, learning[28] all they can. Others "puzzle" over the genome like race-car designers tinkering with an engine whose power they are determined to increase. Yet others collect the incoming "ideas" passed along by their sisters and work together to alter the use of existing genetic parts or to turn them into something new. The "super-mind," to use Ben-Jacob's term, even sucks in lessons from other colonies and, as Ben-Jacob explains, "designs and constructs a new and more advanced genome," thus "performing a genomic leap." Such is the power of what Ben-Jacob calls a bacterial "creative web."[29]

Ben-Jacob has now analyzed thousands of colonies of microorganisms to find out if his creative web hypothesis is true, and if so what makes the collective information processor work. His results indicate that bacteria constantly keep in touch, pool data, then finally sum the product through collaborative decision making and collective ingenuity. In short, bacteria engage in many of the basic activities we associate with human beings.

Ben-Jacob's findings illustrate how a social learning machine's five elements do their thing:

1. Bacterial colonies utilize the **conformity enforcer** of the genome, which, among other things, imposes a common language so that every member of the community responds to the same chemical[30] and molecular vocabulary.

2. *Bacillus* colonies are riddled with **diversity generators.** Bacterial clones (genetically identical offspring of the same mother) can morph into a host of variations. Which form each dons depends on the chemical signals it picks up from the herd around it.[31] In the best of times, when food is overflowing, the colony clumps together for the feast. But divergent appetites and digestive abilities are vital to a gorging group's survival. The bacteria which concentrate on mining a new food source produce a poisonous by-product—bacterial excreta, the equivalent of raw human sewage. Other bacteria adopt an entirely different metabolic mode. To them the excrement is caviar. By snacking heartily on toxic waste, they prevent the colony from killing itself.[32]

   More diversity generators kick in when the colony's glut runs out. We've already seen some of them at work in 3.5-billion-year-old stromatolites. As famine approaches, individuals send out a chemical signal which makes them socially obnoxious, a "body odor" that says "spread out, flee, explore." This prods groups of roughly ten thousand cells to act as scouting parties, sallying forth in a trek which unfolds before the human eye, and creating the forms which had first caught Ben-Jacob's attention—concentric circles, thick fingers flaring from a central core, or outspread rotaries of fractal lace. Meanwhile, other cellular cohorts apparently set up posts in the wake of the outward-bound advance and channel the findings of the explorers toward the center.

3. At this stage individual pioneers are at the mercy of the third element of a complex adaptive system: **inner-judges.** Explorers which find slim pickings have an internal device, the bacterial equivalent of what British theorist Michael Waller has called a "comparator mechanism."[33] This internal biological gauge determines that an outrider has chanced across parched and dangerous terrain. Its mission, in short, has failed. After sending out repulsion signals, the unfortunate reorganizes its genome[34] in preparation for the self-dissolution of lysis—a lonely death.

   On the other hand, a discoverer which encounters a cornucopia of edibles has its inner-judges jostled in the opposite direction. Its metabolism bursts with energy and it disperses an *attractant*—a sweet smell of success.

4. Now the fourth element of the complex adaptive system enters the petri dish: **resource shifters.** Bacteria stranded in the desert are stripped of food, of popularity, and most important from the point of view of the group brain, robbed of imitators. Meanwhile, those which find an overflowing buffet eat their fill and command the attention, emulation, and protection of a gathering multitude.

   Should things prove truly grim, however, and even the most strenuous searchers confirm that food is nowhere within reach, more than just resource shifters come into play. A startling diversity generator we've already seen may rouse to meet the challenge—the mechanism James Shapiro terms the "genetic engineer."[35] As the bacteria refashion themselves in radical new ways, they demonstrate the full scale of what Ben-Jacob calls their "creativity."[36]

Thanks to the synergy of conformity enforcers, diversity generators, inner-judges, and resource shifters, the bacterial colony is capable of what Ben-Jacob calls a "vertical leap"[37] in problem solving, something even humans find difficult to achieve. It would be curious to know what further marvels Shapiro and Ben-Jacob would have observed had their experiments included the fifth of a complex adaptive system's elements:

5. **Intergroup tournaments.** When two myxobacteria groups are in the same petri dish, each grabs its own territory and mounts chemical warfare on the other, sending out poisons designed to inhibit the

other's growth.[38] In seawater, the clashes between native bacteria and intruders from human sewage can be fierce and deadly.[39] Battles between bacterial armies for footholds in the human nose are equally ferocious.[40] Different strains of E. coli, despite being related, will fight to the death for the right to snatch space in their favorite feeding ground—the gut.[41] And when colonies of B. catarrhalis try to establish a beachhead in the human throat, they meet a defender native to the territory, human-friendly C. pseudodiphtheriticum. Combat flares, and the invader is usually tossed out.[42]

Intergroup tournaments up the speed of microbial creativity. In 1982 scientists dedicated to protecting chicks from infection developed a mixture of forty-eight bacterial species which, when fed to the baby birds, garrisoned the gut and fended off salmonella's attempted offensives.[43] The salmonella microbes didn't take this challenge lying down. New strains arose to outfox the guardian bacteria . . . and the scientists who had concocted them.[44] Among the retooled challengers was a super salmonella with the power to resist not only its new antagonists, but to survive the frigid temperatures humans use to preserve foods. In the past, popping a TV dinner into the freezer made it bacteria proof. But now even filet mignon frozen solid as a brick could become the site of a salmonella orgy.

A new innovation had emerged from microbial battle, an innovation of the sort which enriches the fate of a species for eons. Similar intergroup tournaments may have driven ancient bacteria to move from sea to land, or to contrive ways of eating the rock and glass of crustal plates twenty miles[45] beneath the surface of the earth,[46] thus adding abundance to the environment, complexifying the planetary biomass, and transforming ever more of this once-barren orb into food for life.

These are some of the secrets of the nascent global brain. Robert Bakker's work implies that this quintet of principles may have been at work among velociraptors and astrodons 120 million years ago. New finds of early birds* from the same era rouse a suspicion that they, too, may have used the five principles of a complex adaptive system in their group maneuverings.[47] And we will soon see how the pentagram of the learning machine extended its embrace to human beings.

*Confuciusornis sanctus.

5

MAMMALS AND THE FURTHER
RISE OF MIND

210 Million B.C. to 4 Million B.C.

It has, I think, now been shewn that man and the higher animals,
especially the primates, have some few instincts in common. All
have the same senses, intuitions, and sensations—similar passions,
affections, and emotions, even the more complex ones, such as
jealousy, suspicion, emulation, gratitude, and magnanimity; they
practise deceit and are revengeful; they are sometimes susceptible
to ridicule, and even have a sense of humour; they feel wonder and
curiosity; they possess the same faculties of imitation, attention,
deliberation, choice, memory, imagination, the association of ideas,
and reason.

Charles Darwin
The Descent of Man

Memes—habits, new ways of doing things, and other commanding intangibles which migrate from mind to mind—are key to the next jump up in
networking. Memes come in two stripes: *implicit,* those which belong to
the animal brain; and *explicit,* those which depend on human neural add-
ons, the cranial gizmos responsible for syntactic speech. Implicit memes—
the ones transferred by spiny lobsters, birds, octopi, and squid—are
housed in a very old part of the brain indeed. Yet they dominate our lives,
handling everything from our autopilot greetings, quarrels, reconcilia-
tions, unspoken cultural quirks, frustrations, and joys, to the way we drive.
    Mammals appeared approximately 210 million years ago. Vertebrate
paleontologists have largely closed their eyes to the rise of mammalian
sociality. They have a good excuse. Early mammals are almost never found
intact, much less ganged in gregarious heaps like trilobites. Instead, a tri-
umphal dig turns up an isolated bone or two . . . perhaps a single jaw or
claw or tooth. Merely fleshing out the shape of the creature who left these

pitiful remains is almost beyond the grasp of human brains. No wonder so few experts bother with paleosociality. Only Argentina's Jose Bonaparte seems to have found the skulls and a few postcranial bones of half a dozen early mammals who huddled together in a burrow well over 100 million years ago.[1] And even Bonaparte's revelation failed to wrench his colleagues from their narrow focus on new species and their evolutionary family trees to the reconstruction of something equally important: the nature of early societies.

Way back in 1982, John F. Eisenberg stepped bravely from the paleontological pack and summed up his theories on mammalian collegiality in an article he wrote for *The Oxford Companion to Animal Behavior*.[2] Though Eisenberg has apparently abandoned the quest for the origins of social networking since then,[3] he made several important claims which have been echoed by other scientists in more recent years. Regular get-togethers of multicellular creatures must have begun, Eisenberg reasoned, at the latest with the birth of sexuality, some eight hundred thousand years before the first mammal nuzzled its way onto the scene. Said Eisenberg, sexuality forces animals of opposite gender to rendezvous. No meeting, no mating, and hence no procreating. What's more, Eisenberg argued, to guarantee that meandering creatures do not miss their tryst by a month or two, the beings must interlock their moves.

Courtship struts and tournaments set individuals to a public timer, since a battle for the favors of the ladies only works if the contestants and their audience show up at the appointed place on the same date. The abundance of flaring armored skulls and other signs of mating tournaments among dinosaurs indicate that the thunder lizards were well equipped to coordinate socially. Brontotheres, behemoths with the horns and bodies of rhinoceroses, continued this Jurassic interface. But brontotheres were not dinosaurs, they were mammals. Their armaments clearly showed that they were metronomed to a common social pace.

With mammals came another network plug-in: parent-offspring linkage in a high-speed data trade. Many creatures lay an egg and simply walk away, never to meet their youngsters or to offer them care and aid. But among mammals, contact between mothers and their brood was cemented by a calorie relay. Matriarchs shuttled the nutritional mix we know as milk from a specialized gland into their infants' mouths. But they were only able to pull this off if, as Eisenberg says, they could "exchange . . . stimuli which coordinated their activities."

Among love-'em-and-leave-'em egg layers whose broods hatched on their own, life could be short and crude. But nipple-to-mouth dining required that mothers stay alive to supply food. Longer life and drawn-out immaturity stretched the time when young and old were joined together, thus opening an adaptive opportunity. Parents could mind-suckle their young—passing to small fry their experience, the behavioral memes they'd stored in memory.

The resulting communication systems would introduce new forms of flexibility. With ancient lessons and newly minted tricks cartwheeling through the group and over generation gaps, a family or far larger horde could rework its lifestyle swiftly, carving out a paradise in many an environment which otherwise would have been forbiddingly impossible. Mammals could caper out-of-doors in the equator's nonstop summertime or hunker down in caves and furs to weather icy arctic climes.

For 55 million years, small insect-eating mammals eked out their existence in the shadows of the dinosaurs. Most resembled modern rodents, shrews, and hedgehogs. Then 65 million years ago, an asteroid over 6 miles in diameter blasted a 112-mile-wide pustule into the earth's crust north of Yucatán, locking the earth in a shell of atmospheric debris, bringing daylong darkness, icy cold, and fires that raged from sea to sea, and killing the clusters of plants and roving herds which dinosaurs had made their eatery. The last remnants of the megabeasts starved or froze among the ash and char, their adaptive capabilities overwhelmed. But for socially networked animals with larger brains, catastrophe was rife with possibilities. Mammals no longer had to hide in bodies smaller than a dino-snack and in holes and crevasses too narrow for a dino-claw attack. A cold and darkened world challenged mammals to flex their capabilities.

Whales, bats, and rodents had all appeared by 50 million years ago. Since, as mammal expert John Eisenberg says, "behavior patterns do not fossilize,"[4] we have to reconstruct mammalian ways by extrapolating back from what we see today. The indications are that each of the new mammalian types was heavily equipped with conformity enforcers, homogenizers which prod one animal to take its cues from another. Information transmission among social mammals—whether handled by scent, sound, or visual codes—tends to be swift. Rats avoid a strange food until they smell it on the breath of a denmate. Then, assured by the survival of the poison tester, they pounce on the previously suspicious morsel. This imitative skittishness saves lives.[5]

Conformity-enforced codes help the fluffier rodents known as squirrels pool information, using their tails as semaphores to signal trouble and to rally their companions. A special twitch of this fur-fringed flag may mean there's a snake and bring others running to the rescue. A team of squirrels can track and isolate a snake more effectively than one squirrel on her own.[6]

Tail wagging in wolves*[7] seems to be a recruitment signal linked to celebration—one of many conformity-inducing cues canine pack members

---

*The animals of the chase from which wolves descended first appeared in roughly 60 million B.C.

use to match their motions to a common aim. One wild dog cannot bring down an antelope, but a pack working together can. The sleek and panicked prey is defeated not just by teeth and claws, but by the operation of collective brain, the passage of second-by-second strategic signs which keep the hunters well-entrained.

The urge to follow someone else—a consummate conformity enforcer—also speeds the spread of information between primates. Among monkeys, pioneering primatologist K. R. L. Hall noted how a bit of rubbish every beast despises can suddenly go sky-high in popularity. If one animal shows an interest in the detested thing, friends are likely to fall in line and become similarly intrigued.[8] 'Tis another instance of that antique conformity enforcer: imitation. The impulse to follow the crowd also turns perception to a herd phenomenon among baboons. When one baboon howls a warning call, it inspires others nearby to repeat his yell. So a necessary bit of news ricochets rapidly throughout the baboon territory.[9]

Knowing how addicted baboons are to raw meat, primatologist Shirley Strum brought a troop she called "the Pumphouse Gang" a freshly butchered carcass. At first, the baboons were panicked and ran away from the flesh that had arrived in this strange way. Then one adventurous individual tried a bit of the offering. After his troopmates saw one of their tribe dig into the alien fare, the others moved in fast to grab their share.[10]

As among bacteria and bees, there is solid evidence that individual mammal brainpower is often less important than networking. K. R. L. Hall points out that on their own, chimps are more intelligent than baboons, even making their own tools. But baboons have been infinitely more successful than the brainy junior apes. Lone baboons may be rather dumb, but their group creativity is so great that while most of Africa's exotic creatures—including chimps—are being wiped out, baboons are spreading like cockroaches. Their secret is to find the potential bonanza in every new twist introduced by nature or by man. In the dry thorn veld baboons use cattle drinking troughs and thrive in temperature extremes that go from eighty degrees by day to freezing at night. They live along the banks of the Zambezi and in the southern woodland savannas. In fact, they're "the most widely distributed non-human primates" in Africa.[11] Why? Despite their skimpy endowment of solo smarts, baboons have something chimps lack—a vastly superior social web.[12] The average group of chimps is a mere forty or so. Baboons, on the other hand, hang out in crowds three to six times that size. Why does this heightened craving to hobnob give baboons an edge over chimps? Predators usually only manage to snatch one member of a pack. So the bigger the assembly, the less chance any single member has of being attacked.[13] But equally important is a large group's role in expanding its communal database.

Mammals are endowed with both conformity enforcers and a second of the complex adaptive system's elements—diversity generators. Baboon social learning is aided by an itchy creator of behavioral twists—curiosity.[14] Some baboons will toy with nearly anything that comes across their path. Says Hall, baboons "push over slabs of rock," yank at telegraph wires, pry their way through the doors and windows of empty huts and cars, and overturn, crack open, or "fiddle with and try out" nearly anything in sight. This restless test of oddities helps a baboon find new ways to get the most from almost any environment.[15]

Conformity and diversity work together for the betterment of the larger whole. Like bacterial and honeybee scouts, baboons spread out in small groups during the day. The foolhardily inquisitive among them comb the landscape's possibilities. Bacteria pool their discoveries via chemical communication. But baboons gather at night by the hundreds in sleeping clusters, their home base for data processing. In the morning, the troop's males confer and swap their "ideas" about the direction in which the richest food sources can be found. Primatologist Stuart Altmann says that when only two males are involved, one "may 'propose' a direction of march by moving a short distance, then stopping and checking to see whether his partner is moving in that direction. If not, another direction may be proposed."[16] In larger groups, several baboon debaters shuffle a few yards to indicate the route they advocate. The senior males take each suggestion into account, then finally arrive at a decision, give a nearly invisible go-ahead, and the entire troop heads out on the winning path.[17]

Says Jane Goodall Research Center director Dr. Christopher Boehm, "In cases of emergencies (e.g., a river that floods and blocks the most likely direction of travel) this pooling of information can lead to significant conservation of energy for the entire troop. Because such emergency decisions seem to be influenced by males who have extensive experience with the environment, and because each individual's experiences will differ, it is easy enough to imagine that different Hamadryas [baboon] troops might make different tactical decisions about direction of travel under similarly threatening circumstances—for better or worse."[18]

Conformity enforcement and diversity generation once again work hand in hand when one tribe absorbs knowledge from another band. The human inhabitants of a Kenyan army barracks were tired of having their doors ripped open and their cupboards plundered whenever a member of the local baboon troop felt like munching a snack. To solve the problem, the soldiers were prepared to go on the attack, having the overgrown monkeys shot as vermin by the local gamekeeper. Fortunately, the misbehaving primates were the members of the Pumphouse Gang baboon troop Shirley Strum had long been studying. Strum organized a rescue crew to fly her study subjects to a new location. The displaced baboons had no idea of what groceries were available in the unfamiliar locale to which a plane had delivered them, nor did they have even the faintest inkling of how to get at the region's shoots, roots, tubers, buried bulbs, and other forms of food. But they followed native troops and watched to learn their ways, discovering by observation how to harvest tough-but-nutritious plants they'd never seen before, including sansevieria and opuntia so hard to yank from the ground that Strum never figured out how the baboons managed it. In addition, young local males gravitated to the newly arrived strangers in the hope of making themselves at home and finding mates. One applicant for Pumphouse Gang admission dug for salt near a watering hole, something the new arrivals had never seen done before. When the immigrants followed the native's example, they added a vital new skill to their rapidly expanding repertoire.[19]

K. R. L. Hall has said that baboon groups provide "the essential setting for each and every act of learning by the individual . . . the group is the basic unit for . . . learning processes."[20] In short, baboons are more successful than the wily chimpanzees because their troops are better collective learning machines.

Mammals not only network information across distance, they also spread the tendrils of what they've learned into the future, thus penetrating space and time. Elephants, for example, pass behavioral memes from one generation to the next. In 1919, the citrus farmers of South Africa's Addo Park wanted to get rid of a herd of 140 elephants which had been wreaking havoc with their crops. They called in a hunter who shot the elephants painfully, one by one, while their family members watched—and probably cried*[21]—over the dying agonies. After a year, only 16 to 20 elephants still remained. But they had adapted their lives to the hunter's techniques. In a most unelephantlike manner, they had become nocturnal, hiding in the bush until night fell, and not stealing out to feed until utter darkness had shrouded their dining spots. This inventive schedule shift worked—the hunter eventually gave up on gunning down what he couldn't see. Then, in 1930, the elephants were granted permanent sanctuary. There were no more gunshots, no more attacks by murderous human beings. Yet forty-five years later, the elephants retained their reclusive, nighttime way of life. The veterans of 1919 had died off, but the group held on to patterns designed to cope with a danger that had long since passed.[22] Their strategies had leaped the boundary from generation to generation and mind to mind. Implicit memes had shaped the sensibility of

---

*The work of Cynthia Moss has demonstrated that elephants shed tears. Rather large ones, as you might expect.

the elephant community as surely as genes sculpt the rippling canyons of the brain.

Advanced brains were, in fact, one key to the elephants' multigenerational memory. The other was the bond of motherhood and matriarchy. Early elephants, like the humans who would not appear for another 35 million years, possessed a cerebral cortex of substantial size. This is less unusual than it seems. Biological anthropologist Robin Dunbar has shown that the larger the size of a social group, the larger the cortex of each member. Bats, for example, were one of the earliest mammals to evolve. Like elephants, these flying mammals live long lives. (One tagged wild bat in New England survived well over thirty-one years.) Most mothers nurture just one youngster at a time and do it for a lengthy span. A few bat species live in solitude. The cerebral cortexes of these flying hermits are relatively small. Others live in colonies of up to 20 million and have cerebral cortexes of impressive size. Vampire bats, for example, hunker down in a cluster of two hundred or so, yet each mother is able to find her own child when she returns at night from a long flight, despite the fact that her son or daughter is hidden like a lost toddler in a vast amusement park. What's more, before the mother settles down she will seek out the adult "baby-sitter" who tended the children while others were away and repay the favor by regurgitating a share of the blood she's garnered for the day. To top it off, if an unrelated neighbor has had slim pickings in her search for blood, a returning mother will disgorge some of her stomach contents to feed the needy supplicant. On a future night the bat who was aided when she was down on her luck will make her way through the bewildering mob to pay back her benefactress by offering *her* fresh food if she, too, has been starved by snarls in her foraging.[23] A hefty cerebral cortex makes it possible for these bats to pinpoint patrons and trade favors as if in a commodities exchange.[24]

Neither the cerebral cortex nor the network ties of language are unique to humankind. Robert Seyfarth and Dorothy Cheney[25] have shown that though monkeys don't gush a steady stream of sentences chock full of nouns and verbs, they do erupt with symbolic sounds which act as words. Most famous are the vervets, whose distinctive wrrs and chutters warn of killer birds circling the skies, lethal snakes slithering on the ground, and leopards stalking at eyeball height. Each vervet warning must be different, for the response that would allow escape from a cat—going high into a tree—is a surefire way to become an eagle canapé. Even more remarkable is the fact that vervets have more than a single term for each of their dangers. Every call has synonyms—different sounds with the same meanings. One more element of human uniqueness anticipated long before we walked this earth.

Through diversity generators like baboon curiosity and the elephant inventiveness which produced nocturnality, early mammals brought forth implicit "behavioral" memes, improvising tricks which could be passed from brain to brain. Those memes wafted wordlessly through a troop, using conformity enforcers to reshape the mass behavior of the group. At least two components of the complex adaptive system were at work in mammals long before the appearance of the first *Homo sapiens*. We'll soon see how the other three points of the adaptive pentagram snapped into place as well.

6

THREADING A NEW TAPESTRY

65 Million B.C. to 30,000 B.C.

Between 65 million years ago and 35,000 B.C., the collective intelligence of multicellular beasts tentatively stretched beyond local networks toward globality. A bird species salaciously called tits showed up on the scene roughly 10 million years ago. Airborne, they were high-speed spreaders of behavioral memes. It is difficult to tell what kinds of tricks these creatures passed to one another in prehistoric times. But some idea comes from an event famous among animal behaviorists—a modern case of tit-perpetrated crime. During the late 1940s, England's milk vendors replaced cardboard bottle caps with aluminum foil. A few blue tits figured out how to pierce the flimsy metal so they could sip the liquid's crown of unhomogenized cream. This innovation spread so rapidly that seemingly overnight tits the length and breadth of the British Isles were fattening their bellies via dairy robbery.[1] Conformity enforcers had apparently propagated a high-nutrition meme.

Meanwhile, a diversity generator spawned by intergroup tournaments launched new variations on antique avian ploys. Oystercatchers, whose ancestors evolved roughly 50 million years ago, lived on seacoasts and used a flat, knifelike bill to snatch the clams, oysters, and mussels which dug for cover in the soaked slosh left where waves had licked the sand. The birds then pried apart the mollusks' shells and swallowed their inhabitants.[2] Evidence suggests that during the last five hundred years, some Scottish and British oystercatcher flocks were bullied by more powerful oystercatcher mobs out of easy pickings on the beach, leaving the innocent victims without a place to feed. The refugees were forced to flee over savagely unfamiliar terrain. Following the waterline, they winged their way upriver—into lands that spelled starvation for birds adapted to the edges of the sea. But the wandering of the downtrodden opened a paradise not presaged by instinctual memory: irrigated farmlands and wetlands on the river's edge. There were no shellfish on these banks and plots of soil, but there were burrowing insects by the ton.[3] The fugitive oystercatchers used their techniques for excavating sand to dig the mud, altered their diet to inland worms and grubs, and thanks to their fabled imitative learning, added to the database of oystercatcher mind. One small step for a flock or two, one giant leap for oystercatcher kind.

The contagious ways of doing things I've called "behavioral memes" did more than open new frontiers of gluttony. They also knitted separate species into multitalented teams. Pleistocene mammals still live side by side in the Serengeti Plain of East Africa: the zebra, the Thomson's gazelle, and the wildebeest. When the dry season sucks life from their flat pastures in the southeast, these animals make a five-hundred-mile procession north to Kenyan woodlands rich in watered meadow. Zebras lead, each year improvising a new track, but always with the same destination in mind. Conformity enforcers prod a herd of a million wildebeests to follow the striped equines. Then five hundred thousand delicate gazelles, taking their memetic cues from the zebras and the wildebeests, are pushed by imitative drives to join the tail end of the queue. When the immense migration reaches its goal, the diversity generator which provides each species with a unique adaptation turns the mélange into a multilevel harvesting machine. The first arrivals, zebras, crop the roughest and tallest grasses—food too tough for wildebeests to eat. This browsing exposes tender mid-height shoots upon which wildebeests can make their feast. By the time the gazelles show up, the turf is sufficiently low to offer their favorite dish, ground-hugging greenery. The grasses then join the multispecies circle dance, repaying the pruning they've received by sending up fresh shoots and leaves.[4]

Behavioral memes opened jackpots of new land and food to oyster-catchers, tits, and African plant-grazers; and they've done the same for chimpanzees. Chimps became a separate species roughly 6 million years ago, then developed flourishing cultures—inventing memes and passing them down the generational chain.[5] The chimps of Africa's Gombe, Mahale, and Taï forests have all invented tools, but the implements of each band are subtly different and used in different ways. All three groups have learned the handiness of sticks. But only the Taï have figured out that twigs and branches can be used to pull the marrow from the hollow of a bone. Mahale chimps use their hands to nab the protein snacks we call termites and ants—stoically enduring pain as their finger food bites back. The Taï and Gombe residents have pioneered a radical advance—dipping twigs into the fortress entrance of a termite mound and harvesting bugs pain-lessly. Gombe chimps also use sticks to scratch and clean themselves. Their Taï and Mahale counterparts haven't yet invented this method of hygiene.[6] Among the three groups, only the Taï chimps[7] have come up with the ulti-mate triumph of Pan troglodyte technology—fashioning a broad stone anvil, then using a rock hammer to crack a coula nut[8]—a craft requiring up to seven years of practice before a young chimp mounts to full proficiency.[9] It's the shuttling of behavioral memes from adults to their infants which keeps such forms of knowledge warped and wefted through an animal culture's weave.

Tits, oystercatchers, zebras, wildebeests, gazelles, and chimps all profited from the fruits of networking. But it wasn't until 4 million years ago that the strangest networkers of all first raised themselves on awkward feet and looked out on the scene. Their hands were far more dexterous than those of previous clever apes. Chimps had tossed stones and sticks at tigers and others who approached with the goal of dining on chimp meat.[10] But the aim of a chimp was pathetic and the distance it could clear with a two-handed lob from between its knees was barely twenty feet. Protohumans developed an overhand pitch and an aim sufficiently sure to bring a bird down in mid-flight.[11] The new arrivals' fingers were nimbler, too. Chimps primarily use stone tools ready-made by nature. Humans took a more assertive role in stone technology.

Roughly 2.7 million years ago, *Homo habilis* began to crank out implements from stones and bones. In the beginning, this was largely a matter of finding a rock which was already sharp[12] or a cracked rib with an accidental point—techniques on a par with those of chimpanzees.[13] Our shallow-skulled ancestors (their thinking apparatus was half the size of what we carry in our skulls today) figured out how to use crude implements to dig and skin and scrape.[14] The technology spread from group to group, but only reached from Ethiopia to adjoining Kenya, a distance animals like tits would have disdained. A million years later, *Homo erectus,*[15] with a 56 percent larger brain, came up with two new breakthroughs. First, they tapped one stone with another so skillfully that the lines of stress were fractured, dislodging flattened slivers—the first real blades. Later the new hominids devised a second step: tapering both sides of the rocky core left when the flaking was done and creating that hefty multipurpose chopper, chipper, digger, and tree cutter—the hand ax. Now the slog of *Homo* culture at last began to leap. Early flaking traveled from North Africa to the continent's far south. From 1.8 million years ago to roughly 500,000 B.C., stone-flaking somersaulted the thousands of miles to Europe far in the northwest and China[16] in the unimaginably distant east. The hominid collective mind was going global, carried by what archaeologist Clive Walker calls "timewalkers"—those of our ancestors whom travel lust had seized.[17] As humans fanned out, so did other signs that theirs was an expanding global mind,[18] but not one of the mystic kind imagined by believers in the psychic and the supernatural. Over four hundred thousand years ago, fire was popular from Africa to China. Men and women had apparently carried flame to spots over ten thousand miles apart along footpaths of travel.

Chimps vary their tools and their ways of handling food according to locality. But so potent were *Homo erectus*'s conformity enforcers that for 2.4 million years African stone tool–making styles (termed Oldowan) were used from England, Hungary, Germany, and Israel to Peking. Separation did not create major changes, nor did time. The Acheulean-style stone ax was in use from 1.5 million B.C. until a mere four thousand years ago.

The colossally different surroundings into which humans moved eventually spurred diversity generators, too, prodding forth occasional feats of adaptation and flights of gadgetry.[19] Man-the-wanderer trudged through cataclysmic shifts in weather and terrain. Lakes of immense size drowned once-arid plains. New lands opened as the oceans shrank, then hills and valleys disappeared beneath returning waves again. Even the Mediterranean Sea changed and rechanged utterly. This flux provoked occasional bursts of *Homo erectus*'s wizardry.[20] In Southeast Asia many anthropologists are convinced that prehumans of 1.8 million years ago learned to jigger hunting tools not from stone and wood, but from a local plant able to take a sharp, hard point—bamboo.[21] Eight hundred thousand years ago on the Indonesian island of Flores, *Homo erectus* built the first watercraft and went to sea, crossing to the island of Mata Menge, where he focused his connoisseurship on a new treat—the elephantlike pygmy stegodon's meat.[22] In the process, the hungry ancient mariner introduced another modern distinction—overhunting a species to extinction.[23] In northern lands like Spain, China, and Hungary, men used their wits and their technology to fashion ways of coping with the vicious winter cold. One sport which warmed the body was hunting with a wooden spear—a killing apparatus crafted in Germany four hundred thousand years ago.[24] But it wasn't until between 200,000[25] and 100,000[26] B.C. that toolmaking showed signs of full-tilt, regional originality. Producing a faster pace of change required a 20 percent boost in the size of brains—which arrived with our immediate predecessor and near look-alike, archaic *Homo sapiens*.[27]

Chimpanzees and baboons had long since learned to hunt in groups. The meat of colobus monkeys is a favorite among chimps, who still deploy in squadrons to corner their elusive prey. Baboons occasionally invent team tactics to bring down a wild pig or a small gazelle. But after a few years, the crew of baboon males who bring home bacon flies the coop, each member trudging off to seek his fortune in a different troop. When the old gang disappears, the practiced teamwork with which baboons seek out a herd, size it up, pick a vulnerable member, then chase it into the arms of ambushers goes poof.[28]

By thirty-five thousand years ago, our brains allowed us humans to create achievements far beyond the animal ken. Chimps do not forget the tricks of collaborative hunting. Nor did women and men. Chimps had invented weaving (though birds had beaten them to it). Every night chimpanzees laced leafy boughs to make a sleeping nest.[29] But humans pieced together roomier accommodations in which to get some rest. With our delicately flaked stone tools, now far more sophisticated than they'd been 2 million years before, we were able to bring down and skin large prey (woolly mammoths were popular, in those glacial days), cut out a mammoth's ribs, use the colossal struts as framework for large homes, then stretch a quilt of mammoth hides over the curving bones.[30] (One pelt would not suffice to tent a building of substantial size.) Using one tool to fashion another, a skill no bird or chimp seems to have known, our ancestors made threads[31] from sinews and needles from bone. With these they stitched together enough hides to roof and side a structure like the 115-foot-long by 50-foot-wide Paleolithic domicile found in the Ukrainian area of Kostienki.[32] Our forebears could also fashion pelts into elaborate clothes, then encrust their haute couture with beadwork made from carved, drilled, and polished bits of mammoth tusk ivory and of bone.*[33]

Knitting nodes of humans was long-distance trade, which first pulled together the campsites where *Homo habilis* made deals for rare and workable stone a full 2 million years ago.[34] Eventually a crucial aid was added to the give-and-take of craft and raw material: a transmitter capable of threading whole new kinds of intricacy from one mind to another. More than the wrrs and chutters of monkeys, this was a brocade of sound that gathered tufts of meaning into tapestries. Specialists call it syntactic speech: noises linked in structured strings of verbs, adjectives, and nouns.

Some linguistic theorists propose that the germ of sentences may have appeared a full 2 million years ago. One hypothesis suggests that the rise of language coincided with toolmaking.[35] Another says it arose as a substitute for the grooming which holds together a troop of monkeys or apes.[36] However, even the most cautious expert seems to agree that full-fledged language was in place by, at the very latest, 30,000 B.C. Amidst the speculation, one bit of evidence stands out with clarity. Analysis of a 2-million-year-old skull from Koobi Fora in Africa indicates that *Homo habilis* possessed a patch of brain unknown till then in any family tree. This new cerebral curio was Broca's area[37]—an apparatus vital to fluid, nuanced speech. Language, trade, migration, and the imitative memes of men and women, birds, and pasture beasts all laid the groundwork for a burst of

---

*It's popular in anthropological circles to believe that until relatively recently men and women lived together in a state of relative equality. This, according to the remains dug up of mammoth-bone homes in the Ukraine, is not true. Some tents were palatial and others were those of the poor. The "wealthy" were buried with beadwork and carved bracelets of mammoth tusk ivory whose creation had taken thousands of work-hours—not work done, apparently, by the wearer. Others, including the probable bead makers, were buried with a paucity of goods which eloquently tells us of their poverty.

data cabling. Something new was networking beneath our lazy sun. Earth's new way of dreaming and of scheming had begun.

THE NEURAL SLINGSHOT OF THE VERBAL MEME

When Richard Dawkins first published his idea of the meme, he made it clear he was speaking of "a unit of *imitation*" which multiplied in what he called "the soup of human culture."[38] Memes were supposed to be exclusive triumphs of humanity.[39] But memes come in two different kinds—behavioral and verbal. Previous chapters have shown how behavioral memes began brain-hopping long before there were such things as human minds. Because the idea of behavioral memes is new, traditional memeticists may complain that the concept of memetic transmission between animals and of unspoken memes passed between humans can't be true. But the proof of two different sorts of "imitative units" is in the human brain, where each of the two varieties of meme follows a separate trail to a very different storage space.

Human and animal bodies pick up information from pressure gauges in the bottoms of the feet, from nerves which wrap the base of fur and body hairs, from sensors registering the vibrations of bristles in the ear, from the tips of neural fibers groping molecules in the nasal cavity's air, and from light detectors in the eye. All is funneled through the brain's emotional center—the limbic system—a leftover from reptilian and early mammalian days. There, instinct and personal memory set off elation, devastation, fear, anger, and frustration as internal signal flares. Should a batch of input spark emotional ignition, the limbic system routes the hot arrival to the storage lockers of cognition—the cooling vaults of memory. But not all storage lockers are the same. There are two radically different sorts of memory storerooms in the human brain.

The first are antique caches inherited from the animals who came before mankind. They handle visceral memories, things we can't express and yet remain after they're through: the potent feeling of a joy or agony; or our learning to perform a feat of derring-do—ice skating through a triple twirl, shooting a wave, detecting a luscious, deeply buried root, or hammering the tough shell of a nut so that it yields its fruit. These muscle-and-emotion memories are slid to the amygdala, arch sorter of such things, which slings its parcels to a safekeeping net of axons called the striatum. Extra information is packed away in the motor and sensory corridors, the cerebellum, and a widespread nervous system so out of our control that its very name—autonomic—comes from its autonomy, its stubborn independence from our sense of a conscious "me." A wide variety of animals practice wordless habit stashing. It's the core of imitative learning and of body memory.

The result is the behavioral meme, a skill or a strong inkling well beyond the realm of human thought. Yes, we know how to ride a bike. But the finest rally racer can't explain the symphony of neural cues he uses to sustain a simple thing like balance. If we focus consciously on the angle to which we must adjust each of our vertebrae while slaloming through traffic at top speed, we are likely to lose the hang and scrape our head on hard concrete.

Broca's area, the brain enhancement possessed 2 million years ago by the *Homo habilis* known as KNM-ER 1470, helped create entirely new forms of data cabinets, those which house verbal memories. Verbal memes, the kind we can convey by speech, the kind that our storytelling consciousness can spin into debates, myths, tall tales, complaints, or the instructions with which we teach, take a very different route to memory.[40] They slide back to the curved prongs of the hippocampus, which flip them forward to the cortexes of the temporal lobes, accessible to manipulators like Broca's area and to two other verbal twiddlers which emerged in early *Homo habilis*—the supramarginal and angular gyri.[41] These are some of the processors which piece together data for our inner voices and our blathering tongues. They are the brain devices from which verbal memes are wrung.

Spiny lobsters, ravens, zebra, and wildebeests all network what they hear and sense and see. But language and verbal memes have given our collective webs radical new properties. One is a structured mass hallucination, a consensual fantasy,[42] a shared delusion with which far, far down the road we would compete against a group intelligence nearly a hundred trillion generations old. For in the distant future, our survival would depend on fending off the ultimate elder, the swiftest of earth's tricksters—the global brain which first arose among bacteria 3.5 billion years ago.

7

A TRIP THROUGH
THE PERCEPTION FACTORY

This way, that way, I do not know
What to do: I am of two minds.

Sappho

Is there a split between mind and body? And if so, which is better to
have?

Woody Allen

Before we plunge into humanity's tool kit of illusions, here's a question for
you. What in the world is reality? Is it an oh-so-solid thing you can pound
with your fists and rivet with your eyes? Or is it, as postmodernists proclaim, a sheer projection of the social brain?[1] In the Vienna of the late
1920s and early 1930s there throve an internationally famous philosophical
bunch called the logical positivists. Logical positivists took a firm stance
on reality. They said that a key ingredient of knowledge was "sense-data,"
and proclaimed emphatically, in the words of the British logical positivist
J. S. L. Gilmour, that sense-data are "objective and unalterable."[2] Got that?
"Unalterable," as solid as Gibraltar's rock. Good guess, but no cigar!

In the 1980s and 1990s, scientist-philosophers from Switzerland to
Peru rebelled against the positivists and founded a movement seemingly
based on deliberate confusion. No one could agree on where the doctrine
had begun, exactly how, or even on its name. The most common title to
emerge from this movement's spurt of books and journal articles was "radical constructionism." Here's a sample of the way in which the radical constructionists portrayed reality:

> A composite unity by a network of interactions of components that
> (i) through their interactions recursively regenerate the network of
> interactions that produced them, and (ii) realize the network as a unity

in the space in which the components exist by constituting and speci-
fying the unity's boundaries as a cleavage from the background . . .[3]

Hmmmmm, some of us look confused (including me). Well, here's
another shot at it:

> [Reality is] a domain of interlocked (intercalated and mutually trig-
> gering) sequences of states, established and determined through
> ontogenic interactions between structurally plastic state-determined
> systems.[4]

That should clear things up a bit, eh wot? But if the verbiage seems a
wee bit muddled, try this:

> "There can be no such thing as an objective fact"[5] because "the
> observer . . . [is] the source of all reality."[6]

In other words, reality is in the mind of the beholder.

Radical constructionism, as you may have noticed, is often a bungle
of fashionable obscurities. But in this case the fans of plain, unvarnished
common sense don't win the day—the radical constructionists do. Reality
is even more a fabrication than some constructionists dare dream in their
philosophy.

Canadian neurologist Wilder Penfield's studies are often cited by
those who believe reality is something we can grab hold of and bite, a concrete thing we perceive without distortion. When Penfield touched the
naked brains of neurosurgery patients with electrodes in 1933, some
reported vivid, detailed memories.[7] The conclusion many drew and continue believing to this day is that the brain warehouses fine-grained tapes of
past experience. Later analysis showed this conclusion was, well, let's just
say mistaken. Many of the "memories" were confabulations. One patient,
for example, recalled the time that he'd been robbed down to the minutest
detail. There was only one problem—he'd never been robbed in his life.[8]

The reality of the external world registers rather poorly on the
human mind. One eyewitness to Abraham Lincoln's assassination swore
the killer had crawled away on his hands and knees, a second declared
the assassin had stopped to deliver a line of Latin, a third said he'd leaped
fifteen feet, and a fourth said forcefully that he'd shimmied down a flagpole. There was no flagpole at the scene.[9] Even the most highly trained
observers end up mixing fiction with their "facts." Before chromosomes
were discovered, scientists used their microscopes to examine cells in
minute detail, then drew what they had "seen." Not a single chromosome
showed up in their renderings. After chromosomes had become accepted
truths, researchers suddenly peppered their cellular portraits with the

things. Lacking the concept of the chromosome, observers could have sworn the chromosomes were not there.[10]

Other oddities bend, fold, spindle, and mutilate the seeming firmness of "reality." Turn the level of lighting up and down and the sound of a nearby buzz saw seems to rise and fall as well. If you shine a light and play a faint tone over and over again, then turn on the light in utter silence, the mind will hear a tone that isn't there.[11]

Constructionists proclaim that reality is "brought forth through the . . . operations of . . . the observer,"[12] and oh, 'tis true, 'tis true. The image that we see is the product of slicing, dicing, coding, compression, long-distance transmission, neural guesswork, and eventual resplicing. The cutting-and-pasting begins at the frontier where our senses meet the outer world—the eye, which edits and reshapes input rather than faithfully recording just the plain facts of sight. Cells in the retina scrap 75 percent of the light which pours in through the lens of the eye. They diddle mercilessly with what's left, transmogrifying the photons of which light is made into pulses of electrons and bursts of unpronounceable chemicals like prelumirhodopsin. They fiddle with the contrast, tamper with the sense of space, and report not the location of what we're watching, but where the retinal cells calculate it soon *will* be.[13] Though we don't notice it, the eye is constantly flicking back and forth. To pick shapes out of the resulting blur, retinal photoreceptors take enormous liberties. The ones which suspect they've got a handle on what needs to be perceived clamp a blackout on the photoreceptors around them, and keep them from reporting what they're witnessing. Then the domineering cells turn their favored spot of mishmash into the rough outlines of what we'll eventually "see."

Adding insult to injury, the eye crushes the information it's already fuddled, compacting the landslide of data from 125 million neurons down to a code able to squeeze through a cable—the optic nerve—a mere 1 million neurons in size. On the way to the brain, the constricted stream stops briefly in the thalamus, where it is mixed, matched, and modified with flows of input from ears, muscles, fingertips, and even sensors indicating the tilt and trajectory of the head, hands, legs, and torso.

The rearranged gumbo is sent off to the visual cortex, where it is divvied up again. Some areas of the cortex pick out horizontal lines, others work on the vertical, some sort out diagonals for processing,[14] and edge-detectors[15] carve out silhouettes. Each resulting sliver of the visual signal is tucked into a separate storage belt responsible for gleaning a different type of meaning. If you're twirling in a swivel chair, one belt will grab the smear of color whipping past your eyes and retouch it as a freeze-frame.[16] Meanwhile, neurons nooked and crannied throughout the brain will sift the tidbits for interpretation.[17] For instance, cells which signal if an oncoming human is friend or foe will add their best guess to the moving information flow.[18]

Finally, a council of representatives from the superior colliculus, the thalamus, the locus coeruleus, the hypothalamus, and the occipital cortex pool their squabble of conclusions and cast a vote on what the twinges of light impinging on the retina might be. Not until they've agreed on an image do they send it to the left cerebral hemisphere, presenting it to the conscious mind[19] as a panorama accompli. What we see is not the product of direct perception, but a reconstruction bordering on collage artistry.

The reassembly process we call vision is so powerful that back in the 1950s experimenters burdened volunteers with custom goggles whose lenses turned the scenery upside down. If you had been around back then and donned this light-inverting eye gear, at first it would have appeared outdoors as if the ground was above, the sky below, and indoors all the furniture was dangling overhead. Slowly but surely the sensory crew that scrambles, scrunches, and reworks what you see would have taken the topsy-turvy rays of light and rebuilt their image[20] until, voilà, despite distorting goggles, you'd no longer have seen the floorboards hanging above your seat but would have beheld them settled reassuringly beneath your feet.

Vision is not the only sense which uses bickering organs to slap together a pastiche. The human body is composed of a host of separate systems bouncing signals back and forth, often battling clumsily on their way toward compromise. Plato called one form of this struggle the war between spirit, reason, and the appetites. To illustrate it, he told the tale of Leontius, whose path took him past an execution site. One part of Leontius's brain was curious to view the dead bodies. Another cranial constituent was filled with "dread and abhorrence. . . . For a time he struggled and covered his eyes," says Plato, "but at length the desire got the better of him; and forcing them open, he ran up to the dead bodies, saying, Look, ye wretches, take your fill of the fair sight."[21] Researchers on split-brain patients have gained a bit of insight into the battle of brain bits and body parts Plato described.

Split-brain patients have had the connecting lines between their right and left hemispheres cut to prevent epileptic fits. All of us, split-brain patients included, are neurologically wired so the messages from our left eye go to our right brain and vice versa. Anything you show to just the left eye will cross over and be witnessed primarily by the right brain. If you've snipped the massive bridge of tissue connecting your cerebral hemispheres,* the two brain halves won't be able to compare notes. The left

_____
*The corpus callosum.

brain literally won't know what the right brain is doing. But the left brain is not about to admit its ignorance. The problem is that the right brain handles things like getting the big picture. It's the strong-and-silent type. The left brain is the side that handles speech. In one famous experiment, Michael Gazzaniga, director of the Center for Cognitive Neuroscience at Dartmouth College, showed a split-brain patient's right hemisphere a picture of snow and the left hemisphere a picture of a claw. Says Gazzaniga,

> Placed in front of the patient are a series of cards that serve as possible answers to the implicit questions of what goes with what. The correct answer for the left hemisphere is a chicken [which goes with the claw]. The answer for the right hemisphere is a shovel [appropriate for snow]. . . . A typical response is that of P.S., who pointed to the chicken with his right hand and the shovel with the left. After his response I asked him, "Paul, why did you do that?" Paul looked up and without a moment's hesitation said from his left hemisphere [the hemisphere which handles speech], "Oh, that's easy. The chicken claw goes with the chicken and you need a shovel to clean out the chicken shed."[22]

Though the left brain clearly doesn't have a clue about the right brain's preoccupation with snow, it makes up an instant answer and tosses in a nonexistent shed. Even more important, it doesn't let the person in whom all of this is happening know that it's cheating and that the "conscious self" in the left brain actually has no idea of why the right hand was pointing at the shovel so insistently.

Similar splits and bizarre excuses course through you and me. As we sit at our desks and see our coworkers walk by, an emotional federation with three main offices—one in the head, another atop the kidneys, and a third in our pubic apparatus—may well assault us with erotic fantasies. Apparently this interlocked trio of organs doesn't sufficiently respect the penalties for sexual harassment. (Don't bother to keep track of names, but the troublesome triad is called the hypothalamic-pituitary-adrenal axis.) Meanwhile, a hankering for a chocolate-frosted doughnut with custard filling may bedevil our taste buds. The alliance lobbying for this desire probably includes a dynamic duo of neural wheeler-dealers with branches in our stomachs, chests, and brains. (I'm talking about the vagus nerve and the enteric nervous system, but remember, don't keep track.)[23] While all this backstage haggling is going on, we may have a flickering urge to stop work and take the day off. (Lord knows which part of the body-brain combo this desire comes from.) Meanwhile, the poor left brain[24] bravely battles the gaggle of internal tempters and urges us to at least *appear* proper and efficient. After all, this is work, and we're *supposed* to be doing our job.

Then there's the "mind," which makes the decision on which batch of nagging interior natterers to obey. By now it should come as no surprise that said "mind" is actually a committee—let's imagine it as one which meets in a basement strategy room at the White House. Again, don't bother to keep the names of the committee members straight. They're here to give you an idea of how crowded the conference room is. The meeting is kicked off by the prefrontal cortex, near the head of the table, which has been sizing up the situation for quite some time and now presents several strategies for dealing with our rebellious appetites. The anterior cingulate, sitting two chairs further down, weighs the pros and cons of the prefrontal cortex's game plans and passes a directive to the woman wedged in between them—the motor strip—who shuffles through a sheaf of motion memories (that is, behavioral memes), holds a whispering session with a beefy fellow down at the end of the table—the superior parietal region[25]—then quickly lays out a detailed action plan. At this point the president walks into the room, totally oblivious to the proceedings, takes the empty chair at the table's head, and is told the final conclusion. The president then strides out, holds a press conference, and takes full credit for having decided on what to do. Unless the decision involves an operation he prefers to officially disapprove, like inviting the young person of the opposite sex in the cubicle next door to visit the snack wagon with us, pick up a few chocolate doughnuts, then take the afternoon off and visit a posh hotel. In this case the president—our consciousness—blames the whole thing on irresponsible members of the opposition party who have hijacked the congressional districts of our palates and our loins.

Michael Gazzaniga feels that consciousness—a chief executive who seldom gets the final word until after the fact—is centered in the left cerebral hemisphere, where, when necessary, it pulls together messages from the competing factions within us and fashions them into a policy statement of the moment.[26] Which brings us back to the radical constructionist position that "the observer . . . [is] the source of all reality."[27] The left brain's consciousness uses the material it's handed, in Gazzaniga's opinion, to construct two theories of "reality": one of the self and the other of an outside world.[28] Alas, these ever-changing theories can be terribly off base.

Consciousness and its theories are so out of the loop that often we don't have the foggiest idea of what's going on in the easiest-to-reach reality of all—our own bodies. Researchers rigged subjects to a finger-pulse stress detector, then ran their human guinea pigs through a modestly hair-raising situation. When the ordeal ended each victim was quizzed on how he or she felt. "Fine," said many whom the stress detector indicated were extremely rattled. "Upset," said others who, according to their physiological signals, were actually quite calm. Further studies have shown that adolescents who report they are aggressive often aren't, and those who say they're not aggressive are. In one research trial, women were hooked up to a device measuring vaginal blood volume and given an erotic story to read. Many who said they were aroused weren't, and many who said they weren't were. In yet another experimental probe, subjects were wired to a brain-wave-measuring sleep detector and were allowed to doze off. Though many fell into a solid slumber, they usually were certain they'd stayed wide awake.[29]

So the radical constructionists were right. Reality is a fabrication slapped together by an often bumbling inner team. Even the most dispassionate scientist can only see what his bickering body and brain assemblies allow him to perceive. The proclamation that "there can be no such thing as an objective fact"[30] has a great deal of validity.

As we shall discover in the next chapter, the internal assembly steps patching together our reality are filled with open loops demanding input from a crowd of other human beings. At a level far, far deeper than we know, even the most granite of our solidities is largely a collective fantasy.

8

REALITY IS A SHARED
HALLUCINATION

Being here is a kind of spiritual surrender. We see only what the others see, the thousands who were here in the past, those who will come in the future. We've agreed to be part of a collective perception.

Don DeLillo

We are accustomed to use our eyes only with the memory of what other people before us have thought about the object we are looking at.

Guy de Maupassant

After all, what is reality anyway? Nothin' but a collective hunch.

Lily Tomlin

The artificial construction of reality was to play a key role in the new form of global intelligence which would soon emerge among human beings. If the group brain's "psyche" were a beach with shifting dunes and hollows, individual perception would be that beach's grains of sand. However, this image has a hidden twist. Individual perception untainted by others' influence does not exist.

A central rule of large-scale organization goes like this: the greater the spryness of a massive enterprise, the more internal communication it takes to support the teamwork of its parts.[1] For example, in all but the simplest plants and animals only 5 percent of DNA is dedicated to DNA's "real job," manufacturing proteins. The remaining 95 percent is preoccupied with organization and administration, supervising the maintenance of bodily procedures, or even merely interpreting the corporate rule book "printed" in a string of genes.

In an effective learning machine, the connections deep inside far outnumber windows to the outside world. Take the cerebral cortex, roughly

80 percent of whose nerves connect with each other, not with input from the eyes or ears. The learning device called human society follows the same rules. Individuals spend most of their time communicating with each other, not exploring such ubiquitous elements of their "environment" as insects and weeds which could potentially make a nourishing dish. This cabling for the group's internal operations has a far greater impact on what we "see" and "hear" than many psychological researchers suspect. For it puts us in the hands of a conformity enforcer whose power and subtlety are almost beyond belief.

In our previous episode we mentioned that the brain's emotional center—the limbic system—decides which swatches of experience to notice and store in memory. Memory is the core of what we call reality. Think about it for a second. What do you actually hear right now and see? This page. The walls and furnishings of the room in which you sit. Perhaps some music or some background noise. Yet you know as sure as you were born that out of sight there are other rooms mere steps away—perhaps the kitchen, bathroom, bedroom, and a hall. What makes you so sure that they exist? Nothing but your memory. Nothing else at all. You're also reasonably certain there's a broader world outside. You know that your office, if you are away from it, still awaits your entry. You can picture the roads you use to get to it, visualize the public foyer and the conference rooms, see in your mind's eye the path to your own workspace, and know where most of the things in your desk are placed. Then there are the companions who enrich your life—family, workmates, neighbors, friends, a husband or a wife, and even people you are fond of to whom you haven't spoken in a year or two—few of whom, if any, are currently in the room with you. You also know we sit on a planet called the earth, circling an incandescent ball of sun, buried in one of many galaxies. At this instant, reading by yourself, where do the realities of galaxies and friends reside? Only in the chambers of your mind. Almost every reality you "know" at *any* given second is a mere ghost held in memory.

The limbic system is memory's gatekeeper and in a very real sense its creator. The limbic system is also an intense monitor of others,[2] keeping track of what will earn their praises or their blame. By using cues from those around us to fashion our perceptions and the "facts" which we retain, our limbic system gives the group a say in that most central of realities, the one presiding in our brain.

Elizabeth Loftus, one of the world's premier memory researchers, is among the few who realize how powerfully the group remakes our deepest certainties. In the late 1970s, Loftus performed a series of key experiments. In a typical session, she showed college students a moving picture of a traffic accident, then asked after the film, "How fast was the white sports car going when it passed the barn while traveling along the country road?" Several days later when witnesses to the film were quizzed about what they'd seen, 17 percent were sure they'd spied a barn, though there weren't any buildings in the film at all. In a related experiment subjects were shown a collision between a bicycle and an auto driven by a brunette, then afterward were peppered with questions about the "blond" at the steering wheel. Not only did they remember the nonexistent blond vividly, but when they were shown the video a second time, they had a hard time believing that it was the same incident they now recalled so graphically. One subject said, "It's really strange because I still have the blond girl's face in my mind and it doesn't correspond to her [pointing to the woman on the video screen]. . . . It was really weird." In piecing together memory, Loftus concluded that hints leaked to us by fellow humans override the scene we're sure we've "seen with our own eyes."[3]

Though it got little public attention, research on the slavish nature of perception had begun at least twenty years before Loftus's work. It was 1956 when Solomon Asch published a classic series of experiments in which he and his colleagues showed cards with lines of different lengths to clusters of their students. Two lines were exactly the same size and two were clearly not—the dissimilar lines stuck out like a pair of basketball players at a Brotherhood of Munchkins brunch. During a typical experimental run, the researchers asked nine volunteers to claim that two badly mismatched lines were actually the same, and that the real twin was a misfit. Now came the nefarious part. The researchers ushered a naive student into a room filled with the collaborators and gave him the impression that the crowd already there knew just as little as he did about what was going on. Then a white-coated psychologist passed the cards around. One by one he asked the predrilled shills to announce out loud which lines were alike. Each dutifully declared that two terribly unlike lines were duplicates. By the time the scientist prodded the unsuspecting newcomer to pronounce judgment, he usually went along with the bogus consensus of the crowd. In fact, a full 75 percent of the clueless experimental subjects bleated in chorus with the herd. Asch ran the experiment over and over again. When he quizzed his victims of peer pressure after their ordeal was over, it turned out that many had done far more than simply going along to get along. They had actually *seen* the mismatched lines as equal. Their senses had been swayed more by the views of the multitude than by the actuality.

To make matters worse, many of those whose vision hadn't been deceived had still become inadvertent collaborators in the praise of the emperor's new clothes. Some did it out of self-doubt. They were convinced that the facts their eyes reported were wrong, the herd was right, and that an optical illusion had tricked them into seeing things. Still others realized with total clarity which lines were identical, but lacked the nerve to utter an unpopular opinion.[4] Conformity enforcers had tyrannized everything from visual processing to honest speech, revealing some of the mechanisms which wrap and seal a crowd into a false belief.

Another series of experiments indicate just how deeply social suggestion can penetrate the neural mesh through which we think we see a hard-and-fast reality. Students with normal color vision were shown blue slides. But one or two stooges in the room declared the slides were green. In a typical use of this procedure, only 32 percent of the students ended up going along with the vocal but totally phony proponents of green vision.[5] Later, however, the subjects were taken aside, shown bluegreen slides, and asked to rate them for blueness or greenness. Even the students who had refused to see green where there was none a few minutes earlier showed that the insistent greenies in the room had colored their perceptions. They rated the new slides more green than pretests indicated they would have otherwise. More to the point, when asked to describe the color of the afterimage they saw, the subjects often reported it was red-purple—the hue of an afterimage left by the color green. Afterimages are not voluntary. They are manufactured by the visual system. The words of just one determined speaker had penetrated the most intimate sanctums of the eye and brain.

When it comes to herd perception, this is just the iceberg's tip. Social experience literally shapes critical details of brain physiology,[6] sculpting an infant's brain to fit the culture into which the child is born. Six-month-olds can hear or make every sound in virtually every human language.[7] But within a mere four months, nearly two-thirds of this capacity has been cut away.[8] The slashing of ability is accompanied by ruthless alterations in cerebral tissue. Brain cells remain alive only if they can prove their worth in dealing with the baby's physical and social surroundings.[9] Half the brain cells we are born with rapidly die. The 50 percent of neurons which thrive are those which have shown they come in handy for coping with such cultural experiences as crawling on the polished mud floor of a straw hut or navigating on all fours across wall-to-wall carpeting, of comprehending a mother's words, her body language, stories, songs, and the concepts she's imbibed from her community. Those nerve cells stay alive which demonstrate that they can cope with the quirks of strangers, friends, and family. The 50 percent of neurons which remain unused are literally forced to commit preprogrammed cell death*—suicide.[10] The brain which underlies the mind is jigsawed like a puzzle piece to fit the space it's given by its loved ones and by the larger framework of its culture's patterning.[11]

When barely out of the womb, babies are already riveted on a major source of social cues.[12] Newborns to four-month-olds would rather look at faces than at almost anything else.[13] Rensselaer Polytechnic's Linnda

---

*Apoptosis.

Caporael points out what she calls "micro-coordination," in which a baby imitates its mother's facial expression, and the mother, in turn, imitates the baby's.[14] The duet of smiles and funny faces indulged in by Western mothers or scowls and angry looks favored by such peoples as New Guinea's Mundugumor[15] accomplishes far more than at first it seems. Psychologist Paul Ekman has demonstrated that the faces we make recast our moods, reset our nervous systems, and fill us with the feelings the facial expressions indicate.[16] So the baby imitating its mother's face is learning how to glower or glow with emotions stressed by its society. And emotions, as we've already seen, help craft our vision of reality.

There are other signs that babies synchronize their feelings to the folks around them at a very early age. Emotional contagion and empathy— two of the ties which bind us—come to us when we are still in diapers.[17] Children less than a year old who see another child hurt show all the signs of undergoing the same pain.[18] The University of Zurich's D. Bischof-Köhler concludes from one of his studies that when babies between one and two years old see another infant hurt they don't just ape the emotions of distress, but share them empathetically.[19]

More important, both animal and human children cram their powers of perception into a conformist mold, chaining their attention to what others see. A four-month-old human will swivel to look at an object his parent is staring at. A baby chimp will do the same.[20] By their first birthday, infants have extended this perceptual linkage to their peers. When they notice that another child's eyes have fixated on an object, they swivel around to focus on that thing themselves. If they don't see what's so interesting, they look back to check the direction of the other child's gaze and make sure they've got it right.[21]

One-year-olds show other ways in which their perception is a slave to social commands. Put a cup and a strange gewgaw in front of them and their natural tendency will be to check out the novelty. But repeat the word "cup" and the infant will dutifully rivet its gaze on the old familiar drinking vessel.[22] Children go along with the herd even in their tastes in food. When researchers put two- to five-year-olds at a table for several days with other kids who loved the edibles they loathed, the children with the dislike did a 180-degree turn and became zestful eaters of the dish they'd formerly disdained.[23] The preference was still going strong weeks after the peer pressure had stopped.

At six, children are obsessed with being accepted by the group and become hypersensitive to violations of group norms. This tyranny of belonging punishes perceptions which fail to coincide with those of the majority.[24]

Even rhythm draws individual perceptions together in the subtlest of ways. Psychiatrist William Condon of Boston University's Medical School analyzed films of adults chatting and noticed a peculiar process at work. Unconsciously, the conversationalists began to coordinate their finger movements, eye blinks, and nods.[25] When pairs of talkers were hooked up to separate electroencephalographs, something even more astonishing appeared—some of their brain waves were spiking in unison.[26] Newborn babies already show this synchrony[27]—in fact, an American infant still fresh from the womb will just as happily match its body movements to the speech of someone speaking Chinese as to someone speaking English. As time proceeds, these unnoticed synchronies draw larger and larger groups together. A graduate student working under the direction of anthropologist Edward T. Hall hid in an abandoned car and filmed children romping in a school playground at lunch hour. Screaming, laughing, running, and jumping, each seemed superficially to be doing his or her own thing. But careful analysis revealed that the group was rocking to a unified beat. One little girl, far more active than the rest, covered the entire schoolyard in her play. Hall and his student realized that without knowing it, she was "the director" and "the orchestrator." Eventually, the researchers found a tune that fit the silent cadence. When they played it and rolled the film, it looked exactly as if each kid were dancing to the melody. But there had been no music playing in the schoolyard. Said Hall, "Without knowing it, they were all moving to a beat they generated themselves . . . an unconscious undercurrent of synchronized movement tied the group together." William Condon concluded that it doesn't make sense to view humans as "isolated entities." They are, he said, bonded together by their involvement in "shared organizational forms."[28] In other words, without knowing it individuals form a team. Even in our most casual moments, we pulse in synchrony.

No wonder input from the herd so strongly colors the ways in which we see our world. Students at MIT were given a bio of a guest lecturer. One group's background sheet described the speaker as cold; the other group's handout praised him for his warmth. Both groups sat together as they watched the lecturer give his presentation. But those who'd read the bio saying he was cold saw him as distant and aloof. Those who'd been tipped off that he was warm rated him as friendly and approachable.[29] In judging a fellow human being, students replaced external fact with input they'd been given socially.

The cues rerouting herd perception come in many forms. Sociologists Janet Lynne Enke and Donna Eder discovered that in gossip, one person opens with a negative comment on someone outside the group. How the rest of the gang goes on the issue depends entirely on the second opinion expressed. If the second speechifier agrees that the outsider is disgusting, virtually everyone will chime in with a soundalike opinion. If, on the other hand, the second commentator objects that the outsider is terrific, the group is far less likely to descend like a flock of harpies tearing the stranger's reputation limb from limb.[30]

Crowds of silent voices whisper in our ears, transforming the nature of what we see and hear. Some are those of childhood authorities and heroes; others come from family and peers. The strangest emerge from beyond the grave. A vast chorus of long-gone ancients constitutes a not-so-silent majority whose legacy has what may be the most dramatic effect of all on our vision of reality. Take the impact of gender stereotypes—notions developed over hundreds of generations, contributed to, embellished, and passed on by literally billions of humans during our march through time. In one study, parents were asked to give their impression of their brand-new babies. Infant boys and girls are completely indistinguishable aside from the buds of reproductive equipment between their legs. Their size, texture, and the way in which newborns of opposite sex act are, according to researchers J. Z. Rubin, F. J. Provenzano, and Z. Luria, completely and totally the same. Yet parents consistently described girls as softer, smaller, and less attentive than boys.[31]

The crowds within us resculpt our gender verdicts over and over again. Two groups of experimental subjects were asked to grade the same paper. One was told the author was John McKay. The other was told the paper's writer was Joan McKay. Even *female* students evaluating the paper gave it higher marks if they thought it was from a male.[32]

The ultimate repository of herd influence is language—a device which not only condenses the opinions of those with whom we share a common vocabulary, but sums up the perceptual approach of swarms who have passed on. Every word we use carries within it the experience of generation after generation of men, women, families, tribes, and nations, often including their insights, value judgments, ignorance, and spiritual beliefs. Take the simple sentence, "Feminism has won freedom for women." Indo-European warriors with whom we shall ride in a later episode coined the word "dh[=a]," meaning to suck, as a baby does on a breast. They carried this term from the Asian steppes to Greece, where it became "qh^sai," to suckle, and "theEIE," nipple. The Romans managed to mangle qh^sai into "femina"—their word for woman.[33] At every step of the way, millions of humans mouthing the term managed to change its contents. To the Greeks, qh^sai was associated with a segment of the human race on a par with domesticated animals—for that's what women were, even in the splendid days of Plato (whose skeletons in the closet we shall see anon). In Rome, on the other hand, feminae were free and, if they were rich, could have a merry old time behind the scenes sexually or politically. The declaration that "feminism has won freedom for women" would have puzzled Indo-Europeans, enraged the Greeks, and been welcomed by the Romans.

Freedom—the word for whose contents many modern women fight—comes from a men's-only ritual among ancient German tribes. Two clans who'd been mowing each other's members down made peace by invoking the god Freda and giving up (Freda-ing, so to speak) a few haunches of meat or a pile of animal hides to mollify the enemy and let the matter drop.[34] As for the last word in "feminism has won freedom for women"—"woman" originally meant nothing more than a man's wife (the Anglo-Saxons pronounced it "wif-man").

"Feminism has won freedom for women"—over the millennia new generations have mouthed each of these words of ancient tribesmen in new ways, tacking on new connotations, denotations, and associations. The word "feminine" carried considerable baggage when it wended its way from Victorian times into the twentieth century. Quoth *Webster's Revised Unabridged Dictionary of 1913*, it meant: "modest, graceful, affectionate, confiding; or . . . weak, nerveless, timid, pleasure-loving, effeminate." Tens of millions of speakers from a host of nations had heaped these messages of weakness on the Indo-European base, and soon a swarm of other talkers would add to the word "feminine" a very different freight. In 1895, the women's movement changed feminine to feminism, which they defined as "the theory of the political, economic, and social equality of the sexes."[35] It would take millions of women fighting for nearly one hundred years to firmly affix the new meaning to syllables formerly associated with the nipple, timidity, and nervelessness. And even now, the crusades rage. With every sentence on feminism we utter, we thread our way through the sensitivities of masses of modern humans who find "feminism" a necessity, a destroyer of the family, a conversational irritant, or a still-open plain on which to battle yet again, this time over whether the word "femina" will in the future denote the goals of ecofeminists, anarchofeminists, amazon feminists, libertarian feminists, all four, or none of the above.[36]

The hordes of fellow humans who've left meanings in our words frequently guide the way in which we see our world. Experiments show that people from all cultures can detect subtle differences between colors placed next to each other. But individuals from societies equipped with names for numerous shades can spot the difference when the two swatches of color are apart.[37] At the dawn of the twentieth century, the Chukchee people of northeastern Siberia had very few terms for visual hues. If you asked them to sort colored yarns, they did a poor job of it. But they had over twenty-four terms for the patterns of reindeer hide, and could classify reindeer far better than the average European scientist, whose vocabulary didn't supply him with such well-honed perceptual tools.[38]

American physiologist-ornithologist Jared Diamond, in New Guinea, saw to his dismay that despite all his university studies of nature, illiterate local tribesmen were far better at distinguishing bird species than was he.

Diamond used a set of scientific criteria taught in the zoology classes back home. The New Guinean natives possessed something better: names for each animal variety, names whose local definitions pinpointed characteristics Diamond had never been taught to differentiate—everything from a bird's peculiarities of deportment to its taste when grilled over a flame. Diamond had binoculars and state-of-the-art taxonomy. But the New Guineans laughed at his incompetence.[39] They were equipped with a vocabulary each word of which compacted the experience of armies of bird-hunting ancestors.

All too often when we see someone perform an action without a name, we rapidly forget its alien outlines and re-tailor our recall to fit the patterns dictated by convention . . . and conventional vocabulary.[40] A perfect example comes from nineteenth-century America, where sibling rivalry was present in fact, but according to theory didn't exist. The experts were blind to its presence, as shown by its utter absence from family manuals. In the expert and popular view, all that existed between brothers and sisters was love. But letters from middle-class girls exposed unacknowledged cattiness and jealousy.

Sibling rivalry didn't begin to creep from the darkness of perceptual invisibility until 1893, when future Columbia University professor of political and social ethics Felix Adler hinted at the nameless notion in his manual for the moral instruction of children. During the 1920s, the concept of jealousy between siblings finally shouldered its way robustly into the repertoire of conscious concepts, appearing in two widely quoted government publications and becoming the focus of a 1926 crusade mounted by the Child Study Association of America. Only at this point did experts finally coin the term "sibling rivalry."

Now that it carried the compacted crowd power of a label, the formerly nonexistent demon was blamed for adult misery, failing marriages, crime, homosexuality, and God knows what all else. By the 1940s, nearly every child-raising guide had extensive sections on this ex-nonentity. Parents writing to major magazines cited the previously unseeable "sibling rivalry" as the root of almost every one of child raising's many quandaries.[41]

The stored experience language carries can make the difference between life and death. For roughly four thousand years, Tasmanian mothers, fathers, and children starved to death each time famine struck, despite the fact that their island home was surrounded by fish-rich seas. The problem: their tribal culture did not define fish as food.[42] We could easily suffer the same fate if stranded in the wilderness, simply because the crowd of ancients crimped into our vocabulary tells us that a rich source of nutrients is inedible, too—insects.

The perceptual influence of the mob of those who've gone before us and those who stand around us now can be mind-boggling. During the

Middle Ages, when universities first arose, a local barber-surgeon was called to the lecture chamber of famous medical schools like those of Padua and Salerno year after year to dissect a corpse for medical students gathered from the width and breadth of Europe. A scholar on a raised platform discoursed about the revelations unfolding before the students' eyes. The learned doctor would invariably report a shape for the liver radically different from the form of the organ sliding around in the surgeon's blood-stained hands. He'd verbally portray jaw joints which had no relation to those being displayed on the trestle below him. He'd describe a network of cranial blood vessels that were nowhere to be seen. But he never changed his narrative to fit the actualities. Nor did the students or the surgeon ever stop to correct the book-steeped authority. Why? The scholar was reciting the "facts" as found in volumes over one thousand years old—the works of the Roman master Galen, founder of "modern" medicine. Alas, Galen had drawn his conclusions, not from dissecting humans, but from probing the bodies of pigs and monkeys. Pigs and monkeys *do* have the strange features Galen described. Humans, however, do not. But that didn't stop the medieval professors from seeing what wasn't there.[43] Their sensory pathways echoed with voices gathered for a millennium, the murmurings of a crowd composed of both the living and the dead. For the perceptual powers of Middle Age scholars were no more individualistic than are yours and mine. Through our sentences and paragraphs long-gone ghosts still have their say within collective mind.

9

THE CONFORMITY POLICE

Public opinion, because of the tremendous urge to conformity in gregarious animals, is less tolerant than any system of law.

George Orwell

Much as the sage may effect it to despise the opinion of the world, there are few who would not rather expose their lives a hundred times than be condemned [to reproach and scorn].

Charles MacKay

The mind itself is bowed to the yoke: even in what people do for pleasure, conformity is the first thing thought of; they live in crowds; they exercise choice only among things commonly done: peculiarity of taste, eccentricity of conduct, are shunned equally with crimes.

John Stewart Mill

At how early a period the progenitors of man . . . became capable of feeling and being impelled by the praise or blame of their fellow-creatures we cannot of course say. . . . [However], a selfish and contentious people will not cohere, and without coherence nothing can be effected.

Charles Darwin

Why call the first principle of a complex adaptive system "the conformity enforcer," objected a well-meaning colleague. "Doesn't the notion smack of a police state?" Yes. The conformity enforcers pressing perception, behavior, and appearance into a common mold can be far more brutal than we might like to think. And they begin their work at a disturbingly early age.

Jesus Christ, William Wordsworth, and the current New Age Touch the Future Movement in California have portrayed children as avatars of innocence. If so, then innocence is curled in a tightened fist of raw ferocity. In the early 1960s, Eibl-Eibesfeldt found "toddlers . . . hitting, kicking, biting and spitting at one another" no matter what culture he studied.[1] It is unlikely that these toddlers had learned their harshness from parents or from violent movies on television. In many of the societies Eibl-Eibesfeldt scrutinized, television was at best a distant dream. In others parents worked like hell to stop the hailstorms of savagery. If anything, the behavioral circuitry of sadism seems a curse genetically prestamped into us.

The offspring of humans are not alone at inflicting cruelty. A monkey island in Florida was surrounded by friendly alligators. Yes, I know it sounds hard to believe, but according to pioneering primatologist Harry Harlow, the ominous reptiles actually seemed to enjoy mammalian company. The monkey youngsters, however, would wait for one of the gregarious creatures to drift by, grab it by all four legs, flatten it against a cement wall, and proceed to chew on it. To the primate hooligans, this seemed highly entertaining. Harlow implies that the alligators were somewhat less amused.

Gangs of young monkeys will sneak up on a cage in which a mother is kept, pretend to be perfectly innocent until they get within arm's reach, then, when the matriarch's not looking, yank out tufts of her fur. Clusters of baby monkeys allowed to roam free in the lab form gangs that carry out raids for revenge or simply for kicks. Says Harlow, "If they had not learned to cooperatively aggress, there would be no monkeys in the world."[2] In other words, Harlow felt that collective sadism was practice for the intergroup tournaments which pit monkeys against those who would like to dine on them or against rival groups of other monkeys determined to do them harm.

Some form of practice! When Clifford, a juvenile baboon on the savannas of Kekopey in Kenya, injured his leg, he became a target. His age-mates ganged up on him until his mother put a stop to it. But her help only halted the harassment for a short time. Adult baboons disabled by injury undergo the same fate. One male whose leg became damaged found that adults and infants both fled screaming from him, and that the males who had been his buddies now targeted him for attack.[3] Detestation of deformity is not limited to primates. A dominant lizard who has his tail snapped off by a predator will return to the troop over which he reigns only to find that he's now an outcast. The sight of a herring gull in distress often inspires others of her kind not to help, but to attack. Says legendary ethologist Niko Tinbergen, hostility in social creatures is almost universal against "individuals that behave in an abnormal manner."[4]

The tendency is far worse than Tinbergen makes it sound. One young female gorilla in the Virunga Mountains of Central Africa was deeply attached to her brothers and father. She'd sit by her father for hours staring adoringly into his face. Then she became ill. How did the relatives she'd doted on show their concern? They pounded and picked on her.[5]

The tendency to shun the misshapen or ill-fated is not just some dismal quirk of the animal world.[6] An American child whose mother had been killed in a car crash reported that after the accident the other kids in school shunned her. Even her best friend, who comforted her, had to force herself to come near. In the late 1980s, a student at a university in China made the mistake of telling her classmates that her mother had died when she was young. From that point on, they made fun of her mercilessly.[7]

The Kwakiutl Indians of the Pacific Northwest sensed the instinctive manner in which we punish those who experience misfortune, and created rituals to stem the resulting tide of antipathy. The death of a relative shamed a man and would stigmatize him permanently. His only remedy was to reassert his dignity by canoeing away to another tribe and killing a man of his own social standing. If a man's ax slipped while he was chopping wood and he injured himself, his major problem was not the damage to his leg, it was the damage to his reputation. He had to immediately fend off rejection by giving away property and reasserting the sense that he was still a man of standing. If a man lost all of his property gambling with a friend, there were no rituals which could deflect the cruelty of others. Often the only appropriate escape from the viciousness of disdain was suicide.[8]

Byzantine emperors knew the power of disfigurement to trigger repulsion. They cut off the noses of relatives who might have a legitimate claim to the throne, aware that the handicap would keep the nasal amputees from ever commanding respect.[9]

Psychologist and zoologist David Barash feels that our intolerance of the handicapped comes in part from an ancient impulse to distance ourselves from those who may be carrying one of the primary killers of premodern men and animals—infectious disease.[10] There may be merit to his argument. But I suspect the urge to impose physical uniformity springs from the principles which turn a group into a complex adaptive system, a collective intelligence, a learning machine. Remember a networked learning machine's most basic rule: strengthen the connections to those who succeed, weaken them to those who fail. The monkey with a broken limb botched an attempt to maneuver across a tricky landscape. The lizard with the missing tail was insufficiently wily to avoid the teeth of an enemy. The wailing gull may have owed her woes to bad decisions or poor genes. And the noseless member of the Byzantine royal family wore the sign of his incompetence at power games on his face.

Our intolerance of deviations from a physical norm seems bundled into us at birth. Human studies from all over the world show that infants as young as two months old already prefer attractive to unattractive faces.[11] Physical beauty is so rare that it seems to us the exception, not the rule. But not all that seems obvious is true. Ironically, the most attractive faces scientists have been able to construct are composites of photos of as many as thirty-two regular people, their features blended to create a perfect average.[12] Study after study shows that what we rate as beauty is a median, a midpoint, an archetypal mode, the kernel at the core of typicality! Further studies demonstrate that we fawn over those whom we deem beautiful, clustering around them, overrating their intelligence, anxious to be their friends.[13] We flock to what seems special, never knowing it's the essence of normalcy.

Normalizers wire us as modules of the group's machinery. They align us, confine us, define us, synchronize us, and incite us to enforce group uniformity. We've already examined the activation of perceptual coordinators in the first year of human life—from empathy to a fixation on a mother's face and the following of another's gaze. During the second year children fixate on the standards their parents lay down, comparing the objects around them to a social benchmark, and becoming distressed when things deviate from the family ideal. Though fourteen-month-olds are not yet bothered by breaches of propriety, nineteen-month-olds point an accusing finger at the tiniest flaw: a hole in clothes, a chip in the paint on a toy, a spot of dirt on a wall, or, most important, the "bad" behavior of someone else. By twenty months, babies have a rich vocabulary for denigrating deviance—they are incensed when things are "yucky," "broken," "boo-boo," and "dirty." In short, at less than two years old toddlers already show not just the instincts that patrol conformity within themselves, but the weapons which will help them impose it on others.

Protests against imperfection are not just some anal-compulsive trait of middle-class babies in the West. They also show up in children on the Fiji Islands and newly arrived immigrant babies from Vietnam. In 1896, James Sully summed up such phenomena by observing that a child has an "inbred respect for what is customary, and . . . has an innate disposition to follow precedent and rule." Modern psychologist Jerome Kagan wonders what evolutionary advantage this innate tendency to "morality" may have.[14] The answer: it is one of those instincts which make cohesion possible, giving man his most important tool, society.

Older children become far more aggressive enforcers of conformity. The punishments they hand out can be alarming. As long ago as 1992, 21 percent of British schoolchildren had been teased, bullied, hit, or kicked by fellow students. Many had been attacked on their way to or from school and on streets near their homes. Few reported these assaults to their parents.[15] In the United States at about the same time the penalties doled out by kids to their peers were even harsher. Twenty-three percent of schoolchildren had been attacked violently. Some of the children on informal conformity patrol carried knives and guns.[16] Then there were more subtle forms of cruelty. Five to 10 percent of children reportedly have no friends.[17]

Unattractive children or children with strange religious backgrounds, funny names, and unusual ethnic roots are the usual targets for these torments. Kids punish those who do far better than average in school and those who do far worse. One third-grader was cursed with talent—she was outstanding at piano, ballet, and reading. Her classmates hated her. She tried to be pleasant to everyone, but was labeled a snob and treated with the derision her atypical abilities "deserved."[18] In Japan, where the ideal of harmony is king, enforcement of conformity takes on particular viciousness. *Ijime,* bullying or picking on someone who sticks out, is often led by, of all people, the teacher.[19] It's disgusting, but it's the nature of humanity.

Ritch Savin-Williams studied U.S. summer campers and discovered that adolescent leaders were particularly gifted at dishing out ridicule. Female camp trendsetters—praised by gender studies specialists like Carol Gilligan[20] for their warm and gentle cooperation—were particularly wicked conformity enforcers. They did it with the carrot and the stick. A dominant female camper would offer to fix another girl's hair or help her with her choice of clothes . . . both quiet ways of shaping the follower's appearance to fit the mold. But the verbal abuse these teen leaders could mete out to those who failed to conform was so devastating that it agonized even the researchers watching it. One of them, who had been forced to tears on several occasions by the viciousness of the attacks she'd witnessed, said, "Now I know why no one studies junior high school girls. They are so cruel and horrible that no one can stand them! I remember my own adolescence as that way, and this summer was like reliving it. Never again!"[21] Yet when the girls were quizzed about dominance, they claimed to dislike it. Though some of them stomped others with appalling verbal brutality, they abhorred being seen as authority figures because to them it represented being different.[22] And difference among young girls just won't do—conformity has a choke hold that won't let go.

This sort of viciousness can build the backbone of a social structure. Youngsters who lead attacks on the odd often end up running animal troops or human nations. The pubescent Oliver Cromwell roamed the streets of his British hometown clubbing adults he didn't care for with an outsized walking stick, then led the English Revolution of 1648 and ended up as a piously Puritan dictator. As a child, Fidel Castro was a bully and proud of it. In his fifties and sixties, he still enjoyed telling how he once had beaten up another grammar school student because that student was the teacher's pet. In an attempt to stop the battering, the Catholic teacher-priest whomped Castro on the head. The youth turned and pummeled the pedagogue for all he was worth. Fidel gloated that the incident had made him a school hero.[23]

Bullies act as conformity enforcers when they are children, and may become conformity enforcers in adulthood once again. Castro, for example, allows no straying from the norms he sets for Cuban citizens. Nor did Oliver Cromwell, who persuaded the author of a famous work on freedom of the press,* John Milton, to become his censor, and an unforgiving one at that.[24]

The tendency of children and adolescents to force diverse humans into a common mold grows more refined among adults. Max Weber, describing the United States of the twenties, said that to be among the fashionable elite you had to live on the correct street, wear the correct fashions, gush over the correct art, and behave in the correct manner. Otherwise, you would not get invited out. No one would show up to visit you. The threat of social exclusion drove the U.S. upper class to toe the line.[25] In prerevolutionary China "public standards were enforced mainly by gossip, threat of 'loss of face,' (prestige), and ostracism," say anthropologists Allen Johnson and Timothy Earle.[26]

The Utku Eskimo also used social exclusion to enforce conformity. They disallowed angry thoughts, which they were sure could kill. Living on the edge of existence, closeness and cooperation were vital. Anger was regarded as a child's emotion that adults learned to hold back. Those who failed to control their anger were teased, ignored, or tossed out of the band.[27]

Things are not that different in the modern scientific community. Sociological researchers maintain a mask of objectivity. But behind that mask some schools of thought hide ideological goals. When students in these movements report facts that contradict the tenets of their group's creed, they are not praised for the objectivity of their work, but punished for their heresy. They are derided, their papers are rejected by journals, and they are excluded from key symposia—all an indirect way of forcing them "to leave the movement."[28] A similar mechanism of repression is at work in every scientific discipline I know. For numerous scientists, to go against the tide risks academic suicide.

The New Conservatives of the 1980s, like the members of other groups, furtively disciplined their members to parrot the party line. Paul Weaver, an assistant professor of government at Harvard, was a dedicated neoconservative free marketeer who believed passionately in the reigning dogma of his group—that corporations are the salvation of America. So he went to work for one. After two years as a communications planning director at the Ford Motor Company, he became convinced that the corporation could be a self-destructive beast. When Weaver returned to New York with his hard-won insights, his neoconservative friends spurned him. His criticism of the corporation was an affront to the faith.[29]

---

*The anticensorship essay in question was Milton's acclaimed 1644 *Areopagitica*.

Even humor is a conformity enforcer clothed in the garb of congeniality. It focuses on others' weaknesses, disasters, stupidities, and abnormalities. Darwin reports that in the mid-nineteenth century, Australian aborigines would "mimic the peculiarities of some absent member of the tribe" and break into uncontrollable fits of laughter.[30] Even in "gentle" Tibet before the Chinese takeover, Heinrich Harrer, the only Westerner allowed to stay at length in the capital of Lhasa, reported: "If anyone stumbles or slips, they enjoy themselves for hours. . . . They make a mock of everything and everybody. As they have no newspapers, they indulge their criticism of untoward events or objectionable persons by means of songs and satire. Boys and girls walk through the Parcor in the evening singing the latest verses. Even the highest personages must put up with being pulled to pieces."[31]

Thomas Hobbes declared that the man who laughs too much is aware of how many flaws he has and maintains a high opinion of himself by focusing on the imperfections of others. And mid-twentieth-century cartoonist Al Capp observed bitterly that "all comedy is based on man's delight in man's inhumanity to man."[32]

But perhaps the word "inhumanity" is a tad too homocentric. Humor is governed by the animal brain—the thalamus and hypothalamus.[33] Gorillas, like humans, use mockery to punish those who botch their attempts at conformity. For example, two gorilla troops ran across each other on the forested slopes of Central Africa's Mount Visoke. The males strutted and swaggered to show their power. One was young and inexperienced. He charged to show off his boldness, but pulled off the maneuver sloppily. An older rival made his displays with confidence and finesse. The youngsters of the inexperienced male's group followed behind him, "exaggeratedly mocking his awkward displays of bravado."[34]

Social exclusion among another group of nature's innocents, chimps, involves an intolerable willingness to dish out pain. At Holland's Arnhem Zoo one male chimp was attacked by two others so viciously that they removed his testicles, slashed his head, back, sides, and anus, removed several of his toes, and wounded his hands. Yet the next morning he did not want to be separated from his attackers—the key males of the social group in which he'd spent most of his life. Twelve hours later, his wounds had proven fatal. He was dead.[35]

Like chimpanzees and apes, we avoid the deformed and different. In one study an actor was asked to collapse convincingly and dramatically in the middle of a subway car. The apparent victim of a serious problem was much less likely to get help if he had a large birthmark.[36] A questionnaire administered by a psychologist in 1894 showed that tiny breaches of custom—men who wore earrings, people who wore a ring on their thumb or too much stylish jewelry, humans who tried to stand out or who strayed from the pack—roused infuriation.[37] Seventy-four years later an experimental group was given the task of assigning others to jobs that would either pay money or involve receiving an electrical shock. They awarded the paying jobs to those whose personalities matched the majority, and shuffled the painful shocks to those who didn't quite fit in.[38]

A willingness to dish out pain is not limited to the group psychologists' lab. If an American worker in the 1960s and 1970s worked faster than the rest of his group, the other laborers would often "bing" him—snapping their fingers sharply on his arm.[39] If you belong to one of the tremendous number of cultures that believe in witchcraft or the evil eye, your failure to fit in can be lethal. To the Bantu, all evil is caused by a witch, an innocent soul in whom a demonic spirit has taken up residence without his knowledge. To detect the bearer of malevolence the Bantu gather in a circle, chanting softly, as the witch doctor goes from man to man, sniffing each. The tribesmen believe that the volume of their chanting is under the control of supernatural forces. The individual in front of whom the witch doctor is testing the air when the chanting becomes the loudest is the one he picks as the demonic vessel. The unwitting conduit of evil is hauled off and dispatched from this planet by having a stake driven up his rectum. His kraal—the circle of huts in which he and his relatives live—is burned, his family wiped out, his cattle given to the chief, and a smattering of cows and bulls are handed out to the witch doctor as a tip. In reality, the chanting is a popularity contest. And as in the American witch-hunts of the 1600s, suspicion locks in on the person who strays the furthest from the norm.[40]

But which really is more deadly—physical injury or social rejection? When vervet monkeys are attacked by their peers, the bites they get are often trivial—in fact, many don't break the skin. But the punished monkey can go into shock and die.[41] Human children are wounded far more than physically when lashed by humiliation from those who deem them different. Adults assume that youngsters are concerned with problems like the birth of a brother, the prospect of an operation, or a trip to the dentist. But a 1988 survey of 1,814 children by the University of Colorado's Kaoru Yamamoto showed that, in fact, many of the primary fears of nine- to fourteen-year-old children in the United States, Australia, Canada, Egypt, Japan, and the Philippines focused on being shamed or disgraced in front of friends. Yes, children were frightened of the expected horrors: a mother or a father's death, going blind, and seeing their parents fight. But their preoccupations also included being held back a grade in school and wetting their pants in the classroom. Yamamoto's study indicated that even the experts were usually way off base in estimating the importance of the fear of being scorned for abnormality.[42] Ann Epstein of Harvard Medical

School pointed out the most chilling fact of all: that humiliation was one of the most common causes of childhood and teen suicide.[43]

Acceptance is as important to social animals as oxygen and food. The Japanese have incorporated this basic fact of biopsychology into their culture far more overtly than have we. Most Westerners are familiar with one of Japan's most popular adages—"the nail that stands out will be hammered down." The preeminent expert on Japanese culture, Edwin Reischauer, explained a generation ago that the Japanese were painfully concerned about what others might think of them. The most powerful threat parents could deliver to a misbehaving youngster was that "people will laugh at you." The effect, said Reischauer, was "devastating."[44]

The supreme punishment in a Japanese village, explained Reischauer, was ostracism. The inability to trade food and other necessities with your neighbors could seriously threaten your existence on this earth. Something remarkably similar once played a primary role in keeping early New England colonists in line. During the seventeenth century, behavior was policed by taking advantage of the individual's entrapment in a small community. If he went astray, the entire village displayed him in an outdoor pillory in the center of town, lashed him at a public whipping post, branded him so everyone could see his criminality, or made him wear a badge of infamy. He was, in the words of Massachusetts Bay Colony's first governor, John Winthrop, "despised, pointed at, hated of the world, made a byword, reviled, slandered, rebuked, [and] made a gazing stock."[45] Even worse, Puritan settlers were dependent for their meals and feelings of self-worth on the two to three hundred neighbors with whom they shared their labors in the isolation of a dangerous wilderness. As in Japan, exclusion from the community could increase your chance of death dramatically.

By the nineteenth century, on the other hand, land in New England had been cleared of Indians and wild animals. One could easily leave one town and move to another, or better yet, blend into the crowds of a large city. New circumstances called for new conformity enforcers—highly portable ones—like the Victorian sense of guilt. Your parents didn't shame you publicly, they sent you to your room so your "penalty for wrongdoing [would] . . . be exacted internally." The government locked you in a house of penitence—a penitentiary*—where your feelings of remorse would theoretically pummel you without cease.[46] "Each individual," one reformer wrote, "will necessarily be made the instrument of his own punishment; his conscience will be the avenger of society."[47] But the publicly inflicted shame of village life and the more private "conscience" you carried in the

---

*According to *Webster's Revised Unabridged Dictionary* of 1913, originally a penitentiary had been "a small building in a monastery where penitents confessed."

anonymous metropolis both—at heart—came down to the same bottom line: they meant that the people who mattered to you the most were going to reject you, sometimes ferociously. The agony of anticipated scorn usually whips the aberrant tendencies out of us and gets us to conform.

Our instinctual cruelty often pushes peculiar individuals to society's outskirts and sometimes shoves them out entirely, while it squeezes the rest of us into a conscious or unwitting team. Conformity-enforcing packs of vicious children and adults gradually shape the social complexes we know as religion, science, corporations, ethnic groups, and even nations. The tools of our cohesion include ridicule, rejection, snobbery, self-righteousness, assault, torture, and death by stoning, lethal injection, or the noose. A collective brain may sound warm and fuzzily New Age, but one force lashing it together is abuse.

10

DIVERSITY GENERATORS:

The Huddle and the Squabble—Group Fission

In previous episodes we've focused on two kinds of conformity enforcers: those which shape brains to work in harmony, synching our vision, hearing, and attention so that we share the same "reality"; and others which goad us to tailor our public behavior and appearance to the standards of our tribe. Conformity gives the complex adaptive system called a social group stability. But to adapt, the system needs a hefty dollop of something else: originality. The ability to bend, stretch, and create comes not from conformity enforcers but from their indispensable opposites: diversity generators.

Diversity generators show up in many forms.[1] They snake through the solar dust and stars of the inanimate universe, where physicist Paul Davies says there's no telling "what new levels of variety or complexity may be in store." Random fluctuations, explains Davies, "are nature's way of exploring unforeseen possibilities."[2] On the living level there's that old phenomenon called sex. To a cost-benefit analyst, it may look like a time-consuming, energy-swallowing waste of an organism's time.* But the diversity sex produces provides innumerable advantages. One among the many is the ability to make gene repairs on-site. Cells are occasionally bombarded by the bundles of protons and neutrons called alpha rays, by the crazily racing electrons called beta rays, by the thwackingly powerful photons of gamma rays, and, most especially, by the penetrating photons of ultraviolet light. One solid hit from these bits of cosmic artillery can shatter a gene or other twist of DNA. In the face of this barrage, some creatures hang on to chromosomal conformity—splitting in half and giving each daughter a replica of her mother's genes. The solution sounds simple enough, but simplicity has a price: shattered nucleotides on the genetic string pile up until they can short-circuit things. On the other hand, organisms which go through the bother of courting and coupling mix and match

---

*Two hundred years ago, Britain's Lord Chesterfield gave this blunt appraisal of sex's value among humans: "The pleasure is momentary, the position ridiculous, and the expense damnable."

their genes with those of a chosen mate. As the nucleic strands of the happy couple conjugate, each chromosome can discard its damaged genes and replace them with still-perfect equivalents from its mate. The payoff comes when a particle blitz begins, as when the atmosphere fails to protect those living on the planet's skin against the sun's ultraviolet cannonades. Experiments with bacteria show that when slammed by ultraviolet rays, sexually shuffled beasts out-survive the neatly uniform, asexual brigades.[3] In other words, despite the cost diversity generation pays.

Here's a bit of irony—conformity is *strengthened* when it's shored up by its "enemy"—diversity. No life-forms know this better than bacteria, the ultimate diversity creators. Bacteria far beneath the surface of the earth are so skilled at reinventing themselves that they can eat indigestibles ranging from soil to metal to stone—a result of their ability to generate an awesome array of variations on the basic ancestral genome. A mile and a half below the plateaus of Virginia, the deserts of New Mexico, the old volcanic slopes of the state of Washington, and the cavernous ridges of the ocean floor, colonies of bacteria thrive as they've done since the age of the dinosaur, eating the carbon content of rock and metabolizing it with iron or manganese, supping on naturally occurring glass and the compressed mash of multimillion-year-old beasts from which coal and oil are formed, dining on sulfides beneath the ocean's crust, and devouring basalt or raw hydrogen.[4]

Diversity generation wins the jackpot, and it does so handsomely. A full 90 percent of the earth's bacteria have made a happy home beneath the surface of the lands and bottoms of the seas. Never underestimate these subterranean mistresses of the quick-change routine. There are a mere 6,000,000,000 of us, but a full 5,000,000,000,000,000,000,000,000,000,000 of them.[5]

Diversity generators also benefit larger beasts. Macaque monkey females, like some women and men, can't resist the appeal of strangers from outside their group. The fruit of their promiscuity is an influx of fresh genes, whose introduction to the reproductive pool prevents the troop from becoming stagnant and vulnerable to disease.[6] But diversity generators are particularly vigorous among us human beings.

BICKERING, BACKSTABBING, AND THE HATRED BETWEEN BROTHERS

> It is much easier to be friends with a member of the opposite party than a member of one's own party—for one is not in direct personal competition for office with members of the opposition in the way that one is with one's colleagues.
>
> Jonathan Lynn, *The Complete Yes Minister*

Families are as much the matrix of affection as they are the crucibles
of conflict.

David Cannadine

A powerful form of diversity generator prodded hominids roughly 2 million years ago. One wave after another of human ancestors trekked from
Africa to China, Southeast Asia, and Europe, then as early as 176,000 years
ago navigated huge expanses of sea and settled in Australia,[7] where their
wanderings began all over again. The forces behind this scatter are murky
to us. But it is possible to deduce them from the evidence of modernity.

One of the most powerful diversity generators in today's humans
and animals is a force Freud called "the narcissism of minor difference."[8]
For simplicity's sake, we'll refer to it as "creative bickering." Individuals
extremely similar to one another find some petty distinction, then raise
unholy hell about it. To paraphrase Emile Durkheim, a community of
saints will classify a bit of lint on the heavenly robes as intolerable, and will
viciously hound those who aren't lint free.[9] Eventually the supposedly
unkempt may seek out others with a sloppy bent, and wall themselves off
as a separate sect sworn to a messy destiny.

A primitive form of this impulse far precedes us hominids. The
closer insects are to each other in physical form and habits, the more likely
they are to be enemies.[10] The most dangerous foes of ants are other ants.
The major adversaries of South American wasps are other South American
wasps. And, in general, the greatest threat to each social insect group is not
a marauding bird, lizard, or aardvark. It's another cluster of social insects. I
don't mean to say that social insects are more ruthless than human beings.
They greet each other politely, rush to each other's rescue, bow to superiors meekly, and even carry each other around. But cannibalistic ants tend
to feed on those who most resemble themselves. And parasitic ants sup on
hosts who seem their semiduplicates. A major reason: insects with the
same shape, size, and tastes have a yen for the same nest sites, food, and
foraging space, so they're willing to battle for these prizes . . . sometimes
to the death.[11]

Like many forms of turmoil, this conflict between clones has a positive side. If a passel of nearly identical animals is cooped up on a common
turf, it frequently splinters into opposing groups which scramble deter-
minedly down different evolutionary paths. E. O. Wilson, who brought
attention to this phenomenon forty years ago, called it character displacement.[12] The battle over food and *lebensraum* compels each coterie to find a
separate slot in the environment from which to chisel out its needs.[13] For
example, a small number of look-alike cichlid fish found its way to Lake
Nyasa in eastern Africa roughly 12,400 years ago. It didn't take long for
the finny explorers to overpopulate the place. As food became harder to

find, squabbles and serious fights probably pushed the cichlids to square off in spatting cliques. The further the groups grew apart, the more different they became.[14] The details of this process are somewhat speculative, but the result is indisputable. The cichlids rapidly went from a single species of fish to hundreds, each equipped with a crowbar to pry open opportunities others had missed. Some evolved mouths wide enough to swallow armored snails. Others generated thick lips to yank worms from rocks. One diabolical coven acquired teeth like spears, then skewered its rivals' eyeballs and swallowed them like cocktail onions. In the geologic blink of 12,000 years, what had begun as a small group of carbon copies became two hundred separate species—a carnival of diversity.[15] Hence the word creative in "creative bickering."

Creative bickering may have been the spark for *Homo erectus*'s wanderings. Legendary anthropologist J. B. Birdsell built a powerful case for the notion that prehuman clans, like cichlids, hit the limits of their environment,* then budded off in search of untapped openings.[16] Even today, the Yanomamo tribes of northwestern Brazil and southern Venezuela swell until they reach three hundred members or so, then break into arguments between blood relatives. The quarrels often end in violence, convincing an angry group of malcontents to start a new life somewhere else.[17]

The tendency of those alike to fight when times get tough defies a cardinal rule of the theory of the selfish gene—that the closer we are in the makeup of our chromosomes, the more we will work together as a team.[18] The violation is particularly strong in the battle of brother against brother. Species of genetically related ants are, in E. O. Wilson's words, "the least likely to tolerate each other's presence." The same sometimes applies to human beings.[19] On his way through the Alps to spring his surprise attack on Rome, Hannibal ran across two groups of Gauls on the verge of battle. The problem? A pair of brothers were fighting over who would head the tribe. Among the Yanomamo, the biggest clashes are between family members—and between the groups they head.[20] How could evolution favor feuds which current theory says should never be? Creative bickering has been honed by natural selection because, in pitting father against son and brother against brother, it opens up new avenues to genes, clans, cliques, and species. It slices through genetic bonds to generate diversity.

FISSION'S FRYING PANS

When the going gets tough, the diversity generators get going. Earlier in the book I described bacterial colonies whose members found a

---

*In technical terms, they reached the environment's carrying-capacity.

food bonanza, then signaled for companion and stranger* alike to gather 'round. But as the food ran out, the bacteria's chemical communiqués spat a less gregarious message: "keep your distance—get away." There was social value to this bitchiness. It drove colony members to fan out on long treks, probing for richer pickings in the distant wilderness. The scouting expeditions which came up croppers crowed about their victories, chemically signaling "by all means come and settle down near me." In the long run, chemical cantankerousness saved the colony, forcing its members to scour new frontiers in search of opportunity.

The principle of creative bickering pops up on numerous levels of biology. Cues that a once-cozy environment is no longer working make humans chafe until their urge to escape is overpowering. Some people become hostile when jostled from their comfortable routines and hit with the sort of high-anxiety task[21] which crops up as the sledding gets rough. Psychologist Auke Tellegen puts it like this: "People in good moods tend to engage in pro-social behavior; in the absence of joy, people tend to cut themselves off."[22] Show us a depressing film, and when it's over studies indicate that we'll avoid each other.[23] Chip away our quality of life, and our plunge into downward mobility warps not only on the way we act, but the way we hear and see. If a couple is having problems and one member attempts to do something comforting, his mate is likely to interpret the peace gesture as a veiled attack.[24] This paranoid interpretation comes courtesy of instinctual demons planted in our brain loops by the evolutionary triumphs of creative bickering.

Such demons lie in multitudes within us awaiting the slightest opening. They can saw our social bonds apart over irritants so small that the bones of contention seem like hardly anything to get fired up about at all. For example, hit us with air pollution, noise, and broadcasts of bad news[25] and we'll turn on each other viciously. The hotter it gets the more violent we can be. When experimenters deliberately ran a study on a day when the temperature had slid into the nineties, their sweating human guinea pigs—civilized college students—grew far more eager than those cooled down by a glass of water to torture each other with electrical shocks.[26] When some modern humans feel the heat, they vent their frustration with fists,

---

*Standard scientific views have it that there can be no strangers in a bacterial colony, since all members are technically sisters. This venerable view is technically both right and wrong. In reality, bacteria branch out from a common center started by the spore that birthed a tribe. However, each branch of the colony is isolated from other branches genetically. In only a day, every one of those outstretched offshoots of the family line becomes separated from its cousins by seventy-two generations—roughly 1,400 years in human time. In a very real sense, each tendril of a bacterial colony is less a string of sisters than it is a distantly related tribe.

guns, knives, and rocks. Men beat up their wives or women brutalize their men.[27] Prison fights break out in spades when the temperature is swelter- ing.[28] The American race riots of the 1960s took place when thermome- ters hit extremes.[29] What's more, people are generally friendly when their energy is high, but snap, snarl, withdraw, or whine when their internal fuel runs low.[30] So it's easy to imagine that the putrid odor, lack of food, and constant bad news at an overused *Homo erectus* campsite could have kicked up a good deal of backbiting 2 million years ago.

A community of bacteria react to the insults of harsh circumstance by fragmenting in groups whose flight from each other leads to the detec- tion of bounties in the great unknown. A similar irritability was almost certainly among the blows which had splintered prehumans and spattered them across three continents by 780,000 years ago.[31]

CULTURE'S CENTRIFUGE

As they squabbled, broke from each other, and moved further into waste- lands, our ancient ancestors remained connected in a rough but very slow communal brain. During the 2 million years when upright walkers fanned out toward distant coasts, their hand scrapers, choppers, cleavers, and other stone basics stayed more or less the same. Only a smattering of iso- lated groups like the *Homo erectus* of Java and the archaic *Homo sapiens* of southern England[32] managed to miss out on an occasional new sensation like the Acheulean hand ax—an American-football-sized stone core with its blade flaked to razor sharpness from two sides. Then new diversity gen- erators began their rise.

By roughly 40,000 B.C., fresh inventions showered the scene— among them the spear thrower, which more than doubled a spear's range, plus the barbed harpoon and an early form of hook and tackle known as the fish gorge which opened the food stores of the rivers and the seas. New equipment made it easy to follow animal herds and enjoy a slab of steak whenever you had a carnivorous craving. There were social breakthroughs, too. The rain of new hardware tempted scattered families to join forces in large-scale maneuvers. By using beaters in the bush to drive an entire herd of four-legged meat lockers into a trap then descend for the slaughter,[33] or by panicking animals into enormous nets,[34] small groups working together could unlock a cornucopia of previously untouchable abundance—the fil- lets and chops of reindeer, mammoths, bisons, horses, and rhinoceroses.[35] Shared exploits, rites, and festivals knit microgroups together for these new varieties of megahunt. Even the dances at festivals were choreo- graphed to give men practice in coordinated hunting moves.[36] Diversity generators in ritual and technology drew humans into new forms of conformity.

One result was a population explosion. The size of base camps mushroomed. In France, some reached seven acres in size. Encampments had forty-five-foot-long huts with a profusion of hearths, true multifamily apartment dwellings. And these huge structures were no longer temporary settlements of perpetual wanderers. They were permanent.[37] Presumably even then, the most dangerous animals in the vicinity were not lions and tigers but, as in the case of ants and wasps, other swarms of men.[38] This may explain why the Czechoslovakian communities of those days were already fortified with wooden palisades.

Creative bickering slid to warp speed as the ingenuity of humans picked up from crawl to hyperdrive. A host of innovations highlighted the differences between human beings. Decorated clothing, jewelry, body ornaments galore, and splendid hide-covered homes distinguished the loftier members of a tribe from the low.[39] Archaeologist A. Gilman suspects that each newly gathered populace wanted to monopolize its own herd or favorite hunting spot—like a narrow pass through which masses of migrating mastodons or reindeer reliably trotted season after season.[40] Humans needed ways to proclaim their monopoly of such valuable property. Other hunting animals could use urine and musk glands to spray their territories with specialized scents. This was an ability our ancestors no longer possessed. So they found a host of crude but clever substitutes. As early as 130,000 years ago protohumans in Africa had collected colored pigments and rhinestones apparently to use in ceremonies which showed just how different one group was from another.[41] One hundred twenty thousand years ago, the ante went up. At Terra Amata in France, inhabitants gathered a palette of seventy-five tints spanning the spectrum from yellow to red and brown[42]—probably to paint their bodies and make themselves stand out from their brethren and, more important, from their tribal enemies. Seventy thousand years ago, some proclaimed their uniqueness by recontouring the skulls of their young with tight bindings and objects that pressed the head into a bizarre and permanent profile. Later on, others filed their teeth into unnatural shapes.[43] All this demonstrated that our group is not like yours, so if you value your life it would be wise to keep away.

Between 77,000 and 60,000 years ago early humans in Australia began engraving rows of symbolic circles in the local stone.[44] Forty thousand years later symbolic representation reached a breakneck pace.[45] Primordial writing showed up carved in animal bone as early as 27,000 B.C.[46] Sculpture and cave paintings may have driven home the differences between one group and its enemies.[47] In all probability, humans used the growing swiftness of their tongues[48] to paint word pictures, too—trading stories of the day's exploits, exchanging opinions about which step to take next in the hunting and gathering of food, and manufacturing myths with which to grasp the mysteries of an often uncontrollable natural world. To *our* concepts, *our* words, and *our* ritual techniques we added our style of bone calendar making,[49] our mode of carving, and our way of decorating tools as insignia to show who belonged to our group and who belonged to another.

Other differences were likely to have appeared, including one it is currently unfashionable to contemplate—a minor retooling of each band's genes. The legendary psychoanalyst Erik Erikson coined the term "pseudospeciation" to describe the growing sense that group outsiders are subhuman.[50] But pseudospeciation seems to go further than Erikson imagined. Notes David Smillie, "The initial split creates a large genetic difference between the daughter and parent groups."[51] It's easy to see how this could happen. Separating tribes of closely related Yanomamo rapidly set themselves apart by generating new dialects and rituals.[52] In the same way, archaeological remains show that fissioning groups of the Pleistocene generated very different artistry and fashion. Once these elements fell into place, they would have led each group to a separate vision of the ideal Mr. Right. One splinter tribe's picture of male perfection was likely to have differed defiantly from the vision of unblemished masculinity swooned over by the girls who'd stayed at home a mere two valleys away. Everything from romance comic books to scientific evidence[53] indicates that women find it hard to resist their culture's model of the flawless beau,[54] and that those same women shun the weirdos who just can't seem to get the latest group norms right. The result would have been a Pleistocene sexual resource shift. Men who resembled the new culture's notion of Adonis would have lured more fertile females to their beds and sired more children than the schlubbier guys who couldn't seem to kick the uncool habits the old and "infinitely inferior" parent tribe had prized. The resource shifters of status and popularity would have funneled the best food, tools, and homes, fanciest clothing, and most enviable accessories to the new group's paragons. Meanwhile, the conformity-enforcing impulses we share with lizards and chimps would have impelled our ancestors to toss group deviants to the periphery. As a consequence, the children of each Pleistocene group's "beautiful people" would have been healthier,[55] better looking, more popular, and destined for greater adult success (this is true even of the offspring of high-ranking apes and monkeys). By contrast, the fewer youngsters birthed by oddball parents would have been bullied, shunned, and occasionally killed (also true among our primate cousins).

Darwin recognized this process as selective breeding. Selection under any name can put genes through substantial change. Let's hop back to modern times to see the probable result. The Yanomamo, like most tribes, have their own model of he-man magnificence. Men who slaughter the largest tally of humans from competing tribes are rewarded with the greatest number of wives and father far more children than any other villagers. Timid Yanomamo men or those who loathe bloodshed have very few kids at all.[56] Experience with laboratory animals and guard dogs like pit bulls and Dobermans shows that aggression is a highly cultivatable trait.[57] So it wouldn't be surprising if the Yanomamo's selective breeding produced a violent disposition which exceeds even the fairly high human norm.[58]

Among some Eskimos, on the other hand, aggression is frowned on. Men who can't hold their temper are given the cold shoulder. As outcasts, they have a hard time finding a mate. And when it comes time for the traditional method of demonstrating friendship, swapping wives, these wrathful types are sidelined. Just as the Yanomamo breed aggression *in,* the Eskimo breed it *out.*[59] One result: the Yanomamo are constantly at war. Eskimos experience a very different fate: they are blessed with relative peace (though war is such a human universal that even Eskimos, until recently, periodically indulged in the grisly sport).[60]

Anthropologists have noted how a splinter culture's choice of sexual fixations makes some groups tall, some short, and even alters breast and penis shape.[61] What's more, I've sketched in an earlier episode how every culture wires infant and toddler brains in slightly different ways. As groups paraded their uniqueness with distinct dialects, methods, and beliefs, they were likely to have manufactured youngsters who saw the world from starkly different points of view. These principles were certainly at work long before the arrival of glacial sheets and sabertooths. So language, culture, and differing tools would have done for humans what simpler forms of evolution accomplished for Lake Nyasa's cichlid fish—generated an outburst of diversity.

The quibbling, rivalry, and rebellion sparked by creative bickering 2 million years ago eventually energized men and women whose ancestry was in a warm and pleasant clime to conquer the ice floes of the Arctic, the frigid plains of Siberia, the malarial wetlands of southern China, and the bewilderingly varied environments of today's France, Spain, and Germany. Fissioning groups devised ingenious ways to snatch abundance from the grasp of unknown lands and seas. A swarm of new diversity generators radically accelerated these innovations during the age of symbols. Meanwhile, new groups selectively bred genetic innovations. New cultural symbols shaped infant brains which in adulthood would see the world from innovative points of view. A ferment of resulting insights and ideas would have enriched the pan-human brew. Trade made many by-products of these adaptations common human property.[62] Since then the result—cultural evolution—has leaped to dizzying speeds,[63] throwing a spume of ever-increasing options into the communal brain.

11

THE END OF THE ICE AGE
AND THE RISE OF URBAN FIRE

8000 B.C. to 3000 B.C.

> Never doubt that a group of people can change the world. Indeed
> that is all that ever has.
>
> Margaret Mead

There is a power in the push and pull of opposites. Gravity's shackle fights momentum's fleeing force to keep a planet racing on the circle of its course. The battle between flexor and extensor muscles gives a strongman might to lift the front end of a car. Inhalation and exhalation vivify the lungs. Compression and expansion pound the heart to pump a tank-truck-full of blood a day. At Ice Age's end, sociality would draw women and men into conformity compressors of unprecedented size. One by-product would be conformity's antithesis—diversity—pouring ideas of all kinds into the interhuman mind. Old networks would give way to new, hastening the pace at which the fuel of concepts from afar would kindle flares of fire in the furnace of the nascent global brain.

Roughly 130,000 years ago,[1] the diversity generator of creative bickering drove tribes to run an artificial crease down their centers, sorting shoulder-rubbing neighbors into two opposing groups. These primordial forms of fabricated cleavage, known to anthropologists as moieties, were apparently a way to keep the deformities of inbreeding at bay. The members of a moiety were forbidden to marry each other, but were forced to pick their mates from among the members of the competing coterie, no matter how distasteful that prospect eventually may have seemed. For moieties soon showed early man's ability to build mountains of separation from molehills of similarity. Moieties identified with differing forces of the natural world. If one moiety clove unto the powers of the day, its pouting Siamese twin would declare itself an avatar of night. If one chose to be summer, the other marked itself as winter. If one was earth, the other was defiant sky.[2]

Such hairsplitting snowballed in prehistoric times. After moieties came clans.[3] Clan members believed they were descended from an animal or plant whose powers they bore.[4] In a sense each member of a Bear clan was one-half bear as well as one-half man.[5] Clan members were not allowed to kill and eat their ancestral animal or plant—their "totem"— except on sacred occasions.[6] Each totemic dish was one of the dietary staples of the entire tribe. The result: each clan was forced to specialize in a different menu, and to conserve the animals and plants which fed the members of rival clans.[7] Clans of this nature were so universal that they existed in early Australia and among the Siberian immigrants who trekked across the Bering Strait and sired Native American tribes from the Tlingit of the Pacific Coast to the Iroquois near the Atlantic sea.

This was the merest prelude to a coming revolution in creative bickering.[8] Roughly ten thousand years ago, there arose a whole new kind of pressure cooker for the sweet-and-sour opposites of conformity and diversity. This was the neolithic city. Intergroup tournaments like war, ambushes, and raids had long boosted the ingenuity of even bacteria's social brains. A human tribe that did well in the recurrent fray[9] could bully its neighbors mercilessly. A tribe less able to spit forth fresh stratagems was likely to be plundered or simply wiped away.*[10] Cities were a quantum leap in tactical defense—offering almost impregnable sanctuary from the prospect of a violent death.

The founders of Jericho, surrounded by ramparts and three-story-high stone towers, constructed this radically new form of protective complex ten thousand years ago. Jericho's mortarless boulder walls, built when most humans were still living in huts and caves, were 6.5 feet thick and four times the height of a neolithic man. They were surrounded by a trench 9 feet deep and 27 feet wide.[11] The invention of a town wall protected by lookout towers would still be used ten millennia later in the age of Leonardo da Vinci and Michelangelo.

Conformity enforcers such as the need to flock helped cities come to be. The palisades which kept invaders out held citizens tightly in, compressing them into a common mold. All shared a common reservoir of language, mode of thought, cultural reflex, religion, and ritual. Yet urban life

---

*The Patanowä-teri of South America were beaten in war by neighboring tribes, each of whom passed along the word that the Patanowä-teri were easy pickings. The more they were attacked, the weaker the Patanowä-teri became until they were forced to flee from their homes and hide in the foliage by day. Defeated refugees are not considered desirable mates, so it was only a matter of time before the Patanowä-teri would disintegrate.

heightened the generation of diversity. Author Dora Jane Hamblin and Harvard archaeologist C. C. Lamberg-Karlovsky feel that one lure of squeezing like sardines into the urban can was, ironically, the opportunity to express oneself. "A variety of roles," they write,

> awaited the man—and woman—of the city. The city, in fact, depended on variety. Its concentration of numbers was possible only because its residents performed specialized duties that would be supported by the larger society of which the city was a part. But the possibility of specialized occupations, in turn, made the city attractive. No longer was every man forced to be a hunter or a farmer, every woman a mother and housekeeper. In the city—from the time of the very first cities—there were trade goods to be manufactured, commerce to be conducted, shrines to be tended [and] . . . massive construction projects to be undertaken.[12]

There were also walls of plastered brick to be covered with artwork, patterned rugs to be woven, pottery to be made from strips of clay, jewelry to be crafted from limestone, from imported shells, and from the teeth of deer, and makeup to be ground and mixed from red ocher and green and blue minerals brought from far away.[13]

The walls of Jericho couldn't have been built without the division of labor, the fine-tuned organization, and the excess time which dripped like honey from the bustle of a civic hive. Hamblin and Lamberg-Karlovsky add that "the tantalizing variety of city life, offering the possibility of following a personal bent rather than a parent's footsteps, must have been as powerful a lure in 8000 B.C. as in the 20th Century A.D."[14]

The plains of the Middle East on which cities saw first light had been freed from the frigid sparseness of the Age of Ice. The thaw made this vast expanse an Eden.[15] Not only did toothsome plants flourish in quantities modern humans had never been able to find, but deer, wild sheep, pigs, and other roaming cuts of meat were thick as ants on a melon rind. A hunter could bag more barbecue in a day than his glacier-age forefathers, combing lands sucked dry by sheets of ice, could manage to sniff out in a month or two.

Tell Mureybit was a village of stone houses established in Syria ten thousand years ago. The fabled "abundance of farming's harvest" which allegedly allowed the flowering of cities would have been a waste of time in such lush vicinities.[16] The residents feasted on wild grain[17] and ate the animals they brought down with their bows. Even Jericho's monumental builders banqueted on wild seeds and the flesh of charging beasts.

Meanwhile, mountain-dwelling eccentrics tamed tall grasses and sought the secrets of harnessing vegetation's genes.[18] Through breeding, these fanatics nudged the hollow stems of einkorn wheat into bearing not three to six scrawny seeds but between twenty and one hundred bulging starch-and-protein packs. A single stalk could now produce forty times as much food as a dozen or more of its feral progenitors. Some men and women laid out former hunting and foraging lands in plots and cultivated the new crops. Others learned to use the conformity enforcers which hold a grip on social animals.[19] Taking advantage of a beast's need to follow a leader and subordinate itself to a superior, these inventive keepers gentled sheep and goats into following convenient paths. The pioneering pastoralists could now nab a plenitude of meat, milk, and wool without the risky and often fruitless pursuit of the hunt.

Yet Neolithic religion testified to the way in which hunting and the killing power of an animal's armaments still obsessed the early city dwellers. Nearly nine thousand years ago, one out of three rooms in the Anatolian city of Catal Hüyük was a shrine. No sanctuary's decor bore agricultural motifs—sheaves of grain or harvest emblems. The sacred chambers' walls carried mural after mural of men with bow and arrow bringing down herds of fleeing deer. But the most overwhelming symbols paid homage to the animals these fledgling pastoralists were still struggling to tame. Each tabernacle was filled with the skulls of long-horned bulls aligned in overpowering majesty. Horned skulls flared from the walls, were painted full-size on the plaster, and were installed for maximum effect on rigid benches over which the bulls' white weapons swept and curved, aiming points directly at the chests of standing worshipers. The men of the Middle East had only recently domesticated the wild ox, *Bos taurus,* an animal seven feet high at the shoulders whose horns might reach ten feet above the ground. Untamed males of this plains species would step forward from a hard-won harem, lower their heads, and disembowel humans who dared assail them with the irritation of a spear. These confrontations would have been vivid in the minds of the Catal Hüyükans, who still hunted the bulls' wild relatives, aurochs, and stuffed their storerooms with venison by tracking down red deer. The barely subjugated descendants of the *Bos taurus* oozed a strength and sexual potency far beyond that of the human male. Even a modern domesticated bull—whose size has been decreased and whose temperament has been softened by roughly eight hundred generations of selective breeding—is easily provoked to charge with murderous force against the wisp of "mighty" man.

British anthropologist Chris Knight theorizes that Neolithic females deliberately withdrew their sexual favors to prod men into hauling home fresh meat.[20] No wonder the horned heads which could so easily split the rib cage of an owner were painted next to women with their legs spread wide in sexual invitation, pregnancy, or birth, and between plaster wall moldings of round, full breasts yawning with real jaws and teeth where their aureoles should have been while other plaster breasts expressed their obdurate "no's" via bird beaks and animal tusks which stabbed from the centers of their nipples at the viewers' eyes. Other paintings underscored the fierce edge females gain from their allure by portraying yet more open-crotched women, knees splayed outward from their hips—with their arms resting on the bodies of leopards. These images carried contrapuntal meanings. They shrieked the slashing pain of sexual denial while roaring with an appetite for sexual savagery.

Goddesses, the archaeologists call the ladies of the wall paintings and of numerous full-bellied figurines. But we have no way to know if that is what they were. The suggestion of the decor was clear. Bulls had a power to pierce the walls of feminine refusal. This awed male humans with their far smaller penises and infinitely tinier bulk and might. Men, so easily cowed by womanly disdain, could only worship and hope to gain the thrust of a bull's horns and enormous phallus penetrating vaginas with vast over-loads of sperm. It was the bull who could truly make children grow in a grudging damsel's womb.

Despite its evocations of lust, strife, torment, and the wild, religion was used to synchronize the emotions and the symbol set of those who lived within the city's walls. This fervid enticement to cohesion and to the discipline of ritual geared the members of a town to think and work in harmony.[21]

Meanwhile, the diversity generators of environmental peculiarity prodded each city to devise its own approach to survival and to the gathering of wealth. One example came in the contrast between a pair of towns on the Turkish Konya Plains eight thousand years ago. Catal Hüyük was in an area rich in wild herds which included not only aurochs and deer, but the pig *Sus scrofa*. To harvest this abundance of galloping groceries, the Catal Hüyükans were big-game hunters par excellence. Their stone-tool industry specialized in weapons to bring down large prey—long spear-heads, arrowheads, daggers, and the like. But not far away lay the village of Suberde, whose surroundings presented different epicurean opportuni-ties: wild ass, wild sheep, and the mid-to-small-sized roe and fallow deer, augmented by such relatively minuscule game as fox and wolf. The flint and obsidian crafters of Suberde devised weapons very different from those of Catal Hüyük, miniaturized for small-target precision. Even Suberde's arrowheads were an unusual pattern designed for compactness and for exacting accuracy.[22]

Each town offered its particular package of technology, raw materi-als, and technique in the market of intercity trade. Economic competition hit mass circulation ten thousand years ago, when the multimillion-year-old commerce between tribes exploded dramatically. Additional cities sprang up to provide food, shelter, and good service on the paths men used to swap the wares of obsidian-mining and toolmaking towns with the products of those rich in salt, in pottery, in cloth, in copper, in the red coloring material hematite, and in lapis lazuli from hubs in south-central Russia fifteen hundred miles away. The cascade of merchandise drew distant centers into a commercial web. Oasis hubs like Jericho provided accommodations and water channeled from underground wells through stone aqueducts to exhausted and thirsting tradesmen pounding the now-permanent exchange routes; and Catal Hüyük—with its hundreds of worship rooms—offered solace to passing merchants filled with uncertainty and anxious to draw fresh guidance from the gods.

From 5500 B.C. to 2500 B.C., the Middle East's overlapping trade networks for obsidian inched the 1,500 miles from Crete's city of Knossos—eventual birthplace of Greek civilization—to Bahrain, where the Persian Gulf reaches toward the Indian Ocean. But these trade loops were a mere beginning. Obsidian in the ancient Asian city of Hattusas has been traced to Africa's Ethiopia, 2,500 miles away. Iranian cities like the 6,500-year-old Tepe Yahya were pivot points for the transport of goods from India to Mesopotamia.

A city could not survive without the nourishment provided by a mesh of planters and animal domesticators in its outstretched locality. Nor could it thrive without a flow of foreigners delivering essential rarities. The tongue-and-groove relationship between distant minilopolises triggered a cross fire of ideas, methods, and styles, swelling the pool of choices within a city even further than before. Each innovation in thought, technique, or manufacture was tossed into a continent-crossing whirl of notions which cosmopolites could tap, test, adapt, and frequently recast.[23] Conformity enforcers made cultures ape each other imitatively. Diversity generators ensured that the mix in circulation swelled perpetually.

Methods and perspectives spread across amazing distances. Some archaeologists have proposed that the civilizations of Greece and Israel and the agriculture which vivified all of Europe[24] got their start in Catal Hüyük. Even Egypt, some experts declare, imported its practice of soil cultivation from Catal Hüyük and the other Asian cities with which the settlers of the Nile were bartering their wares. Meanwhile, roughly 450 miles away, another complex of prehistoric towns—the Ubaidian minilopolises of southern Mesopotamia—exported the know-how of their cultures up the Tigris, the Euphrates, and the Karkheh-Karun Rivers to the remote coast of the Mediterranean Sea. The later city-states of the Tigris and the Euphrates would, in turn, catapult their wares and myths 1,400 miles east to the Harappan civilization in Pakistan, probably picking up ideas from the early civic snarls of the Indus River in their turn.[25]

The surge of trade's synaptic junctions heralded the coming of a global brain. A city was like a neural ganglion, a center of collaborating ecologies.[26] Catal Hüyük built its edifices and many of its finest works of art from oak and juniper, yet had no trees in its vicinity. All were felled on hillsides far away and floated down the river to satisfy the city's needs. The fir carved into elegant adornments which graced sacred alters and the best homes came from the Taurus Mountains, as did gourmet delicacies like almonds, pistachios, apples, acorns (good not only for feed but as raw material for leather-tanning chemicals and for yogurt making), and berries like juniper and the wine makers' favorite, hackberry. Other mountains closer by provided greenstone, limestone, and volcanic rock. Catal Hüyük's alabaster and calcite came from Kayseri, and its creamy white marble from lands far to the west. Its cinnabar was imported from Sizma, and its shells from Mediterranean beaches many miles and mountain ranges to the south. Salt, one of the greatest lacers of distant cultures into nets of commerce, came from Ihcapmar, whose industry was based on the mineral gifts of a nearby brackish lake. There is no flint on the Turkish plateau where Catal Hüyük is located, yet the citizens of the town used the finest varieties of this stone to manufacture daggers with serrated blades and everyday utensils to start the kitchen fire or to scrape skins for the leather industry. The Catal Hüyükans ignored the multicolored pebbles of quartz, agate, and chalcedony lying on the ground nearby in favor of more exotic semi-gems like rock crystal, carnelian, and jasper from such unfamiliar locations that archaeologists have not yet pinned their sources down.

All were fodder for Catal Hüyük's most value-added artisans. Some maestros made ordinary implements into status symbols for far-flung elites, creating the finest in weaponry, in sickle blades for wealthy farmers, in chisels and in gouges for the upscale carver of wood and bone, and in prestige knives with finely sculpted wooden hilts and pommels of the choicest chalk. Then there were the luxury items from the town's master obsidian polishers: bowls of awesome grace, makeup trays, and rings. Adding to the extravagance was the output of artistes in shellwork: wonderfully colored necklaces, armlets, bracelets, and anklets. Meanwhile, virtuoso carvers worked with horn and bone, covering boar tusks with geometric designs, and fashioning fine-tipped makeup applicators, cups, spoons, and ladles, wrist guards for bowmen, and hooks and eyelets to fasten belts. Pottery, so prominent in other cultures, was apparently a sop for the poor. Mass-produced into bargain containers of all kinds, its forms imitated the workmanship of status items with traditional prestige—those painstakingly crafted of wood and basket weave.

The opulence of high-priced craftworks and the drabness of cheap clay goods bespoke the presence of the complex adaptive system's resource shifters, which shuttle wealth and influence to those whose work appears majestic and away from those whose contributions seem mundane.[27] Though early cities like Catal Hüyük had up to thirty-five acres of straight-lined, low-slung, flat-roofed, brick housing complexes, each nuclear family's two-room suite was completely self-contained, walled off from the others for privacy and equipped with its own kitchen and sitting and sleeping nooks. Despite the nearly identical layouts of their apartments and the fact that each dwelling was sandwiched wall to wall between those on either side, the citizens of Catal Hüyük took petty divisions between groups of neighbors a good deal further than did the still-wandering tribes of Mesolithic hunters, with their simplistic separations between moieties and clans.

In tribal society, each member had been a generalist. Even the shaman presumably had known the arts of tracking prey and of other disciplines essential for survival day to day. But a city provided the luxury of concentrating on a single field, then of walling oneself off with those who shared one's métier. Catal Hüyük's shrines, as we've said before, filled one out of every three apartments in the sprawling complexes. The plethora of priests were a pampered elite whose services the masses were apparently prepared to pay for at a more than handsome rate. They lived in their own part of town, inhabited more spacious buildings than their fellow citizens, and deigned to purchase the finer things in life from the bead maker and weaver sweating in the marketplace. The complex adaptive system's resource shifters shoveled privilege to these intercessors with divinity. They did it through the urge compelling ordinary men and women to offer homage and their savings for services proffered by an aristocracy. None of the two hundred priestly quarters excavated during the first two years of the Catal Hüyük dig (which, admittedly, only covered a few neighborhoods in a very large town) showed any signs that priests were forced to do the chores which lesser folks performed in their kitchens, fields, and homes. Priests had no sickles for reaping crops and grain, yet they were gluttonous gastronomes, savoring fourteen different kinds of delicacies—from wheat, barley, and peas to apples, almonds, beer, and wine, along with the meat of game and in all probability honey and such elegant milk products as butter and cheese. The servitors of the supernatural had no looms, yet their homes were rich in cloths and draperies. They possessed no implements for crafting finished stone, yet their household treasures included ceremonial weapons of polished obsidian graced with intricate carved handles depicting such whimsies as the entwinement of a pair of snakes. Priests admired themselves in obsidian mirrors. Their jewelry contained beads with holes too fine for penetration by a needle made of modern steel. These perks of holiness were the most lavish since the dawn of human artistry.[28] No wonder country tribes deserted "indigenous culture," headed toward the towns,[29] and sought to make their fortunes in previously unheard-of ways.

Like the Ice Agers who'd sealed themselves in arbitrary moieties, pioneering urbanites subdivided into pockets of exclusivity. Weavers lived in one district, potters in another. This cliquishness swiftly upped the subtlety of the collective brain. We discussed in a previous episode how each fissioning group forms its own ideals and dialect, its own view of the world, its own emotional stance, and even its own gene pool. Fission takes a different shape when contained within a city's walls. Groups may set their boundaries, but they do not march off to start afresh in distant territory. Instead, they remain in town, each adding to the repertoire of tactics bubbling in the public mind.

Biologist and cultural observer Lewis Thomas might have called the contribution boiled up by each group to demonstrate its uniqueness a newborn hypothesis[30] available should others fail. But subcultures do not meekly wait for the day when their chosen system may save the day. Instead, all compete like soccer teams for top rung in power and prestige.[31] This claustrophobic contest jolts city life into a soaring progress that would have dizzied earlier Stone Age bands. But more about the surprises which would pop from subcultural gamesmanship in later episodes. Let's return to nitty-gritty here.

Judging from what we know of early Sumer, Greece, and Rome, the spokesmen of the gods in Catal Hüyük surely had an attitude restricted to their clique, a priestly hoard of creeds which justified their wealth, aloofness, and disdain for toil. Rival subcultures, those of the butchers, bakers, and tanners, would have clung to their own realms of expertise, their own emotional postures, and their own peculiar ways of getting at the "truth." The group in charge would have dominated mass consciousness, dictating which manners of perception, dress, and speech were allowed and which were thoroughly outré. If a crisis arose and the reigning elite seemed helpless to turn catastrophe aside, other microgroups would have vied for number one. The winner—often the subculture whose mode of operation could best defang the threat—would become the next maker of the group's mind, the arbiter of what was chic and what was not, of how to think, talk, walk, and decorate oneself. The more subcultures, the greater the playbook of vying strategies, and the sharper a community's combinatorial craftiness.

Through all of these advances, from the axons of trade to the dendrites of diversity, synapses were forming for a faster interhuman brain than any deployed before by multicellular organisms. As early as 6,000 B.C., we can see the birth of a new modernity—the evolving blueprint for whole new forms of future interactivity.

12

THE WEAVE OF CONQUEST
AND THE GENES OF TRADE

◻ ◻ ◻

In the first nanoseconds of the big bang two of the most fundamental opposites of all revealed themselves—attraction and repulsion. The repulser of explosion begat a rush apart which hasn't ended to this day. Its aftermath is an "expansionary universe" separating stars and galaxies at breakneck pace. Then there are the forces of attraction—those which pulled together quarks in threesomes, linked atomic shells to produce molecules, then sucked masses of these interlocks into the swirls we see as galaxies, stars, and human beings. Physicists are still debating whether attraction or repulsion will have the final say. But the fact is, repulsion and attraction are not battling to the death, but twining in continuous tango.

The success of a society depends on the dance between its repulsers and attractors, its huddle and its squabble, its elements of competition and of cooperation. One of our most powerful attractors is an instinct often overlooked in treatments of human history. It's the principle of reciprocity. Bacteria give each other information, and even change forms to eat what others find poisonous. To pay for this cleanup effort, the bacterium whose environment is cleansed turns more raw material into food for its decontaminator. An alga will live in the protective tendrils of a fungus, paying its "rent" by turning sunlight into fungus fare. Two lionesses will share food. If one is dying, the other will still bring her meat in token of past favors—repaying what looks very much like the gift of lasting friendship. Fair exchange holds together alliances of male baboons. A group of elite males frequently confronts a youth gang trying to woo and corner one of its most valuable assets—a female in heat. Another squad of male adults will come to the rescue and help chase the juvenile delinquents away. However, the deliverers expect that the favor will be returned someday. And it is, or the winning coalition will not last.[1] Male baboons are unexpectedly good hunters. When one comes home with a haunch of meat, he tends to keep it to himself. But a female baboon can sometimes call on the good deeds she's done the hunter in the past for the right to join in his repast.[2] Males hold babies or take care of youngsters, thus racking up the right to call on female help. But the coziest rule of repayment in chimp and baboon societies is "I scratch your back, you scratch mine"—one good grooming session deserves another.[3] Back rubs are even used as currency:[4] a subordinate chimp can groom his "overlord" into ecstasy and be repaid with the right to mount one of the head honcho's females.[5]

Homo sapiens would vastly enlarge the scope of this social adhesion device, engaging in forms of reciprocity whose long-distance reach went far beyond that of any other animal. Australian aborigines traditionally trekked a hundred miles or more to meet a rival band and swap stingray spears, axes,[6] grinding stones, necklaces, girdles, shells, hair belts, dilly bags, boomerangs, and an early version of the news couched in the entertainment format of stories and of songs.[7] In Bougainville, the tribesmen of Petats exchanged women's hoods to obtain pots. Then they traded their pots to the people of Lontis for taro. The clan leaders of Lontis in turn took the pots and swapped them elsewhere to acquire pigs. But even the pigs were just one link in a long chain of barter. The Lontians had "purchased" the porkers to trade them further down the line for shell wedding jewelry.[8]

The Puyallup-Nisqually tribe of northwest America's Puget Sound had names for ten different forms of trade, including *fgwis*—a straight swap of one item for something of the same sort; *obetsáleg*—putting down a payment for something which as yet doesn't exist; and *ábaliq*[9]—"when you take someone to somebody so you won't be afraid of them any more." Then there was the "silent trade" in which one people would leave its goods at a traditional drop point and disappear, waiting overnight for another tribe whom they might never see to replace the offering with exotic specialties before they, too, disappeared into the forest again. This was going on in Herodotus's day between the Carthaginians and the mysterious peoples on Africa's far west coast, and was still a major way in which the olive-skinned citizens of North Africa obtained ivory and gold from the black denizens of the malarial jungles to their south as late as the fifteenth century.

The silent trade's ubiquity hints that the practice's roots may have gone far back into prehistory. The Siberian Chukchee used the silent trade to swap consumer items with the inhabitants of Alaska. Africa's herding and farming Bantus used it to trade with their pygmy neighbors hidden in the bush. In New Caledonia shore dwellers would trudge to a prearranged rendezvous where they'd lay down piles of dried fish and seafood by the side of the path, then wait for their arm's-length inland partners to take what was offered and leave in their turn a pleasing assortment of tubers similar to yams.[10] And fourteenth-century Islamic traders who had ventured into northern Asia's "Land of Darkness" used the silent trade to obtain ermine and sable from furtive primitives. Despite the anonymity, haggling was part of the process. If a Muslim businessman felt that the pelts he'd been left weren't worth the items he'd deposited on the ground the night before, he'd refuse to accept them. The next night the invisible natives would either up the ante by adding to the pile of furs or would take their wares and walk.[11]

It is conceivable that our talent for long-distance reciprocity* resides not just in our learning, but in our suites of genes. A long-standing evolutionary principle called the Baldwin Effect says that once an advantageous behavior has been embraced by a population, it will gradually reshape the genetic string of the species which has adopted it, resulting in biological rewiring. The mechanism for this transformation is simple. The spiny lobsters of nearly 300 million years ago lived where glacial cliffs plunged down to the sea. In winter, this hangout became treacherous indeed. Some lobsters learned to line up in single file and trek to a warmer clime. Those creatures which went along with the crowd saved themselves from a glacial freeze and survived to reproduce. Those which stayed behind were frozen out of the mating game when they died in slabs of ice. Generation after generation, the killing cold chiseled recalcitrant genes away until finally what began as innovation became instinct. The environment continued this sorting process until only those with a genetic hankering to follow fashion and vacation in a cozier locale still multiplied. The result: permanent fixation of migratory autumn cavalcades. Time's lessons shaped a genetic template which automated the travel plans of generations far down the line of geological time.

There are numerous hints that trade, like the spiny lobsters' seasonal parades, may have been stamped by the Baldwin Effect into human DNA. Animal behaviorist Frans de Waal feels that humans offer each other presents (and expect returns) much more often than other primates do. This tendency shows up just a few years after birth, when children are often driven by instinct more powerfully than by what they've learned.[12] Children who've been fighting give each other gifts to make up. Reinhard Schropp found that in a German kindergarten, children also used gifts to become close to kids with whom they hadn't had ties before.[13] Studies in

---

*The concept of reciprocal altruism was introduced by Dr. Robert Trivers, at Rutgers University, in 1971. The view basically says that an individual will only be generous if he or she is motivated by the hope of a payoff equal to or greater than what he or she gives up. In other words, humans and the genes driving them are unalterably self-centered and greedy. Since Trivers's original formulation of the concept of reciprocal altruism, the theory has become accepted dogma among evolutionary anthropologists, evolutionary psychologists, and mathematical games modelers. Dr. Trivers proceeds from an individual-selectionist model. My account utilizes a group-selectionist model in keeping with the complex adaptive systems theory of mass behavior put forth in this book. Reciprocal altruism, by the way, is consistent with many of the views expressed here.

social psychology show the same possible trading instinct at work in college students. For example, one experiment revealed that if a student brought a stranger a soda while the two were filling out a form, the one who'd received the Coke bought twice as many raffle tickets from his benefactor when the paperwork was over as did similar experimental subjects not softened up with a gift. So strong was the need to pay back an unexpected kindness that the ticket buyers opened their wallets and coughed up the cash even when, as they admitted to researchers later, they could hardly stand the person who had bribed them with beverage.[14]

We might suspect that this attempt to balance the books through reciprocity was just the product of civilized parenting . . . if it weren't for the fact that every society that's been studied, no matter how advanced or primitive, has a principle of give and take. Aztec emperors, for example, used to slit their flesh and offer the gods their blood. They were paying for the glories they'd received and putting a down payment on triumphs yet to come. This was definitely not the result of Western influence. The civilizations of the Americas had been out of contact with those of Europe for well over eleven thousand years.[15] The Indians of America's Pacific Northwest, equally isolated from cross-cultural contamination, used to blow a pinch of tobacco off their palm as a gift to the spirits, and to request a gift in return. The Chibcha of Colombia were saddled with deities who set far higher rates for their services, demanding everything from cotton cloth to gold.[16] The most expensive blessings came from the god of the sun, to whom the Chibcha offered up their children in sacrifice. When the conquistadores first marched into Chibcha territory, the Indians, convinced the alien beings were sons of the solar deity, literally threw a hail of children at the Spaniards' feet. Far more than any other animals, we are the species who live by the rule of "to get you have to give."

Natural selection would easily have favored genes* for long-range reciprocity. Imagine that you are a late–Stone Age farmer. How do you handle the fact that your soil in the lowlands only yields low-fat crops whose constant ingestion would either bore you to death or give you malnutrition, while someone a thousand feet up the mountainside can raise fat-saturated seeds you'd find scrumptious, but that are driving *him* nuts with their monotony? How do you handle the fact that a tribe in the forest glories in hunting something both you and the mountain dweller can't get your hands on—meat—that folks living 200 miles away have the best stones for making tools which ease the job of building and harvesting, and

*Complex behaviors require not just one gene, but a team of them. Such genetic interweaves go under a variety of names, among them "complementary gene sets," "polygenes," and "multiple gene effects."

that 150 miles in another direction live folks in a territory rich in the salt needed to keep your body running? According to economic anthropologist Melville Herskovitz[17] and numerous others, you make networks of friends and exchange presents. Every gift you hand out obliges your friend to give you something back someday. You try to stay even so no one ever feels ripped off. Some of the friends you make are well beyond the boundaries of your tribe. Those who don't care to trudge the roads to distant trading spots have more meager diets than you do. They have a harder time attracting wives and protecting their children from malnutrition and disease. You have many women and a plethora of offspring. In the long run the genes of the stay-at-homes are weeded out while those of you, the enthusiastic traveler and bargainer, grow fruitful and multiply.

On a grander scale, if you belong to a society whose inhabitants possess the genes it takes to connect with a massive marketing net, you'll be blessed with a well-balanced diet. If those in the society around you are genetically predisposed to shun goods-shuttling you'll be cursed by an insufficiently varied bill of fare, be weakened by reliance on weaponry made with whatever comes to hand, and eventually will find that the territories of your band are snatched by folks more able to hopscotch commodities. *Homo isolatus* will be replaced by *Homo commercialis*.

In fact, things have worked out very much this way. Barter bound stone and iron miners in the highlands of Guatemala to food producers on the plains and shores near the Gulf of Mexico and linked together Mesoamerica's architecturally magisterial Olmec civilization 3,300 years ago. Trade pulled together a canoe-based trading circle of Trobriand Island tribes spread over hundreds of nautical miles in more recent times.[18] And we'll soon see the wonders trade accomplished for the early Greeks in roughly 600 B.C. During the last 3,000 years, long-distance trading cultures have steadily conquered, displaced, absorbed, or erased indigenous clans less enthusiastic about peddling wares across wide stretches of both sea and land.

Cities like Catal Hüyük may well have bred significant enhancements into the genetic fastener of a long-distance trading instinct. The currently dominant view in evolutionary psychology is that our instincts, now called "mental modules," were stamped into our DNA during the "Environment of Evolutionary Adaptedness,"[19] a roughly two-and-a-half-million-year hunter-gatherer phase which ended before the climax of the last Ice Age. Since then, our preprogrammed heritage has supposedly been locked in stone (or in amino acids). However, the facts don't seem to bear out this contention. Genes change far more speedily than most evolutionary psychologists realize. Explains University of Washington evolutionary ecologist John N. Thompson, "Genomes are structured and continually restructured through . . . species interactions [and] gene-for-gene coevolution." Thompson adds that "dozens" of these genetic transmutations have been known to take place in a mere "100 years."[20]

Behold the refinement of the LA gene which confers the ability to digest milk on adults. Some people, notably those of northern Europe, have it. Others—like East Asians and Polynesians—don't. It's particularly handy in wintry climes, where the sun frequently refuses to reveal enough of its radiance to generate vitamin D in human skin. This is a deficiency which cow's milk neatly cancels out.[21] However, humans, as we've seen, probably didn't domesticate animals from which they could derive dairy drinks until after the first cities were founded. Which means the gene for adult milkshake tolerance did not appear until well after the walls of Jericho were erected and *Bos taurus* was taught to toe the line. Other genes have arisen during this geological wink of time.[22] One is the sickle-cell anemia gene which protects black African peoples against malaria.[23] Still more are found in the immune shields which defended the European conquerors of the Americas from scourges like measles and smallpox. This heritage of disease resistance seems to have begun in the last five thousand years or less and developed to its fullest just within the last millennium. One clue to the immunological recency: measles is thought to have jumped to humans from the rinderpest of domesticated cattle. It was the dense-packed urban environment which turned it to a killer. In the grisly manner evolution favors, the measles virus massacred those in European cities who had no genetic resistance and left only the fortunates whose genes were able to adjust the immune system to mount an appropriate defense. These protective genes then grew robust within the following generations, making a profound mark on the face of history. The genetic acquisition of immunity was the greatest weapon of the conquistadores and colonialists, who wiped out an estimated 70 million Native Americans* with the unseen weapons of their germs.[24]

Trade, social organization, and combat may well have sorted genes with rigor, pampering those able to handle increasingly sophisticated human interactions, and punishing those unable to play the networking game. During the ten thousand years from the rise of Jericho to that of the welfare state, disasters which favored the newly fit and which winnowed out genes not up to the challenges of modernity struck over and over again. They struck in the form of war—a variety of misfortune which would inspire human ingenuity to create offensive weapons and clever stratagems able to undo the invincibility of city walls. Jericho would fall

---

*In fact, there are no "Native Americans." Even the Sioux, the Hopi, and the Navajo are descendants of immigrants, some of whom arrived over eleven thousand years ago.

and become a wasteland for thousands of years. So would the early cities of the Indus Valley's Harappan civilization. To the best and most cleverly organized went the spoils—one of which included survival. Then there were the postagricultural plagues, which continued to decimate populations from biblical times through the glory days of Athens, the height of the Renaissance, and the Age of Reason, up to the influenza pandemic of 1919 and the spread of antibiotic-resistant tuberculosis, staphylococcus, Hansen's disease, AIDS, and a host of others today. Humans have been out-foxed over and over again by a collective mind far older and nimbler than any we've developed to this point—the 3.5-billion-year-old global microbial brain.

During epidemics, the rich have nearly always outsurvived the poor. In some cases they've even benefited, as did the founder of the Krupp fortune, a wealthy burgher during the Black Death of 1349 who bought up scads of properties left vacant by plague-eradicated families for mere pennies, and whose descendants prospered off his callous canniness well into the twentieth century.[25] Krupp's windfall shows how those who master the art of social integration are privileged to protect themselves from the probability of death. Krupp had money, a bonus shoveled toward those who specialize in the skills of mass sociality. Other virtuosos of large-scale connectivity include politicians (masters of conformity enforcement, even when their primary devices for cohesion are horse trading, persuasion, coercion, corruption, coalition building, and tyranny), warrior-leaders (masters of social mobilization for intergroup tournament victories), merchants (welders of intergroup links), and priests (coagulators and keepers of the social norms).* Experts in web building are given larger, more hygienic, safer living spaces, more generous and nutritionally varied allotments of food, and servants to help them avoid such crippling daily chores as heavy lifting or grinding meal for beer and bread. (Kneeling over a grinding stone and dragging it back and forth against a slab beneath it was the method used by the low-scale women of nearly every early civilization until

---

*Social integrators—actors, politicians, persuaders, salesmen, marketers, admen, demagogues, religious leaders—unknowingly look for the basic cues which will trigger an emotional synchrony in their audience, whether it be an audience of one or of many. Hitler was probably the twentieth century's greatest master of this art—he practiced his moves in front of a mirror and even used Albert Speer's architecture and ritual mass spectacles to trigger the sort of basic neuro-emotional patterns described originally by ethologists like Niko Tinbergen. Hitler had a genius for zeroing in on what ethologists call "supernormal stimuli"—exaggerated signals which switch on primal emotional reflexes. Fortunately Roosevelt and Churchill, though less dramatic in their ability to dig out the animals in the brain, were also able to handle this feat via the use of a more subtle symphony of supernormal stimuli.

the invention of the water mill. It made good flour; but grinding destroyed the tissue in a woman's joints and deformed her skeleton permanently.)

The elite have nest eggs and even extra homes to see them through hard times. (True in the days of Rome and of Europe's great plagues—see Boccaccio's *Decameron* to get a sense of how it worked.) This means if disaster strikes, those whose genes allow them mastery of integrative skills are (and were) the best placed to survive.

Plagues came over and over. So did war. Each ran humanity through a selective sieve, culling out the socially unskilled from the high-level wheeler-dealers—politicos, bureaucrats, warrior-generals, celebrities, aristocrats,*[26] businessmen, and priests—who had mastered the art of network maintenance and enrichment. In the end, those who had most successfully meshed into the massively integrated ecosystem of a metropolis—that knot of town and surrounding countryside tied to numerous other junctures of its kind—were the ones who triumphed and survived. Well over 75 percent of the dwellers in industrial nations have now been urbanized.[27] And inhabitants of underdeveloped areas are moving with unprecedented speed from bush and farm into the megalopolises.[28] Would some mental abilities be favored and others suppressed by five hundred generations of this sorting process? Five hundred generations were enough to create massive changes in Lake Nyasa's cichlid fish. And fifty years sufficed to rearrange proboscis genes in soapberry bugs so the insects could pierce the rinds of fruits newly introduced in the American South.[29] It is very likely, then, that ten thousand years of urban living have left their mark on the genes of women and of men.[30]

One of the genetic propensities which may have been fine-tuned by this process is on display in children. Studies show that toddlers and very young kids gravitate toward and defer to those who are the best social

---

*The follies and injustices of hereditary aristocrats whose wealth was based on land are well known. What goes unnoticed was their role in organizing and administering the central machinery of mass agriculture. The Teutonic knights who conquered the Slavic realms known later in history as Prussia restructured the region's farming totally, generating sufficient bounty to turn former tribal subsistence farmers into export creators whose product fed the pipelines of international trade. The aristocrats of Tudor England put enormous sums into drainage projects which turned former fens and swamps into highly productive land. The importance of mass-agricultural organizers—whether they are aristocrats or plutocrats—to the economy of the countryside became clear in Rhodesia, where white farmers who had employed large numbers of black helpers were driven off the land. The former employees took over the acreage, stripped it of timber to feed their kitchen fires, and used primitive, each-family-for-itself techniques which drained the richness from the soil, impoverishing both the land and those who had taken it over. The former underlings to "rich whites" were far worse off when it was all over than they had been before the liberation which had made each a small property holder free of "wage slavery."

organizers. Little nippers adept at turning rambunctious companions into an orderly team not only win the most popularity contests, but become the focal point of play groups and the leaders of gangs of friends.

A seemingly valid argument has been raised that city-centered peoples would be *ill* nourished and tremendously *unfit*. How, then, could they have ratcheted up our genetic heritage? Champions of this view cite the unequivocal truth that once mankind converted whole hog to agriculture, the archaeological record shows a spread of malnourishment and its attendant diseases, diseases unknown to hunter-gatherer societies. There are two counterarguments. Towns were the spawn of trade, not, as I've mentioned before, of agricultural surplus. The first cities were founded on gathering plant abundance and on hunting where the prey was so plentiful one didn't need to shift one's abode every week or two to follow it about. What's more, it took a long time for wild game to disappear from the serving dishes of the earliest urban inhabitants. Bones of undernourished grain eaters don't show up until thousands of years after the walls of Jericho were built. Meanwhile, those in cities could spread like kudzu on a fertilized lawn. In battles they could outnumber and overwhelm hunter-gatherers. Their only competitors were nomads who stressed animal herding and made plant gardening a modest trimming to their way of life—cattle-breeding predators like the Indo-Europeans or, later, the Huns and the Gauls (otherwise known as the Celts). However, even nomad leaders like Genghis Khan were masters of social integration, organizing followers for the large-scale enterprises both of war and peace. Though they stayed on the move, the Mongols were far more urbanized than even most history fans realize. They traveled in virtual mobile homes: tents of lath and felt built on wheeled wagons. Ibn Battuta, the Islamic voyager, reported that when he visited the Mongols on their native steppes in the 1300s, these portable housing units—complete with grill windows—were gathered in "vast cities on the move."[31]

But nomads eventually could no longer menace the meshworks of metropolises. The proof is in the pudding. Urbanized, highly networked agricultural societies have taken over the world. Hunter-gatherers and wandering herders have long since been forced to exist on the fringes, scrounging in lands so poor others were not interested in exploiting them. Now those who are a part of the megalopolitan webwork have found uses for even fringe lands, and indigenous cultures are in danger of final extinction. This would be a tremendous loss for our understanding of human diversity. But why are these low-integration cultures considered more "indigenous" to this earth than those other products of the planet's evolution, cosmopolitan societies? Especially considering that the earth itself, in the form of evolution, seems to have awarded its prizes to the interwoven for at least ten thousand years?

CONQUEST AS A DATA CONNECTOR

Reciprocity was by no means the only human agglomerator and information exchanger. There was also a data connector whose power to produce
two-way information flows is filled with irony—war and conquest. When
the Indo-Europeans (succulent details about them coming up) conquered
from western China to the Atlantic, they took lives and labor, subjugating
the peoples in their path. What did they give? The chariot and their language—two of the most important staplers of human society for the next
three thousand years.[32] When, in the Middle East, those who'd been born
with the Indo-European gifts learned to use them, they stitched together
empires—those of the Hittites, the Hyksos, the Assyrians, the Babylonians, and the Persians. In the Far East, such warrior peoples as China's Chou
(1111–255 B.C.) and the horse-mounted Siberians we know as the Japanese[33] quilted together the cities and lands of thousands of small tribes and
city-states as well. Invaders dealt out something no one wants and should
ever have to accept—slaughter. And these warriors took, oh, how they
took. But most subduers kept their subjects alive. The empire builders
among them spread the tendrils pulling us together to this day. They broke
the barriers separating minigroups by standardizing languages, writing systems, laws, trade, weights and measures, and by building roadways over
which their troops could march and on which merchants, pilgrims, and the
curious could follow.

Like trade, the hunger of some societies for growth and for the subju-
gation of the less powerful is not uniquely human. Some ants are entrepreneurs, striking out on their own, finding an unexploited niche, laying eggs,
and raising their own employee pool.[34] Gradually, the new nest turns from
a mom-and-pop operation (without pop) to a major corporation as the
growing staff of underlings develops a caste structure and a division of
responsibilities. The sheer growth of the community produces advantages.
As colony size increases, so does the safety (and the expendability) of the
individual. The group can shift resources from a frantic fecundity[35] to
other forms of production . . . or of usurpation. Large insect colonies
benefit from improved defense. Small colonies have to live in vulnerable
lean-tos. Large colonies manage to construct fortresses. (Shades of Jericho!) What's more, large colonies have evolved the luxury of defending
their ramparts with castes of biologically remodeled soldiers—huge, well
armored, and well armed. Small ant colonies, where everybody has to do a
little bit of everything, cannot afford to produce six-legged battle tanks.

Large groups can also spare foot soldiers aplenty. If challenged, they
can mount massed counterattacks. Should a small colony throw all of its
members into a last-ditch defense, it would risk losing its entire population

in minutes. For a large colony, the loss of a massive fighting troop is just a minor setback.

Megacolonies provide other useful comforts. The air-conditioning systems of beehives control for temperature and humidity, increasing worker health, productivity, and hive survivability. Teams of carefully coordinated bees bring water from distant parts, relay it to indoor workers who slap it on the roofs of cells, ceilings, walls, and floors, while others at the hive entrance fan outside air into the corridors with their wings, producing a pleasant cooling effect. In winter, large hives can form a cluster of bees clinging solidly to each other, providing insulation and warming the mob by moving their muscles to generate heat.[36] Inhabitants of smaller groups are forced to endure the freeze of winter and the summer heat, a serious threat to their mortality.

Larger groups can also capture bigger, meatier prey. Because of the enormous numbers in their platoons, army ants the size of a match can bring down pigs. No wonder celebrated entomologist E. O. Wilson tells of insect empires which control huge territories, monopolize food-rich trees, and hunt down and eliminate rival nests or turn their conquered neighbors into slaves.[37]

Among humans, this sort of conquest often triggers an information swap, one in which it's hard to tell whether the memes of the victors or those of the vanquished will come out on top. The Mongols took China in 1215 A.D., conquered its landmass, and killed off as much as a third of its population in the process.[38] Then the defeated peoples' culture conquered the conquerors, who learned to live in cities, to use Chinese firearms and navies, to rule through a semi-Chinese-style administration, and to tax peasants instead of turning farmland into pasture for their horses. Despite their despotism, the Mongols expanded China's use of paper money (a financial web enhancer the Chinese had only used sporadically), encouraged private enterprise, opened free road passage throughout the empire, extended use of the system of official post roads and rapid long-distance communications, and introduced new ways of easing trade which made domestic commerce and international export and import a relative breeze.[39] They also popularized a dish previously almost unknown in China, yogurt.[40]

Conquest made it possible to extend an information network in yet more peaceful ways. India was a largely tribal culture until roughly the sixth century B.C. Then the improvement of the plow led to agricultural surplus, a cash economy, and the rise of businessmen. A new breed of ruler discovered it could tax the commercial classes, use the money to build a professional army, and set out to subjugate every neighboring territory in sight. Expansion was built into the system. Kautilya, the Indian analyst of statecraft and teacher to kings, argued in roughly 300 B.C. that it is the duty of a ruler to make war "whenever a king has at his disposal instruments of force adequate enough to ensure victory." Kautilya stated the goal unabashedly: the potentate was obliged "to make acquisitions" of territory.[41] The result was the Mauryan Empire, which launched a far more peaceful march: that of Buddhist emissaries who followed trade routes into China, Southeast Asia, and eventually most of the Far East. Thanks to a foundation of carnage, a bloodless overthrow performed by Indian ideas wove half a continent into commonality.[42]

Conquest is the needle which stitched together virtually all of the "great nations" which we know today—allowing such multitribal hodge-podges as Germans, Russians, Arabs, Japanese, English, and French to convince themselves that they have always been *ein Volk*—one folk with a unique bloodline and history. In fact, before their conquerors arrived, each of these patchworks had consisted of thousands of squabbling "folks" of different histories and different genes—their bloodlines sometimes almost as far apart as could be.

To manufacture unity, the imperialists of history followed another animal pattern, that of the dominance hierarchy—the strangely unjust principle which sometimes uses brutality to bring individuals or collections of groups together in a stable and ultimately peaceful form. It's ironic that one of our strongest forces of cohesion should be something so unpleasant as our will to lord it over others and that this ace-attractor should be egged on by repulsers—our animosities and our savageries. But that's one way the global brain has grown.

Reciprocity and conquest would slowly reweave something nucleated cells had lost a billion years ago—the ability to swift-swap information across continents and seas. For 3.5 billion years, the bacterial brain had been upgrading *its* worldwide web, and frequently showed how its collective "wits" could turn befuddled herds of humans into livestock for its feasts. But we were making progress. Next we'll see how reciprocity and conquest knitted a culture known as ancient Greece.

13

GREECE, MILETUS, AND THALES:

The Birth of the Boundary Breakers

3000 B.C. to 550 B.C.

It is not the consciousness of men that determines their being, but, on the contrary, their social being determines their consciousness.

Karl Marx

Neolithic centers like Catal Hüyük left only wordless clues to the new forms of diversity boiling in the cauldron of late–Stone Age cities. The time has come to jump five thousand years from towns of the Ice Age aftermath to municipalities which have left us detailed tablets and scrolls—artifacts whose calligraphy fills in the blank spots left by archaeology. There we can see new factors focusing the frenzies of the nascent global brain. Far below the surface mapped by standard histories, these forces tapped the power lines of neurobiology, built new arenas where instincts could chorus in harmony, and placed a spotlight on formerly mute emotions which, in turn, transformed society. The catalysts for transformation would be three: freedom to escape group boundaries; ideas; and the games subcultures were about to learn to play. We'll see in coming chapters how this triad altered utterly the workings of mass mind.

While some citizens of that western Asian flank now known as Turkey mixed mud to erect the housing developments of Catal Hüyük, their neighbors rowed over the horizon of the Aegean Sea, came across Crete and the thirty islands of the Cyclades, put down roots, and stayed. Sixty miles of sea did not stop the Neolithic data rush: the settlers kept up with developments back on the Anatolian plains, importing the new art of wheat cultivation and cattle raising, then joining the trade loop of obsidian which helped make Catal Hüyük's stone craftsmen rich. Archaeological evidence suggests that at the same time other Anatolians paddled their high-prowed boats to mainland Greece, where they found accomplished sailors who had been plying the waters, fetching obsidian from distant isles, and fishing tuna since roughly 12,000 B.C.

Judging from linguistics and from archaeological remains, eastern immigrants and seafaring salesmen maximized the genes of reciprocity— keeping the interlocked communities of the Cyclades, Crete, and mainland Greece awash in Syrian-style seals to secure doorways and storage jars, and in jewelry which followed fashions set by Mesopotamian Ur. But pre-Hellenic civilization remained Middle Eastern in more basic ways. Catal Hüyük, for example, had been decorated with wall-to-wall paintings starring women with wide-open thighs. The Cretans zestfully embraced this female fixation, but covered the loins which had entranced the folks in the old country and targeted their adulation on paintings of fully clothed women whose standing bodies made a bold display of very naked breasts. Flaunting the need of each group to distinguish itself from its benefactors, Greek mainlanders abandoned brush strokes for carved figurines. What's more, Greece's nude *femmes* lay instead of sitting or standing, and coyly crossed their arms to hide their mammaries. Just as in Lake Nyasa's cichlid fish,* biology continued serving up diversity using the slice-o-matic of creative bickering.

The pre-Hellenes had walled hilltop towns, but their defensive foresight did them little good against the connective force of conquest. They were clobbered by cattle-herding Indo-Europeans, subduers who swept down from the Caucasus in waves, the first around 2200 B.C., the second just two hundred years later.[1] Vanderbilt University's Robert Drews writes that the Indo-European invasions were "well planned and well organized, and their leaders knew where they were going and what they would do when they got there. The[ir] . . . object in leaving their country was to take control of societies that were vulnerable and that could be profitably exploited."[2] These land-scudding inventors of the chariot carved up the Helladic sea-lovers brutally. However, they brought gifts, among them the language which would become what we call Greek.[3] Being patriarchal,

---

*In 1998 Kelly K. Kissane, whose research focuses on a common North American genus of spider, *Dolomedes*, completed a study of the fishing spider *Dolomedes triton* and reported a resolute tendency to create differences between cliques among these eight-legged subjects. Groups with the same ancestry tend to harrumph themselves into difference by embracing incompatible mating rituals. Though the aquatic arachnids may continue to live in each other's vicinity, the clumps of quibblers refuse to mate with an outsider who does his mating dance "all wrong." This "reproductive isolation" is precisely the kind which leads to genetic separation between groups. In the case of modern humans, it also leads to cultural diversity.

the conquerors banished womanly charms from the local artwork* and imposed a fixation on more masculine things: a man fighting a centaur, men hunting a lion with the help of dogs, and—the real key—statuettes of horses, those sources of speed and power which lifted heroes from the ground on which mere women walked. For the Indo-Europeans had snatched the battle edge with which they'd overrun large chunks of Turkey, Iran, India, Europe, and western China by being the first to tame the "mighty steed."[4] With this hoofed transfer device they'd spread their vocabulary[5] from Xinjiang to Scandinavia, leaving a legacy which would eventually sprout as English, German, Sanskrit, Hindi, Persian, Latin, French, Italian, Spanish, Portuguese, Dutch, Swedish, Danish, Armenian, Irish, Welsh, Albanian, Latvian, Lithuanian, Russian, Czech, Polish, and Serbo-Croatian, among others.[6] As you might expect from their artwork, the Indo-European method for breaking barriers and imprinting this data wasn't exactly nice.

It would appear that the Indo-European warrior tradition frequently dictated sending out battle parties of men to slay the males in the territories they seized, then to impregnate the newly widowed or newly fatherless women.[7] Lines attributed to eighth-century B.C. poet-farmer Hesiod may give an idea of the results. He describes the origins of a testosterone-soaked, post-invasion warrior tribe thus: "And she conceived and bare Aeacus, delighting in horses. Now when he came to the full measure of desired youth, he chafed at being alone. And the father of men and gods made all the ants that were in the lovely isle into men and wide-girdled women." The Greek descendants of the settlers from Anatolia were likely to have been Hesiod's (or should we say Aeacus's) "ants." In India, war records left by descendants of the Indo-European savagers (the *Mahabharata* and *Rama-yana*)[8] portrayed the native population as monkeys. Any local males who'd evaded Indo-European slaughter would presumably have been put to work doing what ants and monkeys did best—laboring for the sons of their weapon-loving masters. The purpose for which the defeated women would have been used may explain Hesiod's emphasis on their wide-girdled hips.

Back to Hesiod for more on the product of the resulting insemina-tions: ". . . The sons of Aeacus . . . rejoiced in battle as though a feast."[9] Little wonder, given their father's probable origins.[10] This particular batch of legendary sons became a tribe called the Achaeans, the slaughter-obsessed heroes of Homer's *Iliad* and the dominators of the Peloponnesus. Says Hesiod of the Achaeans' downtrodden subjects: "These were the first

*The feminine fascination which the first Hellenes had brought across the waves disappeared for twelve hundred years, but apparently persisted underground. According to religious historian Ioan Petru Culianu, it surfaced later in goddesses like Hecate and Artemis.

who fitted with thwarts ships with curved sides, and the first who used sails, the wings of a sea-going ship." Captained by Indo-Europeans, the craft of the conquered natives became not only vessels of trade but conveyors of the sort of pillage Andromache would mourn at Troy. Later, the launches would be used for an even more historically significant task—the waterborne weaving of the Mediterranean mind.

Thanks to the "joy in battle" Hesiod refers to, no one was safe. This may be why the Greek hill forts of 1200 B.C. where superstars of plunder like Achilles, Odysseus, and Agamemnon were said to have lived[11] were so heavily buttressed. The twenty-four- to fifty-seven-foot-wide walls of these Mycenaean palaces and their outer ramparts were constructed of stones so large that later generations were convinced only a race of one-eyed giants, the Cyclops, could have wedged each boulder into place.

A thousand years later the *old* invaders were faced with new marauders. These were yet another northern people, the Dorians—every bit as pitiless as the Indo-Europeans had been. The interlopers were illiterate, but they'd managed to perfect a pole vault in materials science—the art of making swords from iron. Indo-European Greeks still perforated gizzards with weapons of a material tried and true by two thousand years of satisfied warriors, good old-fashioned bronze. However, a stubby stabbing sword was the best the orange metal's strength would allow. A fighter equipped with a long, slashing blade of iron could cut down a bronze-armed warrior before the outclassed champion could lunge close enough to puncture anything but air. Among other things, the heavy iron sword's reach allowed an invader to fight from horseback. The Bronze Age defender was forced by his runty weapon to dismount and fight on foot—a crippling disadvantage. Even if he remained mounted, the bronze-equipped battler would not have been able to get within sword's length of his opponent without being cleaved like a cut of meat. So despite the massive defensive structures the Mycenaean hunters of man and beast had built, they were overrun. Many used their subjects' ships to flee and do their *own* conquering, planting cities on the Turkish fringe of the Aegean Sea, seizing the native municipalities, killing their men, appropriating their women, and impregnating the females with more than just Mycenaean culture. Plutarch gives some sense of this old Indo-European tradition in action. The Mycenaeans, he writes, "brought no wives with them to the new country, but married Carian girls, whose fathers they had slain. Hence these women made a law, which they bound themselves by an oath to observe, and which they handed down to their daughters after them, 'That none should ever sit at meat with her husband, or call him by his name'; because the invaders slew their fathers, their husbands, and their sons, and then forced them to become their wives." The new lords and masters named the territory they'd taken Ionia. The flight which had pushed the

Mycenaeans to commit atrocities would soon produce fresh nodes of connectivity.

Meanwhile, back on the Hellespont and the Peloponnesus, the illiterate Dorians wiped out the budding alphabet—Linear B—used by the Mycenaean kings' record keepers and reduced the beginnings of a civilization to what seemed like civilization's end. It took four hundred years before the mainland Greeks would invent a new writing system and get their cultural act together again.

By this time, all manner of tribes, clans, and families had staked their claims to real estate. Some of them must have wondered if it had been worth the effort. Greece's crowd of mountainous crags left only stingy valleys in the creases where they met. The land within these rocky folds was so poorly clothed with soil that Plato later called the place "a fleshless skeleton." Grain would not grow readily here. But olives and grapevines could suck moisture from the lower depths. The families who had been first to seize the best and greatest amount of land—the Eupatrids[12]—lived off the surplus produced by their exorbitant acreage. Others were not so lucky. As they multiplied, they split their tracts amongst their sons.[13] These lots (named because each son picked a lot to see which swatch of terra firma would be his) grew smaller, so small that their owners were gradually impoverished. To help the unfortunates over hard times, the Eupatrids lent them funds, then collected their debtors' strips of soil when the harvest didn't live up to expectations. Soon a poor farmer's collateral for his loans became his body, and a defaulter became his lender's personal property.

This was one of many times when "the wings of a sea-going ship" would rearrange the evolution of human life. Farmers who couldn't pay off their borrowings ended as their lenders' slaves. Others gambled on the risky business of adventure, took to the sea, and grabbed fertile footholds in coastal enclaves whose goods could bring a hefty price back in the newly flourishing city-states of home, which badly needed grain[14] and craved exotic luxuries. The colonists spread their web twenty-three hundred miles from Mainica in Spain to Tanais in Russia. Conquest once again carried cohesion's threads.

Mainland cities were not the only ones sending out boatloads of the land-and-profit hungry to weave a mesh of trading colonies. Nor were they the first to be jolted forward by the rush of data which comes from the resultant upgrade in network power. By the seventh century B.C. the Asian Greeks of Ionia were hell-bent on rooting excess citizens wherever they could find a promising beach. One of those leading the pack was a rampant Ionian reciprocity-spreader called Miletus.

The city of Miletus imported wool from the local Anatolian countryside, ran it through textile mills, built an impressive clothing industry, and sowed eighty colonies from the Black Sea and Egypt to Italy. With her sixty colonies to the north, Miletus traded cloth ware for flax, timber, fruit, and metals. What's more, Miletus was part of a web of hundreds of additional Greek colonies. These were perched, as Plato said, "like frogs" on the shores of Europe's and western Asia's central seas. So Miletus's commerce stretched from Russia to France, Spain, and North Africa, and it sucked in goods from the exotic east. Miletus's resulting wealth became "a scandal"[15] throughout the Mediterranean. But the real bonanza would be in something more important than material goods.

When Stone Age hunter-gatherers had first clustered in cities, *Homo sapiens* had unexpectedly acquired the freedom to choose a major part of their identity, becoming a priest, bead maker, carpenter, or the odd etcetera. Yet they had apparently remained yoked to clans and tribes. Seven thousand years later, Miletus still ranked its inhabitants in three tribes. Across the Aegean where things were less advanced, a backward folk called the Athenians identified themselves as members of one of four tribes, each of which was slivered into ninety clans.[16] Tribes would still be vital to Rome's voting system over half a millennium later, when Mario, Sulla, and later Julius Caesar would run for consul.

As time went on, the meaning of "tribe" was no longer clamped entirely to birth. Newcomers to a Greek city were often assigned a tribe at random—no matter who their forefathers might have been. But once you had your tribal label, it was unchangeable. Despite vocational liberties, preordained identities still stuck to you like glue.

During the seventh century B.C., intercontinental nets of city-states embedded in an even larger mesh of trade were fast eroding the old fetters of affiliation and breaking down old boundaries. Thales of Miletus exemplified the heady nature of the new fluidities.

In roughly 640 B.C., a Phoenician[17] couple decided to settle down in Miletus. Their son, Thales, became a speculator in vegetable oil futures or the rough equivalent thereof—a form of investment choice unheard of in pre-urban days. During the winter when idle oil presses were money losers for their owners, Thales "gave deposits for the use of all the olive-presses in Chios and Miletus, which he hired at a low price because no one bid against him. When the harvest-time came, and many were wanted all at once and of a sudden, he let them out at any rate which he pleased, and made a quantity of money . . . his device for getting wealth is of universal application, and is nothing but the creation of a monopoly." (This description comes to us courtesy of a financial reporter named Aristotle.)[18] However, Thales had long since been speculating in several other ways. For starters, he had taken the plunge into politics, befriending his local dictator—Thrasybulus—and whispering foreign policy recommendations in the tyrant's ear. Since Thrasybulus was a major figure on the intercontinental stage, such whisperings made a difference. Thales became the mover behind an attempt to coax the Ionian cities into a defense alliance against the Persians.[19]

A rival king, Croesus of Lydia, had something else in mind. He hired Thales to reroute the Halys River so that it would clear a path for his troops. Thales was also reportedly a friend and adviser to both Solon and Lycurgus, the great lawgivers of Athens and Sparta, putting him in position to help shape two of the most influential governmental systems of antiquity.[20] If Plutarch is correct, Thales not only visited Solon in Attica, but entertained the traveling Athenian back home in Asia Minor.[21] As a political consultant, Thales was a relatively new form of data connector—a cable zapping knowledge from one society to another, a linkage in the wiring laid by conquest and reciprocity.

Probably the least-recognized data connector is the need to keep up with the mental twists of your competitors and enemies. To win, you have to fathom the tricks of your foes so well that you can use them in your sleep. You copy your adversary's database within the very core of your own brain. "Know the enemy . . . and you can fight a hundred battles with no danger of defeat," said a distant wise man of Thales' era, Sun-tzu.[22] Thales was a knower of the highest degree.

Miletus's opponents included the conquering Persians—who'd mastered the accumulated lessons of two thousand years of Levantine imperialism, including the art of war, the logistic organization of masses of men, and the administration of huge, culturally jumbled territories.[23] Insights from competitors synapsing into the Milesian mind also included innovations from that other unpredictable superpower, the reciprocity-mastering Lydians, creators of silver and gold coinage—a financial barrier piercer the Milesians grabbed and made their own, becoming lenders throughout the Mediterranean basin and the land bridges to the heart of Asia and the Ganges River valley. Both Persia and Lydia were on Thales' beat.

The days were over when a new idea like that for the Acheulean hand ax had taken seven hundred thousand years to worm its way through the far-flung tribes of prehumanity. The travel perks of an intercontinental troubleshooter like Thales made him a one-man data link able to zap new notions between cultures almost instantly. Thales mused on Persian astronomy with such intensity that legend says he walked into a well while pondering the stars. He also reportedly dropped in on Egypt, learned its mathematical and geometric systems, used them to calculate the heights of the pyramids, then brought them home, where some of this high-tech software would later be attributed to Euclid. Using Egyptian computing techniques and material on the heavens from the Mesopotamians, Thales is said to have predicted the eclipse which accompanied the May 585 B.C. battle between the Lydians and the Medes. He thus demystified for the Ionians an event which brought the soldiers of Lydia and Media to their knees, filled with dread that if they drew more blood, the gods would be displeased.

Thales also helped initiate the concept of secular philosophy, reducing the perverse magic of a persnickety universe to elements graspable by reason. Among other things, he tossed aside the habit of explaining everything via mythology and generated a down-to-earth theory of how this cosmos had come to be. The world, he declared, had self-assembled from water and from "psyche." The macrogods of sky were banished in favor of microgods one could finger in a stone or chew in a leaf of grass . . . for, as Aristotle put it in one of his references to Thales' cosmogony, "all things are full of gods"[24]—an early way of saying that each object has its own inherent properties.[25]

For those who needed something with more oomph than metaphysics, Thales' Milesian neighbors offered poetry, prose, and history. As time went on, a minisociety, an advanced subculture, a whole new kind of tribe, would grow around each of these inventions. In fact, as different schools budded off to argue for the centrality of some hairsplittingly reworded point of view, the choice of groups to join would soon become bewildering. The couplings and scufflings of these factions were poised to reinvent the operating system of mass mind.

Pondering the blush before the dawn, Thales offered a piece of advice which would be attributed to several philosophers after him: "know thyself." He didn't say which of the many possible new selves he meant. As a financial wiz, a guru on policy matters, a transnational thought conductor, an arch reciprocator, a Phoenician, a Milesian, an Ionian, a Greek, an heir to Middle Eastern culture, a speaker of an Indo-European tongue, a replacer of theology, and an inventor of philosophy, the man could count up quite a few . . . had he really wanted to.

14

SPARTA AND BABOONERY:

The Guesswork of Collective Mind

1200 B.C. to 600 B.C.

Make no little plans. They have no magic to stir men's blood
and probably themselves will not be realized. Make big plans;
aim high in hope and work, remembering that a noble, logical
diagram once recorded will never die, but long after we are
gone will be a living thing, asserting itself with ever-growing
insistency. Remember that our sons and grandsons are going to
do things that would stagger us. Let your watchword be order
and your beacon beauty. Think big.

Daniel Burnham, Chicago architect
(1864–1912)

Since the age of Jericho ten thousand years ago, trade had turned the pace
of human interaction from glacially slow to brisk. Now, in 600 B.C., men
like Thales breezed back and forth across state boundaries, twisting
together threads from alien cultures and tinting them with original
thought. Other professional thinkers would soon use these strands to
weave new visions of reality. But some of the audacious would go much
further, taking what Thales and his ilk produced to drastically recraft societies. Before we tell the story of these social architects, it's necessary to
digress. For the deeds of men like Solon and Lycurgus will reveal the full-
ness of their meaning only after we explore some further secrets of a mass
intelligence.

A complex adaptive system is a "nested hierarchy"—a net whose nodes are
each a part of a larger entity. Each of these larger "superorganisms," in
turn, is a node in a far larger web. And each is also a hypothesis.

Every brain cell in a newborn is a guess. If the spot to which it migrates and the function it adopts turn out to be necessities, it stays and even gains in "popularity"—other cells massage it with nerve endings begging for what it has to give. If it is a motor neuron geared to make a tongue click like the African San language's popping "!" and everyone chattering 'round it pops away, it will grow vigorous and stay. If it's sandpapered by the baby talk of English burblers whose syllables never make their palates snap a "!", the cell prepared to make a tongue click will shrink, then die away.[1]

The cell is a feeler for the person in whom it gets its start. And that individual is a mere stab in the dark for the groups of which he or she is part. Should he try twenty ventures and see every one of them slide down the tubes, he'll be shunned by other humans, spurned by money, hounded by the torment of his failures, and will probably die at an early age. Even if he dodges homelessness and alcohol, he may be done in by depression, which persuades the immune system to lower its weapons when bacteria storm the gates.

The hopeful baby turned an adult wreck is a probe sent out by the group to explore its possibilities. His path helps teach his group the routes it must elude. But if his attempts turn to gold, then love, health, and recognition may be his destiny. He has taught the group a better way. Right guess and you win. Wrong guess and you lose. Or in that unholy refrain I keep drilling into you from the words of Jesus, "To he who hath it shall be given. From he who hath not, even what he hath shall be taken away."

Meanwhile, the individual depends for his success on the greater group's hypotheses. Should the tribe or nation of which he is a part march toward a utopia which turns out barren, his band will shrivel round him, and he will shrivel, too. But should the collective mind time its trek so it arrives as fruits abound, it may gain power over groups from far around. And from its fate those other groups may learn their ways to navigate. This spray of antennae, test shoots, and guesses gives the complex adaptive system its genius: its ability to slip out of tight spots and into opportunities, even if it means inventing ways to pole vault over barriers, or better yet, to turn obstacles into possibilities.

Each node in a collective brain represents a different approach available to the larger mesh of mind. Individuals and subgroups are disposable rovers, sensors for an interlaced intelligence. Way back when, we saw a bit of how this works in bacteria. Here's how individuals and groups operate as hypotheses in the collective brainwork of baboons.

Baboon societies differ vastly from each other, so vastly that they once befuddled primatologists. Seemingly definitive studies on baboon behavior came out in the 1960s, only to be contradicted by other, equally authoritative studies performed a few years later. Why did one scientific team find that baboons behaved in one way, while another discovered that baboons acted in a different manner entirely? Because the researchers had studied different baboon societies, each of which had adopted its own particular strategy. For example, *Hamadryas* are the democrats of the baboon world.[2] Each male has a family which he keeps rigidly in line. But there's relative peace and freedom in the band of 130 or more. When it comes time to decide which direction to move, the younger males offer body-language suggestions to the decision-making elders and usually "steer" the contingent as it roams.[3] Troops of *cynocephalus* baboons, on the other hand, are smaller, rougher, and more despotic. Primatologist Hans Kummer compares them to "monolithic armies." Males fight it out to see which few will come out on top. Winners receive the privileges of an oriental potentate—a disproportionate amount of food and most of the sexual attentions of the females. The resources of the group are under the victors' control. And the ruling elite can be brutal in fending off the influence of its juniors. The difference between *cynocephalus* and *Hamadryas* may be partially determined by genes. But that's by no means always the case.

The Pumphouse Gang was a troop of *cynocephalus* baboons originally located in Kekopey, Kenya, which relays of researchers began to study continuously in 1970.[4] Anthropologist Shirley Strum, in the book retelling her thirteen years with the Gang, shows many of the ways in which individuals and groups can become test pilots for speculative strategies.

In the Pumphouse Gang males were quiet types, incidental footnotes to the real social scene. The power that counted was wielded by the females. This female-centered structure was—like social structures of all kinds—an educated guess about the best way to ride destiny's roiling tides. It offered the advantage of relative peace (though baboon females can be brutal to each other) and stability. But when circumstances change, advantages can often prove a liability.

Female baboons are conservatives at heart. Males, with their rambunctiousness, tend to be explorers of untested possibilities. During the Pumphouse Gang's first years under the eyes of scientists, a lone male occasionally got lucky, came across a helpless antelope, grabbed it, killed it, and brought it home. The ladies in the troop became increasingly voracious for meat, and soon began to grab at rabbits and gazelles. However, their success rate was abysmal. Then came the real value of male restlessness and strength. As often happens when a group goes through a change, an unlikely leader kicked the whole thing off. His name was Rad. He was young, strong, and an individualist who didn't so much follow the troop as run ahead or lag behind, exploring to his heart's content. A meat craver of the highest order, Rad had been following the best hunter in the troop for years, learning his elder's tricks while being denied a share of his kills. Finally Rad set off on his own, stalking groups of Thomson's gazelles and scouring the grasses for a hiding fawn. He'd eat his prey in private, then return to the troop covered in blood—an inadvertent advertisement of his expertise. When the mighty hunter took off again, the other males spread out and followed from a distance, trying to figure out how Rad was pulling off his flesh-acquiring feats. One day when Rad was hunting in his solitary way, he dashed into a gazelle troop and lunged at a fleeing fawn. He missed, and caused the herd to bolt . . . directly into the arms of the surprised males who'd been peeking at the action from behind the crest of a nearby hill. Rad and his observers learned a critical lesson—that by using teamwork, they could panic the gazelles into an ambush. Thus was group hunting born.

While the female baboons stayed at home, each trying to improve her skills at snagging an occasional animal without aid, the males were mastering squad maneuvers, trekking two hours to find the most vulnerable prey, scoping out the terrain to maximize their stratagems, and learning to focus on the herd riddled with the greatest number of easy-to-catch infants. What's more, the males pulled off a social breakthrough. Under normal circumstances baboons are so selfish that a mother won't share food with her own children. But the team hunters grudgingly doled out their catch—something unheard of in the standard baboon repertoire. They looked the other way when females grabbed a portion of the groceries they'd brought home. And even mothers let their infants join in finishing what they'd snatched.

Here's where the hypothesis of female-centered power turned out to be disabling. The sophisticated hunting team of males eventually left the Pumphouse Gang and moved on, each departing in his own time and for a different destination. The females—backbones of this matriarchal troop—remained. Though they continued their fruitless solitary hunting, they never learned to set forth on scouting expeditions and to utilize group strategies. The gazelles in the neighborhood could relax again. Pumphouse's meat lovers had been reduced to vegetarians.

One guess down, three to go. Another Pumphouse subgroup, again led by the young and the restless, came up with a far easier way to feast. Adolescent and postadolescent males went off in bodies of six or seven to indulge in a new discovery: farm raiding. For the stodgy—females and older males—trespassing in a cultivated field or garden was utterly outré. They turned their backs on such dangerous asininity. The daredevils who'd gone off to filch ripening ears of corn had momentarily tossed themselves out of the competition for sexual favors, but the chance they took paid off. Their rich source of nutrition allowed them to bulk up, making them physically larger than the rivals who had "wisely" stayed at home. This advantage would later bring the young risk takers the female adulation they had seemingly foresworn.

The first females to join in were the postpubescent crowd. Networks of alliance proved critical to information uptake—for these gang molls had been longtime friends and partners of the renegades who'd invented the profitable new sport. Next came the occasional older female who'd been mating regularly with one of the crop-stealing males. But once these ladies saw the richness of agricultural pickings, they were utterly swept away. Says Strum, they not only pitched in vigorously, but "often took the lead," stuffing their stomachs, cramming their cheek pouches until they bulged ridiculously, grabbing as many corncobs as their hands and underarms could carry, and, like shopping mall looters, making an awkward getaway. Meanwhile, the well-established oldsters stuck with their traditional routines and for the most part stayed away.

Three months later, the crop thieves—male and female alike—set up a separate sleeping camp. Says Strum, "Older sisters, nieces and nephews, even a few of the mothers of some of the delinquents joined in." Families separated. Most mothers stayed with the master troop while their children went off with the adventurers. A few discontented females abandoned their children to join the rebels, among them the lowest-ranking mother in the Gang, who was angling (says Strum) to exit her bottom slot and snag a more prestigious lot. Her juvenile sons stubbornly remained behind. At first, some members slipped back and forth between the two assemblies, unable to make up their minds. But as the units' sleeping quarters moved further and further apart, loyalties grew more firm. Then, as is often the case when groups divide, violent skirmishes broke out. The well-fed upstarts, apparently sensing their growing upper hand, were usually the ones to start the fight.

Farm raiding proved a boon. Males gained far more time to laze around. Females not only acquired the leisure for extra socializing, but won an even more significant prize. In the old days they'd eked out their subsistence by grubbing the savanna dirt for deeply buried bulbs. Back then it had seemed normal to give birth only every eighteen to twenty-four months. But fattened on the fruits of croplifting, the females were now able to produce new babies every twelve. This put the farm raiders in position to outbreed their reactionary Pumphouse rivals nearly two to one.

As yet there were no clear winners in the occasional battles between the old group and the new, but the innovators had yet another edge. Not only could they spend less time gathering food and more time marshaling their strength for each coming fray, but their fertility might easily tilt the balance conclusively someday, since overwhelming fighting numbers can bring invincibility.

Yet a third group had discovered an even bigger (and more dangerous) windfall—the garbage dump at the Gilgil army barracks a fair distance away. Not only was the human refuse of gourmet quality, but the location came complete with amenities: running water and convenient sleeping sites in the wilderness nearby. To top it off, the army camp's wives and children, bored and having no televisions to watch, found the daily baboon escapades irresistible. They handed out food to the youngsters and the mothers in the troop. Here's where the hazards came in. First off, the garbage dump was surrounded by electrified barbed wire. Getting through it meant accepting significant injuries. And once the baboons realized that the humans were rooting them on, some of the bigger males took the liberty of breaking open the doors of storehouses, pillaging women's gardens, barging into homes, overturning furniture, prying open cupboards, and snatching food from shelves. This the humans did *not* appreciate. One day a seething officer convinced the local game warden to take out his rifle and handle the problem. Only one baboon was shot. Meanwhile, the farmers being raided back at Kekopey had set their dogs on *their* baboon pests, killing more than ten. Had things gone on, Shirley Strum felt she would soon see the Pumphouse Gang and its offshoots annihilated entirely. At that point, she obtained a plane and spirited the groups away.

The Pumphouse Gang had split into three entities. Each was a separate hypothesis, a distinctive gamble on destiny. The value of stealing human food became obvious when Strum's baboons were inspected medically. The garbage harvesters were the most robust and healthy of the three. The farm raiders came next in the vigor of their vital indices. In the worst condition was the backward-looking team which had been eating the food Mother Nature provides.[5] Though nutritionally the garbage and farm strategies were way ahead, both were leading to a showdown with human beings. One of the three conjectures would eventually prove its superiority. Others might be clobbered by catastrophe. Through this trio of competing strategies, the interknit communities would have advanced their schooling. But thanks to Strum's rescue we'll never know which practices would have won the day.

There was no Shirley Strum to stop the group tournaments which educated the neural net of ancient Greece.

LYCURGUS'S GAMBLE ON FUTURITY

The rise of men like Thales ushered in the coagulation of social groups not around tribes or baboon-like discoveries, but around the probeheads of philosophies. Societies and the subgroups within them were not shaped entirely by accident and environment.* Some high-placed citizens thought their way through to the kind of state they felt would be ideal. Two examples were Athens and Sparta. One was allegedly given its shape by the legendary lawgiver Lycurgus,[6] the other by yet another savant whom Thales had once counseled, Solon. One was a land-rooted[7] military society built with all the no-nonsense, no back-talk discipline of an overgrown army encampment. The other was a seagoing trading empire which thrived on the intake, debate, and export of opinions. Sparta faced resolutely inward;[8] Athens faced without.[9] The Spartans and Athenians were allies, copartici-pants in Greek society, and simultaneously enemies, jostling for control of Greece in its entirety. In their story lay many of the new properties of com-plex adaptive systems which meshworked cities ushered into being.

The extent to which Sparta embodied a radically different hypothesis from those embraced by its Greek neighbors was attested by the historian and short-term resident of Sparta Xenophon, who states outright that it was "not by copying other states, but by deciding on an opposite course to the majority" that made the Spartan "country outstandingly fortunate."[10]

In the days of the Trojan War, the city of Sparta had not existed. Its future territory had been home to the Mycenaean kingdom of Lacedaemo-nia, seat of Helen of Troy's cuckolded husband Menelaus, whose urge for revenge gave Homer a clash worth singing about. Then came the Dorians, sweeping from the north around 1200 B.C., wiping out the culture that they found, then mysteriously departing. For two hundred years the archaeological record shows a blank. Then came a second Dorian sweep. This time the conquerors stuck around, giving birth to the first thoroughly Spartan polity.[11]

The Dorian band which overran the southern Peloponnesus clamped the lid tight on the peoples it conquered. Following the tradition that had led their distant relatives to establish the caste system in India, the invaders, who renamed themselves the Spartiate, dictated that only they could be citizens. Those they'd squashed were the usual "ants and mon-keys," a subrace to be used like harnessed beasts. One subjected class, the *perioeci,*[12] were, much like India's Vaisya caste, allowed to call themselves freemen, to administer their seaside cities, and to carry on trade and man-ufacture—a necessary source of income for an arrogant overclass. Others were less fortunate. These *helots*[13] were worked "like asses under a heavy burden," according to the early Spartan poet Tyrtaeus.[14] They toiled like the lowly Indian *shudras* to raise food for the mess tables of their masters.

---

*It's entirely possible that *no* human societies have been shaped entirely by accident and cir-cumstance. However, the archaeological record does not reveal what role, if any, leadership, innova-tion, and ratiocination may have played in the crafting of collective modi operandi before roughly 3000 B.C., when the Sumerians invented writing.

The Spartans were only a small minority. But they held full sway over the city-state's group brain—allocating its resources, speaking for it internationally, making its decisions about everything from daily life to matters of war and peace, and, to top it off, forcing the majority to live under the Spartan banner in a state bearing the Spartan name.

In 700 B.C. revolution became the byword nearly everywhere in Greece. Aristocrats waxing fat with wealth had gone on a power spree. Those on the lower end of things were anything but pleased. Nor were they particularly quiet about their newfound miseries. The authorities felt a desperate need to pull order *(Eunomia)* out of the chaos of riot and incipient anarchy. Some used tyrants to solve the problem. Sparta took another tack. Lycurgus—son of a king, king for a short time himself, and now the guardian of a king too young to rule—reputedly traveled to Crete, Asia, and Egypt, analyzing their governments.[15] In the face of crisis, he set off for Delphi,[16] then returned home claiming that Apollo, speaking through the Delphic oracle, had given him something which was becoming all the rage—a constitution—one which threw enough bones to quiet the howling underdogs among the Spartan citizenry.[17]

Shaping Lycurgus's system and the peaceful assent of the public to its decrees was a ferocious revolt by the underdogs *below* the underdogs, the *perioeci* of the enslaved province of Messenia—which made up roughly half the territory Sparta called its own.[18] This made the obvious even clearer than before. There were only nine thousand Spartans.[19] To prevent overthrow by their enslaved majorities, they were going to have to garrison themselves and permanently militarize.[20]

The Spartan center of the day was not exactly what a good Greek would have called a city. It was a collection of towns surrounded by a common defensive wall.[21] Thucydides wrote that when only Sparta's ruins were left, "future generations would never believe that its power had matched its reputation . . . without any urban unity, made up as it is of distinct villages in the old style, its effect would be trivial."[22] Sparta was banking on a hypothesis which has some moderns in its grip—that urbanism is an evil. Meanwhile, Athens, Sparta's future competitor, was destined to be a city extraordinaire.

Sparta tested yet another guess which still seems much in vogue today—that we humans do our best functioning in ancient tribal ways. Children were educated and men were housed communally—as in common nineteenth- and twentieth-century utopian visions. Under Lycurgus's constitution, all were theoretically viewed as equals—another goal of utopias from the revelations of Christ to the French Revolution of 1789. These practices were throwbacks, says the historian W. G. Forrest, to tribal Dorian days. Unlike Athenians, who used old Ionian tribal markers primarily for ceremony, the Spartans continued to organize many of their basic activities around the three Dorian tribes from which they'd apparently come: the Hylleis, Pamphyloi, and Dymanes.[23]

Not only was Sparta the proving ground for ideals of village living and adhesion to the tribe, but it was the test case for yet a third hypothesis about the perfect life. When the other cities of Greece had planted overseas colonies, many of their aristocrats had turned from farm management to manufacture and trade. Sparta spurned industrialism and stuck with soaking her wealth by force from the vanquished peoples in her immediate vicinity.[24] This created an atmosphere in which, as Plutarch retells it, the citizenry saw "working at a craft and with moneymaking as only fit for slaves."[25]

Back to Lycurgus. When it came to wagering on hypotheses, the lawgiver placed his chips on the enforcement of conformity. He slashed the importance of private homes and ordered that from now on all meals would be served in mess halls "to reduce to a minimum disobedience of orders." So if you were fortunate enough to benefit from Lycurgus's cerebrations, your food rations were measured out without regard to your appetite. If you wanted to add to your meager diet, you were required to organize a hunting expedition and bring down your own meat. Wine—the only thing a decent Greek would drink—was restricted just as thoroughly. Getting tipsy was not allowed. Men could not bunch off in groups of friends, but were required to eat with the mix of members assigned to them. Table conversation was limited to one subject: the latest "noble acts" performed by worthy citizens.

Lycurgus did not want his people wearing themselves out with the silliness of love and sexuality. Says Xenophon, "He made it a matter of disgrace that a man should be seen either when going into his wife's room, or when leaving it. For by having intercourse under these circumstances their desire for one another was bound to be increased. . . . Besides he would no longer allow each man to marry when he liked, but laid it down that they should marry when at their peak physically."[26] The marriage itself was not exactly romantic. The bride was "carried off by force," had her hair cut short, was dressed in men's clothes, and was left "on a mattress alone in the dark."[27] Explains Plutarch, "The bridegroom came . . . after dining with his companions," removed his bride's ceremonial belt, "then, after spending a short time with her, he left for his usual accommodation soberly to sleep with the other young men." The groom continued "sleeping with his comrades" from that point on, and the first baby was often born "before the father had seen his wife by the light of day."[28]

Possessiveness was out of the question. If a man spotted another whose body and mind he felt were superior to his own, it was his duty to invite this model of masculine perfection to impregnate his wife. Similarly, if a bachelor spied a beauty who seemed particularly disciplined and rugged, he was required to ask her husband for permission to impregnate her. The point of all three rules was to produce babies with the right stuff militarily.[29]

Life was as hard for women as it was for men. Unlike in other Greek city-states, female children were fed every bit as well as males. And unlike their counterparts elsewhere, they weren't expected to hide in their homes spinning wool and making clothes. Says Xenophon, Lycurgus left clothes weaving to slaves, and "required the female sex to take physical exercise just as much as the males; next he arranged for women also, just like men, to have contests of speed and strength with one another, in the belief that when both parents are strong their children too are born sturdier."[30] Women ran and wrestled against the boys, threw the discus, hurled the javelin, handled horses, and drove carts.[31]

Under Lycurgus's rules, obedience to authority dominated even the "democratic" government. The country was run by a council of thirty aristocrats, the Gerousia, chosen for life in a rigged election. Judges decided who had gotten the most votes by hiding themselves from the voters and coming to their own conclusions about which candidates had allegedly roused the loudest applause. To pacify the masses, there was a citizens' assembly—but post-Lycurgan amendments placed it under a virtual gag order. Only the council of thirty could propose legislation and topics for consideration. Apparently if these patricians didn't care for the way their edicts were received, they could shut down discussion by using their right to make "withdrawals."[32] According to the post-Lycurgan amendments, there was to be no "crooked speech," only a yea or nay, after which the Gerousia withdrew and made its decision anyway. To be fair, on occasion the citizens' assembly slipped out of hand and managed to get its opinions across. But this was a rarity primarily limited to moments of impending catastrophe.[33]

The Spartan state fashioned by Lycurgus was known for its "military fitness and efficiency" and for its "austerity."[34] But it also seems to have had more than its share of intolerance for modernity. According to Plutarch, it banned the use of coined money, which as we've noted was revolutionizing the connectivity of the international scene. It dictated exactly what tools and what tools only could be used to put up a house—the old standbys of ax and saw. Employing any new construction implement was against the law.

Wrote Critias, the dictates of Lycurgus tested "endurance to the limit."[35] *Agoge,* the Lycurgan military discipline ruling life, began with childbirth. The elders of the tribal subgroup inspected each newborn and determined if it was up to Spartan snuff. If not, it was left to die. If, however, it made the grade, its fate was still severe. Each toddler was allowed six years with his mother, then was uprooted from home and conscripted into the educational "squadron." He would serve in this soldierly unit for the next fourteen years of his life. Here he would work his way upward through the ranks in a system W. G. Forrest calls "increasingly brutal and brutalizing."[36] The title given to the aristocrat in charge of the educational apparatus was telling enough: "Trainer-in-Chief." Under this brusque figure was what Xenophon calls "a squad of young adults equipped with whips." Severe punishment was considered the path to "respect and obedience." To harden the six-year-old draftees as they passed through their school (or was it boot camp?) years, Lycurgus insisted they go barefoot at all times, even on the rockiest terrains, and be issued only one piece of clothing to see them through both summer sun and winter cold, so they'd become inured to extremes of temperature. Their unappetizing and inadequate food rations were designed to keep them permanently hungry, giving them a fourteen-year-long "taste of what it is not to have enough."* If they wanted more, they were forced to steal it; and if caught they were given many a lash. Yet those who managed to purloin "as many cheeses as possible" were looked upon with honor. All this was crafted to make each pupil an expert in the skills of the soldier who "must keep awake at night, and by day must practise deception and lie in wait, as well as have spies ready if he is going to seize anything." Not that the boys were given privacy to plan their thefts: the youngsters of Sparta had a commander watching over them twenty-four hours a day. To draw the noose a little tighter, any citizen who walked by when the Trainer-in-Chief wasn't on the spot was duty bound to issue orders to the students and to punish them harshly if each dictate wasn't precisely obeyed.†

When they hit adolescence, Greek boys in other states were given their freedom and allowed to roam at will. Not in Sparta, where it was "appreciated that at this age youths become very self-willed and are particularly liable to cockiness" (per Xenophon again). The Spartan system overloaded these would-be rascals with labor, and declared that any who tried to ditch part of his burden would be condemned to second-class citizenship for life. What's more, teen ebullience was crushed by rules that would make modern youths protest publicly, burn buses, and overturn cars: "Even in the streets they should keep both hands inside their cloaks, should proceed in silence, and should not let their gaze wander in any direction, but fix their eyes on the ground before them." At their military-style communal meals boys were only allowed to speak when spoken to. Xenophon was impressed with the resulting "self-control." Said he, "You would

---

*These words come from Xenophon.

†On the other hand, men were not permitted to take young boys as lovers, something Xenophon, when recounting the phenomenon, said was so utterly un-Greek that he doubted his readers would believe him.

sooner hear a cry from a stone statue or succeed in catching the eye of a bronze one."

The theme of young adulthood was cutthroat competition—but competition between groups, not individuals. Three young leaders were allowed to select a hundred each of their peers, explaining to those who did not make the cut exactly why they were not up to par. This produced an in-group which felt it was its job to rat on those below them, and a resentful cluster of rejects who did everything possible to catch their "superiors" in breaking the rules of "honor" and of "bravery." No matter which side you were on, you constantly had to watch your step. One slip and you were not only disgraced, but went directly to the whip. In Xenophon's words, the children were virtually "at war."[37] They had to stay "physically fit," since they could "come to blows whenever they met." On the other hand, if one of them kept on fighting when an adult ordered him to halt, he would be brought before the Trainer-in-Chief and given "a stiff fine" to show that "anger must never prevail over respect for the law."

Perhaps the ultimate display of the system's results came in a ritual of passage eventually required for Sparta's young men. Teenagers competed to see who could endure a flogging the longest. Some of them died trying to come out first.

Though historians W. G. Forrest and J. M. Moore declare that many of these rules were elaborations of old Dorian tribal tradition,[38] the fact is that they were not adopted by any other city in mainland Greece. Plutarch said that Lycurgus had begotten a system which trained humans "to be like bees, always attached to the community, swarming together around their leader, and almost ecstatic with fervent ambition to devote themselves entirely to their country."[39] Aristotle was more harsh. He wrote that the Spartan system "turned men into machines."[40] However, the constitution of Lycurgus was a thought-out choice in which a people concurred. It was a wager on the values of authority, obedience, war, and uniformity—a conscious hypothesis about the best way to make it through the challenges the future would bring. How valid this communal guesswork was, only coming centuries would tell.

15

THE PLURALISM HYPOTHESIS:

Athens' Underside

3000 B.C. to 399 B.C.

> A removal from one set of people to another . . . will often include a
> total change of conversation, opinion, and idea.
>
> Jane Austen

Meanwhile, above the Peloponnesus in Athens, another hypothesis was
taking shape within the nascent global brain. This one would increase not
merely the speed with which information flowed, but the way in which it
would be used to fuel the flame of creativity. Sparta was a heartless breeder
killing off diversity. Athens was diversity's warm and cozy hatchery. As a
result, Athens and Sparta tested a fistful of opposing policies: among them,
authoritarianism versus libertarianism, internationalism versus isolationism, and totalitarianism versus democracy. The winner would not be the
contestant anyone was likely to foresee.

The peopling of Athens' hills came relatively late. In 3000 B.C.,
roughly five thousand years after the inhabitants of Jericho had laid down
their walls and three thousand after Catal Hüyük had gone into full swing,
unknown peoples settled on the northwest slope of a hill we know as the
Acropolis and left behind their high-grade burnished pottery. How long
they'd been there before they made their fancy kitchenware, archaeologists have as yet been unable to ascertain.

The spot was blessed by several advantages. Its Acropolis had a direct
water supply, unlike the hilltops on which other settlers relied for their
defense. The surrounding region was shielded by four mountain ranges.
And its landmass jutted into the sea, allowing it to caress the trade-bound
winds of interchange. Catal Hüyük had long since traded with the isle of
Crete, and Crete is the place from which the early Athenians pulled their
heaviest influence. While the Thebans looked back to a land-wandering

Oedipus and the Spartans to a power-packed Hercules, the legendary planter of Athenian culture was reputed to have been Theseus, a transport-obsessed inventor who had solved the mystery of the Cretan maze, defeated Knossos's Minotaur, then devised a flying machine (more about that aircraft later). By 1200 B.C. the Acropolis was a heavily buttressed hill fort whose walls, unlike those of other Greek cities, would never be breached by Dorian assailants.[1] While Sparta's helots would be cut off from their heritage, Athens' cultural thread would not be snapped.

Legend has it that in roughly 1250 B.C., Theseus united the 12 pivotal towns of Attica's 140 villages into one city whose merger was commemorated with a wall of great circumference. Archaeologists place the date of centralization at closer to 700 B.C.[2] The newly consolidated polis could have lived, like Sparta, off the grain produced by its surrounding territories. But the soil was fit for barley, and Athenians craved wheat—available in quantity only from Greece's southern Russian colonies. To earn the dough that fed their appetite, many of Athens' aristocrats ceased living on their land, moved into town, and made new fortunes founding industries. A taste for foreign bread would eventually make Athens the center of the web of commercial outposts from which Miletus had also sucked its wealth. However, this import-export approach initially looked like a losing proposition, for Sparta eclipsed Athens in fame and power by what seemed an infinite number of degrees.

In Sparta, there was very little pick and choose. Your dinner, your dining place, the nature of the table you sat at, the hall in which you noshed, the companions at your table, and the subjects you discussed were ordained by custom and authority. Athens was the opposite. A man could select from a catalog of social clubs which reflected his personality—whether it leaned toward materialism, snobbery, intellectuality, or emotionality. If you wanted to show off the glory of your ancient lineage, you could join one of the *gennetai* who monopolized the priesthoods. If you were seeking comradeship, mutual aid, and a major hoo-ha on the holidays, you could sign up for a lower-class religious fellowship, a secular burial society, a benefit association, a trade group, or a hobby club, then join together with your adopted "brethren" in feasts, chat fests, and business gatherings. If getting bombed, trashing the town, and occasionally mucking about in politics with aristocratic chums caught your fancy, you could join a high-class drinking club (a *symposion*). If you wanted to get in shape after evenings of debauch, you could enroll in the sports club at a *gymnasium*. For the mystically minded, there were sects galore. For intellectuals, there were schools gathered around great thinkers. No matter what your disposition, there was a famous mentor within whose circle you could hone it royally and know that despite the oddness of your views, you were among friends who eagerly agreed with you. Even pirates had their official societies, complete with bylaws sanctioned and regulated by government authorities.[3]

This gave the diversity generator the opportunity to upgrade a component critical to the development of global mind, one which had appeared back in Catal Hüyük in more primitive form—the interest group, the subculture, the specialized clique, the gestator of new ways of thinking, new preoccupations, new emotional stances, new techniques, and new beliefs. Through subcultures the concepts coming in from across the seas would show their power to make of cities something no tribal society ever could have been. Each new conviction flowing from abroad entered the marketplace of ideas, vied for buyers, and if it became a hit, gathered 'round it a fan club, its own minisociety. Behind this process lay forces rooted in the strangeness of emotion and biology.

THE PSYCHOBIOLOGY OF SUBCULTURES

> Our brains differ as much as our bodies. Indeed, they may differ
> more. One part of the brain, the anterior commissure . . . varies
> seven-fold in area between one person and the next. Another part,
> the massa intermedia. . . , is not found at all in one in four people.
> The primary visual cortex can vary three-fold in area. Something
> called our amygdala (it is responsible for our fears and loves) can
> vary two-fold in volume—as can something called our hippocampus
> (involved in memory). Most surprisingly, our cerebral cortex varies
> in non-learning impaired people nearly two-fold in volume.
>
> Dr. John Robert Skoyles

Thanks to Plato, we have what purport to be records of the conversations of a human Cuisinart of concepts, an eclectic sage whose roughly fifty-year-long intellectual life bracketed the Periclean Golden Age (443–429 B.C.). This all-purpose conceptual chopper and blender was that son of a socially high-placed family, Socrates. Experts and neophytes agree that it's impossible to tell how many of the words Plato ascribes to this self-appointed gadfly were authentic and how many were simply Plato's way of getting his own notions into the public eye. But one thing *is* generally accepted as accurate—the names of the folks from whom Socrates extracted opinions before shredding them with the quiz mastering which now bears his name (Socratic dialogue). The cast of characters palavering with Socrates in Plato's *Dialogs,* says learned reasoning, was too well known in Athens for Plato to have fudged.

Just who were the fonts of learned conversation whose wisdom Socrates whipped and whirled? Socrates' interlocutors were frequently famous thinkers from distant cities, each of which specialized in a different

manner of plucking goods from its surroundings and injecting them into
the circulatory system through which the trade of the Mediterranean and
the Black Sea swirled. Socrates was a student of Anaxagoras, who came
from the Ionian city of Clazomenae on the coast of today's Turkey. He was
also a disciple of Archelaus, another Ionian import. The Socratic dialogues
Plato "chronicled" included those with Protagoras from the Balkan city of
Abdera, Hippias from Peloponnesian Elis, Parmenides from Italy's Elea,
and Gorgias from Sicily's Leontini. Each visiting intellect had been shaped
by contact with a unique group of surrounding tribes, and by the exigencies
imposed on city structure, domestic habit, and vested interest by distinctive forms of enterprise. One result: each arrival presented a philosophy
which appealed to a very different configuration of the human mind.

To understand how philosophy couples with the mind's biology, let's
track the complex adaptive system's best-concealed constituent to its hiding place. The five elements of the complex adaptive system are conformity enforcers, diversity generators, inner-judges, resource shifters, and
intergroup tournaments. Inner-judges may be the most unusual of the
crew, for they are physiological built-ins which work deep inside the body
to transform a bacterium, a lizard, a baboon, a me, or a you into a module
of a larger learning machine. The basic rule of learning machines is one
we've already seen: turn on the juice to components which have a grip on
the problem at hand and turn off the power to those components which
just can't seem to understand. Inner-judges help decide whether the components in which they reside will be enriched or will be denied, then they
aid in carrying out the sentence. The irony is that these evaluators, prize
givers, and executioners are *built into their victims* biologically. On the
microlevel, inner-judges work through "programmed cell death"—apoptosis—a molecular chain reaction deep within the genes which ends in cellular suicide.[4] In higher animals the inner-judges dole out interior
punishments which range from overdoses of stress hormones[5] to emotional miseries. Or they grant internal bonuses of zest and confidence to
those of us fulfilling our group's needs.[6]

When we feel like kicking ourselves around the block or curling up
and disappearing, our condemnation comes from inner-judges like guilt
and shame. What's a good deal harder to realize is that behind the scenes
our inner-judges sicken us and dumb us down quite literally. If they sense
we're a drag on the collective intelligence, inner-judges downshift our
immune system and neurochemically cloud our ability to perceive. They
induce a narcotic haze by swamping our system with endorphins, the
body's self-produced equivalent of morphine.* And they flood us with glu-

*"Endorphin" is a contraction of the term "endogenous morphine."

cocorticoids which kill off both brain cells and lymphocytes—critical cells in our fight against disease.

Inner-judges measure our contribution to the social learning machine by two yardsticks: (1) our personal sense of mastery; and (2) the hints we get from those around us telling us whether they want us eagerly or couldn't care less if we disappeared like a blackhead from the face of decent society.

Mastery is a useful gauge. It measures whether we're coping with the trials tossed our way, and whether our example can help steer others in their trip through choppy seas. Popularity is an equally practical yardstick. It measures the extent to which we're feeding others' physical, organizational, and/or emotional needs.

Nestled deep within our neuroendocrine complex,[7] inner-judges operate on a sliding scale. By adjusting our mix of neurotransmitters like serotonin, dopamine, norepinephrine, and acetylcholine, or the balance between the gloomy right and sunny left side of the brain,[8] they shift us from fear to daring, from misery to happiness, from grouchiness to charm, from timid silence to expansive speech, from deflation to elation, from pain to ecstasy, from confusion to insight, and from listlessness to lust or to the resolute pursuit of goals.

Some of us are born with inner-judges whose verdicts are perpetually harsh. The result is depression, shyness, and heightened susceptibility to pain. Others arrive from the womb with inner-judges preset to treat us generously, endowing us with energy, few inhibitions, a deep sense of security, and little sense of guilt or shame.[9] But most of us are in the middle— our inner-judges sentence us sternly or magnanimously depending on the snugness with which we fit our social network's needs.

Those born with inner-judges excessively lenient or severe have taught us much about the secrets of mental and emotional diversity. Harvard University researcher Jerome Kagan has probably never heard the term "inner-judges," yet he may have done more than any other psychologist to uncover their capabilities. To understand what Kagan hath wrought, a background briefing is in order.

The early-twentieth-century psychoanalytic thinker Carl Jung, says Kagan, originated the concept of introverted and extroverted personalities. Jung also believed that each had a slightly different brain structure. Kagan feels that in his own way, he has proven Jung right. He's found that 10 to 15 percent of infants are born with a tendency to be fearful and withdrawn, while another 10 to 15 percent are born with a flair for dauntless spontaneity. During the last few decades of the twentieth century, Kagan performed numerous experiments and accumulated large amounts of data demon-strating his concept's validity.

He refers to facts like these:

- In studies of Japanese and American newborns, some infants took the removal of the nipple from their mouths calmly, while others went into emotional fits.[10] The babies as yet had had no opportunity to learn these reactions from their parents. The tendencies were those they'd brought with them from the isolation of the uterus. At four-teen months, the babies who'd been easily upset at birth were still so oversensitive that they often broke out crying when the sight of a stranger loomed.[11] On another test, babies who became upset at birth when they were switched suddenly from water to a sugar solu-tion squalled hysterically at the age of one or two when their mothers left the room, but babies who had taken the change in beverage casu-ally did not.[12] In addition, a study of 113 children showed that those who had a hard time handling the unexpected when they were one year old were still shy and withdrawn by the time they reached six.
- This tendency toward variation in personality was not limited to human beings. According to Kagan, it appeared in dogs, mice, rats, wolves, cats, cows, monkeys, and paradise fish. Some of these ani-mals were fascinated by novelty. Others were terrified by anything the least bit out of place.[13]
- Fifteen percent of cats steered clear of strangers and even avoided attacking rats. This was remarkably close to the percentage of humans frozen by anxiety attacks.[14]

Kagan traces these differences to genes, which can help set off a life-long domino effect in the brain. The production of a key manufacturing enzyme for the stimulant norepinephrine, says Kagan, is controlled by a single pair of genes, making norepinephrine levels highly heritable.[15] Nor-epinephrine—which is also a potent stress hormone—shows up very early in the development of the embryo,[16] making the hippocampus oversensi-tive to the unfamiliar, and hyperactivating the amygdala, which jolts us with the warning signal we call fear. The hippocampus and amygdala—as we've seen earlier—are central shapers of the memory bank we call real-ity. They are also key to the inner-judges' machinery.

Genes are not the only reasons 30 percent of us pop out of the pla-centa with our inner-judges preset to render us outgoing or fearful and shy. An Italian scientist, Alessandra Piontelli,[17] has studied identical twins from their first weeks as embryos to their childhoods. Using a series of hour-long ultrasonographs, she's watched how siblings compete within their mother's womb. One twin will become dominant and the other subordinate. The dominant will hog up the most comfortable space, leaving only a cramped corner to its brother or sister. As one-year-olds, the pair will show the same characteristics. One will be active, an extrovert, the other passive, an introvert.[18] The outgoing, domineering twin will enthusiastically attempt to communicate with adults, even if its only language consists of eye contact, smiles, googles, wriggles, crawls, and runs. The other one-year-old will make a face of silent distress and often toddle fearfully away, attempting to hide as if it were still seeking protection from the kicks of its more boisterous and expansive twin during their days in utero. Both twins have the same genes, but prebirth experience makes a walloping difference.

Now here's the rub. Many of us are conceived as twins. Roughly 150 million[19] people alive today are victors in a competition with a brother or sister who never made it past the early embryonic stage. We helped kill them off well before birth. And all of us, as fetuses, rapaciously plundered our mothers' reserves, battling her defenses for all we were worth. Despite the bounty she naturally shared with us, we used biological weapons like placental lactogen to raid the pantry of her glucose stores and destroy the nerves and muscles with which she rationed our blood flow.[20] So our social experience—be it good or bad—began in utero and shaped the very way we formed. In addition, we marinated in chemicals our mother secreted to handle her own crises and her joys. Her stress hormones were capable of premolding us as emotionally troubled infants. Her hormones of happiness would have had the opposite effect. But one way or the other, we each emerge scrambled in our own peculiar way.[21] It's from this scramble that the inner-judges' preset at extremes are made.

Later in life the products of a prebirth norepinephrine cascade are timid children, who, in carefully controlled studies, are alert to slight changes in tones or brightness of light that other children miss. In other words, these children literally see and hear their world in ways others would not recognize. According to Kagan, the constitutionally frightened are endowed with a limbic system hair-triggered to curse them with a sense of imminent catastrophe. As a consequence, shy children attempt to escape punishment by hiding from everyday events which threaten to torment them hideously. Uninhibited children, on the opposite end of the scale, have underaroused limbic systems and demand a deluge of entertainment to dodge boredom's intolerability.[22] Their craving for excitement can sometimes wear their parents to a frazzle.

Kagan's shy children are condemned to solitude and pain by hanging judges in their own biology.[23] Kagan's uninhibited kids are gifted with indulgent inner-judges predisposed by the limbic system to offer such unearned rewards as boldness and social dexterity.[24] But most of the animals and humans Kagan has studied avoid these two extremes.[25] Seventy percent remain in the middle, their inner-judges handing out positive and negative verdicts according to the rules of the learning machine.

Nurture and nature both play a role in shaping the way the inner-judges of adults behave. Culture affects the manner in which mothers raise their children, and that, in turn, helps shift the balance between boldness and vulnerability. Seventy-five percent of American one-year-olds studied are upset when their mothers leave the room and abandon them with a stranger, but only 33 percent of West German babies seem disturbed. Thanks to their country's child-rearing style, the inner-judges of West German infants cut them a bit more slack to handle things independently.[26]

Class—a factor that existed 25 million years ago[27]—also has an impact on our interior sentencers. Lower-class mothers say they cherish physical closeness as much as mothers in the middle class. However, testing indicates they tend to ignore their babies' needs and leave their infants untouched and unheld a high proportion of the time. This produces more timid, fearful children than those nurtured by middle-class mothers whose approach is often far warmer and more benign.[28]

Healthy inner-judges shift a creature from inhibition to boldness depending on the signals hinting at its value to society. Babies receive these signals from the interest their mothers and others show in what they try "to say." When babies invite you to play, you can see their inner-judges wilt them if you frown or thrill them if you grin and bend down to clown around. When a mother comes back into a room after leaving a baby with a stranger, some calm down quickly, some don't, and others are relatively indifferent. The easily comforted tend to be the ones blessed with an engrossed mom who listens to her baby's sounds, watches its face and eyes, and twines her coos and burbles with her infant's in emotional duet.[29]

The complex adaptive system algorithm for collective intelligence gives inner-judges a vital role to play. Early in this book, two networked learning machines trotted to the turf—the neural net and the immune system. In each, elements helping unravel the problems at hand were hyper-energized and became the centers of attention. A successful node in the neural net attracted an eager concentration of connections from the crowd around it. Immune system lymphocytes with a lock on the invader's weak points shot out signals which a myriad of other cells obeyed. Lymphocytes and neural-net processing components without a clue, on the other hand, went into self-imposed retirement.[30] They shunned fuel, retreated from the team, and in essence shut themselves away. Inner-judges granted the most valuable players pep and enfeebled bumblers. In animals like humans, inner-judges serve the same purpose. Meshes of our own brain tissue and cocktails of glandular chemistry give the most valuable players among us zest and pull those of us who blow it toward depression and obscurity.

Nature is often cruel and sometimes kind when she makes us modules of mass mind.

Jerome Kagan has drawn conclusions with wide-ranging consequences from his work. He feels he has demonstrated that what a parent does is not the ultimate in child raising.[31] Because human infants react differently to the same potentially stressful events,[32] several children growing up in the same household will each perceive a vastly different environment. What's more, a baby wailing unstoppably at three A.M. trips a different frame of mind in parents than a cuddly infant who sleeps through the night and whose tears can be turned to smiles almost instantly.[33] Other researchers agree with Kagan that an infant partially shapes its parents' behavior, molding the contours of what it and it alone perceives as the nature of its family.[34]

Kagan's work also implies that those with inner-judges whose verdicts are at the extremes play a powerful role in expanding a culture's capabilities. Kagan studied the biographies of T. S. Eliot, Franz Kafka, Alfred North Whitehead, Alan Turing, and Nobel Prize–winning neurologist Rita Levi-Montalcini. His conclusion: an inhibited temperament can nudge a child toward a life of creative scholarship.[35] Kagan cites a study of pioneering architects and mathematicians which revealed that they'd been rejected by their peers when they were teenagers (indicating that they'd probably been hypersensitively shy). Their later success was motivated, among other things, by a desire to gain power over those who had once mocked them mercilessly.[36]

Apparently in some cases, once such people have taken their revenge and proven their merit to society, even the most dour of their inner-judges are forced to grant a reprieve and wipe some of the fearfulness away. For a 1990s version of this neuroendocrine about-face, see the career of the once-timid Bill Gates, who, like the withdrawn John D. Rockefeller[37] a century before him, became a pitiless competition crusher, a fearsome match in almost any game.

Even more important for the evolution of communal mind, as we'll soon see, is Kagan's belief that infant personality type influences the subculture with which a child identifies later in life. What's more, Kagan's studies of creativity imply that inhibited babies may grow to be the prophets around whom subcultures congregate.

Athens did not kill off its introverted, physically weak, and oversensitive newborns who would have made poor warriors. Nor did it regiment them mercilessly as youths and eradicate their sense of individuality. Over to Xenophon again: "The time when boys develop into youths is the very moment when . . . [Athenians] remove them from tutors, remove them from schools and have nobody in charge of them any longer, but leave them independent."[38] Independent to make choices. Independent to find others with aberrations like their own. Independent to seek or build a subculture in which their otherwise oddly shaped dispositions might find a welcome home.

Sparta after Lycurgus was straitjacketed in a single cultural mode. It picked and chose at birth who would live and die, then continued to winnow its ruling class as they grew up, using rough competition to pound its youths into one mold of acceptability. There was no room for musing poets like the pre-Lycurgan Spartan Alcman—reputed to have been the inventor of love poetry. When Sparta needed a rhymer in roughly 650 B.C. to write the fiercely war-soaked anthems which would stir it to military victory in the second Messenian War, it was forced to import a rhapsodizer—Tyrtaeus—from Athens. Athens, on the other hand, allowed the development of subcultures galore. Each of these subcultures provided a haven for a different kind of temperament. Each struggled to make its mixture of emotion and ideas a defining force in the city-state's collective perception. And each worked to impose its chosen techniques for bending fortune to its will.

Just as inner-judges turn the individual into a component of his society's intelligence, the subcultures around which abnormal individuals gather help a society switch its settings and either take on leadership or blend in with the multitude. In other words, subcultures help adjust the inner-judges of entire states, making them nodes in yet a larger thinking entity—the web which would eventually become a networked brain for all humanity. In Athens, subcultures and the nourishment of both boldness and oversensitivity would literally change the course of history.

16

PYTHAGORAS, SUBCULTURES, AND PSYCHO-BIO-CIRCUITRY

570 B.C. to 399 B.C.

All the world is queer save me and thee, and even thou art a little queer.

> Attributed to a Quaker,
> speaking to his wife

My school yard was like something out of *Lord of the Flies*. The bully ruled.

> Reed Konsler

I'll probably regret saying this, but . . . for me kin have always been bad news. Warmth and hope came from strangers as they became friends, mentors, allies, etc., while family is the shared trait of those who diminish my happiness and augment my griefs. I know in my bones that blood is not thicker than water.

> David Berreby

People who were suffering silently found that they were not suffering alone.

> Melody Beattie

There had been little room for a sensitive, vulnerable, and inward-looking male in a tribal community. Should he be sufficiently determined, he might become a shaman. If he was shy to the point of panic and had the rare luck to be in a tribe which allowed such a thing, he might seek refuge as a berdache,[1] a man-woman who dresses like a female and becomes a wife. However, if he was in a tribe like Brazil's Yanomamo, he'd end up shunned for his cowardice, have no wives, and hence no progeny. As a result, his genes would disappear from his group entirely.

But there was a new slogan in the urban Greece of the fifth century B.C.: "know thyself." On the surface, it seemed to imply rugged self-sufficiency. However, its subtext preached a hunt for connectivity. "Somewhere," it whispered between the words, "is a subculture into which you fit. Seek it and adopt it as a home for your identity." Fugitives from the normal, driven by off-kilter inner-judges and by the disdain of others, had what researchers call "low stakes in conformity."[2] They could ferret out a freakish idea, then use it to declare their independence from those who had rejected them. What the folks you'd grown up with had called weirdness could be your pass card to a brotherhood of strangers who shared your "insane" sensibilities. In the interurban web, you could for the first time choose a group which fit the contours of your neurobiology. But while the culture which had spat you out was probably local, the new form of subculture which welcomed you in was often a vortex of transnationality.

To see these whirlpools of psychobiology spiraling in the greatest number we'll have to find the right city. And Athens was the rightest one in sight. Foreigners like Zeno—who laid out what would be recognized for thousands of years as the three basic domains of philosophy: logic, ethics, and physics—arrived in Athens from Italy's Elea carrying the seeds of an intellectual sport for thrill-seeking underexcitables, those whose perpetually parched limbic systems thirsted for neural thrills. Zeno's contribution was the mental rough-and-tumble Aristotle called the dialectic. Socrates gave this gift a local twist and presented it as his own "Socratic method," within whose social confines a variety of convention piercers found abode. Athens' haven for more conventional extrovert daredevils came from across the Thracian Sea with the Sophists, who peddled swashbuckling in politics, in the law, in public duels of rhetoric, and in the hornswoggling of international diplomacy. (Sophists also offered the best all-around education of their day. But that's a matter for another time and day . . . or better yet, for the footnotes.)*[3] The word "sophist" actually meant a knowledge

*The Sophists have been given a bum rap by history, largely due to Plato, who was jockeying with them for influence. Sophists were among Plato's favorite villains in his write-ups of the Socratic dialogues. He went so far as to call them intellectual "prostitutes." However, the great Sophists—of whom there were many—were walking universities. They taught a full curriculum of fields from science, literature, and philosophy to government, diplomacy, and, of course, oratory. Here's how Protagoras, a superstar Sophist whose visits to Athens sent the town into a pop-star-style tizzy and whose ideas made substantial contributions to Western thought, sums up the Sophist curriculum and his variation on it: "If Hippocrates comes to me he will not experience the sort of drudgery with which other Sophists are in the habit of insulting their pupils; who, when they have just escaped from the arts, are taken and driven back into them by these teachers, and made to learn calculation, and astronomy, and geometry, and music . . . ; but if he comes to me, he will learn that which he comes to learn. And this is prudence in affairs private as well as public; he will learn to order his own house in the best manner, and he will be able to speak and act for the best in the affairs of the state."[4]

carrier,[5] and was equivalent to the current term "professor." Sophist experts like the Abderan inventor of the traveling seminar, Protagoras, and the touring Leontinian lecturer Gorgias charged tidy sums for their public presentations, but would also make themselves available to visit your home and give you private lessons for a mere $100,000 or so.[6] They equipped you to argue your case persuasively before audiences which would have made an introvert cringe. Meanwhile, other cosmopolites wafted into town from the Italian colonies carrying Pythagoreanism, a refuge for limbically bruisable introverts and others of the contemplative sort. Then there were the homeboys—Athenian-born Cynics, magnets for those drawn to self-denial, pain, and the desolate caverns of the soul. Like most bands of social rejects, the Cynics offered a belief system with which their followers could turn the tables on their tormentors, damning the high and mighty who had mocked their childhood sensibilities. Fourth-century-B.C. Cynics preached purgation from the "deformations" of Athens' "modern" state and urged return to nature's purity.

But we're getting ahead of our tale. In the previous chapter, we glimpsed the manner in which the diversity generator cranks out neurological extremes. Now it's time to see how the conformity enforcer hooks together the resulting irregulars—creating networks whose connecting pins punched through the parochial barriers of the ancient polis and continue to perforate the barriers of the modern state.

At heart is the fact that like is attracted to like—which amounts to a good deal more than common sense. Thanks to the findings of mid- and late-twentieth-century science, it is now apparent that our itch to be with those who seem simpatico goes far back in evolutionary time. For example, it shows up in sponges, creatures which first arose roughly 550 million years ago. Sponge cells, like most organisms, live in structured communities. Normally the amoeba-like creatures hunker together indivisibly. But run a sea-fresh sponge through a sieve into a bucket of water, and you jolt the inhabitants from their fastenings, forcing them to wander homeless and alone. All you'll see in your bucket is a cloudy liquid, but that cloudiness is much more than it seems. Despite their forced separation, the animalcules seek each other frantically and finally clump together again. Now sieve one red sponge and another of a yellow hue into the same container. The differently colored refugees will spurn desegregation and madly seek out others of their kind. Bigoted red cells will shun yellow and vice versa. Within three days, the reds will have reassembled with their fellow ruddies in one clump and the yellows with their lemony sisters in another.[7]

Our bodies, too, are steered by this microorganismic legacy. Put cells from the eye and the liver in water. The liver cells will gang up with other liver cells; the eye cells will chill with others from the eye.[8] Cells continue to seek soul mates when they're not pureed. In the growing brain of an embryo, neurons reach for partners with whom they'll spend a lifetime in embrace. Each scouts out fellow nerve cells which share its electrical rhythms, then taps into a cellular clique which pulsates to its clannish beat.[9] (Reggae fans in one corner, please; rappers, rockers, and classical lovers in another.) And so it goes on up the bio-chain. Little as exterior pigmentation means to the muscle power, swiftness, or brainpower of a cichlid fish, the color-conscious swimmers congregate according to their shade of physio-body-paint.[10] Chimp mothers raising youngsters clot with other toddler-moms, and adult males hang out with other male adults.[11] Sapient humans also follow this primal rule. A Detroit survey of 1,013 men showed that whites tended to choose whites as best friends, Protestants to choose Protestants, Catholics to choose Catholics, Republicans to choose Republicans, and working class to choose working class.[12]

Individuals, especially those with temperaments which defy the norm, are pulled together by two kinds of similarity—their emotional wiring and an important by-product, the extent to which they see things eye to eye. Experiments show that humans are drawn to those who share their attitudes on religion, politics, parents, children, drugs, music, ethnicity,[13] and even clothes.[14] They'll do everything from standing closer to their kindred-in-belief[15] to marrying[16] them in preference to someone other factors tag as a more likely candidate for matrimony.

Beliefs are not just rallying flags, but symbols of emotionality. And emotion-flooded souls—like those filled with misery—love company. Social psychologist Stanley Schachter told one group of college girls they'd receive a painful shock. He explained to another how enjoyable the electrical surge would be. Then he gave the girls a choice of spending the time before their voltage dose in a waiting room with a bunch of other young women about to undergo the same amps and watts or in a room by themselves. Twice as many of those who thought they were about to be tortured wanted to nestle in the comfort of a similar-fated gathering.[17]

We mammals are uncannily good at gravitating toward those who share our hidden joys and woes. This talent for emotional homing crops up among beavers, wolves, and even deer.[18] In the rhesus monkeys Harry Harlow studied, it's particularly astonishing. When it came to mating, those who'd been raised in isolation fell for others also brought up in quarantine. Those who'd spent their youth in cages wooed other victims of captivity. Now here's the topper. Some of the monkeys had been lobectomized. Though none were handed pictures of each other's brains, those with similar neurosurgery managed to sniff each other out. So subtle were the differences detected by the simians that even researchers couldn't spot them without a careful study of medical and rearing charts.[19]

Humans are much the same.[20] Children whose gifts or disabilities make them seem bizarre, for example, manage to find each other and to congregate.[21] Among our kind it's called validation. Without others on our wavelength the strangeness of our emotions can make us feel we're losing our minds. Jerome Bruner explains one of the prime reasons we despair. Some cultures, he says, forbid discussion of experiences which other cultures wallow in quite heartily.[22] Should we be stranded in a milieu which taboos the moods we live from day to day, our essence is expelled from permissible reality. Aggregations of others like ourselves can offer a nest in which we finally feel worthwhile. But more goes on than mere emotional rescue. Once gathered, those of us who are akin often escalate our similarities, aping each other until we've feedback-looped our common thread into an insignia of our group's superiority.[23] Drop this psychosocial filament into the solvent of a multi-urban brew—et voilà, you have the formula for a boundary-breaking subculture glue.

The big fastener-fest hit in roughly 600 B.C., when prophets and philosophers took advantage of the new trans-territorial fluidities. Until then notions like religion had been used to demonstrate who was related to whom and who was a member of which gene-based tribe.[24] The house of Jacob in the valley of the Jordan had shown its genetic solidarity by worshiping Jehovah, while the tribe which invented the Greek language had flaunted its lineage through fealty to Zeus. But the users of the new mucilage epoxied together *transcontinental* circuits of genetically mixed humanity. These clans were based on neurohormonal fraternity—a solidarity of sensibilities. Among the large-scale integrated patch makers of the fifth and sixth centuries B.C. were figures like China's Confucius and Lao-tzu (both active around 525 B.C.), Zoroaster in Persia (628–551 B.C.), and Plato in—where else—Greece (428–348 B.C.). We'll reveal Plato's dangerous side in later chapters, since some of the attitudes he proclaimed spell trouble for the twenty-first century. But our story really begins in pre-Platonic times.

The éminence grise behind Plato, his modern followers, and their adversaries was an arch cementer of transnational subculture. Today we associate him primarily with a theorem about triangles, but this future-shaking figure was Pythagoras.

FAUSTIANS AND FLOCKERS—A TALE FROM THE CRYPT OF PSYCHOBIOLOGY

Introverts are often tossed outside the locked gates of conventional society. The quirks produced by their cerebral oversensitivities turn them into "geeks"—round pegs excluded from the square holes of acceptability.

Even when they're toddlers, preschool teachers shudder at their strangeness, stigmatizing them with terms like "negative" and "loner."[25] Other kids and adults jettison them from their presence or refuse to look their way.[26] Meanwhile, extroverts are grabbed in warm embrace. As hard as introverts may campaign for leadership, it's the extroverts who are usually handed the top slots.[27] The hormones of power propel extroverts to take out their aggression on those beneath them. And oversensitives become their punching bags.*[28]

Then things go from bad to worse. Introverts are stuck with aiming their aggression at themselves.[29] One of the few strategies for overcoming this conundrum is to bail out and find a group or build a philosophy in which you're no longer pushed to the periphery.[30] No one was more in need of an emotional home than were the introverts of the age of urban interface: the free-floating rejects of ancient Greece's city-states.

Introverts break down into two types—the flockbound and the Faustians. Faustians cross the boundaries of the system[31] and wrestle with forbidden mysteries.[32] Flockers bury themselves in a bevy of others like themselves[33] and follow the certainties preached by an authority.[34] Faustians take off on odysseys,[35] occasionally returning with fresh visions,[36] Promethean flames around which to shape subcultural societies. Flockers crowd into the warmth provided by the Faustians' discoveries.

One of the pacesetting researchers into introversion and extroversion, Hans Eysenck, paved the way as far back as 1939[37] for the discovery of the Faustians—introverts who in spite of their inhibitions are surprisingly good adventurers. The vast majority of airplane pilots are extroverts. Most introverts freeze at the idea of rudder-and-sticking an aircraft. However, Eysenck's followers discovered a pocket of introverts whose success in pilot training outdid that of every other group.[38] These were introverts Eysenck had previously called "stable." Their opposites he'd dubbed "neurotics."[39] As research progressed, more and more clues to the Faustian nature of the "stable" introverts emerged. The most important evidence arrived in 1972, when social psychologists Jeff B. Bryson and Michael Driver found that some introverts shy away from complexity and others dive into it with great avidity.[40] However, the psychological community was preoccupied with other questions, and didn't see the emerging pattern. In 1998, I subjected sixty years of research on these topics to reanalysis, detected the shape of something new, and called it Faustianism. See if you notice its silhouette, too.

The star players on China's championship women's volleyball teams have been, surprise, surprise, introverts,[41] achieving a glory their extro-

---

*Even among lower primates, the savageness of attacks on those who are shy are devastatingly visible. Some of the most vicious of these assaults are carried out not by males, but by females.

vert teammates envied. Extroverts lust the most fervently for money, status, and power,[42] but it's the less-materialistic introverts, of all the ironies, who make the best businessmen and entrepreneurs.[43] Business often involves not only risk, but the building of a miniculture. Entrepreneurs, in particular, have to craft a microsociety of employees who champ at the bit to challenge giants . . . and who can pull it off. There are good reasons complexity-seeking introverts like those discovered by Bryson and Driver do well at such social engineering. First off, they dive eagerly into theory, penetrating the surface to root out cause and effect.[44] They also have a flair for left-field solutions to problems which leave more conventional types perplexed.[45] And they tend to think with both mind and emotionality. Imaging of their brains shows that when they're pondering, the cerebral web of feeling called the striatum lights up[46] like Tokyo's Ginza district at midnight.*[47]

Extroverts, on the other hand, tend to flee emotion, barricading themselves behind the left brain's abstractions and externalities.[48] By freezing consciousness out of their emotional right hemisphere, extroverts often blind themselves to the unusual. They frequently turn their backs on exploratory reconnoitering.[49] The extroverts' left-cortical tyranny throttles access to areas which house emotional[50] and hunch-based abilities. But it's the wedding of the nonrational with the intellect which studies show produces the most accurate anticipation of upcoming pitfalls and strategic openings.[51]

What's more, complexity-seeking introverts embrace potentially useful strangers[52] whose alien nature tosses their fellow citizens into a tizzy. Complexity-seeking introverts can also calm the rhythms of their brain at times when others panic[53]—a gift that puts them among those charismatic heroes who keep their heads when all others about them are losing theirs. But the bottom line of complexity-seeking introverts' success is probably their tenacity. Extroverts jump into diversions glittering with excitement, run through them quickly, and make a slew of mistakes. Introverts are slow and steady. They take on long and monotonous tasks, but make few missteps and plow through to the bitter (or sweet) end.[54] Research shows that this sort of persistence is the raw stuff of greatness.[55] And greatness is something Faustian introverts frequently achieve.[56]

Pythagoras (580–500 B.C.) was the ultimate Faustian introvert. His father was a gem engraver in a day when artisans were near the pinnacle of society. Like most introverts, Pythagoras did not feel particularly welcome

---

*Even the sheepish followers among introverts show more activity in the thinking areas of the brain—the cerebral cortex and its frontal lobes—than do extroverts. Among extroverts, on the other hand, images of the striatum during thought look more like the subdued lighting of a suburb, and the cortex is less energized.

on his home island of Samos. One reason may have been a common introvert trait—he was a bookworm in a city where single-minded intellectuals
were, at best, an oddity.[57] Another may have been that as a child Pythagoras
was admired by adults for his precocity[58]—a form of attention which can
make you highly unpopular among your merely normal peers.

The regime which came into power when Pythagoras turned eighteen wanted to beef up what Thucydides called "a powerful navy" so it
could trounce the Persians and "reduce many an island."[59] Pythagoras left
town, certain "that under such a government his studies might be
impeded, as they engrossed his whole person." Those studies would have
been more than just impeded if Pythagoras had been drafted to man an oar
in a battle galley. So the young thinker began a thirty-seven-year journey
in search of fellow knowledge addicts, including two citizens of Miletus
who took a shine to him—Anaximander, father of astronomy, and our
old friend Thales,[60] known today as the father of a discipline for which
Pythagoras would later coin a name—philosophy.[61] If Pythagoras resembled the cross-country hitchhikers of the mid-twentieth century scrutinized by psychologist Stephen L. Franzoi,[62] he was intuitive, emotionally
open, impulsive, independent, enjoyed complexity and change, and had a
strong interest in diving into others' souls: the perfect type to dredge the
essence from the sages of faraway lands. Like a 1960s hitchhiker, Pythagoras is said to have spent nearly four decades traveling the width of the
Mediterranean from Gaul to Phoenicia sucking knowledge from mystics
and from priests. He touched on the African realm of Egypt—famed for
the secret knowledge of its sacerdotes' beliefs—then dove into the
shamanic depths of Asia,[63] hitting Persia (and its magi), Arabia, and reportedly reached India[64]—the country

- where the founder of Jainism, Vardhamana, was piecing together a
  new monastic religion which condemned violence toward animals
- where Goshala Maskariputra was piecing together the self-denying
  Ajivika sect
- where two new Upanishads[65] had just been "discovered" in the forests
- and where Buddha, twenty years younger than Pythagoras, was, like
  the Greek tourist, soaking up the spiritual zeitgeist

Both the founder of a religion which would someday permeate the East
and the visitor whose ideas would saturate the West drank deeply of
Brahmanic[66] mysticism, reincarnation, vegetarianism, asceticism, and, in
Pythagoras's case, political elitism.

When Pythagoras returned to his hometown at fifty-six with an
almost fully developed philosophy, his old fans among the elders gathered
eagerly to hear his exploits. However, says Iamblichus, "Not one proved

genuinely desirous of those mathematical disciplines which he was so anxious to introduce among the Greeks." In fact, the folks back home absolutely refused "to learn" the lessons with which the intellectual adventurer had returned. So Pythagoras took off for religion-saturated Delos and for authoritarian Sparta "to learn their laws." Then he went back to Samos and demonstrated the depth of his introspective propensities. According to the Syrian philosopher Iamblichus, Pythagoras "fashioned a cave . . . in which he spent the greater part of the day and night, ever busied with scientific research, and meditating" until he had "unfolded a complete science of the celestial orbs, founding it on arithmetical and geometric demonstrations."

Still, no one in Samos would listen. Pythagoras's townsmen ignored his theories and sent him off on diplomatic missions or buried him in bureaucratic natterings. So once again the master of the hidden worlds which introverts best know escaped his mother polis. "Disgusted at the Samians' scorn for education," says Iamblichus, Pythagoras put down roots in the Italian Greek colony of Croton, where he hung out his shingle as a guru and soon attracted roughly six hundred dedicated disciples and another two thousand devotees who attended his orations in the city's auditorium.[67]

Pythagoras turned his budding cult into exactly what flock-seeking introverts most need—the haven of a tyranny. Ambiguity is a tension-provoker[68]—and the sheep among introverts go bonkers at the strain. Indecisive grays savage the flockers' limbic systems, so they desperately need the tranquilizer of a world spelled out in black and white.[69] Pythagoras's cult was a cleaver slicing ebony from ivory. It severed insiders from outsiders with the ax of a loyalty oath. It sheared away uncertainty and squabbles over internal status through the sharing of all earthly goods, the wearing of identical clothes, the adoption of an unvarying manner ("never yielding to laughter,"[70] among other things), and a monastic prohibition against consuming meat, wine, eggs, or beans.[71] To enter, one endured an initiation ritual of purification via abstinence, massive study, and best of all for those who needed to deactivate their limbic overexcitability, five years of silent obedience to every command, no matter how seemingly ridiculous.[72] Once this half-decade of self-smothering was ended, one could enter the inner sanctum of "esoterica"—students permitted to sup "the secret wisdom of the Master himself."[73] Ironically, for a sheepish introvert this final headlock in the fold is exactly what it takes to make his formerly smothered sense of freedom and achievement bloom.[74]

Pythagoras's philosophy is frequently tossed into the same vat as Greek mystery religions and the occult practices surrounding Egyptian gods.[75] The normally reserved classical historian Michael Grant calls Pythagoreanism "weird," "occultist," "irrational," and just plain "crack-brained."[76]

Socrates, the favorite of historians like Grant, was the Athenian alternative—a magnet for extroverts whose thoughts were disconnected from introspective emotionality. One of the most famous of Socrates' followers (and possibly lovers)[77] was Alcibiades, a thrill seeker (meaning an underexcitable extrovert) to the nth degree. He was known for breaking what Will Durant swears was "a hundred laws" and injuring "a hundred men."[78] Alcibiades pursued every pleasure one could wring from the pool of Athenian excess. He kept a stable of horses which consistently were winners of the Olympian races.[79] His house was stacked high with luxuries. His finances were perpetually overdrawn. His popularity among Athens' high-priced prostitutes was unparalleled. And his wisecracks to men of power were the stuff of legend. (An example: Athens' preeminent figure of political command, Pericles, tried to put Alcibiades in his place by remarking that he, too, had been able to roll supposedly witty remarks off his tongue when he was young. Snapped back the unchastened youth, "Too bad I couldn't have known you when your brain was at its best.")[80]

Then there was Alcibiades' most infamous act of hooliganism—knocking the noses,[81] ears, and prominently displayed penises off of nearly every sacred statue of Hermes in town, and completing the entire deed in just one night. As Alcibiades grew older, he grew bolder. He and his sidekicks started a war with Syracuse . . . which Athens promptly lost. The national humiliation got Alcibiades tossed out of town. So he signed up as a military leader for Athens' archenemy, Sparta, and made the citizens back home tear their hair out by the roots.[82] They also eventually tore out the root some felt had caused this whole fiasco—Alcibiades' teacher and mentor, Socrates himself. Alcibiades' misadventures spurred Athenians to hemlock the quizzical codger and philosopher into the afterlife.[83]

Saint Augustine managed to nail down the difference between Socrates and Pythagoras when he wrote: "Wisdom consists in action and contemplation . . . Socrates is said to have excelled in the active part of that study, while Pythagoras gave more attention to its contemplative part."[84] Pythagoras focused on obsessions with which introverts are blessed—theory, mystery, scholarship, and creativity. While Socrates centered his attention on left-brain hairsplitting, Pythagoras specialized in such right cortical gifts as music, on whose study he left indelible marks. Musical talent comes, in part, from those abnormal prenatal ablutions in which some of the fetuses of future introverts are bathed[85]—prenatal hormones which produce men and women who don't fit standard gender molds. These are the very sorts of personalities which conformity-enforcing normals pummel to the outskirts of society. And they are the kinds of introverts who, with "low stakes in conformity" and a good education, seek out the sympathetic pocket of a tradition-bucking subculture.[86]

If Iamblichus is correct, Pythagoras administered a virtual personality test to make sure his followers were introverts. He "was much more anxious that they should be silent than that they should speak." In other words, he sought shyness[87]—the major hallmark of introversion. During his admission interviews, Pythagoras also looked for overexcitable limbic alarm systems which would render his followers "astonished by the energies of any immoderate desire or passion," and which would spark an instant aversion to qualities of thrill-seeking extroverts like "ferocity," "intemperance," and "savage manners." What's more, Pythagoras wanted a swift but unrebellious mind able to grasp his teachings readily. In short, he demanded high IQ accompanied by the prime flock-seeking-introvert qualities of "gentleness and mildness"[88]—the signs of absolute docility. Pythagoras's followers didn't question his commandments, but tranquilized their wills with a phrase borrowed from slaves, *"Autos epha ipse dixit,"*[89] generally translated as "He himself has said it." In other words, "it's true because Pythagoras said so."

In blinding white robe and Asian trousers, Pythagoras preached reincarnation—the Indian reason for avoiding meat (swallowing an ancestor might prove deleterious to the soul) and for another Indian-style proscription, never killing an animal unless it posed a danger to one's fellow man. Pythagoras's sect increased its appeal by opening its doors to women—a rarity in the Greece of that day, and a downright illegal act in Croton. Yet Timon of Athens called Pythagoras a fisher of men, a phrase that would take on a powerful resonance later on in history.

The syllabus at Pythagoras's spiritual boot camp seems sound enough: mathematics, music, and astronomy. But geometry and arithmetic were used to regiment minds—by giving them Spartan calisthenics in discerning the dictatorial forms beneath the surface of reality. If a single Greek ever designed a philosophical hideout for those attempting to flee the painful spikes of exterior corporeality, it was Pythagoras. Little wonder that he's credited with inventing the word "kosmos," whose derivative, "cosmic," would become popular with those seeking refuge from externality in the twentieth century.

True to the rules of the intergroup tournament, Pythagoras mustered the members of his social cocoon to impose their totalitarian system on the entire Croton state. Iamblichus claims the sage "liberated . . . Croton, Sybaris, Catanes, Rhegium, Himaera, Agrigentum, [and] Tauromenas," establishing "laws which caused the cities to flourish" while rooting out "partisanship, discord and sedition [in other words, disagreement] . . . from all the Italian and Sicilian lands."[90] Some say Pythagoras and his disciples managed to hijack all of southern Italy.

Unfortunately (or perhaps very fortunately indeed), Pythagoras picked the losing side in a battle between Croton's aristocrats and the adherents of a relatively new movement called democracy. The democrats rioted, burned down the headquarters in which Pythagoras's disciples were hived, murdered a few, then chased the rest beyond the city's walls. Pythagoras himself soon perished, though whether at the hands of angry citizens or by starving himself to death is a matter of uncertainty. However, his movement lived on vigorously, demonstrating how an influential ancestor can continue to raise his voice in the communal mind.

Despite the looting of Pythagoras's personal command center, new communities and brotherhoods dedicated to the master's teachings cropped up in Greek territory for the next three hundred years, first in southern Italy, then in Athens,[91] Thebes, and Phlious. Pythagorean trailblazers developed advanced mathematical and scientific concepts based on those which Pythagoras had introduced—the notions of odd and even numbers, prime numbers, an early algebra, a round earth, an eclipse which occurs because the earth slides between the moon and sun,[92] and many more foundation stones for those who would later conquer the outer world by withdrawing from it: scientists. Aristotle reports that Pythagoreans like Philolaus, through a path of peculiar reasoning, even concluded, among other things, that the earth revolves around the sun and not vice versa.

Pythagoras's more Faustian followers were renowned as social and political leaders throughout the Mediterranean and the Aegean. Pausanias, in his *Description of Greece,* describes how the battle-turning fourth-century-B.C. Theban leader Epaminondas was sent to study with "Lysis, a Tarentine by descent, learned in the philosophy of Pythagoras the Samian."[93] Herodotus felt that the Lacedaemonian custom of not wearing wool in the temples or using it in burial ceremonies had been borrowed from the Pythagoreans.[94] The Greek historian Diodorus[95] refers in passing to Prorus of Cyrene as a Pythagorean. The stamp of Pythagoreanism* is all over Euclid's geometry[96] (written in the Alexandria of roughly 300 B.C.). Plutarch attributes many of the ideas of Rome's second king, Numa Pompilius (ca. 700 B.C.), to the Pythagoreans.[97] And the Pythagorean politician Archytas (whom we shall meet again in the next chapter) was elected "strategos" (dictator) of Taras seven times running. Despite the dedication with which he handled his political duties,[98] Archytas also found time to write in-depth treatises on Pythagorean philosophy, on the mathematical underpinnings of music, on mechanics, and (according to Aristotle) to invent two devices which would change humans' ability to engineer their world—the pulley and the screw. No wonder Diodorus praises the Pythagoreans as "the cause of so great blessings" not just to a few, but to all "the states of Greece."[99]

*And of Pythagoras's friend and mentor, Thales.

Finally there came the series of stepping-stones which would lead to the Pythagoreans of modernity. Pythagoreanism revived in the Roman Empire during the second to fifth centuries A.D.,[100] popped up afresh during the Renaissance, wound its way through everything from Montaigne,[101] Shakespeare, Milton, and current literature to modern mathematics, and thrives in the form of philosophical, religious, and "pagan-revival" movements at the street level, on college campuses, and even on the Internet today.

Some have considered Pythagoras a mystic sage, others a founding father of secular inquiry. Copernicus swore his solar-centric system was sturdily planted in Pythagorean tradition (it was). Galileo was called a Pythagorean by his fellow Italians. And Leibniz openly dubbed himself a devotee of the Pythagorean heritage.

However, Pythagoras's most important legacy would be this: he had pioneered a new sort of subculture—one which could cross the boundaries of nations; one in which those whose personalities did not fit could find a home.

17

SWIVELING EYES AND PIVOTING MINDS:

The Pull of Influence Attractors

479 B.C. to 1990 A.D.

ANY assumption can be made, but not all assumptions are created
equal, and from their deductions you will know them.

John Edser

It is not easy to explain to the lay mind the extremely intricate
ramifications of the personnel of a Hollywood picture organization. . . . The chief executive throws out some statement of
opinion, and looks about him expectantly. This is the cue for the
senior Yes-Man to say yes. He is followed, in order of precedence,
by the second Yes-Man—or Vice-Yesser, as he is sometimes called—
and the junior Yes-Man. Only when all the Yes-Men have yessed,
do the Nodders begin to function. They nod.

P. G. Wodehouse

Each epoch presents new puzzles. The culture or subculture built around
the most relevant answer comes out on top. All groups wish for fame and
fortune. It comes to those whose combination clicks the tumblers of a
shifting lock.

Athens put her chips on pluralism. She was a breeder of cross-
cultural diversity. Sparta bet on the one-size-fits-all hypothesis—a cradle-
to-grave, cookie-cutter conformity. Subcultures also represent hypotheses.
Socratics challenged old rock solids and romped through rebellious
multiplicities. Pythagoreans, at least while Pythagoras was running the
show, imposed obedience to just one master and hammered home uniformity. Which hypothesis, the courting of many voices or the push for

only one, would prove the rightest of them all? The answer would emerge through trials of circumstance whose tides continually rise and fall.

Remember the five elements of the complex adaptive system: conformity enforcers, diversity generators, inner-judges, resource shifters, and intergroup tournaments? In the last few chapters we've examined inner-judges—bioresponders which shut us down and slam us in a shell or open us up and let us romp like hell. Now let's see how intergroup tournaments and resource shifters slyly change the focus of mass mind.

From 500 B.C. to 300 B.C., three tiers of intergroup tournament ran simultaneously. The Persians battled the Greeks for dominance over the lands looped 'round the Aegean Sea. On the next level down, a gaggle of city-states jockeyed for position in that skein of polities known as Greece. And below, within, and around these contests, subcultures vied to shape the way men feel and think. Each level echoed the same motif: pluralism versus uniformity, centralism versus parallelism, democracy versus tyranny—all variations on the themes of conformity enforcement and diversity generation.

The major intergroup tournament of the early fifth century was a war that raged for more than fifty years. In computer terms, the contest pitted a serial processing system—Persia—against a parallel distributed processing system, Greece. A serial processor lines up all decisions and runs them through a unit at its center—in this case, the emperor Xerxes or his appointed general, Mardonius. A parallel processing system depends on a mesh of independent decision-making centers which pool their attempts at wisdom—as did the Greek city-states. Persia's Xerxes mobilized between three hundred thousand and a million soldiers and sailors,[1] then penetrated deep into a Greek homeland defended by just seventy-five thousand Hellenes. Xerxes' troops burned Athens to the ground, yet to everyone's surprise, in 479 B.C. he met defeat. When Plataea—a land battle—sent the Persians packing, the Spartans went back home to savor the victory. But the Athenians continued to command the naval alliance which would keep the snarling Persians in their place and would open the entire Mediterranean to Greek trade. The seventy years that followed were Athens' Century. Power in this period did not pour from Sparta's specialty, marching soldiers across hill and dale,[2] then returning home, closing the shutters against foreigners,[3] and practicing for the next bout of war. Dominance was a product of sea-lane commerce[4] and open boundaries. The politician of the day was Athens' Pericles. To wire the far-flung nodes of Greeks into a single commercial, military, and political network, Pericles and his predecessors bashed the heads of many a recalcitrant "allied" city-state.[5] In some towns Athenians slaughtered, evicted,[6] or enslaved the citizens, not an act of tolerant generosity. But in most, Pericles encouraged the growth of democracy.[7]

At home in Athens, Pericles was pluralism personified. His friends were immigrants from all over the Greek world. He enjoyed gathering them together and taking them to the theater of Dionysus to see the latest exercises in free speech . . . like tragedies by Euripides which trashed the ancient gods and turned tradition on its head. At the leader's side were Phidias, the sculptor, Socrates the idea-mincer, Anaxagoras the protoscientist from Clazomenae,[8] and Pericles' mate in sex, childbearing, and enterprise—not his wife, but his mistress, Aspasia, a brilliant Milesian who ran one of the world's first intellectual salons.

Pericles' Golden Age produced the flowering which would lead to Athens' place in history and would crest in the marble-columned buildings and literary works which cornerstone the civilization of the West. But despite the permanence of these monuments, the intergroup tournaments weren't over—in fact, they never are. War broke out between the Greek states in 459 B.C. Athens managed to stomp it out. In 446, Sparta attacked Athens with a military thrust which the Athenians parried handily. From 431 B.C. onward the Spartans again pitted themselves against the Athenians (and vice versa) in the Peloponnesian War. Athens fought Sparta with one hand while with the other she carried on her naval empiring, her refinement of wealth, and her explosion of theater, sculpture, and philosophy. Then, in 404 B.C., Athens lost to Sparta catastrophically.[9]

Sparta now controlled the cables of the inter-Greco web. Athens resurrected a ghost of her former prosperity, but her once-raucous pluralism grew pinched and dry.[10] This would become apparent in the dominion of a new philosopher, one whose work would etch Pythagoras's authoritarianism deep into the Western mind.

From the defeat of the Persians to the end of Athenian hegemony, each tectonic shift in politics reset the balance between personality types, the subcultures they forged, and the resulting dominance of philosophies. But to see the reasons why, it's time for another dive into the depths of biopsychology.

FOLLOW THE LEADER: ATTENTION STRUCTURES AND THE MONKEY BUSINESS OF PHILOSOPHY

> The first responsibility of a leader is to define reality.
>
> Max De Pree

> He was indeed the glass
> Wherein the noble youth did dress themselves.
>
> William Shakespeare

Resource shifters shovel a society's goods from losers to winners in a manner notable for lack of mercy. When Athens had maintained the choke hold on the Delian League, she'd raked in tribute from her subject states and used the loot to build such glories as the Parthenon. Now that Sparta was in charge, it, too, sucked taxes from its tributaries till they screamed. But the most potent of the resource shifter's prizes isn't money; it's influence, the joystick moving the collective eye.

The hunger for attention and influence is far more primal than most realize. Even "smart molecules" compete to "have their say."[11] Among these are the receptors studded like rivets across the surface of a cell's membrane. Witless as these twists of atomic beads might seem, receptors have decision-making powers, a primitive IQ.[12] If a passing particle fits snugly in the trap these molecules wear on their exterior, their bells of recognition ring, clanging a "eureka" to the multitudes—the fellow receptors spread out on the cellular skin.

Should the hallooing receptors' find be a needed rarity, other lookout molecules will stop what they're doing and join the hue and cry, alerting the totality of the cell. The choiring of a bacterium's receptors, for example, whips the creature into motion, propelling it toward food or flicking it from threat. But a receptor community's tastes shift rapidly. A "discovery" which was hot stuff an hour ago may now seem cold and boring as can be. Should a receptor strike a form of pay dirt whose provision has become routine, the yell it sends will lose what Cambridge University's Dennis Bray calls its "infectability"—its ticket to fame and popularity. Other receptor molecules will ignore its "yikes, I found it" almost totally.[13] In the struggle for attention, timing is everything.

When complex adaptive systems pioneer John Holland was devising a general theory of intelligent machines he realized the need for hierarchy[14]—a resource shifter of tremendous potency. The higher you are on the totem pole, the more food,[15] sex, and perks[16] are tossed your way. But the real prize is your right to be the focus of all eyes—to captain mass perception, pilot the "common sense," and decree the concepts chorused by the folks deemed sound and sane. Among apes, baboons, or chimpanzees, most males are hunched and nervous, radiating insecurity. There's one spot to which their pupils keep on swiveling. In it is a lordly animal,[17] his fur erected in a regal mane.[18] Females groom him,[19] youths court his favor, and infants loll at his feet.[20] Some go into a semblance of a groupie's swoon. One young mountain gorilla studied by Dian Fossey stared for hours every day into her troop leader's face. If he deigned to toss a glance her way, a shudder of pleasure rocked her body visibly.[21] Those outside the primate regal's court also look to their lord for cues.*[22] Cues which indicate whether there's a hail of punishment or reward about to come their way. Cues to whether they should sit or stand—for when the master

---

*Or they look to their lady—dominant female primates are also perceptual autocrats.

settles his haunches, many follow his exalted lead.[23] The leader's power is so great that the smallest change of expression can send his toadies scurrying to please.[24] What he does, they do. They ape his style and even his emotionality.[25] His political approach, be it despotic or indulgent, becomes that of the troop.[26] His strategies, even for basics such as how to snag food and when to eat, become the rage.[27] He orchestrates a cluster of eyes, ears, hands, and primate brains, hooking his "inferiors" into what primatologist Michael Chance has labeled an "attention structure"[28]—a collective gaze.

The human attention structure shows itself among infants, four-year-olds,[29] Mayan rulers,[30] and the members of street gangs.[31] At a cocktail party in the United States where all are alleged equals, men of lesser power show emotion nervously and look for conversational cues from those who telegraph supremacy. Superiors wear their hash marks unobtrusively, radiating the animal signals of calm, control, and dignity, gifts of serotonin,[32] the biological ambrosia which flows even from a winning crayfish's sense of mastery.

There are other tiny clues which sort out who's on top from who is not. High-status *Homo sapiens* take the liberty of touching those below them, granting an instant of camaraderie. Those on a more humble plane seldom dare try this familiarity.[33] The primate to whom all eyes are glued gazes into the middle distance, underscoring that he need not look to others for his cues. Dominant humans also follow this instinctual rule, focusing intensely on their listeners as they themselves pontificate, but allowing their glance to wander when their lessers let an opinion fly.[34] Top dogs act as helmsmen for a conversation. We let them direct its length and tone.[35] When they speak, they do so with conviction. We fill our sentences with words like "uh," "like," and "I think," betokening uncertainty.[36] But not to worry. Studies show that those high on the social scale do most of the talking anyway.[37]

Some attention structures are matters of the moment, cobbled together by strangers who've just met for the first time. But most involve an equivalent to what Adam Smith called "stored labor"—his phrase for the eons of innovation captured in such simple objects as a brick, a spade, or a high-tech turbine engine's ceramic blade. A presidency, a Ph.D., a title of nobility, and even instant celebrity sit atop a pile of perceptual capital accumulated from the days of mammoth-tusk tent making, city building, kingdom crafting, and construction of the institutions of modernity. A professor's title taps 6 million years of primate instincts and the authority heaped up by institutions of higher learning during the forty-five generations since the University of Bologna struggled to build its reputation in the eleventh century.[38] Even the title of Ph.D.—doctor of philosophy— gains its impact from stored attention piled high during the twenty-six hundred years since Pythagoras coined the term "philosophia."

Attention begets attention—it's the first rule of publicity. It's also basic in biology. And it's the process through which stored attention is amassed. When a slime mold amoeba senses that food is running low, it signals potential famine through a plume of cyclic AMP. One warning in the wilderness is a sorry thing. Indifference is its likely destiny. Only when a mob of chemical cries pulses forth in synchrony do the remaining amoebas drop what they're doing and come to see what's happening.[39] The piling up of signals focuses the lens of collective perception in a honeybee colony too. Each flower patch has campaigners shanghaiing others to waggle-dance rallies promoting their find. In the end, the hive will sum up the numbers each team of dancers has won over, then direct its energies to the food source whose proponents have topped the popularity poll.[40]

Ants are ace demonstrators of stored attention and its irresistibility. A wandering worker who's found a crumb of picnic cake will return home with her tail a-dragging. She's marking the path with a scent trail, a chemical road sign to the newfound eatery. One scent trail will get you nowhere. But things pick up a bit when a second ant trips across the discovery, then adds her chemical recommendation to the lone "I found it" lingering forlornly on the ground. Now that there are two scent trails, other ants may check out the route and should they be pleased, leave their signals on the return trip to HQ. As the overlay of scent trails grows intense, its come-hither finally turns immense. Ants rush madly to get their crack at a treasure which has gained celebrity.[41]

Insect scent trails wear off after hours or days. But humans lay attraction markings in more indelible ways. We warehouse them in rituals, institutions, and "immortal" works which sing a great man's praise. Homer's paeans to Achilles and the virtues of Greek bravery were imitated, taught, and written about endlessly, building a trail that grew more potent over each succeeding century. Alexander the Great read *The Iliad,* tried to follow in Achilles' path, and by repeating it in his own way, boosted its pull immeasurably. Julius Caesar couldn't resist the lure of Homer's and Alexander's heroic scent and left his aping of Achilles and Alexander in the ten books he wrote* publicizing not only his own military victories, but those of Alexander as well. Napoleon—a voracious reader—was intoxicated by the mounting odor of success and hero-worshiped Achilles, Alexander, and Caesar, too. His conquests added to the mounting trail of attraction cues. Young Adolf Hitler fell under the spell left by Achilles, Alexander, Caesar, and the Corsican leader of the French, and once again followed the seductive path all four had left. Homer's song of Achilles'

---

*Caesar wrote seven books of his *Commentarii de bello Gallico* (generally translated as The Conquest of Gaul), three books of his *Commentarii de bello civili* (The Civil Wars), and is alleged to have written *De bello Alexandrino* (The Alexandrian War).

killing spree gathered so much stored attention in three thousand years that it's made an everlasting mark on history.

This sort of multimillennial hoard of perceptual allure also gets compressed into a quality called prestige.[42] Prestige is the attention stored in titles like doctor, president, prime minister, and king. Much as we deny the fact, we cozy up to those who've got prestige and shy away from those who don't.[43] We put our confidence in those who've commandeered it, and shun the same message—even if delivered in the very same words—from the mouths of lesser folk.[44] If someone from the lower ranks comes up with a notion that makes sense, we often deep-six it in our memory, then bubble it back to consciousness a few weeks later dressed as our own idea or that of an authority.[45] We rearrange our sensory input to erase the flaws of those above us and exaggerate, or just plain fabricate, those of the folks below.[46] Even our eyes cooperate in the charade. We literally see those of high position as physically taller than they really are—adding as much as two and a half inches to a dignitary's height.[47]

Like apes and chimps, we copy the mannerisms of our masters, down to their speech patterns, choice of clothes, automobiles, and style of homes. In meetings, when they rock slightly forward and back, we inadvertently rock in imitation of their ways.[48] This is not just postmodern Western lunacy. In Uganda, when "the king laughed, everyone laughed; if he sneezed, everyone sneezed. If he had a cold, everyone else said he had one. If he had his hair cut, so did they."[49] In pre-twentieth-century China, courtiers chuckled when the emperor did, but if his face grew sad they showed immediate despondency.[50] The Kwakiutl chief Yaqatlenlis once made a boast which captured a basic truth of mass psychology: "I am known by all the tribes over all the world . . . all the rest . . . try to imitate me."[51]

We're pulled into the wake of those "above" us by the social instincts Thomas Carlyle described when he compared the literary critics of the 1820s to sheep.[52] Put a stick in the path as a lead sheep goes by, wrote the sage, and the beast will jump to get over it. Remove the stick, and each remaining sheep in line will hop at precisely the same spot . . . even though there's no obstacle left to hurdle! Things haven't changed much since then. It's widely recognized in the publishing industry that if the key critics at the *New York Times* fall in love with an author, critics all over the country will follow their lead. On the other hand, if these lead sheep say a book is worthless, a cascade of commentators from Portland to Peoria will arrive at the same conclusion. Imitative learning evolved among spiny lobsters nearly 300 million years ago. This is one of its updates in the world of human beings.

Reality is to some extent a hallucination which we generate collectively.* In the "attention space" of this fantasy landscape, the high and

---

*See chapter 8, "Reality Is a Shared Hallucination."

mighty glom up more than their share of real estate. They possess the greatest acreage of our gossip and our media coverage. As role models, they roam the private attics of our musings and our dreams. They are the lead sheep in the herd of humankind—the eyes, the ears, and shapers of collective mind.

The shift of attention to those on top—whether they be neurons,[53] experts, or empires—is central to the operation of collective brain. Both countries and subcultures compete for influence constantly. The tug of those who triumph at subculture games and intergroup tournaments is so strong that members of out-groups struggle to be members of the in.[54] The more a troop of langur monkeys loses its battles, the more its members desert and join the winning team.[55] We are no different. Some Jewish children in the 1930s goose-stepped and Heil-Hitlered with great glee.[56] Some concentration camp victims barked like their guards and sewed Nazi badges onto their shabby prison clothing.[57] When the Muslim Moguls conquered India, Islamic fashions became the rage.[58] The same has happened in a more recent day and age.[59]

From 1981 to 1986, the securities traded by Japan went from $15 billion to $2.6 *trillion,* leading to a new saying in Japan: "It took Britain one hundred years and the US fifty years to become the richest country in the world, but it only took Japan five."[60] By the mid-eighties, Japan had more money controlling more resources in more countries than any other nation in the world.[61] Its foreign exchange reserve was the world's largest.[62] It was the world's leading exporter and source of loans.[63] The shares on Tokyo's stock exchange topped in value those on the Big Board in New York.[64] Japan possessed 54 percent of all the cash in the world's banks.[65] The top twelve global banks, in fact, were wholly Japanese.[66] The average Japanese per capita income had surpassed that of U.S. citizens,[67] who still spoke of themselves mistakenly as the wealthiest in the world.

Japan owned six hundred companies in Britain and two hundred in France.[68] Analyst Ezra Vogel predicted a new Pax Nipponica.[69] In a sense, it had already come. Japan was the world's biggest donor of foreign aid,[70] spreading money and the influence it bought through Pakistan, Egypt, Turkey, Mexico, Panama, South America, Vietnam, Burma, the Caribbean, and sub-Saharan Africa. More important was something most of the experts missed. As of 1988, Japan's general trading companies, its *sogo shosha,* had become global nerve centers "controlling the greatest centralized commercial network of production, distribution, and communication facilities existing in the world."[71] The *sogo shosha* possessed an unparalleled ability to work with cultures others found baffling. In other words, they were consummate masters of networking. These trading giants raised money from virtually every land, shuttled it to a developing nation, offered that country out-of-the-box installation of industries being phased out of

Japan itself, threw in technology, management expertise, and even the tools for building megaproject consortia, then provided marketing for the final goods. Little wonder that *Le Monde*'s Bruno Thomas called the Japan of the 1980s "the center of the world economy."[72]

All eyes swivel to the nation at the top. English-language authors cranked out more than ten thousand books on the subject of Japan. A slew of them taught how to organize, industrialize, and strategize in the Japanese way. During the 1980s, Sony founder Akio Morita gave Americans regular sermons on how to conduct their affairs, and Americans listened eagerly.

Veteran foreign correspondent Henry Scott Stokes reported that Tokyo's young "regarded the world as a 'tabula rasa' upon which they will write a large design. That design is the Japanese flag."[73] Thanks to the tricks of the attention structure, this was no idle dream. Brazil aped the Japanese model for development—locking out foreign investment and imports, and using government to guide business.[74] In Africa, the head of Senegal's Center for Research on Internal Development, Joseph Ki-Zerbo, urged his continent to follow the Japanese lead, taking what it wanted from the West, and closing its doors to all the rest.[75] Sub-Saharan Africa's grassroots consumers needed no such prodding, since Japanese goods were among their favorite novelties.[76] Keiretsu-like companies popped up in Taiwan, South Korea, the Philippines, Thailand, India, Brazil, and Argentina. In Britain, where Japanese-run auto plants outproduced local factories dramatically, English industrial managers reorganized operations to ape those of the Japanese.[77] In France, Peugeot spent $200 million to train workers in what Pacific Rim specialist Daniel Burstein called "Japanese-style assembly techniques."[78] Even ultranationalists in the former USSR talked of "following the Japanese model"—combining economic reform with one of Japan's dirty little secrets, obsessive ethnic purity.[79]

Tokyo's charisma recalibrated the minutiae of worldwide daily life. The number of sushi bars in New York went through a tenfold increase from 1980 to 1989.[80] In Burma, the biggest-selling T-shirts had slogans in either English or Japanese.[81] In Asian cities like Hong Kong and Bangkok, Japanese television stars, pop songs, and soap operas were rapidly replacing American pop cultural exports.[82] *Time* and *Forbes* declared that Tokyo's fashion and architecture had become trendsetters for the world.[83] Japanese designers hogged the limelight at the 1987 Paris fashion show.[84] Japan's retailers set the pace even in their product presentations and use of video directories. Some Japanese food stores became so wary of having their secrets stolen that they put signs in their windows asking customers not to take pictures or draw sketches of the displays. In the global web of retail commerce it was said that what Japan does today, the rest of the world will imitate in two years' time.[85]

For Western scientists, Japan became a place of pilgrimage. American politicians went hat in hand to Tokyo so often that Tennessee governor Lamar Alexander admitted he'd done the grueling trans-Pacific trek more frequently between 1983 and 1987 than the easy hop to Washington.[86] In fact, by 1988, thirty-six U.S. states had established permanent offices in the city of the $6 cup of coffee. When our country's politicians and former presidential candidates left office, it was the Japanese to whom they sold their clout.[87] In 1990, Peru elected a second-generation Japanese president—Alberto Fujimori. The move paid off with a nod of acknowledgment from the troop leader. Fujimori was granted an audience with Japan's emperor and was promised foreign aid.[88] Too bad that he could not, in good chimp, ape, and baboon fashion, repay the favor by scratching the emperor's back.* Meanwhile, as the Philippines forced out American military bases, Manila turned to Japan to save its faltering economy. Japan became the country's largest investor, while the United States, traditionally the country's "older brother," fell to number four. Philippine prime minister Corazon Aquino, whose relatives had been killed by the Japanese during their occupation of the island nation, spoke cheerfully of the days when Japanese soldiers had handed out chocolates to Philippine children, and fondly remembered the songs she had learned when she was a child during the Japanese occupation.[89]

The Singaporeans named the Japanese the people they most admired, citing their spirit, their strength, their work ethic, their willingness to sacrifice, and the fact that their cars and technology were number one.[90] A Korean studying for his M.B.A. was asked what country he esteemed above all others. He answered bluntly, "Japan is the future."[91] In 1989, Russia's perestroika and China's economic reforms brought down the Iron Curtain and Americans took the credit. But in the Asian block, that credit was given to Tokyo.[92] When Eastern Europe opened to foreign investment, Japanese auto- and steelmakers led the way in shaping the region's new economy.[93] Suzuki, for example, built a $200 million auto factory in Hungary, then used the country's low-cost labor to manufacture cars for sale in the newly unified European community.[94] The Mexicans at the end of the eighties wooed Japanese investment more eagerly than that of Americans. The lure of Japan to upper-class Mexicans became so strong that the well-heeled, including the country's president, sent their children to Liceo Japones[95]—Japanese schools in Mexico City. Even the Iranians

---

*Though most baboons, according to Hans Kummer, crave to scratch the back of their troop leader, only those in the very top positions are granted the favor. Others are forced to content themselves by gazing with what Kummer describes as religious awe at his majesty's mane and sweeping the ground with their fingertips as if they were "redirecting" their ache to groom their master.

saw trading with Japan as a convenient way to escape the twin devils of the United States and the former Soviet states.[96]

By 1993, Japanese authors and other authorities had become increasingly outspoken in their opinion that the American model of capitalist development was obsolete, and that the new economic paradigm had been made in Japan.[97] The country most eagerly listening to this advice was the contender-to-watch in the twenty-first century—that empire of a billion educated, canny, and driven worker/soldier/consumer/innovators known as the Chinese. Just as monkeys ape the leader who has climbed to their troop's apogee, so human groups look to winners for superior strategies.

Molecular sensors, bacteria, crustaceans, ungulates, primates, and intellectuals—we are very much the same. We act as nodes in a neural net. Thanks to resource shifters, we follow the lead of victors in the mass perception game.

PLATO—A SPARTAN IN PYTHAGOREAN GARB

Greek philosophers of the fifth and fourth centuries B.C. also followed the resource shifter's rules: shovel influence to the winner; get everyone to ape the guy (in those days it was always a guy) on top. While Athens was master of the Aegean hierarchy, the leading philosopher, Socrates, was thick as thieves with Athens' leader, Pericles. When Sparta became king of the heap, the new philosopher-in-vogue turned ethics, metaphysics, and epistemology into a mirror of the Spartiate. Forty years later the Spartan approach would cease to fit the landscape of a difficult century. Then another kind of political leader would seize the attention structure and once more swivel the pupils of philosophy.

Plato was born three years after Athens had been pulverized by plague and one year after the death of Pericles, guarantor of Athens' Golden Age. Good-looking, athletic, awarded for military bravery, an outstanding student in everything from music to mathematics, a woman-charmer skilled at love poetry, a playwright who penned a series of four tragedies, the young Plato couldn't decide whether to make his name in poetry or politics. Then, through Socrates (a family friend),[98] he found philosophy. As we've seen with Alcibiades and his efforts to take Syracuse, Socratic thought could be used politically. Plato was twenty-three when his home city fell at last to Sparta. Athens' glory was eclipsed and its civic pride was smashed, as was its economy. Reigns of terror replaced elections during the coup of the Thirty Tyrants—funded with a hundred talents from the Spartan enemy. One of the bloodthirstiest of the Thirty Tyrants was Critias, a relative of Plato's who was also a good friend. A year later, when the Tyrants were deposed, two of Plato's favorite uncles, Critias and

Charmides, lost their lives in an attempt to fend off the returning democrats. Then the restorers of the vote—a generally peaceful bunch—killed
off Plato's mentor, Socrates.

Plato was fed up with Athens' liberal education, its openness, and
especially its democracy. He fled to Megara, Cyrene, and Egypt in disgust,
visiting eminent Pythagoreans like Philolaus and Eurytus, then retraced
Pythagoras's footsteps through the audience chambers of African and Asian
priests.[99] Drawing him was a view whose founder—a Spartan admirer like
himself—had imposed obedience and uniformity.

In periods of upheaval, authoritarian certainties offer to restore
what's lost[100]—a bedrock beyond crumbling. Plato found a path in
Pythagoras to the hardest of rock solids. This earth was but an impure
shadow of archetypes lodged in a space more permanent and potent than
this shifting plane. Every college student who takes a course in philosophy
knows this Platonic concept from the story of the cave. What most don't
know is that its roots were in Pythagoras. Aristotle, however, was not
fooled. He treats Plato's concept of archetypal forms as Pythagorean pla-
giary. Diodorus puts it in more positive terms, calling Plato "the last of the
Pythagorean philosophers."[101]

Perhaps, then, it should come as no surprise that when Plato finally
set up his own school for the sons of the wealthy it bore the stamp of
Sparta and Pythagoras. The Academy's suburban grounds were bought for
Plato by highborn friends. The Academy's expenses were underwritten by
Spartan-supported tyrants like Syracuse's Dionysius II. And the Academy's
students were known for their ultra-aristocratic manners and finery. The
campus bore a telling motto over its portal: *Medeis ageometretos eisito* (Let
no one without geometry enter here). The point couldn't have been
driven home more soundly if the arch-geometer Pythagoras himself had
chiseled it there.

Like Pythagoras, Plato carved out vast new territories for the intellect. But his most disturbing heritage would be totalitarian. The utopia he
portrayed in his *Republic,* his *Statesman,* and his *Laws* is stretched on an
unbending Spartan rack. Eugenics chisels conformity at birth.[102] A
nomenklatura of philosopher aristocrats, backed by a secretive night council, clamps down on those whose appetites might carry them from virtue
and austerity. Poetry, music, and literature—the loves of Plato's early
days—are filled with dangers to the young, and must be censored vigorously. Plato's republic is a cold and moralistic place, much like Stalin's Russia in the 1930s, the ayatollah's Iran in 1979, or the Republic of the Taliban
in 1999.

Though *The Republic* gets most of the attention, it's in *The Laws* that
Plato really lets his aping of the new top power—Sparta—show. Here he

paints an "ideal" nation, one which copied the Spartan dream. Asks an Athenian interlocutor at the beginning of *The Laws:*

> Ath. And first, I want to know why the law has ordained that you shall have common meals and gymnastic exercises, and wear arms. [Spartan practices of the purest kind.]

> Cle. I think, Stranger, that the aim of our institutions is easily intelligible to any one. . . . All these regulations have been made with a view to war, and the legislator appears to me to have looked to this in all his arrangements. . . . He seems to me to have thought the world foolish in not understanding that all are always at war with one another; and if in war there ought to be common meals and certain persons regularly appointed under others to protect an army, they should be continued in peace. For what man in general term peace would be said by him to be only a name; in reality every city is in a natural state of war with every other, not indeed proclaimed by heralds, but everlasting.[103]

The intergroup tournament is always among us. Like Sparta, each state should be disciplined with the severity of a military camp. Plato's influence would rise and fall throughout the ensuing centuries. His approach would be emotionally in tune with twenty-five hundred years of absolutist zealotries. And in many an important sense, he would be the father of what will menace us in years to come—the fundamentalisms of a brutal new modernity.

PLURALISM BY THE SWORD—
ARISTOTLE AND THE WEBSPLICER OF MACEDON

When an information-intake-centered RAM approach was needed, Athens had stayed on top. Then came the day when men hungered for a strategy of ROM (no data input, just an output of commands), and Sparta had become the cock of the walk. The next shift paved the way for a kingdom perched on the boundary between Greece and its Persian enemy. The royal palace of this border land was one where input and output coexisted of necessity. Athenians, Spartans, and Persians mingled in the chambers of its kings.[104] Far from the reigning centers, this nation was considered by the Greeks "barbarian." But Sparta had won power then lost it, and now Athens' Second Empire was sputtering. Even the Persians weren't making out half as well as they were wont to do.[105] Times of social disintegration open vacuums which new integrators fill. When the ruffian nation rose, so did its chosen political leader and his tutor, the court philosopher. The teacher was Aristotle. His pupil: Alexander. Both became switchworkers of a new attention skein.

Like Alexander, Aristotle grew up on a land bridge and built bridges with his philosophies. He joined the best of the Socratic, the Pythagorean, and the Platonic, yoking diversity to conformity as partners in the spring-works of a golden mean. He questioned nature vigorously, a quasi-Socratic enterprise. He applied Pythagorean science and, like the Pythagoreans and the Platonists, built where Socrates had specialized in pulling things apart. He rejected authoritarianism, but mentored a commander whose power would shake destiny. Platonists aped Sparta. But soon Mesopotamians, Indians, Afghans, Hebrews, Phoenicians, Egyptians, and even the Romans would imitate the Hellenists of Macedon.

The best of Aristotle's heritage would thrive in science and plural-ism. The worst of Plato's would survive in intolerance and bigotry. Which leads back to a question raised many chapters ago—in the final tally which hypothesis would win, Athens' openness or Sparta's harsh rigidity? The answer would be both and neither, for the contest never ends. Rome chose Athenian democracy in the centuries of its rise, then switched to Spartan harshness when it faced barbarian onslaught and decline. The chaos of the Dark Ages triggered a retreat to Sparta's fortresslike mentality. Then Europe calmed in the twelfth century, open trade returned, and merchants once again interlinked societies. The Renaissance and the Age of Reason exulted in Athenian-style information flow. But in our brave new century, it is hard to see which way the tilt will go.

18

OUTSTRETCH, UPGRADE, AND IRRATIONALITY:

Science and the Warps of Mass Psychology

3000 B.C. to 20?? A.D.

The British neurobiologist Charles Sherrington spoke of the brain as an enchanted loom, perpetually weaving a picture of the external world. . . . The communal mind of literate societies—world culture—is an immensely larger loom. Through science it has gained the power to map external reality far beyond the reach of a single mind.

Edward O. Wilson

And we are here as on a darkling plain
Swept with confused alarms of struggle and flight,
Where ignorant armies clash by night.

Matthew Arnold

Good ideas are not adopted automatically. They must be driven into practice with courageous patience.

Admiral Hyman Rickover

Nations and their leaders battle for control over the common sense. But other contestants struggle behind the scenes—subcultures, clusters of like-minded folk who live within and ooze between societies. Dr. Gilbert Ling's life story is shot through with the threads which tie the ancients to modernity. Yet it also uncovers a thought-throttler hidden in the workings of the modern global brain—a legacy from Catal Hüyük, Miletus, Athens, and Sparta, where subcultures first learned to play their sometimes crooked games.

Ling was born and educated in China and received his graduate education in the United States in a field developed largely by Germans. His intellectual heroes included Lao-tzu, Confucius, Socrates, Joseph Priestley, Benjamin Franklin, Michael Faraday, and Japan's Saburo Ienaga. Ling became influential in America, Europe, and Russia, then fell afoul of authoritarian mind-closers—ironically those within science itself. Should he overcome the obstacles hurled in his path, he feels strongly that his research could aid us in an intergroup tournament with a global brain which we managed to beat for a short time, but is threatening to leave us in the dust again—that of our bacterial brethren.

To understand the puzzle and the stakes, let's move back a bit in history. When contemporary scientists like England's Peter Russell, France's Joel de Rosnay, Belgium's Francis Heylighen, and America's Gottfried Mayer-Kress—all of whom we met in the prologue of this book—predict the coming of a global brain sometime in the twenty-first century, their minds are often fixed on electronic communication. They see satellite-relayed beams, fiber-optic cables, and next-generation technologies as neuronal axons in a cyber-cortex spanning continents and seas. But trade and stone-tool technology-sharing produced early versions of these far-flung axons over 2 million years ago.[1] The pace at which these interconnects grew after 3000 B.C. was nothing to sneeze at. The Egyptians used the Nile to unite one long strip of towns into an empire which endured fourteen times as long as the United States has been a nation. Riverboats cruising the 4,160-mile-long river provided remarkably quick and easy intercourse between one Egyptian center and another.[2] Foreign merchants carried Egypt's interconnectivity all the way to Asia. In 800 B.C. the Phoenicians used agile craft capable of handling turbulent waters to sail the Mediterranean and the Atlantic waves, shuttling goods and ideas from Britain and Spain to Argos and Assyria, 2,400 miles away.[3] Shortly before the time of Christ, Greeks[4] and Arabs tapped the monsoon winds[5] to power their water craft on yearly trips to India and China,[6] then hauled back wondrous cargoes of pearls, Malabar pepper, perfumes, and jewelry. The Romans quickened the growing connective skein by building roads, copying Phoenician ships, and refining Egyptian and Persian forms of long-distance data exchange into a 170-mile-per-day long-distance mailing service called the *cursus publicus.*[7] Saint Paul, a master of his era's form of internet promotion, used Rome's communication systems to send his Epistles throughout the eastern Mediterranean, then took advantage of the ready availability of passenger berths on ships[8] to travel from city to city, carrying his gospel as far as Rome and possibly even Spain.[9] Paved highways and post-horses were so effective that young Constantine, according to Edward Gibbon's *The Decline and Fall of the Roman Empire,* left the palace of Nicomedia in the middle of the night, traveled by relays of fresh post-horses through Bithynia, Thrace, Dacia, Pannonia, and Italy, "and amidst the joyful acclamations of the people, reached the port of Boulogne in the very moment when his father was preparing to embark for Britain."[10] Gibbon fails to tell us how many days this 1,600-mile trip took, but he is clearly impressed with its breakneck speed.

Meanwhile, the Chinese pulled together over 3 million square miles of Asia by founding their empire, another whose antiquity makes the nations of Europe look like flashes in the pan. Land routes and river transport ribboned together this vast entity. Then came such enterprises as the 600-mile-long Imperial Canal—a project started in roughly 580 A.D. by the Emperor Sui Wendi[11]—which provided express access for 500-ton freight barges from the Yangtze to the Yellow River.[12] In the Americas, the Incas used roadways, llama caravans, and canals to pull together a domain 2,700 miles long by a few hundred miles wide.[13] Data flow was vital to such massive meshes of city and countryside. The Romans spread a common language and culture which made international communications a snap. The Chinese spread a common alphabet to do the same. The Incas instituted a uniform accounting system based on knotted ropes.[14] And all three imperial powers smoothed things further by conformity enforcing such web fasteners as standard weights and measures, and, in China's case, decreeing a common span between wagon wheels so that vehicles from one province could travel the wheel-rutted highways of another with ease.[15] Meanwhile, Chinese merchants carved a Silk Road through the Takla Makan desert to carry their wares to Rome,[16] where the great matrons of history like Livia one-upped each other with their shimmering imported clothes. To pay for these Far Eastern luxuries, the Romans used slaves abducted from numerous continents to scrape metal from the silver mines Rome had snatched in its conquests. So heavy was the semiglobalization of commerce that Rome ran a ballooning negative balance of trade with the commercial tigers of India, Arabia, China, and Spain.[17]

Ideas surged and blended as these connective tissues quickened in vitality. India soaked up artistic traditions from descendants of the generals of Alexander the Great, who had carved out kingdoms in the Hindu land's northern territories. In turn, India's fairy tales became European common currency.[18]

After Rome fell in 410 A.D. to the Goths, the Mediterranean sealanes which had fed the citizens of Rome and Gaul with grain from Egypt fell to pirates and were cut. The roads bringing Cornish tin were snipped by bandits and by pillaging tribes. Europe would not rebuild its interconnectivity to Roman levels until the late eighteenth century. But sprouts of reconnection grew from 1100 A.D. on, when pathways into which roads had crumbled were sufficiently crime free to allow a fresh rebirth of continental trade.[19] Then came a crucial turning point. In the fifteenth and six-teenth centuries, the human information network finally spanned the planet. Portugal and Spain spread the threadwork of oceanic exchange to Africa, India, and the Americas, introducing the sweet potatoes of Mexico to the Philippines,[20] the chili peppers of Peru[21] to Hunan,[22] and carrying African culture from the Congo River to a Caribbean which would soon echo to the conga drum.[23] At the same time, Europe's long-dead postal service was resumed. Now Erasmus could become a snail-mail junkie, using a stream of letters[24] to unite a community of international eccentrics into a movement which downgraded religion's death ride to heaven and exalted earthly humanity. The moment was appropriate. For the first time since the rise of multicellularity, there was no longer just one global brain but two—one microbial, the other that of man- and womankind.

Homo sapiens were not the only ones to gain from this growing global weave. Bacteria and other microorganisms traveled the trade routes of Greece and ancient Rome poking about for new possibilities. And they found them, spawning waves of plague in Athens and the empire the Cae-sars ruled. An infectious epidemic hitchhiked from the Himalayas and India to Constantinople in Justinian's reign and killed off ten thousand of that glorious city's inhabitants a day. Smallpox took to the ocean waves with the conquistadores of Spain and gorged on a Native American popula-tion which had no barriers of immunity, sweeping an estimated 90 percent of Indians to their grave. And tuberculosis, a disease which had first trekked across continents with primitive hunter-gatherers in pre–Ice Age days,[25] carried both the parents of Gilbert Ling away.

Ling begins the telling of his own tale with China's Boxer Rebellion, an event born of the worldwide web at the end of the nineteenth century. The development of ships driven by steam in 1807 along with the creation of breechloaders (1838), Gatling guns (1862), and light artillery[26] freed Western nations like England and the United States to escalate intergroup tournaments—both those of conquest and the more peaceful ones of com-merce. China was brimming with export goods the West demanded, but the Middle Kingdom showed little interest in the woolens and other com-modities Europe and America offered in return. The result was a serious balance-of-payments deficit, one the English made up by peddling a prod-uct manufactured in its Indian provinces—opium.[27] Sales were brisk— eight million pounds of the narcotic were smuggled into China during just one year (1836)[28]—and American merchants like Warren Delano, Franklin Delano Roosevelt's grandfather, wanted in on a good thing. Many of America's elite families had made their initial fortunes shipping sea otter pelts from North America's frontier to Chinese ports. Now the Coolidge and Forbes clan followed the British lead and took up drug dealing in Can-ton. The Chinese counterattacked against this traffic in addiction with a series of crusades they couldn't win. One was the Boxer Rebellion. When it was over, China was milked for "reparations." The United States made up for such unconscionable exactions by providing money for a university in Peking and by establishing fellowships which would allow over one hundred of China's top students to study in the U.S.A.

Battle had proven a synapse builder, and in 1945 one of those synapsed was Gilbert Ling, who showed up at the University of Chicago during the Christmas season to go for his graduate degree. Ling chose to focus on the field of cell physiology, one which would carry him further into the meshwork of minds which binds together the collective intellect of humankind.

Cell physiology began in England in the 1600s when a young scientific jack-of-all-trades got his hands on a new lens-based device with many a peculiar property. The magnifying lens had first appeared in ancient Assyria,*[29] had been forgotten, then had been reinvented in Florence in 1280[30] as an aid to reading for the privileged classes—clerics and noblemen. Within five years, wearing spectacles had become a craze. Early in the 1600s, an obscure Dutch spectacle maker, Hans Lippershey, discovered that by combining several lenses in a tube he could craft a long-distance viewing device which he believed would allow the Netherlands' military defender, Prince Maurice of Nassau, to spot attacking Spanish armies long before they appeared clearly to the naked eye. Lippershey's secret weapon—the telescope—soon leaked to Paris, Frankfurt, and London. Galileo, a Pisan teaching university courses in Florence, made one of his own with some beefed-up properties. In an act of diversity generation which literally horrified fellow scholars and churchmen, he turned his telescope away from the fields of potential battle for which it was originally designed and aimed it audaciously at a territory sacred to religion and philosophy—the heavens, the night sky. There he discovered that the plan-

---

*It is frequently claimed that the "ancients" utilized the lens, but this is a bit of an exaggeration. The Assyrians knew that by removing slices from a glass sphere, they could obtain a magnifying device. The Greeks did not employ such bulging bits of glass to enlarge minute or distant objects. Aristotle merely discussed the ability of the "burning-glass" to start fires in his *Posterior Analytics*. In his play *The Clouds,* Aristophanes also toyed with the mischief one could do by using "a crystal lens" to melt from a distance the wax letters some unsuspecting victim was reading. By Cicero's time, the glass was an object through which one saw things, but not a device with which one was able to magnify. For Ovid and Plutarch, a glass was merely a mirror. The one genius in the Western bunch seems to have been, of all people, the foppish Roman emperor Nero. Said Pliny the Elder, writing in 23–79 A.D., "Emeralds are usually concave so that they may concentrate the visual rays. The Emperor Nero used to watch in an Emerald the gladiatorial combats." Alas, no one in antiquity seems to have followed Nero's example. Even the Chinese, often credited for using spectacles, apparently wore flat panes of colored glass, not true lenses, and did it for show, not to improve their vision.

ets were not orbs on crystal spheres moved by angels, but craggy globes of matter much like ours. Another university lecturer in Bologna diversity-generated in reverse. Marcello Malphighi turned the telescope around, peered into the large end, and used it to look down, examining such hidden things as the inner layers of skin, the bumps on the tongue, the capillaries carrying blood, and the outer tissue of the brain's cerebral cortex. This was the beginning of microscopy.

As snail mail became more efficient, it upped the speed with which pre–Age of Reasoners could data-swap. Malphighi was a member of one newly formed local chat group—the Accademia del Cimento. Englishman Robert Hooke was a member of another which had used the mails to go intercontinental—the British Royal Society. Hooke picked up on Galileo's innovations, built his own telescope, used it to discover a new star in Orion and to sketch the planet Mars, then turned it on its head, transforming it into a microscope with which he could examine the invisible intricacies of snowflakes, mosquitoes, feathers, and fungi. He hit the jackpot when he pointed his lenses at a slice of cork, for here he spotted what he described in his best-selling book *Micrographia* as "the first microscopical pores I ever saw." Because they reminded him of the chambers in which monks slept, Hooke called these microrooms cells.

The human network continued its arch. Two hundred years later, a pair of German labmates, Matthias Schleiden and Theodor Schwann, took the work of Hooke and transformed it to a theory which would elevate our understanding of bacteria, beasts of burden, and almost all biology. Every plant and animal was composed of cells, the southern German and the Prussian said.[31] And cells were water-filled cubicles, walls enclosing a passive micropond.[32]

This led to a mystery—molecules of sodium were locked out of a cell and potassium was locked in, a fact which defied the laws of physics. How could two chemicals which should have spread with equal density on either side of the cell membrane be kept from gushing through the cellular barricades? In the 1860s, a German wine merchant turned physiological chemist—Moritz Traube[33]—proposed that the membrane had some sort of "atomic sieve" pushing one molecule out and trapping the other inside. Alas for Traube's notion, twentieth-century work with electron and X-ray diffraction showed that his idea was full of holes—but holes of entirely the wrong kind[34]—pores too large to be the sieves of Traube's imaginings. Faced with this fact, modern biologists upped the power of Traube's sieves and turned them into bilge pumps studding the cell membrane.

Meanwhile, Gilbert Ling built an edifice of evidence indicating that one could take large steps ahead by recognizing that the passive liquid in a cell's interior was active as could be, and because of its busy nature, no wall pumps were necessary.[35] Ling's proposals would eventually amount to what Cambridge University's Lancelot Law Whyte called "one of the most important and advanced contributions to the understanding of the structure of living systems which I have seen."[36] They would provide what Ling feels is "the only verified, unifying theory of cell physiology."[37] And Ling himself would be hailed by the Nobel Prize–winning discoverer of vitamin C's oxidative properties, Albert Szent-Gyorgyi, as "one of the most inventive biochemists I have ever met."

This is where the word of caution comes. Science is accelerated by the games subcultures play. It thrives on subcultural tournaments set off by creative bickering over trivialities. One side argues for its point of view; another marshals facts to prove its opposite. But when debate downshifts to power politics, irrationality can strip the gears with which science gains its speed. This idea-lock claims victims, among them: Gilbert Ling.

NETWORKS, NETWORKS EVERYWHERE
AND NARY A DROP TO DRINK

Complex adaptive systems are geared not by reason, but by biology. This can be a blessing and a curse. Take resource shifters, which can build you up, then knock you down again. Because Gilbert Ling's ideas were considered extraordinary, they brought the good things of this world tumbling his way. When Ling finished postdoctoral work at the University of Chicago in 1950, he was given funds to teach and do research at Baltimore's Johns Hopkins Medical School. Then came a proposition from the University of Illinois, an offer which would allow him to forgo teaching responsibilities and plunge into his cell research full time. Ling's work drew an escalating level of popularity. When Philadelphia's Eastern Pennsylvania Psychiatric Institute established a Department of Basic Research, it invited Ling to design and run his own laboratory. Next came the chance to build a lab at Philadelphia's Pennsylvania Hospital. Attention begets more attention. Says Ling, "Bright young science students flocked to my laboratory to work toward a higher degree or otherwise participate in the exciting research."[38]

Ling's work soon penetrated the global data flow. He recalls lecturing "all over the world," receiving standing ovations as he went. In 1957, he reports that he was visited by "the Soviet scientist, A.S. Troshin, another anti-membrane-pump-cell physiologist and Director of the huge Leningrad Institute of Cytology," who had apparently been following Ling's work avidly.[39] Along the way Ling's concepts of saltwater biophysics knitted together with a Nobel Prize–winning thread of theory created by Isidor Rabi, an immigrant who had come to the United States from the Austro-Hungarian Empire at the age of three.[40] The result was one of modern biology's most powerful tools—the nuclear magnetic resonance

(NMR) imagers used today to probe the body and the brain. Once Ling's ideas were acted on further, he was confident that the bacteria which had killed his parents would take a beating. Armed with a better theory, humans would be able to build new tools with which to battle the microbial invaders anxious to turn the cells of which we're made into their private banquet halls. Here's a taste of why Ling may be right.

The liquid of a cell's interior is anything but random slosh and splash. An atom is a network of quarks and leptons—there are three quarks and a lepton in even the simplest atom that we know. Molecules are networks of even more complexity. Each has plugs and sockets on its exterior. These can produce a hypersensitive mesh which physicists call a "cooperative state."[41]

For example, Ling points out that a water molecule is not the placid stuff of our imaginings. Its one oxygen and two hydrogen atoms are arranged in such a way that one end of the molecule is charged positively and the other charged negatively. The positive charge of one water atom docks with the negative charge of another, impelling water molecules to knit together in vast skeins.[42] A loose molecular weave keeps water liquid; a tight one turns it into ice. But the water, proteins, nucleic acids, and carbohydrates of a cell's interior are webbed in an in-between state we call life.[43] Ling's Polarized Multilayer Theory of Cell Water[44] helps explain the how and why.

Says Ling, "Jell-O is almost all water and yet in Jell-O, water can 'stand up' as no normal pure liquid water ever can. This ability of the water in Jell-O to stand—which ice can also—indicates that the water-to-water interaction in the Jell-O water has been altered by the only other component present, gelatin." The key to this transformation, according to Ling, is the material from which gelatin is made—collagen, a protein which pulls coats of water molecules around itself as if it were dressing in layer after layer of winter clothes.[45] When water-bundled proteins Velcro together, one has Jell-O.[46] Or one has the flesh of which you and I are made—for proteins cloaked in woven water give our muscle and our fibers strength and shape.[47] So powerful are Ling's proteins at holding on to their water-molecule snowsuits that they can be chopped up, then centrifuged with a force a thousand times that of gravity and not lose a drop, despite the fact that water makes up 80 percent of their raw material.

Ling's electrostatically connected water molecules sheathing long, thin protein braids are as easily swayed as Carlyle's sheeplike critics—a little influence applied in the right place goes a long, long way. Each water molecule shifts the interior balance of its neighbors' electrons, and every water molecule is able to pivot its electrostatic charge. The result is a typical crowd response: when conditions change, every molecule of H_2O swivels in the same direction. The result, in Ling's words, is a "functionally coherent and discrete cooperative assembly." If a relatively small but insistent molecule called a cardinal adsorbent*[48] steps up to one of the many podiums (docking sites) along the protein chain, it galvanizes attention, making all the water molecules swivel their "heads"—the polarity of the electrons in their shells—simultaneously. This changes the chemical properties of the assembled multitude dramatically.[49] But, hey, that's life— quite literally. When the molecular crowd disperses, a cell is dead.

Ling's concept of meshed molecular audiences driving each other to cooperative alertness[50] may have seemed strange when he proposed it in 1962. But since then, other researchers have shown that a cell's interior is a hive of structure and activity, replete with a framework of architectural girders and a bureaucracy of messengers, receivers, traffic directors,[51] administrators, and gatekeepers.[52] The interactions of these cellular constituents make the workings of the largest human city look simple by comparison. The result is one of the most intricate complex adaptive systems ever seen.

For example, some researchers propose that the microtubules of the cell's frame (its cytoskeleton) have the computational power of a computer.[53] These living beams and crossbars operate according to complex adaptive system rules. Assembly centers called centrosomes mass-produce identical new tubules—an act of conformity enforcement. Each microtubule juts out of the centrosome's center like a porcupine quill. But the direction in which each thrusts is random, since each is a probehead sent to explore the needs of the cell's interior space. This helter-skelter poking about is an act of diversity generation.

Microtubules are made of a highly unstable protein—tubulin. If a new microtubule's loose end fails to find a connection point, its tubulin self-destructs. If the microtubule *does* hook up with a docking site—say a wandering chromosome—proteins in the welcoming partner trigger the microtubule's raw material to retain its strength. The chemical system giving the tubule vigor or melting it back from whence it came is a primitive inner-judge.[54]

Those microtubules which find their way to connecting points critical to the operation of the cell become superstars in the cytoskeleton. Resources are fed to them by economic centers in the cell's interior near and far. Some of these goods are influence—including the ability to

---

*The best-known cardinal adsorbent is ATP, adenosine triphosphate, the key energy carrier within the cell. In trying to describe the massive chain reaction produced by the relatively small twitch of a cardinal adsorbent, Ling has said that "the water molecules dance with the protein, and ATP is the conductor."

reshape the cell, to make calculations on the cell's behalf, to organize activities inside the cell, and to give orders to the cell's nucleus.[55] Other payoffs arrive in the form of building materials.[56] The impulses moving other cellular components to pass assets to successful microtubules are resource shifters.

Cell biologists Marc Kirschner and John Gerhart are awed by the result. They call the complex adaptive system of microtubules "an exploratory system . . . [with an] unlimited range of possible configurations," a learning machine so precocious that it literally "facilitates . . . evolutionary change."[57]

Cells are probeheads too, but they act on behalf of the organism of which they are the building blocks. Like microtubules, competing cells are subject to complex adaptive system rules. They search the possibilities around them and compete for connections,[58] for the power to make copies of themselves, for building material, and for fuel. The intergroup tournament between cells has a potent impact on the life or death of the microtubules which have competed to form a single contestant's skeleton. If the cell within which the microtubule resides is a fetal neuron and finds connections, the cell thrives and the microtubules it contains survive.[59] If the fetal neuron fails to find a hookup, its inner-judges kill it off,* and the microtubules—no matter how successful they were inside the cell—eat dirt (or are eaten by bio-cleanup fluids). The same complex adaptive system rules power the immune system, organisms, bacterial colonies, and human societies—as Ling would learn at first to his delight and later his dismay.

WHEN LIBRARIANS BURN THE BOOKS

Ling demonstrated with a series of compelling experiments that cells do not need membrane pumps to rid themselves of sodium.[60] The busy meshwork of the cellular inner city excludes sodium quite easily. Do away with membrane-pump theory, and drugs could be conceived which would do far more than current pharmaceuticals to battle a bacterial foe, or so Ling was convinced. But before this possibility could be tested, Ling would run into conformity enforcers of the old Spartan kind—those which get a hammerlock on the workings of mass mind.

Many scientists supporting the theory of the sodium pump took exception to Ling's work. In 1967 the president of the Federation of American Societies for Experimental Biology planned a debate at the federation's annual meeting. To extend its outreach, the public airing of the

---

*A cell with no takers for its services commits suicide via apoptosis, programmed cell death.

exchange was to be broadcast on closed-circuit television. Dr. Ling agreed to pit his evidence against that of his critics. Six proponents of the membrane-pump theory were invited to participate. Though the symposium was a year away, says Ling, every one of his debaters claimed he had other commitments for the appointed day. Then, in 1969, Ludwig Edelmann, a graduate student at Germany's University of Saarland, read one of Ling's books,[61] performed an ingenious experiment on guinea pig heart muscles, and obtained results which strongly supported Ling's theories and called into question the concepts of the membrane-pump school. Edelmann was told by his superiors that in spite of his evidence, membrane-pump theories were correct and Ling's had been disproved. Ignoring this pronouncement from his elders, Edelmann presented his results before a group of referees. The year was 1972, and Edelmann received no refutations of the validity of his findings. But in October of 1973 he was told he no longer had a job. Says Edelmann, "One referee wrote me a private letter suggesting not to follow a fixed idea [Ling's idea, to be more precise] further and to look for another scientific field."[62] Edelmann's position in his discipline had been jeopardized at least in part because of his support for Ling's work. Eventually, he was forced to switch from biophysics to electron microscopy.[63]

In 1971 a student of biophysical chemistry at the University of Pennsylvania became curious when Ling's alternative to the sodium-pump model was mentioned briefly in class, then was ignored for the rest of the semester. When he brought the matter up with his professor after class, the pedagogue explained nervously, "If I consider Ling I'll hear repercussions, and my position is threatened. So, I won't consider Ling—I have a wife and children. . . . Even now I can't get most of my stuff published."[64]

Ling fought back against this muzzling of scientific diversity. In 1973 he was appointed to a seven-person committee established at the request of Texas congressman Bob Casey to examine the creativity-smothering peer review processes used by government agencies.[65] (Ling cites a bit of advice given to those applying for grants to the National Institutes of Health: "The author of a project proposal must learn all he can about those who will read his proposal and keep these readers in mind constantly as he writes."[66] In other words, forget fresh insights—conform, or better yet, kiss up.) And in 1975 Ling was able to testify on the obstacles being tossed in his path to a congressional committee.[67]

However, one of Ling's adversaries made a far more vital move. In 1973, he took over the chairmanship of the Physiology Study Section within the National Institutes of Health, the source of Ling's funding. In a stroke, Ling discovered that the money which had supported his efforts for nearly twenty years was slated to disappear.[68] He was saved at the last minute by two NIH officials who saw the merits in the objections he filed.

But the ad hoc group erected to maintain Ling's research air supply was fragile, and lacked the power to finance other researchers who might have wished to follow the leads Ling was uncovering. Ling's subcultural base was dwindling. The resource shifter had been commandeered and was now taking away what it had previously offered up so generously. Ling was disturbed by what he calls "the terrifying power of coercion" leveled against him. Then he recalls being "deeply shaken when (virtually) all my graduate and postdoctoral students suddenly left en masse." Later, in an apparent attempt to demonstrate their allegiance by aping the top dogs in the winning camp, most would publish works attacking their former mentor, Dr. Ling.

Meanwhile, the flow of money to membrane-pump-theory-based projects grew ever more substantial.[69] The complex adaptive system was following its primary dictum: "To he who hath it shall be given; from he who hath not, even what he hath shall be taken away."

In a masterly subcultural coup like those we'll see in the next chapter, the supporters of the sodium-pump model created the impression that there were no dissenters from their point of view. In 1975 a key review article appeared purporting to summarize all of the year's studies on sodium-pump theory.[70] In it, reports Ling, "I.M. Glynn and S.J.D. Karlish from the 'Mecca' of cell physiology, the Physiology Laboratory of Cambridge University . . . cited 245 references, all in favor of the sodium pump hypothesis." They mentioned none of the research against.[71] As Ling sees it, "Documents after documents of experimental evidence against the sodium pump hypothesis were . . . made to look as if they had never existed. So were the identity of those scientists who had made these critical contributions." By projecting an illusion of unanimity, the sodium-pump faction silenced those who might have raised objecting voices to their views.

In 1988, Ling was forced to close his lab at Pennsylvania Hospital and become a scientific refugee.[72] Actually he was more fortunate than most. Because of his role in developing nuclear magnetic resonance imaging, Ling was offered haven on Long Island at the headquarters of the Fonar Corporation, a maker of NMR devices run by Dr. Raymond Damadian,[73] originator of whole-body NMR scanning, winner of a National Technology Award, and developer of a cellular theory complementary to Ling's, one which also discredits the theory of the sodium pump.[74] Though Ling was often forced to buy chemicals and lab animals out of his own pocket, his isolation from the mainstream allowed him to lay out what could be an important mathematical expansion of his theories and to continue research on the cellular mechanisms of cancer and on drug activity.[75] However, Fonar could not afford to underwrite a full program of research, one able to train a new generation of scientists, a new subcultural cohort, in the pursuit of fresh discoveries. Ling continued publishing new studies—one hundred of them from 1966 to 1997—however, they increasingly appeared in just his own journal, *Physiological Chemistry and Physics and Medical NMR*. And Ling's main contact with the general public was reduced to a web site.[76]

In China Ling had learned the values of Western science. Now, he says, "it is excruciatingly painful to witness the West's magnificent contribution to human civilization dragged in mud by some of the highly respected members of the scientific community."[77]

Ling had benefited in his early years from the axons of a global brain spread through China, Russia, Germany, England, the United States, and all points in between. This was no longer the sluggish collective mind which had clung to the same stone tools for millions of years before making a major change. It was one whose shifts took place in generations, decades, or, in the case of the about-face which had turned Ling into a pariah, could flash across continents in days. But despite the switch from foot power to modems and planes, the underlying mechanisms remained the same. Biology had long since knit together humanity's global brain. The result was a planetary mind which even in its finest form is riddled with irrationalities.

Ling—a victim of that brain's subculture games—contends his loss will cripple us in yet a larger intergroup battle, that between drug-resistant bacteria, viruses, and humans. If new breeds of *Mycobacterium tuberculosis, Streptococcus pneumoniae, Staphylococcus aureus,* vancomycin-resistant enterococci,[78] and the human immunodeficiency virus (HIV) have their way, the struggle between the microbial global brain and ours may be the Great War of the twenty-first century.[79] And Ling's ideas, though suppressed today, may later prove a necessity.

Whether Ling's theories are right and those of his opponents are wrong or vice versa, ultimately each belief, individual, clique, movement, and nation is a hypothesis in a larger pondering process. Through the interplay of hypotheses—their battles, their rapes, and their miscegenated offspring—the mass mind grows and learns, even when it does so by taking a mistaken turn.

19

THE KIDNAP OF MASS MIND:

Fundamentalism, Spartanism,

and the Games Subcultures Play

1932 A.D. to 2030 A.D.

When I, or people like me, are running the country, you'd better flee, because we will find you, we will try you, and we will execute you.

Randall Terry, Operation Rescue

I represent a Political Party (The Creator's Rights) that would not hesitate to use nuclear weapons . . . and all the potential slaughter that entails . . . in defense of a State's Right to secede in order to restore God's plan for government.

Neal Horsley, candidate for governor of the State of Georgia, the Creator's Rights Party[1]

The one who blows up the enemies of Allah by blowing up himself as well . . . is, Allah willing, a martyr.

Br. Abu Ruqaiyah,
"The Martyrdom Operations"

We are the millennial promise. Get used to us.

Mary Matlin,
American conservative activist[2]

As we'll soon see, the battle to win out over microbial foes is one which could make late-twentieth-century atrocities like those in Rwanda and Kosovo look very small indeed. But the subcultural struggles retarding science's advance are minor maladies of mass mind compared to a set of twenty-first-century subcultural clashes in which Sparta and Athens remain vigorously alive.

Today's cyber-era Spartans are bone crushers of conformity. They are the fundamentalists of both the left and right. Some are godly; some are secular. Religious extremists, ultranationalists, ethnic liberationists, and fascists[3] fall on the fundamentalist side of the line. Brooking no tolerance of those who disagree, they invoke a golden past and a higher power, both of which demand submission to authority. The worst shoot, burn, and bomb to get their way. Their opposites are Athenian, Socratic, Aristotelian, diversity-generating, pluralistic, and democratic. They pay lip service, and often a good deal more, to such slogans as "I may disagree with what you have to say, but I will defend to the death your right to say it." The groups which follow this pattern are more diffuse than their rivals; in fact, the right to be loose-knit is part of their philosophy. They include liberals, democratic socialists, libertarians, and many free-market capitalists. These champions of human rights use the word "freedom" to liberate the individual, not to hammer home the triumph of a chosen collectivity.

The Athenian strategy moves to the top when things are going well. Spartanism grabs the throne when the world is going to hell.

RULES OF THE SUBCULTURE GAME

Give me a lever long enough, said Archimedes, and with it I can move a world. Many are the levers in the games subcultures play.

Back in 1932, a social and political psychologist named Richard Schanck carried out a field study of "Elm Hollow," a literal horse-and-buggy town in rural New York State cut off without a bus stop or a passenger railway.[4] Nearly to a man and woman, Elm Hollow's Baptist residents spilled forth hatred of the era's godless sins: card games involving the use of the king, queen, and jack, alcohol, and Satan's incense burners—the products of tobacco leaves.[5] Strange thing was that once Schanck had gained a bit of trust, he was often pulled behind closed doors and shuttered blinds by some lone figure hungry for a game of gin, a hard-cider-drinking binge, and a cigarette or two. The hell-tempter sneaking off the straight and narrow wasn't a town hoodlum, but one of its respected citizens.

Despite a population of less than five hundred, Elm Hollow was riddled with subcultures: insiders versus outsiders, Baptists versus Methodists, those who lived near the freight station versus those who lived near the post office, and members of the local lodge versus those who hadn't joined. But of the town's competing factions, one had trussed up the community.[6] Thirty-five years before Schanck showed up with his sharpened pencils, an influential Baptist minister had died. To all intents and purposes, this worthy's memory no longer should have thrived. As Schanck puts it, "The church has been remodeled, every one of his parishioners, save one, is gone, and in fact, in the realistic sense, nothing except the property upon which the building rests remains the same."[7] But the minister had left more than just a minor legacy—he had provided the platform for what the researchers called a "personality tyranny." The heritor of the deceased minister's stored influence was his daughter, Mrs. Salt, a woman who puzzled the researchers enormously. Forty-two percent of Elm Hollow's Baptist parishioners declared publicly that her word demanded "extraordinary respect."[8] She "was not liked," writes Schanck, yet "she dominated" the attention structure of the community.[9]

How, Schanck wondered, did Mrs. Salt manage to hold her sway? The answer was in imitative behavior and the fear which keeps us sheep from going astray. Mrs. Salt controlled what you did and did not say. Hence Elm Hollow chorused with almost unanimous piety. In private, Schanck heard the choked-off sound of heresy. Numerous Baptists secretly hankered for their nip of alcohol. Even the most vehement behind-the-scenes believers in the good of sipping bourbon from time to time, when caught in the spotlight of neighborly attention, echoed the religious party line. Even the new minister, who'd only been on the scene a year, admitted privately to liberal views. Yet on the pulpit he preached fire-and-brimstone services.

How completely the anointed had commandeered collective perception became apparent when Schanck asked the closet dissenters how *other* people in the community felt about face cards, liquor, a smoke, and levity. Hoodwinked by suppression, each knew without a doubt that he was the sole transgressor in a saintly sea. He and he alone could not control his demons of depravity. None had the faintest inkling that he was part of a silenced near-majority.

Here was an arch lesson in the games subcultures play. Reality is a mass hallucination. We gauge what's real according to what others say. And others, like us, rein in their words, caving in to timidity. Thanks to conformity enforcement and to cowardice, a little power goes a long, long way.

FROM OPEN HAND TO SOCIAL FIST— THE CLENCHED SOCIETY

> Things fall apart; the centre cannot hold;
> Mere anarchy is loosed upon the world,
> The blood-dimmed tide is loosed, and everywhere
>
> .   .   .   .
>
> The best lack all conviction, while the worst
> Are full of passionate intensity.
>
> .   .   .   .
>
> And what rough beast, its hour come round at last,
> Slouches towards Bethlehem to be born?
>
> William Butler Yeats, "The Second Coming"

In our flexible, reengineered economy . . . we are unmoored—from
our pasts, our neighbors, and ourselves.

Patrick Smith

Frantic orthodoxy is never rooted in faith but in doubt. It is when we
are not sure that we are doubly sure. Fundamentalism is, therefore,
inevitable in an age which has destroyed so many certainties by
which faith once expressed itself and upon which it relied.

Reinhold Niebuhr

Groups under threat constrict.[10] They do it to gain leverage and force.
Toss bacteria onto a surface so hard that feeding becomes almost impossible, and they'll abandon individual freedom, pull together their
members, and form a tight-knit phalanx which can, according to physicist-
microbiologist Eshel Ben-Jacob, carve through the obstacles around it like
a blade.[11] Human groups in times of trouble stiffen up their unity,[12]
squelch ideas,[13] rally 'round their leaders,[14] and spit out those who fail to
ape the top dog faithfully. Group members project their own forbidden
emotions onto others, and in their ferocity become enforcers for the
group's norms. They spot the smallest sin among their fellows and punish
it intolerantly.[15] In biology, emergency measures like these have a tendency
to cut two ways. In short jolts they produce bursts of power. But used in
the long run, they destroy.[16] The oneness which gives society the punch of
a bayonet produces over the course of time a paralyzing rigidity.

This lesson may prove critical for the twenty-first century. Old ways
of life are crumbling, and the victims of disintegration hunger for something new around which to cohere. As I write, Russia has ceased functioning as a state. It cannot collect its taxes, pay its workers, or protect its
citizens from crime. In this chaos, money is worth nothing and ordinary
citizens cannot buy the food they need to survive. Siberian workers unpaid
for over a year told television crews from Britain's Independent Television
Network News[17] that if they could get their hands on enough rubles they'd
forget buying groceries, purchase a rifle, go to Moscow, and take revenge.
The hungrier, angrier, and more desperate they become, the more they
will be ripe adherents when some new "liberator" arrives.

In China between 70 million and 100 million citizens—more than
the entire British population and, more important, more than the members of China's Communist Party[18]—adhere to the teachings of a New
York–based martial arts master named Li Hongzhi.[19] Like some other fundamentalists, Hongzhi promotes his own books as the path to truth, claims
the ability to heal miraculously, and preaches the hatred of outsiders,
denouncing homosexuals, rock music, and television. The world is about
to end, polluted by these sins, says he.[20] According to Reuters, "rapid

change and upheaval" have driven vast hordes of Chinese into cults which, in the words of high-ranking Chinese nuclear physicist He Zexiu, constitute "a real danger to society."

During the late twentieth century, the seams of European and American society were also fraying ominously. Waves of migration brought Turks into Germany, North Africans and sub-Saharan Africans to France, Pakistanis and Albanians to England, and Mexicans, Puerto Ricans, South Americans, Asians, Lebanese, Palestinians, and a host of others into England, Canada, and the United States. Drowned in a sea of strangers, many natives of the West were groping for a handhold, a new form of identity.

Americans were thrown from jobs which had been in their families for generations—everything from auto work to farming.[21] Their expectations for the future were tossed aside as economic boundaries melted and Western workers were forced to compete with those of China, Malaysia, Indonesia, Thailand, India, and the Philippines.[22] In the new environment, Koreans could take advantage of their average worker's superior education in mathematics and the sciences,[23] Indians could slam home their aptitude for computer programming, the Chinese could utilize their Confucian tradition of higher education, their genius for grassroots and elite technologies,*[24] and their willingness to turn agricultural laborers into industrial and technological slaves.[25]

Meanwhile, U.S. jobs had disappeared in areas like Michigan and migrated to new territories like North Carolina, sucking the life from towns and severing the roots of citizens in communities like Flint, Michigan, which had once seemed fixed eternally. General Motors alone had laid off sixty-five thousand workers between 1993 and 1998.[26] As jobs shifted from lifetime positions to serial career transitions, even homes riddled with luxuries—central air-conditioning, computers, cellular phones, and extra TVs—were bedeviled by new forms of layoff anxiety.[27] Then came the economic scare of the late 1990s, when Asia was first to hear the sucking sound of a looming credit catastrophe. Bank after bank discovered that roughly 60 percent of its assets were tied up in loans no one would ever repay. While the International Monetary Fund scrambled to plug the holes with $6 billion here and $10 billion there, the total vacuum mounted to over a trillion dollars[28]—more than half the size of the federal budget of the United States.[29] Some foreign economies imploded. During the nineties, Japan tottered economically.[30] With her went Malaysia, Indonesia, Thailand, Taiwan, Hong Kong, and Singapore. Washington State's apple farmers, wheat growers, and timber harvesters had grown to depend

---

*Chinese technologies consistently eclipsed those of the West until just the last few centuries.

on the markets of the Far East. So had Oregon's electronics producers, Maine's fishermen and paper products makers, Colorado's industrial machine crafters, Alabama's chemical manufacturers, and Nebraska's food processors and leather makers,[31] not to mention the oil producers of Venezuela[32] and of the Arab states. With no place to sell their goods, many small entrepreneurs and international corporations stared disaster in the face. More jobs were downsized and yet more who had known security were tossed from their accustomed place.

Sponge cells sieved apart frantically regain their hold by constructing fresh communities. Chimps and rhesus monkeys clump together on the basis of similarity. Displaced humans find common ground by rallying around a common sentiment, a shared sense of experience, a point of connection whose expression offers hope and certainty. When the formulae of the center prove powerless to save one's soul from chaos, the new hypotheses to which men cling may be peculiar "truths" which come from the periphery.[33] Often these beliefs stress the Spartan fist instead of Athenian creativity.[34]

The collapse of global boundaries had wrenched many from their comforts, but it had also enlarged the ways in which those stripped of self could seek new brethren and new ideologies. On the Internet Americans, Japanese, and British who found their solace in worshiping nature could make common cause with ecoterrorists like the Germany-based Earth Liberation Frontists and their ilk, self-styled "warriors" who attacked fur farmers, tossed firebombs, and set up booby traps for loggers to restore Earth Mother's sanctity. Those convinced that the Aryan race was biblically destined to rule creation could cluster 'round extremist web preachings. Muslim terrorists who felt like driving a car bomb into a crowd of civilians could read on the web site of Australia's *The Call of Islam: A Comprehensive Intellectual Magazine* why this form of "martyrdom operation . . . will be rewarded in paradise."[35] And even Serbs from Chicago and British Columbia to Belgrade, whose extremist brethren were carrying out mass killings in Bosnia and Kosovo, could share the justification for this ethnic cleansing via cyberspace. It had become easier than ever for new subcultures to inject fresh (though sometimes unsavory) hypotheses into the collective mind.

More than ever, the newly dispossessed needed a sense of social warmth,[36] of a home, a family, a nest filled with friendship, caring, and stability.[37] Postmodern Spartans were masters at catering to this need. Through schools, clinics, youth groups, political committees, missionary programs, prayer clubs, parenting classes, safe houses, family nights, societies for the oppressed, volunteer organizations, charitable projects, and other nests of nurture, extremist movements offered those who'd been cut loose instant roots and genuine nourishment. They also proffered haven from confusion and uncertainty,[38] for all they did was anchored in eternal bedrock—God's or Gaea's verity.

Nothing grows a subculture faster than opposition to assault.[39] Confront two squabbling cliques with a common danger and the two will join to face the adversary.[40] Give humans a sense that death is in the air and their individualistic views will ebb in favor of the creeds clasped by the collectivity.[41] What's more, they'll go on the attack against those who challenge commonplaces they themselves had doubted in untroubled days.[42] Aggression drowns confusion's agonies.[43] The German youth of 1914 were lost in anomie. World War I was anesthetic and salvation—redemption from a hell of private pain. Ernst Troeltsch testified to the healing power of warfare when he made the following wildly popular speech:

> The first victory we won, even before the victories on the battlefield, was the victory over ourselves. . . . A higher life seemed to reveal itself to us. Each of us . . . lived for the whole and the whole lived in all of us. Our own ego with its personal interests was dissolved in the great historic being of the nation. The fatherland calls! The parties disappear. . . . Thus a moral elevation of the people preceded the war, the whole nation was gripped by the truth and reality of a suprapersonal, spiritual power.[44]

If a crisis seems indecipherable, its victims are condemned by their inner-judges to a shutdown, the helpless nail-biting of anxiety. But if the causes of a crisis seem explainable, the result is a surprisingly healthier emotion—fear. Fear is a form of arousal which prepares us to fight back rather than give up. And it drives us toward group unity.[45] This is true even if an actual solution is nowhere to be seen.[46] The leader's ability to create the illusion of a handle on the situation matters more than the reality. And the idea of a demonic foe is often at that handle's core.[47] Postmodern fundamentalists are masters at the craft of enemy creation and the manufacture of a siege mentality. Their world abounds in villains—one-worlders with black helicopters, Satanic secular humanists, Beelzebubian New Agers, homosexual conspirators, Zionist bankers, Illuminati, Trilateral commissioners, Great Satans, real estate developers, mink farmers, abortionists, and genetic engineers. Bringing the sense of battle to fever pitch is the myth that we are on the brink of the mother of all subculture wars, that of the final days in which an avenging Nature, Allah, or Jehovah will wipe this old, iniquitous world away. The impure and unbelieving (that means you and me) will die in manners horrible to contemplate. When all is stripped and cleansed through nuclear flame, greenhouse flood, or the bloodbath of the scimitar, the righteous will finally take their place at the right hand of God, of Nature, or of racial destiny.[48] Unfortunately, extremists armed with weapons of mass destruction can turn eccentric visions of apocalypse into self-fulfilling prophecies.

Imagine what might happen if a group believing as the Spartans did that death is a noble thing and that adherence to authority is the ultimate salvation of mankind gains control of long-range missiles tipped with atomic weaponry.[49] The dedication of those who know that a violent end will bring a kingdom of blessedness could lead to human suicide. This is exactly what one of the first of these groups to get its hands on chemical weapons, Japan's Aum Shinrikyo,*[50] had in mind. In 1995, Aum Shinrikyo attempted to fill Tokyo's subway system with poisonous sarin gas, and maintained a research and development lab working on weapons designed to literally exterminate humanity. Wrote Robert Jay Lifton, professor of psychiatry and psychology and director of the Center on Violence and Human Survival at New York's John Jay College of Criminal Justice, "Aum Shinrikyo . . . had in mind nothing short of worldwide genocide. They wanted to murder everybody in the world."[51]

Here's how the rain of brimstone forecast looked at the turn of the twenty-first century. Iran, Afghanistan, Chechnya, and Sudan were already in Islamic fundamentalist hands. Though the fact had gone largely unpublicized in the West, by the end of the 1990s the Moro Islamic Liberation Front had established what it called the Autonomous Region of Muslim Mindanao on the second-largest island of the largely Catholic Philippines, had appointed its own governor, Nur Misuari, and was fighting to gain even more territory.†[52] Afghanistan's Taliban had stripped women of their jobs and rights,[53] required that they cower in their homes wrapped in black with their windows curtained and painted over, wear no shoes a man could hear, and possess no reading material outside of pamphlets promoting official religious views.[54] Ladies of learning and skill who transgressed were beaten with rifle butts in the streets. The Taliban's Ministry for the Propagation of Virtue and Suppression of Vice—modeled on Saudi Arabia's religious police—was known for its stonings, its amputations, its hangings from mobile cranes so a body could be paraded high above a city's roofs, its deaths by crushing under a wall, and a long list of other "divinely ordered"

*Aum Shinrikyo was a Pythagorean-style subculture gone wild. Its leader, Shoko Asahara, had the usual goals of a Faustian introvert. He was an outsider assembling a cultural home around himself. In the words of a classmate, he was "trying to create a closed society like the school for the blind he went to. He is trying to create a society separate from ordinary society in which he can become king of the castle."

†Mindanao has a population of over 13 million and includes fourteen Filipino provinces.

atrocities.[55] Citizens deemed guilty of perjury had their tongues cut out. Even men were allowed an education in only one subject: the Taliban version of the Koran.

Using the slogan "Film and music leads to moral corruption," the Ministry for the Propagation of Virtue and Suppression of Vice ordered that Afghanistan's citizens destroy their televisions, VCRs, satellite dishes, and other devices of depravity.[56] Meanwhile, Taliban troops pursued a war of extermination against heretics, specifically the Hazara Shiites in their northern territories.[57] The slaughtered were victims of Allah's order to eliminate unholiness. Human rights organizations reporting on this murder of women and children saw it differently—they called it simple genocide.[58]

All predictions were that Pakistan's 130 million people (more than the population of England and France combined)[59] would be the next to experience fumigation à la Taliban. Pakistan's new leader, it was said, would be someone like the bomber of U.S. embassies and financier of worldwide holy war Osama Bin Laden—the man who had issued this simple order about Americans: "kill them wherever they are." The Taliban takeover of Afghanistan had been secretly financed by the Saudis,[60] and now Saudi Arabia was reportedly doing it again, pumping funds into Pakistan to underwrite yet another fundamentalist coup. The suspected source of the cash was none other than Saudi crown prince Abdullah, who, according to British reporter Robert Fisk, was turning Saudi Arabia, the allegedly "moderate" American ally, "into an anti-American nation in front of our eyes. Of course," added Fisk, "we're not told about that."[61] Nor were we told that Saudi Arabia had long been yet another fundamentalist theocracy.

Meanwhile, the groups who would purge Pakistan of its sinners were hardening their homicidal skills through bloodbaths in the Punjab, where they chanted "Death to America" and "Death to Democracy." Between guerrilla raids, these holy warriors headed out to spray graffiti slogans on the town walls in their training camps' vicinity: "democracy leads to secularism" and "Jihad [holy war] leads to dominance of Islam." North of Pakistan's second-largest city, Lahore, was Muridke, home to the Army of the Pure, which daily drilled recruits in slitting throats, dynamiting bridges, and rocket-attacking with precision blows. The Army of the Pure's goal was death to heretics—Indians, Americans, and Jews. John Stackhouse of the *Toronto Globe and Mail* delivered the following message from this arm of Allah's will made real:[62]

> "We will go to America with the gun," vows Sultan Atiqur Rehman Allehadi, who quit the Pakistani air force and now guides younger

men in the Army of the Pure. "First we will ask them [Americans] to take up Islam. If they don't, then we will use the gun."*

Should Pakistan fall to the fundamentalists this would give Islam's ultramilitants a heroic Roto-Rooter with which to extirpate American impiety. Zealots would control Pakistan's nuclear arsenal, called by its cheering supporters when it was first tested in 1998, "the Islamic Bomb."[64] The result, said Australian journalist Greg Sheridan, could be "devastating."[65]

Then there was the Saudi-backed Gamaa al-Islamiya (Islamic Group), which had turned the hinterlands of Egypt into a killing field, slaying tourists, assassinating policemen, stabbing moderate writers, and massacring Christians who'd been in the land of Egypt since long before Muhammad's birth. Russia, China, and the Central Asian republics—Turkmenistan, Dagestan, Uzbekistan, and Tajikistan—feared they were next on Allah's hit list.[66] Those fears turned real in 1999 when Chechen guerrillas, led by a Jordanian[67] and reportedly financed by the Saudi Arabian Osama Bin Laden,[68] opened a war to establish a fundamentalist Islamic state whose borders would range, as Russian prime minister Vladimir Putin put it, "from the North Caucasus to the Pamirs."[69] On a Western Internet discussion group, an Islamic militant who spoke for many laid out the Word as he sees it: "The oppression and aggression of the United States of America and its protectorates has no end. So it is time that they pay, and it is time that their people taste . . . torment."[70] Meanwhile, America, target number one, slumbered in a state of ignorance.[71]

On the other side of the religious divide stood the new fundamentalists of racial purity and Christianity, whose leaders skillfully manipulated conformity enforcement mechanisms which impel us to pick on those who

---

*Even moderate Muslims living in the West were enjoined by their would-be spiritual leaders to adopt key aspects of this powerfully intolerant point of view. Said the imam of Jerusalem's Alaqsa Mosque during a Friday sermon reprinted for worldwide distribution:

Surely, the Muslim in this world has a mission. This mission is to guide humanity to Islam. Therefore, it is inconceivable for a Muslim to accept to coexist with non-Muslims without working to guide them. It is also prohibited for a Muslim to be in a non-Islamic society without actively working to remove its evil and replace it with the mercy of Islam.

As the hadith clearly indicates, Islam is a unique entity and hence it has no correlation with other entities. In other words, Islam is the Haq (truth) while everything else is Batil (false) and not that Muslims should isolate themselves. However, the concepts of Muslims and there [sic] actions should be molded Islamically as to function for the removal of evil from the society without integrating into it.[63]

Fortunately the peaceful can interpret this call to action as a mandate for persuasion, not for violence.

are different.[72] Typical among these was a group called Russian National Unity,[73] which claimed an army of seventy thousand, snapped up members at the age of nine, enrolled them in "military patriotic clubs," dressed them in fascist uniforms, gave them the insignia of the swastika, taught them the Nazi salute, drilled them in combat maneuvers, honed their marksmanship, and identified the target of their weapons in their anthem:

> We are going down the straight road
> to believe, desire, and dare
> death to the cosmopolitan dogs and Jews and masons!

National Unity's founder, karate coach Alexander Barkashov, quoted yet another tune when giving interviews on the dangers of world Jewry and of the United States:

> Plunge your knife in the vampire's throat
> And the world will become good again.

Many parents welcomed National Unity's training for mass murder. With crime rampant and all order gone, at least groups like Unity kept their children off the streets and lived up to the promise enunciated by one local Unity leader, Andrei Dudinov, who declared that the movement would "drag" the young "away from drugs and vodka."

Meanwhile, like their Nazi models in the early 1930s, Russian Unity's leaders were taking advantage of their country's feeble democracy. They won elections in Vladimir, curried favor with the militia of Krasnodar (which let Unity use its facilities), and earned such approbation from the 101st Internal Brigade and from the 22nd Airborne Brigade that government soldiers trained Unity troops in Stavropol. Unity contingents were welcomed by the militia of Voronezh to officially patrol the town, and were under contract with the police to perform similar services in Moscow itself. National Unity leader Barkashov, quoting a book he says is "a must for any intelligent man," Hitler's *Mein Kampf*, explained that "he who controls the streets controls the politics."

As of 1998, there were roughly eighty-five similar fascist groups in the former Soviet Union with enough sympathizers to support 140 newspapers. London's *Economist* declared that it was just a matter of time before the proponents of the iron fist would seize the reins of government. The result, at its best, would be

> a kind of extreme nationalism: intensely prickly and Pan-Slavic, anti-Semitic, hostile to foreigners, and eager to reabsorb the Slav heartlands of Ukraine and Belarus within the Russian fold. . . . The armed forces and the KGB would be raised again to special eminence. The press and television would be corralled. Russia would become an

angry place—neither democratic, nor prosperous, nor kind to its neighbors. It is a nightmare scenario.[74]

For a preview of how such "angry nationalists" would behave, the assembled journalists in a 1998 issue of the magazine *World Press Review* urged us to look at the margins of the Slav territories, where rehearsals for the new millennium were already taking place. The future, they said, was Serbia's Slobodan Milosevic writ large: ethnic cleansing, mass rapes, mass graves, and blood in the streets; cities and towns destroyed; everything we'd learned to loathe about the Balkans, but this time with a nuclear brigade.

Then there was the United States, where the Covenant, Sword & Arm of the Lord Christian Identity enclave in Bull Shoals Lake, Arkansas, had planned the bombing of Oklahoma City's Murrah Federal Building in the 1980s, and where a Christian Identity hanger-on named Timothy McVeigh, guided by phone calls to another Christian Identity stronghold in the Ozark Mountains, had followed up on the Oklahoma bombing strategy and killed 168[75] enemies of Yahweh's chosen Aryan race.[76] Not far away, the Bible-toting Christians of Oklahoma's bunker-and-arsenal-riddled Elohim City were reputed to have armed every man, woman, and child to the teeth. The Aryan Republic Army was on a rampage of Midwest bank robberies designed to finance triumph for the one true form of Christianity. Sandpoint, Idaho's Promise Ministries[77] were charged with planning similar holdups, the bombing of a newspaper, and the demolition of an abortion facility. The Phineas Priesthood provided encouragement to a network of lone killers "cleansing" in the Bible's name. Christian Reconstructionists insisted that women who undergo abortion be hanged or stoned to death.[78] The Aryan Nations movement[79] was determined to seize the mountains of the West as a new Canaan[80] for pure Anglo-Saxon whites.[81] The North American Militia of Southwest Michigan was accused in court of having targeted the Battle Creek Federal Building, an IRS office in Portage, Michigan, and, in the words of one militia member, "anybody in higher ranks of politics," including the state's two senators.[82] At the same time, self-styled eco-"warriors" were setting up clandestine "direct action cells" devoted, as one anonymous cell organizer put it, to ever "bigger and better actions . . . with more severe amounts of damage being done to the target. This, of course, includes arson."[83]

America was the land where Eric Rudolph, the accused bomber of abortion clinics and of the 1996 Olympic Games in Atlanta, roamed free in the mountains of North Carolina, supported by the local populace.[84] Ours was the nation where two young men in Wyoming had tied a twenty-one-year-old gay college student to a fence post and beaten him to death, encouraged by Christian guardians of righteousness who said all homosexuals must die,[85] where a Buffalo, New York, obstetrician had been killed by a sniper as he stood with his wife and children in his kitchen, and where this murder had been cheered by the organizer of the Christian Gallery, a web site dedicated to such causes as "secession via nuclear weapons" and the replacement of the two-party system with "God's plan for government." The Gallery had posted the medical practitioner on a cyber-hit-list of "child-killers" whom it declared to be in the "war zone."[86] In the hinterlands of fifty states,[87] extremist Christian militias drilled their detonation, reconnoitering, and shooting skills as thoroughly as their doppelgängers in Pakistan and in the former Soviet states, aiming for the day when the use of violence would eradicate the blemish of secular democracy and allow the True Believers to structure the ways of men according to the dictates of an unforgiving Deity.[88] The Spartan premise Plato had stated in his *Laws* was very much alive again: "In reality every city is in a natural state of war with every other, not indeed proclaimed by heralds, but everlasting."[89]

The threat of today's Spartan storm troopers turned some defenders of democracy toward what in better times could have been dismissed as paranoid delusion. However, when one is up against a real-life conspiracy, paranoid thinking can produce accurate predictions of reality.[90] Those who despise alarmism forget the lesson of 1933, when a Keystone Kops political party which had been able to boast a mere twenty-five thousand members in 1925 declared itself ruler of the German nation and swept democracy away. Democrats use words. Zealots use weaponry. With guns and bludgeons, twenty-five thousand rapidly de factos into a majority. Hitler seemed preposterous to his opponents. There was no way, they felt, that someone with such outlandish views could take over the nation of Goethe, Schiller, Beethoven, Handel, and Brahms. Lack of paranoia lost these defenders of reason their lives, their homes, their children, and their wives. The Germany they were sure could never fall to lunatics soon plunged beyond the fringe of sanity. It was the sane, the sober, and the complacent who were erased from Germany's collectively enforced reality.[91]

MONKEYWRENCHING THE WORKS

When conformity enforcers overwhelm diversity generators, all of us are in trouble. Spartans—fundamentalists, militia groups, fascists, and ultranationalists—can freeze the machinery of collective mind. A shutdown of urban diversity devastates that exercise of collective acumen we call an economy.[92] Christian Fundamentalism has been shown by the research of sociologists Alfred Darnell and Darren E. Sherkat to retard the learning of children raised within its grasp.[93] Darnell and Sherkat sum up a common Fundamentalist attitude in the following words: "No schooling is better than secular schooling." Then there's the paralysis of thought which outright battle brings. When World War I erupted, Sigmund Freud was horrified by the sudden "narrow-mindedness shown by [even] the best intellects, their obduracy, their inaccessibility to the most forcible arguments."[94] Such closings of the mind may explain why authoritarians are prone to ignore it when their approaches flop. They often goose-step from one year to another rigidly glued to backfiring ways.[95]

Here are the implications of a neural net simulation carried out at Tokai University in Kanagawa, Japan.[96] If citizens combine civility's self-restraint with social freedom, the neural net in which they're twined spills forth solutions of the finest kind—those which produce a host of winning strategies. If citizens rage out of control when left alone but are held in check by a police state, the neural net goes blank. All power seeps into the hands of a central commander. This maximum leader, implies the research, can be a previous fringe figure and an utter incompetent. What's worse, complexity pioneer Stuart Kauffman has shown with yet another mathematical model how a "Stalinist" chief and his morality enforcers can give the collective brainwork a lobotomy.[97]

The Kanagawa simulation implies that fundamentalist strategies of beating up on others for their sins rather than controlling one's self drags us all toward authoritarian ferocity. The trick is to reassert the right to self-restraint and to the definition of one's own boundaries of behavior and of privacy, boundaries which include consideration toward those who disagree with us—the virtue which we call civility.

The consequences of imposing control on others rather than controlling one's self show up in the traditional American South, where culture pushes a disproportionate number of men to discipline those who wrong them while letting their own instincts rip. A University of Michigan study demonstrates that when they are threatened, the testosterone level of old-style southern males shoots up to almost three times the level of males in northern society.[98] Northern males have a greater grip on their emotionality. One result: "Forty-three percent of the murders in the United States occur in the sixteen southern states," according to an *ABC-TV Weekend News* report digging into the phenomenon.[99] In an interview on the broadcast, Southeastern Louisiana University historian Samuel C. Hyde explained that "here in the South violence is not merely an accepted response, it's an *expected* response."[100] Both Hyde and sociologist Dr. Edward Shihadeh[101] agreed that when one's honor was violated, a southern male had to retaliate, even if it meant the end of his own life. If the simulations of neural net researchers are correct, this helps explain why the southern swath of the country is the so-called Bible Belt.

Other research indicates that the wider the horizons of a culture, the more its citizens are likely to exercise self-control.[102] This self-control, in turn, tweaks a group brain to top strength.[103] In large-horizon situations among ants, each worker is a law unto herself, exploring the landscape in the way that she feels best. This allows the collective mind to test a maximum of new possibilities. In small-horizon ant colonies, each worker seems restricted by the presence of others, puts the clamps on her rambunctiousness, crimps her path, and with her tightened focus becomes a thread in a search web which zeroes in on just one thing.[104] In human groups, a society open to distant horizons also produces those who will probe for the next great uplift, a soar into the wonders of dreamworked possibilities.

Fundamentalism and its fellow militancies are one way in which authoritarian small-town subcultures like that of the South, of Muridke, of Elm Hollow, or of Voronezh assert themselves in the national and the global body politic. This was true of ancient Greece, in which liberty-loving Athens was a big city for its time, while Sparta—home of the muzzled mind—was an encampment of remarkably small size. It's true of the United States, where Christian Fundamentalism has been called a suburban upheaval of nostalgia for small-town certainties.[105] The authoritarian small-town mind-set seems to have pervaded the Nazi Party, whose leaders were so parochial that few had ever traveled internationally.[106] Small-town closed-mindedness bedevils the Middle East, whose fundamentalisms hark back to the days when Ibn Khaldun's desert nomads came in to "purify" the corrupt cities through a control-the-other-rather-than-yourself sweep with a broom of sword and flame.[107] Small-town authoritarianism even reigns in the West's major metropolitan centers. It holds sway in small, parochial enclaves like Boston's Southie, New York's Howard Beach, and L.A.'s Orange County. The inhabitants of these neighborhoods frequently have never seen the centers of their own hometowns. Yet they periodically wage subculture wars against their cities' cosmopolites—the movers and shakers whose connections snake across continents and seas.[108]

Open-minded internationalists are part of a massive social brew. The folks who've seldom traveled outside their neighborhood are part of a tiny social stew. Sometimes a society needs the cutting focus of small-town single-mindedness. But more often it benefits from pluralism's many points of view, from carrying many arrows in its quiver, not just one or two.

As in Elm Hollow, the new Spartans are poised for a kidnap of mass mind. Each fundamentalist army has its weapons ready: the Internet, Armies of Virtue, and instruments of massacre. And each feels chosen to impose its purged and regimented paradise on all humanity. But the mass mind needs its Faustian introverts, its oddballs, kooks, and deviants, its challengers of holy Mother Earth, of sectarian righteousness, and of traditional ways. It needs its internationalists, cross-cultural floaters, homosexuals, abortion supporters, cosmopolitans, explorers, and imagineers, those who extend their reach beyond old boundaries and open new frontiers.

The blasphemers the fundamentalists feel God-bound to eliminate are the mass mind's option makers, its catalysts of new hypotheses. In an atmosphere of debate, new approaches thrive. But when bullets replace words as disagreement-settlers, the mass intelligence nose dives. Totalitarian Spartans must be stopped by pluralists. This is a task which falls to our century's Athenians—upholders of diversity in a sea of harsh conformity.

20

INTERSPECIES GLOBAL MIND

◧ ◧ ◧

If you want to stride into the Infinite, move but within the Finite
in all directions.

Goethe

The global brain is not just human, made of our vaunted intelligence. It is
webbed between all species. A mass mind knits the continents, the seas,
and the skies. It turns all creatures great and small into probers, crafters,
innovators, ears and eyes. This is the real global brain, the truest planetary
mind.

The honey badger uses the black-throated honey guide as its surveillance craft. The bird cruises the forest hunting a beehive, then, chattering
and darting, leads the badger to its find. The clawed beast tears apart the
tough walls of the hive, and both honey guide and badger share in a wax-
comb-and-honey repast.[1] When black-capped chickadees spot danger and
shriek their mobbing call, ten other bird species recognize the signal and
prepare to flee or to defend their all.[2] Off the shores of Costa Rica, tuna
and dolphins hunt together, meshing information to increase the harvest of
their prey. Seabirds watch their movements carefully and follow in their
wake, waiting for them to churn a school of fish to the surface so they, too,
can partake. Fishermen scrutinize the movements of the seabirds and from
them learn where the meaty tuna and the dolphin churn.[3]

A mere five hundred thousand years after the earth took form there
set forth upon this orb a global brain. It was the product of the birth of life
and sociality—a network of mass communities—those of cyanobacteria,
which were among the first cellular beings. The microbial brain rapidly
learned new lessons and added to its arsenal. From it evolved a super
weapon, a sliver of fifty or sixty genes wrapped in a protein capsule capable
of boarding and grappling. This was the virus, the bacteria's collaborator
and its foe. Viral assaults devastated bacterial colonies—yet they tested
bacterial intelligence, tweaked bacterial ingenuity, and amplified bacterial
skills. Viruses also pried loose genetic pages from the creatures they
attacked and inserted them in the DNA library of those they visited next

while on their predatory rounds. Thus they became couriers through which bacteria swapped molecular pamphlets of new tricks and old collective memories.

Later cells learned to build a different form of collectivity—the multitrillion-celled cooperative which forms a plant, an animal, a you, or a me. Three and a half billion years ago bacteria had their global brain up and running—running very fast indeed. By swapping genetic bits and reengineering themselves they could create upgrades in hours or in days. Multicellular collectives, on the other hand, were no longer able to stitch borrowed genetic snippets into their DNA and instantly switch form and strategy. Instead of creating new ways of being in blinks of time, they were limited to evolving in geologic yawns and heaves.

In a sense, microbes saved the day. Roughly 300 million years ago, colonies of the bacteria *Blattabacterium cuenoti* worked out an arrangement of mutual advantage with primitive cockroaches. The cockroach easily became starved for nitrogen. However, it had a body of fat in whose molecules nitrogen was trapped. The bacteria took up residence inside the roaches, fed on the fat molecules, peeled off the nitrogen, and offered it up in usable form to their cockroach hosts.[4] In exchange the cockroach furnished new forms of haven and mobility.[5] When another species—a protozoan—entered the cockroach gut, it submitted as its admission ticket the ability to digest cellulose. If the bacteria-bearing roach chewed and swallowed wood, the protozoan in its digestive system turned the chunks of sawdust into food. Some roaches evolved into termites, and from a triple-species web[6]—insect, bacteria, and protozoan—a lumber-munching way of life was born.[7] Meanwhile, viruses teamed up with parasitic wasps to penetrate the immune defenses of live caterpillars on which both fed their young. The wasp transmitted the virus from generation to generation as a snippet of its own chromosome.[8] Our body, too, has a knowledge base it gains from plug-ins to the microbial brain. At first, we used microbes accidentally. Bacteria in our intestines provided us with skills we didn't have. They manufactured pantothenic acid, a vitamin without which we would stop growing, develop skin sores, and end up prematurely gray. "Friendly bacteria" also fed our needs for folic acid and for vitamin K.[9] Bacteria produced our body odors. This may sound pretty ghastly, but our aroma contains pheromones with which we communicate. The fragrance we give off has been shown to play a crucial part in our sex lives[10] and to bring the menstrual cycles of women who live together into synchrony.[11]

The 100,000,000,000,000 noninfectious bacteria which inhabited each of our bodies[12] helped protect us against those which would pounce and sicken us.[13] By struggling to maintain their turf, they made it difficult for disease-causers to gain a foothold deep inside our inner organs and our breathing passageways.

Then we learned to use bacteria more deliberately. We employed lactic-acid-making bacteria to turn milk to yogurt and to cheese, acetic-acid-producing bacteria to turn wine to vinegar, and carbohydrate-fermenting bacteria to preserve our food through pickling.

But our information mesh with other species went much further. We studied the hunting ways of the leopard, the jaguar, the eagle, wolf, and bear. We gave our clans their names, pretended they were our ancestors, imitated them in our rituals, wore their skins, and carefully, painfully, we learned their hunting skills, the silence of their stalk, their patience as they lay in wait, and the stratagems with which they tracked, confused, and trapped their prey. Eventually we built our ramparts, improved our tactics and our weaponry, and seized the hunting territories of our predatory teachers. We gave the land, recast and reformed, to plants and animals which had networked into our communities—wild grasses turned to crops and goring bulls now turned to oxen. The grasses which fattened under our tutelage to become wheat and corn and rice thrived by using us to clear their land, to water them, to convolute the plains to feed them rivers and to channel them vast lakes, to find new ways to give them nitrogen, to rearrange their genes, and to fight their killers off. We made religions and vast industries to minister to the super grasses' needs and service their fertility. Dogs, sheep, cows, pigs, chickens, and cats also snagged us with the lure of their docility. They prospered and together the net of they and we—the "master" humans and the animals and plants which exploited our powers and our brains—remade the world to fit our needs. Our partners thrived in multitudes, and so did we.

The wolf was the mother of all Rome. The eagle is still an emblem of the United States, Poland, Albania, and Germany. The lion stands for England. By emulating these lords of plains and skies our ancestors invented modes of conquest whose spoils they turned to empire—and to farmland—in later times. But our ancient opponents, the noble beasts of prey, had only the most rudimentary form of collective mind. Their societies, too crude and local, went into retreat. Eagles and lions relied on an old form of adaptation—the natural selection which culls the individual. We had the swift flow of interlaced ideas—they the sluggish isolation of seldom-changing genes.

When the current of human concepts became fast and planetary we were finally ready to take on our ultimate aide de camp and adversary—a worldwide-webbed intelligence which telegraphs its knowledge more instantaneously than do we, a group brain without parallel in creativity, the microbial mesh which links quadrillions of individuals into its processing machinery. In the 1900s, France's Louis Pasteur, England's Joseph Lister, and Germany's Robert Koch discovered that microbes were the source of much disease. What they failed to realize was the cleverness of the newfound enemy. Bacteria are quick to dodge the onslaughts of our drugs and immune systems. By lengthening or shortening bits of apparent junk DNA,[14] they re-tailor their membrane and their interior workings so they can shift appearance, roles, and tactics, evade our defenses, then a few generations down the line switch back again. Viruses hack into our genetic code to undo our immune system. They incorporate bits of DNA similar to the regulators[15] inside our immune cells, bits which allow them to bamboozle our internal defenders by giving them disinformation in our immune system's own vernacular. Others, like cytomegalovirus and Kaposi's sarcoma-associated herpesvirus, steal genes outright from their prey, then use them to churn out subversive proteins which twist the control knobs of their target's cell duplication, intercellular communication, and immune gadgetry.[16] The bacterium responsible for the sexual disease chlamydia has stolen twenty genes from higher organisms during its evolutionary history.[17]

Some bacteria work in multispecies teams, reading each other's chemical whisperings. When *P. aeruginosa* colonize human lungs, they pave the way for the more fatal *Burkholderia cepacia* to move in and take over. *Burkholderia* wait until they detect the signals with which *P. aeruginosa* talk among themselves, then rush in to multiply.[18] Bacteria swap scraps of DNA with creatures from other microbial species or simply scarf up the DNA fragments from dead cells and try them out for efficacy.[19] Some of the tricks their new genetic combinations allow them to play include: producing enzymes which pick drugs apart;[20] and sending up a flack of decoys which lure our drugs away.

In trying to defeat the microbial foe we have learned how to make its intelligence an extension of our own. Experiments with *E. coli* in 1958 helped uncover the double helix which makes DNA tick. By the 1970s, we'd further learned to harness our bacterial enemy. We hunted for new ways to study genes, and found we could use bacterial enzymes to snip them from their chromosomes. Then we tucked the gene that interested us into a living bacterium—*E. coli,* a microorganism we nourish in our intestines every day. Occasionally *E. coli* attack—churning out poisons which sicken us, invading the mucosal lining of the gut in which they normally reside in peace, or simply clinging to the intestinal wall in a manner which gives us diarrhea, weakness, cramps, or bloody stool. Tamed and confined in the lab, however, we turned *E. coli* into shoemaker's elves, diligently cranking out identical replicas of themselves . . . and of any gene we'd spliced into their chromosome. So productive were these bacterial laborers that less than a quarter teaspoon* of the creatures could churn out 10 billion copies of, say, an intriguing human gene between the time a

---

*Actually, a milliliter is all it takes—a fifth of a teaspoon.

researcher left her lab at night and her first sip of coffee in the morning.[21] Once we had gene copies to spare, we could go to work to see exactly what they do, or use them to catch a thief or find the father of a child or two. Scientists got credit for the resulting discoveries, but it was a bacterium doing the grunt work in our genetic laboratories.

Plasmids helped the process mightily. These are boarders in the bacterial cell, walled off with their own DNA, which itch to multiply from time to time. When two bacteria slide up against each other, a plasmid will build a pipe between the pair and move a part of itself through,[22] hauling precise forgeries of a few of the old host's genes as it goes. Then it will create a clone of itself and of its genetic contraband in the new home. We used plasmids to inject a gene into a cell like that of *E. coli,* thus turning it to a factory.[23] Or we used it to insert a gene into a human cell in gene therapy. Completing the new package we called biotechnology was PCR—the polymerase chain reaction—which rapidly copied DNA fragments using an enzyme[24] we burglarized from the heat-resistant bacterium *Thermus aquaticus.*

This paved the way for the drug industry in the 1990s to create new pharmaceuticals by harnessing bacteria, plasmids, and their products as gene sequencers, splicers, and reshufflers, and even as assembly line producers of complex proteins.[25] A thousand biotech firms were operating in the United States in 1996, and a feeding frenzy had developed among drug companies to either buy a biotech operation or partner with one or more. Molecular biology became heavily dependent on recombinant DNA technologies, many of which relied on bacteria to do their dirty work. We used bacterial enzymes to help monitor glucose levels in diabetics.[26] We employed bacteria to assemble new materials—like a substance which can be turned from a liquid to a gel and back on cue, useful properties for delivering medicines or live cells to target sites inside the body of a suffering human being.[27]

We tapped bacteria's wisdom in many other ways. The bulk of the labor force in our sewage treatment plants consisted of microbial teams grouped into tiny spheres.[28] On the outer husk were bacteria which ate the sewage and other unpleasantries. In the core were others which feasted on the poisons excreted by the waste eaters.[29] As of 1997, most of the world's farmland was planted with crops in which microbes had been used to insert new genes. The result was larger harvests courtesy of fruits, vegetables, and grains resistant to herbicides, insects, and disease. Meanwhile, food processors experimented with a chemical weapon bacteria manufacture to destroy their foes—bacteriocins. Our goal was to prevent food spoilage and food poisoning.[30]

During their existence on this earth bacteria have built a hefty database, and we've been plugging into it at full speed. Plans were afoot in 1998 to genetically engineer bacteria as pipe protectors, sending them deep into plumbing and air-conditioning systems to fend off corrosion. Ironically, these defenders would guard the pipe against the chief hasteners of its destruction—other bacteria which use metal as fast food.[31] Biotechnologist Judy E. Brown discovered a bacteria at Yellowstone National Park which may, in the future, be employed to rid soil of contamination by aluminum.[32] The Department of Energy created an entire research arm—the Microbial Genome Program in Germantown, Maryland—to borrow secrets from and/or to harness bacteria.[33] One of its tasks was to sequence the genes of a remarkable bacterium called *Deinococcus radiodurans*[34] which can withstand radiation levels that would kill any other life-form on this earth. The goal was to reengineer the bug so that vast armies of the creatures could be put to work cleaning up nuclear wastes.[35] Meanwhile, the effort to bacterially reengineer plants like potatoes and corn so they could be turned into biofactories, cranking out vaccines and other drugs, was sailing along. As we turned the corner on the year 2000, attempts to use bacterial tinkerers to convince these same plants to produce fabrics blending cotton with natural polyester or to mass-produce plastics was foundering, but still hadn't given up the ghost.[36] Astonishingly, when many of these efforts began, the bacterially based revolution in genetic engineering which had spawned them was less than fifteen years old. With a few more lessons in multispecies cooperation, who knew what we humans might unfold?

Our bacterial knowledge helped us further network with the data banks of numerous other entities. We used the bacterial tricks of genetic engineering to rework cows and goats so they'd produce pharmaceuticals in their milk.[37] We did the same to retool lab animals so they'd spew out drugs in their urine.[38] Sheep were among the animals we employed to produce "recombinant protein therapeutics." One example was Alpha-1-antitrypsin, a cystic fibrosis treatment which was almost impossible to obtain by normal means. However, a team at the Roslin Institute in Edinburgh, Scotland, managed to insinuate the gene necessary to produce the stuff into the milk-making machinery of a sheep named Tracy in 1988. By 1998, she'd had eight hundred granddaughters, many of whom were pumping out the rare healing substance in their daily milk.

The udders of microbially reengineered goats poured forth antithrombin III, which prevents unwanted blood clots. (Retroviruses—our archenemies when they cause AIDS—turned out to be among our best allies in reconfiguring goat genes so they'd pull off this feat.)[39] On the way was milk frothing with a vaccine for hepatitis B and with monoclonal antibodies, a hopeful candidate for cancer treatment. The journal *Science* went so far as to predict that farmers might go from eking out pennies in old-style agriculture to making a handsome profit in the twenty-first century by turning their efforts to "pharming"[40]—raising pharmaceutical-producing herds and crops.

In the mid-1990s a Yale University developmental neurobiologist, Spyridon Artavanis-Tsakonas, managed to extract some key genes from the lowly fruit fly which proved remarkably similar to those of women and men. (Think about it—thanks to our common ancestry with all things low and high, you and I have genes like those of a humble fly.) These magic fruit fly genes were developmental programmers, fountain-of-youthers which could prod cells into acting with the growth power of embryonic tissue. More important, the insect genes showed promise of being able to regrow human body tissue which disease or accident had wiped out forever. The potential practical applications were mind-boggling: healing shattered bones, mending fractured spines, resurrecting missing muscle, seducing barren brain cells into once again emitting the critical neurotransmitters missing in Parkinson's patients, re-creating the lost marrow of cancer sufferers during chemotherapy, stopping strokes dead in their tracks, and healing those with cancer of the skin. Other researchers ransacked the genetic wisdom of the mouse and rat, pilfering rodent genes which could potentially regenerate the lost nerves of those with Lou Gehrig's disease or replace the unusable blood vessels of those with damaged circulatory systems.[41]

In the 1980s, we were challenged to an intergroup tournament by HIV. "Know the enemy," said Sun-tzu. We learned more about viruses in our war with AIDS than we'd ever known before. From 1981 when the disease was first identified to 1998, 150,992 studies on AIDS were published in peer-reviewed journals—each a contribution to our knowledge of biology. This involved new work in recombinant antigenic peptides,[42] new disease detection techniques,[43] innovative studies of viral structure,[44] new genetic approaches to the study of viral ancestry,[45] fresh insights into the operations of the immune system,[46] the development of bacterially based recombinant gene techniques to create varieties of vaccines a generation ahead of those in the past, and the creation of new drugs like nucleoside analogs (among them, AZT) and protease inhibitors. The research effort was so intense that by 1998, investigators David Baltimore and Carole Heilman declared, "Scientists know more about HIV—the human immunodeficiency virus that causes AIDS—than any other virus."[47] Johns Hopkins University School of Medicine's John G. Bartlett and Richard D. Moore saw even broader implications: "The therapeutic advancement achieved [via the study of AIDS] since 1995 has few parallels in the history of medicine, save perhaps for the revolution sparked by the introduction of penicillin."[48] Conflict had been a data connector, forcing us to copy our enemy's database into our own.

The knowledge gained could prove vital for yet another face-off with the microbial army, one whose loss might devastate humanity. Back in 1918 and 1919, an influenza called the Spanish flu wiped out 20 million people in a matter of months—more than all the victims of the recently concluded First World War.[49] The origins and treatment of this pandemic long remained a mystery. Then a worldwide scientific team was assembled after the Hong Kong flu of 1968 to find out where influenza viruses hid between assaults on human beings. Robert Webster of the St. Jude Children's Research Hospital was among the members of this research team. Webster suspected that the viruses might be using birds as airborne storage and transport vehicles, so during the 1970s he studied the droppings of waterfowl flying through his backyard in Memphis on their way from Canada and discovered that the indelicate smears of waste harbored influenza A[50]—the shape-shifting viral type which had triggered the Spanish flu. Webster networked with colleagues from Europe and Asia to check out his findings. For the next twenty years the team surveyed migratory wild birds in Germany, Canada, and Alaska, and bird flu in the pigs of Italy. Their detective work revealed that influenza viruses are worldwide transportation experts.

Fifteen varieties of influenza subtype A travel thousands of miles in the guts of migratory birds without disabling these aerial carriers.[51] The microbes take advantage of such commuters between continents and hemispheres as mallard ducks, Peking ducks, pintail ducks, and dabbling ducks, geese, swans,[52] and gulls. The viruses which voyage in the intestines of wild fowl are also adept at moving from birds into horses, seals, and whales. In the marine mammals, they can continue their intercontinental voyages by sea.[53] When they land in a promising spot, they use their talent for mix and match, and easily infect monkeys—and hence, in all probability, human beings.[54]

Research indicated that influenza viruses had used three different species to concoct the biological mix which led to the Asian flu pandemic of 1957[55] (killer of seventy thousand Americans) and to the pandemic of Hong Kong flu in 1968 (eradicator of another thirty-six thousand Americans). Viruses soared from Arctic tundra down the Asian continent in the bellies of migrating birds. They disembarked from their flights in the farmyards of China, where they transferred to the innards of the homesteads' chickens. From there, it was a short hop to the farms' pigs, in which viruses able to infect humans had long before established residence.[56] When two viruses invade the same cell, they trade and shuffle genes, creating offspring so novel they can baffle the immune system of human beings.[57] The visiting avian viruses had apparently picked up secrets of camouflage and strategy from the human viruses homesteading in the pigs. Then the visitors had switched from pigs to people and had struck in their new garb and with their new weaponry.[58] When we *Homo sapiens* counterattacked using drugs and our immune systems, the clashes only spurred the microbes' creativity, impelling the viral raiders to take on forms even more clever and deadly than before.[59]

Thanks to this early research, when Hong Kong officials first spotted an unusual poultry-killing virus in 1997 they realized how easily it might make the passage through the guts of other animals to the human race. So they called in Robert Webster and sent the bug for inspection to a U.S. lab. In May a three-year-old Hong Kong boy was hospitalized with a fever. Five days later he died of what appeared to be "influenza pneumonia, acute respiratory distress syndrome, Reye's syndrome, multiorgan failure, and disseminated intravascular coagulation."[60] The toddler's school class had been raising baby chicks. Hong Kong's medical authorities became alarmed and sent fluid from the dead boy's throat for inspection to a team of viral specialists at Erasmus University in Holland,[61] who identified the culprit as H5N1, a type of virus which Robert Webster's work had led him to understand quite well.[62] Webster's lab confirmed the findings, and Webster says the villainous virus turned out to be "nearly identical . . . to the H5N1 virus that was isolated during a March 1997 outbreak of influenza that killed 70% of the chickens on three farms in Hong Kong."[63] But this wasn't just any old chicken flu. The virus had bypassed pigs and gone directly from Hong Kong's chickens to one of its citizens, thus arriving in a form for which neither the human body nor modern medicine had developed a defense. The modus operandi of this flu virus was so unfamiliar to the immune system that according to Webster, it could have wiped out "half the world's population"[64]—3 billion victims, a figure decidedly immense.

The first of these viral direct-jumpers was stopped in its tracks when Hong Kong's government ordered all the poultry in the area killed. But the appearance of the Hong Kong bird flu posed a disturbing possibility. Said Webster, the next fatal worldwide influenza pandemic was now a "certainty."[65]

The global mesh of humans and that of microbes were in a deadly race—the sort of intergroup tournament which accelerates a complex adaptive system's pace. Webster and the Department of Virology and Molecular Biology he heads at St. Jude's went on the alert, working with the World Health Organization to study the 1997 Hong Kong bird flu in an attempt to blunt the force of the next worldwide microbial outbreak. By now, Webster was co-leader of an international task force which had set up a special lab at Hong Kong University "to elucidate the origins of and molecular changes associated with the transmission of this avian H5N1 virus to humans."[66] Like AIDS researchers, Webster and his colleagues used an interspecies team in their efforts to understand and stop the viral mechanisms which could cause the next great plague.[67] They employed bacterial products for gene sequencing,[68] plasmids and laboratory mice to test the promise of "gene gun DNA immunization,"[69] the cells of chicks, guinea pigs, and African green monkeys to culture vaccines,[70] a careful study of the flight paths of birds and the ways of pigs, and insights into viral structure learned from the battle with HIV.

Our interconnection with other species was matched by that with which we meshed with other humans to scan each continent. One hundred ten surveillance centers worldwide collected influenza samples each year and sent them to a central location for analysis, the World Health Organization picked the three strains most likely to produce the next season's flu, and a vaccine was made up in time for the onset of winter.[71] Meanwhile, the U.S. government kept up a worldwide viral watch via the Centers for Disease Control's National Center for Infectious Diseases[72] and the National Institutes of Health's Pandemic Advisory Committee. The World Health Organization, headed by a Norwegian, maintained a Division of Emerging and Other Communicable Diseases Surveillance and Control to keep an eye peeled on every continent for outbreaks of new bacterial and viral illnesses.[73] When Hong Kong restructured its poultry market system—separating chickens from ducks and other birds for the first time— it was in part a result of recommendations based on what Webster had learned in Tennessee and published in a Viennese journal.[74] However, another strong recommendation made by Webster and his colleagues called for even more "intensive global influenza surveillance." Our worldwide efforts could not afford to fall short of their goal, for if there were a twenty-first-century flu pandemic, Webster and his fellow experts shuddered at the probable death toll.[75]

With human minds from China to Scandinavia interchanging their ideas on everything from simple alterations in farming practices to the structure of the AIDS virus and the intricacies of its reproductive tricks, we may finally be the first multicellular creatures who can take on the microbial global brain and win. The irony is that in the process, bacteria will be our allies as well as our foes. For the global brain is not just a human interconnect, it is a multispecies thing. And that, in fact, is what it's always been.

It is said that we have enraged nature by tearing at the pattern of her tracery, and for this transgression we shall be punished mightily. But we are nature incarnate. We are made up of her molecules and cells. We are tools of her probings and if, indeed, we suffer and we fail, from our lessons she will learn which way in the future not to turn. For all that lives and all that ever has is part of a collective brain, a neural net of the most sprawling kind . . . an evolution-driven, worldwide, multibillion-year-old interspecies mind.

21

CONCLUSION: THE REALITY

OF THE MASS MIND'S DREAMS:

Terraforming the Cosmos

1000 B.C. to 2300 A.D.

Humankind was put on earth to keep the heavens aloft. When we fail, creation remains unfinished.

The Kotzker rebbe

Let us roll all our strength and all
Our sweetness up into one ball,
And tear our pleasures with rough strife
Through the iron gates of life:
Thus, though we cannot make our sun
Stand still, yet we will make him run.

Andrew Marvell

While others predict the impact of computer networks on mankind's futurity, this book has chronicled the rise of global mind from earth's primordial seas. Implantable communicators and quantum computers[1] will vastly change the way we interface in this dawning century. Yet eons of evolution have long since networked our emotions and biology.

In the 1960s, observers of the Kalahari Desert's !Kung bushmen came to a rather startling conclusion about the hunter-gatherer way of life. While those of us in modern societies worked forty- to fifty-hour weeks to eke out a living, the !Kung were able to supply themselves with food, shelter, and clothing by putting in a good deal less than twenty. Our "laborsaving" technology was just a new form of wage slavery. The real laborsavers were the hunter-gatherers.[2] Later analysis revealed a flaw in this argument. While the !Kung spent only a few hours several days a week hunting for meat (men's work) or gathering mongongo nuts (a task restricted to women), they invested a considerable amount of time in information processing.[3] !Kung tribesmen sat up all night in meetings to resolve disputes or to discuss where the next watering hole might be found now that the current one was drying out. The anthropologists who'd initially portrayed a !Kung utopia had acted as if the work delegated in modern society to newspapermen and -women, TV producers, CEOs, administrators, bureaucrats, lawyers, and magistrates was nothing but leisure activity.

In fact, both for the !Kung and for postindustrial humans, a great deal of information exchange *does* take the form of entertainment. Even child's play, an apparent lark, is a primary form of practice for advanced data processing.[4] However, the networking, interpretation, and reshaping of knowledge, no matter how much pleasure it gives, is a serious affair. For it is a survival tool without which we would cease to live.

During the 1980s, the Gabbra people—herders of cattle, goats, camels, and sheep in northern Kenya—ran into weather conditions which were interpreted very differently by the young Turks of the tribe than by their elders. The time came to decide where the season's rains would fall many months hence and where the grass that fed the livestock would rise so the long trek toward new pastures could commence. The younger leaders had never lived through one of the area's major droughts. They argued persuasively for heading toward the Chalbi lowlands not that many miles away—a pleasant walk which, with a bit of pleasant talk, would take an easy matter of several days. Grass had usually been plentiful in these lowlands during the young men's times. The Gabbra's patriarchs, on the other hand, put their trust in a legacy compiled by their fathers and their fathers' fathers before them, a traditional method of calculating weather cycles which indicated very bad times ahead. In fact, the ancient system of prediction augured catastrophe.

While the youngsters argued for the obvious and easy route, the elders spoke of a far more difficult itinerary—a long trek north through barren land to southern Ethiopia, a country whose armies had killed Gabbra tribesmen systematically from 1913 to 1923.[5] The perils of this course were plenty, but the elders were convinced the risks were a necessity.

Some members of the tribe followed the young leaders to an easy destination. Others joined the elders in a tough and grim migration. The rains did not come for the next two years to their customary place. The tribal groups which had gone with the younger leaders "lost 95 percent of their cattle, 60 percent of the goats, 40 percent of the sheep . . . five percent of their camels"[6] and were forced to beg aid workers for famine relief. The expedition which had followed the elders emerged from the drought with its pride and with its herds largely intact.

A complex adaptive system had exposed part of its decision-making machinery. At work was a social process common to sensors on a cell membrane, to bacterial colonies, to groups of crustaceans, to troops of baboons, and to squabbling moderns of the cyber-age. Two subgroups with two hypotheses. Two roads, both taken. But only one subgroup's unity, wealth, and way of life survived. With this triumph something else lived on—the best guess of the tribe's mass mind.

There are numerous technologies with which we'll soon upgrade our interconnectivity—from smart clothes and digitized pens to information-sending-and-receiving shoes[7] and computers which divine our interests by watching the dilation of our pupils, then go out as personal servants to crawl the World Wide Web for finds to surprise us, to entertain us, and to help us through emergencies.[8] But the web of inventions about to alter our lives will work all the better if we understand the interconnects built into our physiology. The global brain has a pulse and power grander than its constituent beings. We are modules of a planetary mind, a multiprocessor intelligence which fuses every form of living kind.

Current evolutionary theory holds that an individual is "fit" only if he or she can maximize the number of his or her offspring. Even a brilliant thinker like Richard Dawkins says that the ultimate individual is not you and me, but a gene within us driving us remorselessly, and that that gene is selfish to the nth degree.[9] Such contemplations leave out the universal nature of networking. Less than a second after a false vacuum burped this cosmos into being, entities like quarks and leptons precipitated, separated, and set up boundaries which gave them their identity. Yet all were laced together in spite of their autonomy. When the strong force, the weak force, and the electromagnetic force failed to hold them, there was always gravity. The cohesive forces are more intricate in social systems, but the principle is the same: you can run, but you can never get away. You can put distance between yourself and the center of your nation or your family, but you can never totally cut your lines of connectivity. Even when we turn inward, an army of invisible others speaks through our thoughts, twists our emotions, and populates our privacy. We are wired as components of an internet which literally shapes our brain, orders what we'll hear and see, and dictates what we'll comprehend as reality.

This book has been the story of the information nets which gave us birth and of the twists those webs have taken as we and a horde of allies and enemies have struggled for ascendance on this earth. We've moved from the attractive forces piecing together molecules to the pull which per-suaded the bacteria of 3.5 billion years ago to live in megalopolises with populations larger than the total number of humans who have ever been. We've seen how these earliest of our single-celled ancestors built their massive colonies while sending and receiving information globally. We've watched as our more advanced progenitors achieved the benefits of size and skill which came with multicellularity. We've witnessed the price multicellular creatures paid when they lost their ability to communicate minute-to-minute via data-saturated strands of DNA. We've seen the invention of a new information-cabler—imitative learning—nearly 300 million years ago among the spiny lobsters of the Paleozoic age. We've seen how later data strands like words and symbols upped the speed of a new mass mind—that of spreading humankind. We've watched the rise of Stone Age cities and their mesh of commercial interlinks. We've noted the hastening of data flows created by war and rivalry in the sixth century B.C., when the philosopher Thales pioneered international consultancy. We've beheld the emergence of subcultures—havens for those of different tem-peraments, different limbic system bents—and have seen how the games subcultures play enhanced mass mind in ancient Greece's day.

We've seen how a collective intellect uses the ground rules of a neural net:* shuttling resources and influence to those who master prob-lems now at hand; stripping influence, connections, and luxuries from those who cannot seem to understand. We've seen in the tales of Sparta, Athens, and Gilbert Ling how mass minds work to churn out, test, and sometimes crush hypotheses, and how they do it through the contest between social collectivities. We've seen how the groupthink imposed by leaders like the preacher's daughter, Mrs. Salt, and even more, by human sheepishness, creates facades of unanimity, and how fresh options spring from introverts who build subcultures on the margins of society. We've seen how subgroups vie to commandeer the larger group's perceptual machinery, piloting masses of men and women on their next flight through the storms of destiny. We've glimpsed the Spartan fundamentalisms striv-ing at this minute to highjack the shared visions of humanity. And we've seen how the brain we think belongs solely to our kind achieves its goals by tapping the data banks of eagles, wheat, sheep, rodents, grasses, viruses, and lowly E. coli.

This has been not only the saga of social data processing, but the tale of how higher animals lost their worldwide mind, then slowly won it back again. It has been the tale of two global brains in conflict—that of microbes, and that of women and men. We've seen how the data swaps of science, despite their flaws, have finally allowed us to approach the swift-ness of our microbial foes. Since these bacterial and viral relatives are often allies and just as often enemies, 3 billion human lives could be lost if we don't maximize our group mind's speed and creativity.

We've seen how the group brain uses opposites—instincts which tear us apart and equally powerful instincts which yank us back in stride again. Diversity generators drive us to be different. Conformity enforcers compel us to agree. We taste the fruit of paradox in creative bickering. The

*Or any complex adaptive system.

referees are sorters . . . sorters of three kinds: inner-judges planted in the tissues of our bodies and our minds; resource shifters couched in mass psychology; and intergroup tournaments determining which tribe or species wins the contests between social teams.*

Life, as Aristotle knew so well, is a matter of avoiding the extremes. Conformity enforcers are necessities. But they are mass-mind throttlers when they grab hold totally. Diversity generators are equally essential. But taking them too far can destroy a civil culture and devastate once-vigorous centers of humanity. Inner-judges—from depression and elation to those hidden in the tissues of our psychoneuroimmunology—make us effective modules of a collective thinking machine. But they can also turn us into preachers of mass murder or morose and suicidal beings. Resource shifters can work to motivate or can be pirated to pour the wealth too heavily on those who manage to usurp authority and who block the contributions of allegedly lesser beings. Intergroup tournaments work well in business, but can destroy men by the millions when weapons are their chosen means.

The mass mind, like its members, has its dreams. These aspirations can be turned to hard, cold fact if they're pursued not for merely hours or years, but for millennia and centuries. Humans have dreamed of flying since at least the days when the myth of Daedalus was first told in ancient Greece three thousand years ago. But it would take the workings of a global brain 150 generations to turn this airy fantasy from fiction to reality. Italy's Leonardo da Vinci did some interspecies borrowing—studying the motions of birds in flight—then took notes on pads of a material created in China—paper[10]—and drew potential aeronautical machines. The Chinese innovation which had helped da Vinci popped up once again when two sons of a French papermaker discovered that by filling a flimsy bag made of their father's product with hot air, they could cause the bag to fly. In 1783, the papermaker's sons—the Montgolfiers—sent two friends drifting over Paris in the first manned long-distance† flight. America's Benjamin Franklin suggested adding a propulsion engine, a visionary impracticality at the time. In 1799 England's Sir George Cayley conceived of a fixed-wing vehicle with a tail assembly for horizontal and vertical stability and stuck

---

*In a sense, these sorters are all Charles Darwin's pickers and choosers—his natural selectors. Yet they add to Darwin's insights some harsh realities. Not all selectors are elements of the exterior environment. The inner-judges, in particular, are *inside* our biology. Why? Because individuals are parts of a larger survival machine. And just as individuals compete, so do the groups of which we are a part. Groups compete not only with their might, but with their smarts. Each of us is a hypothesis in a larger mind. And unfortunately, not all hypotheses work out. Hence some of us are disabled by what previous centuries called a broken heart.

†The friends the Montgolfiers sent skyward—Pilatre de Rozier and Francois Laurent—drifted a full 5.5 miles on their maiden voyage through the air.

with his vision long enough to build a successful glider fifty years later in 1849. Germany's Otto Lillienthal—jumping from an enormous mound outside of Berlin—made two thousand glider flights from 1867 to 1891, carefully recording each result except the last, which killed him in a glider crash. Before his final plummet, Lillienthal published a working set of data on the most effective curvature of a wing, on the means to achieve stability, and on the manner of turning the tail assemblies drawn by Cayley into practicalities. Ohio's Wright Brothers used Lillienthal's results to design a powered heavier-than-air machine which could be made to turn rather than to simply fly an uncontrolled straight line.[11] Finally, thanks to the development of new materials of all kinds, a human-powered aircraft flew the flight path the mythmakers of Daedalus had conceived—from Crete to the Greek island of Santorini—in 1988, three thousand years after the initial telling of the tale of Icarus's wax wings.

This book's exploration of the mass mind's history is like the notes Lillienthal made on one of his early glider flights. It's intended as a small move in the progress toward another ancient dream—not the dream of flying, but the dream of peace. We will always cling to common threads yet stake out grounds for squabbling.[12] Such is the way the global brain does its thinking and creating, its testing and imagining. The more we can play out our necessary contests civilly, the closer we will come to turning spears to pruning hooks and swords to plowshares—purging the global brain at last of blood and butchery. But we will need to pursue that dream for several more centuries. If each of us contributes one small step to this long march of history, we will finally achieve what no god but the will within us can bequeath—a peaceful destiny.

In 1998, Tel Aviv University's Eshel Ben-Jacob moved from the study of microbes into the creation of microprocessors and gears so small thousands could be placed on the point of a pin. To what will this nanotechnology in its many forms lead? Here's one highly speculative possibility. Imagine the day when we plug molecular computers[13] into the bacterial genome, then let the critters multiply. Imagine swarms of these bacterial microcyborgs interacting in creative webs, continually upgrading their collective software via genetic and inventive algorithms. Imagine them climbing the ladder of paradox at our behest to find new ways of conquering ancient problems from cancer to the carnage of war. A single colony of cyanobacteria in the wild—like that which is responsible for the creation of a stromatolite—has a population a thousand to a million times that of the entire human race. With a million times more participants than there are cells in a human brain, what could a nanocyborgian bacterial population of learning and creating hybrids do? Would they make our manner of searching possibility space—through the interacting intelligences of billions of humans spread out over 2.5 million years—look clumsy and lumbering?

Would they unfold their fractal branches into realms beyond imagining? Could they suck sustenance from subatomic plankton—the seething stew of particles*[14] flickering in and out of existence in the vacuum depths of space?†[15] Could swarms of nanocomputerized bacteria climb and dive through nebulae, resculpting the Einsteinian landscape of space and time? Could they crawl through wormholes and discover ways to harness the energies of the vast flares spewed by black holes at the centers of some galaxies?‡[16] Could they become our descendants' exploratory engines, the antennae and energy gatherers for the human race? From global brains, could we go to universe-hopping megaminds? One small step for bacteria, one giant leap for all mankind?

Ancient stars in their death throes spat out atoms like iron[17] which this universe had never known.[18] The novel tidbits of debris were sucked up by infant suns which, in turn, created yet more atoms when their race was run. Now the iron of old nova coughings vivifies the redness of our blood. Deep ecologists and fundamentalists urge that our faces point backward and that our eyes turn down to contemplate a man-made hell. If stars step constantly upward, why should the global interlace of humans, microbes, plants, and animals not move upward steadily as well? The horizons toward which we can soar are within us, anxious to break free, to emerge from our imaginings, then to beckon us forward into fresh realities. We have a mission to create, for we are evolution incarnate. We are her self-awareness, her frontal lobes and fingertips. We are second-generation[19] star stuff come alive. We are parts of something 3.5 billion years old, but pubertal in cosmic time. We are neurons of this planet's interspecies mind.

---

*Technically this particle froth is known as "vacuum energy density."

†Plans are already in the works to use bacteria in space. Bacterial colonies will be vital to sustaining human colonies on the moon and on other planets. Among other things, bacteria will be assigned the task of recycling sewage and other wastes and of purifying the air.

‡An active galactic nucleus, as the black-hole-powered energy generator at the heart of a galaxy is known, pours out the power of 10,000,000,000,000 suns.

NOTES

PROLOGUE: BIOLOGY, EVOLUTION, AND THE GLOBAL BRAIN

1. Derrick De Kerckhove. *Connected Intelligence: The Arrival of the Web Society.* Toronto: Somerville House Books, 1997: 186.

2. Peter Russell. *The Global Brain Awakens.* http://www.peterrussell.com/GBAPreface.html. Downloaded August 1999.

3. Gottfried Mayer-Kress. "Messy Futures and Global Brains." http://www.santafe.edu/~gmk/MFGB/MFGB.html. Downloaded August 1999.

4. Derrick De Kerckhove. *Connected Intelligence:* jacket copy.

5. B. Y. Chang and M. Dworkin. "Isolated Fibrils Rescue Cohesion and Development in the Dsp Mutant of *Myxococcus Xanthus.*" *Journal of Bacteriology,* December 1994: 7190–7196; James A. Shapiro. Personal communication. September 24, 1999.

6. W. D. Hamilton. "The Genetical Theory of Social Behaviour." *Journal of Theoretical Biology* 7:1 (1964): 1–52.

7. Peter Russell. *The Global Brain.* Los Angeles: Jeremy P. Tarcher, 1983.

8. Robert Wright. *The Moral Animal: Why We Are the Way We Are: The New Science of Evolutionary Psychology.* New York: Vintage Books, 1994: 186.

9. William McDougall. *An Introduction to Social Psychology.* Boston: John W. Luce, 1908.

10. Walter B. Cannon. *Bodily Changes in Pain, Hunger, Fear and Rage: An Account of Recent Researches into the Function of Emotional Excitement.* New York: Appleton, 1915.

11. Robert E. Thayer. *The Biopsychology of Mood and Arousal.* New York: Oxford University Press, 1989: 108.

12. David Livingstone. *Missionary Travels and Researches in South Africa.* New York: Harper & Brothers, 1860, quoted in Daniel Goleman, Ph.D. *Vital Lies, Simple Truths: The Psychology of Self-Deception.* New York: Simon & Schuster, 1985: 29.

13. W. D. Hamilton. "Altruism and Related Phenomena, Mainly in Social Insects." *Annual Review of Ecology and Systematics* 3, 1972: 193–232.

14. David C. Queller, Joan E. Strassman, and Colin R. Hughes. "Genetic Relatedness in Colonies of Tropical Wasps with Multiple Queens." *Science,* November 1988: 1155–1157; Thomas D. Seeley. *The Wisdom of the Hive: The Social Physiology of Honey Bee Colonies.* Cambridge, Mass.: Harvard University Press, 1995: 7.

15. Hans Kummer. *In Quest of the Sacred Baboon: A Scientist's Journey.* Princeton, N.J.: Princeton University Press, 1995: 303–304.

16. D. S. Wilson and E. Sober. "Reintroducing Group Selection to the Human Behavioral Sciences." *Behavioral and Brain Sciences,* December 1994: 585–654.

17. David Sloan Wilson. "Incorporating Group Selection into the Adaptationist Program: A Case Study Involving Human Decision Making." In *Evolutionary Social Psychology,* ed. J. Simpson and D. Kendricks. Mahwah, N.J.: Lawrence Erlbaum, 1997: 345–386.

18. Richard Dawkins. *The Selfish Gene.* New York: Oxford University Press, 1976; J. Philippe Rushton, Robin J. Russell, and Pamela A. Wells. "Genetic Similarity Theory: Beyond Kin Selection." *Behavior Genetics,* May 1984: 179–193.

19. Rene A. Spitz. "Hospitalism: An Inquiry into the Genesis of Psychiatric Conditions in Early Childhood." *The Psychoanalytic Study of the Child,* vol. 1. New York: International Universities Press,

1945: 53–74; Rene A. Spitz, M.D., with Katherine M. Wolf, Ph.D. "Anaclitic Depression: An Inquiry into the Genesis of Psychiatric Conditions in Early Childhood, II." *The Psychoanalytic Study of the Child,* vol. 2. New York: International Universities Press, 1946.

20. M. E. Seligman. "Learned Helplessness." *Annual Review of Medicine* 23, 1972: 407–412; William R. Miller, Robert A. Rosellini, and Martin E. P. Seligman. "Learned Helplessness and Depression." In *Psychopathology: Experimental Models,* ed. Jack D. Maser and Martin E. P. Seligman. San Francisco: W. H. Freeman, 1977: 104–130.

21. Marvin Zuckerman. "Good and Bad Humors: Biochemical Bases of Personality and Its Disorders." *Psychological Science,* November 1995: 330.

22. K. Sokolowski and H. D. Schmalt. "Emotional and Motivational Influences in an Affiliation Conflict Situation." *Zeitschrift für Experimentelle Psychologie,* 43:3 (1996): 461–482.

23. See, for example, James S. House, Karl R. Landis, and Debra Umberson. "Social Relationships and Health." *Science,* July 29, 1988: 541.

24. Bert N. Uchino, Darcy Uno, and Julianne Holt-Lunstad. "Social Support, Physiological Processes, and Health." *Current Directions in Psychological Science,* October 1999: 145–148.

25. B. M. Hagerty and R. A. Williams. "The Effects of Sense of Belonging, Social Support, Conflict, and Loneliness on Depression." *Nursing Research,* July–August 1999: 215–219.

26. Erkki Ruoslahti. "Stretching Is Good for a Cell." *Science,* May 30, 1997: 1345–1346.

27. See, for example, B. Lown. "Sudden Cardiac Death: Biobehavioral Perspective." *Circulation,* July 1987: 186–196; C. M. Jenkinson, R. J. Madeley, J. R. Mitchell, and I. D. Turner. "The Influence of Psychosocial Factors on Survival after Myocardial Infarction." *Public Health,* September 1993: 305–317.

28. See, for example, Michael Davies, Henry Davies, and Kathryn Davies. *Humankind the Gatherer-Hunter: From Earliest Times to Industry.* Kent, U.K.: Myddle-Brockton, 1992.

29. For numerous explanations of this sort, see: David P. Barash. *The Hare and the Tortoise: Culture, Biology, and Human Nature.* New York: Penguin Books, 1987; Richard E. Leakey and Roger Lewin. *People of the Lake: Mankind and Its Beginnings.* New York: Avon Books, 1983.

30. J. J. Lynch and I. F. McCarthy. "The Effect of Petting on a Classically Conditioned Emotional Response." *Behavioral Research and Therapy* 5, 1967: 55–62; J. J. Lynch and I. F. McCarthy. "Social Responding in Dogs: Heart Rate Changes to a Person." *Psychophysiology* 5, 1969: 389–393; N. Shanks, C. Renton, S. Zalcman, and H. Anisman. "Influence of Change from Grouped to Individual Housing on a T-Cell-Dependent Immune Response in Mice: Antagonism by Diazepam." *Pharmacology, Biochemistry and Behavior,* March 1994: 497–502.

31. Charles Darwin. *The Descent of Man, and Selection in Relation to Sex.* New York: D. Appleton, 1871: 146–148.

32. For additional support of the group selectionist point of view, see Daniel G. Freedman's interpretation of Sewall Wright's shifting balance theory (Daniel G. Freedman. *Human Sociobiology: A Holistic Approach.* New York: Free Press, 1979: 5).

33. Thomas Welte, David Leitenberg, Bonnie N. Dittel, Basel K. al-Ramadi, Bing Xie, Yue E. Chin, Charles A. Janeway Jr., Alfred L. M. Bothwell, Kim Bottomly, and Xin-Yuan Fu. "STAT5 Interaction with the T Cell Receptor Complex and Stimulation of T Cell Proliferation." *Science,* January 8, 1999: 222–225; Polly Matzinger. "The Real Function of the Immune System or Tolerance and the Four D's (Danger, Death, Destruction and Distress)." Downloaded from: http://glamdring.ucsd.edu/others/aai/polly.html. March 1998.

34. Glaucia N. R. Vespa, Linda A. Lewis, Katherine R. Kozak, Miriana Moran, Julie T. Nguyen, Linda G. Baum, and M. Carrie Miceli. "Galectin-1 Specifically Modulates TCR Signals to Enhance TCR Apoptosis but Inhibit IL-2 Production and Proliferation." *Journal of Immunology,* January 15, 1999: 799–806.

35. M. E. Seligman. "Learned Helplessness." *Annual Review of Medicine;* W. R. Miller and M. E. Seligman. "Depression and Learned Helplessness in Man." *Journal of Abnormal Psychology,* June 1975:

228–238; William R. Miller, Robert A. Rosellini, and Martin E. P. Seligman. "Learned Helplessness and Depression." In *Psychopathology*, ed. Jack D. Maser and Martin E. P. Seligman.

36. Joseph V. Brady. "Ulcers in Executive Monkeys." *Scientific American,* October 1958: 95–100.

37. For excellent discussions of the executive monkey study's failings, see: C. Robin Timmons and Leonard W. Hamilton. *Drugs, Brains and Behavior.* Previously published by Prentice-Hall as *Principles of Behavioral Pharmacology.* Updated and available online from Rutgers University. http://www.rci.rutgers.edu/~lwh/drugs/. January 1999; Paul Kenyon. *Biological Bases of Behaviour: Stress and Behaviour.* Plymouth, U.K.: University of Plymouth, http://salmon.psy.plym.ac.uk/year1/STRESBEH.HTM#Everday and Executive Stress, January 1999.

38. M. E. Seligman. "Learned Helplessness." *Annual Review of Medicine;* W. R. Miller and M. E. Seligman. "Depression and Learned Helplessness in Man." *Journal of Abnormal Psychology;* William R. Miller, Robert A. Rosellini, and Martin E. P. Seligman. "Learned Helplessness and Depression." In *Psychopathology*, ed. Jack D. Maser and Martin E. P. Seligman; Daniel Goleman, Ph.D. *Vital Lies, Simple Truths: The Psychology of Self-Deception.*

39. V. C. Wynne-Edwards. *Animal Dispersion in Relation to Social Behaviour.* New York: Hafner, 1962; V. C. Wynne-Edwards. *Evolution through Group Selection.* Oxford: Blackwell Scientific, 1986.

40. V. C. Wynne-Edwards. *Evolution through Group Selection:* 87.

41. See: Helena Cronin. *The Ant and the Peacock.* New York: Oxford University Press, 1991: 282–283; David P. Barash. *Sociobiology and Behavior.* New York: Elsevier Scientific: 1977: 70–75; and the blunt statement by geneticist Lawrence Hurst that Wynne-Edwards's views are just plain "fundamentally wrong." (Lawrence Hurst. "You Scratch My Back . . ." *New Scientist* 12 (September 1998, http://www.newscientist.com/ns/980912/review1.html. January 1999.)

42. P. Ward and A. Zahavi. "The Importance of Certain Assemblages of Birds as 'Information-Centres' for Food Finding." *Ibis* 115 (4) 1973: 517–534.

43. David Sloan Wilson. "Incorporating Group Selection into the Adaptationist Program: A Case Study Involving Human Decision Making." In *Evolutionary Social Psychology,* ed. J. Simpson and D. Kendricks: 345–386.

1. CREATIVE NETS IN THE PRECAMBRIAN ERA

1. The National Computational Science Alliance. Cosmos in a Computer. http://www.ncsa.uiuc.edu/Cyberia/Cosmos/CosmosCompHome.html. January 1999.

2. Alan H. Guth. *The Inflationary Universe: The Quest for a New Theory of Cosmic Origins.* Reading, Mass.: Addison-Wesley, 1997: 93.

3. Christian de Duve. "The Birth of Complex Cells." *Scientific American,* April 1996: 52.

4. Eshel Ben-Jacob. Personal correspondence. January 25, 1999.

5. James A. Shapiro. "Thinking about Bacterial Populations as Multicellular Organisms." *Annual Review of Microbiology.* Palo Alto, Calif.: Annual Reviews, 1998: 88.

6. L. J. Shimkets. "Social and Developmental Biology of the Myxobacteria." *Microbiology Reviews,* December 1990: 473–501; James A. Shapiro. "Thinking about Bacterial Populations as Multicellular Organisms." *Annual Review of Microbiology:* 88. Then there's always bacterial sexuality—by no means a solitary process. (C. Bernstein and H. Bernstein. "Sexual Communication." *Journal of Theoretical Biology,* September 1997: 69–78.)

7. S. A. Sturdza. "Expansion Phenomena of Proteus Cultures. I. The Swarming Expansion." *Zentralblatt für Bakteriologie,* December 1975: 505–530.

8. S. Esipov and J. A. Shapiro. "Kinetic Model of *Proteus Mirabilis* Swarm Colony Development." *Journal of Mathematical Biology* 36, 1998: 249–263.

9. C. Allison and C. Hughes. "Bacterial Swarming: An Example of Prokaryotic Differentiation and Multicellular Behaviour." *Science Progress,* 75:298, pt. 3–4, 1991: 403–422.

10. R. M. Harshey. "Bees Aren't the Only Ones: Swarming in Gram-negative Bacteria." *Molecular Microbiology,* May 1994: 389–394.

11. O. Rauprich, M. Matsushita, C. J. Weijer, F. Siegert, S. E. Esipov, and J. A. Shapiro. "Periodic Phenomena in *Proteus mirabilis* Swarm Colony Development." *Journal of Bacteriology,* November 1996: 6525–6538; J. A. Shapiro. "The Significances of Bacterial Colony Patterns." *Bioessays.*

12. R. M. Harshey. "Bees Aren't the Only Ones." *Molecular Microbiology:* 389–394; S. Moens, K. Michiels, V. Keijers, F. Van Leuven, and J. Vanderleyden. "Cloning, Sequencing, and Phenotypic Analysis of laf1, Encoding the Flagellin of the Lateral Flagella of *Azospirillum Brasilense* Sp7." *Journal of Bacteriology,* October 1995: 5419–5426.

13. C. Allison and C. Hughes. "Bacterial Swarming." *Science Progress:* 403–422.

14. E. O. Budrene and H. C. Berg. "Dynamics of Formation of Symmetrical Patterns by Chemotactic Bacteria." *Nature,* July 6, 1995: 49–53; C. W. Douglas and K. A. Bisset. "Development of Concentric Zones in the Proteus Swarm Colony." *Journal of Medical Microbiology,* November 1976: 497–500.

15. James A. Shapiro, "Bacteria as Multicellular Organisms." *Scientific American,* June 1988: 84, 89.

16. A. Pierro, H. K. van Saene, S. C. Donnell, J. Hughes, C. Ewan, A. J. Nunn, and D. A. Lloyd. "Microbial Translocation in Neonates and Infants Receiving Long-term Parenteral Nutrition." *Archives of Surgery,* February 1996: 176–179; R. Sanchis, G. Abadie, and P. Pardon. "*Salmonella Abortusovis* Experimental Infection Induced by the Conjunctival Route: Clinical, Serological and Bacteriological Study of the Dose Effect in Female Lambs." *Veterinary Research* 26:2, 1995: 73–80.

17. C. Allison, H. C. Lai, D. Gygi, and C. Hughes. "Cell Differentiation of *Proteus Mirabilis* Is Initiated by Glutamine, a Specific Chemoattractant for Swarming Cells." *Molecular Microbiology,* April 1993: 53–60.

18. G. P. Salmond, B. W. Bycroft, G. S. Stewart, and P. Williams. "The Bacterial 'Enigma': Cracking the Code of Cell-Cell Communication." *Molecular Microbiology,* May 1995: 615–624; R. E. Sicard. "Hormones, Neurosecretions, and Growth Factors as Signal Molecules for Intercellular Communication." *Developmental and Comparative Immunology,* spring 1986: 269–272; N. H. Hussain, M. Goodson, and R. J. Rowbury. "Recent Advances in Biology: Intercellular Communication and Quorum Sensing in Micro-organisms." *Science Progress* 81, pt. 1 (1998): 69–80.

19. J. Muñoz-Dorado and J. M. Arias. "The Social Behavior of Myxobacteria." *Microbiologia,* December 1995: 429–438.

20. James A. Shapiro. "Thinking about Bacterial Populations as Multicellular Organisms." *Annual Review of Microbiology;* V. Norris and G. J. Hyland. "Do Bacteria Sing? Sonic Intercellular Communication between Bacteria May Reflect Electromagnetic Intracellular Communication Involving Coherent Collective Vibrational Modes That Could Integrate Enzyme Activities and Gene Expression." *Molecular Microbiology,* May 1997: 879–880.

21. K. M. Gray. "Intercellular Communication and Group Behavior in Bacteria." *Trends in Microbiology,* May 1997: 184–188.

22. A. J. Wolfe and H. C. Berg. "Migration of Bacteria in Semisolid Agar." *Proceedings of the National Academy of Sciences of the United States of America,* September 1989: 6973–6977.

23. Eshel Ben-Jacob. "Bacterial Wisdom, Gödel's Theorem and Creative Genomic Webs." *Physica A* 248 (1998): 57–76.

24. E. Ben-Jacob, A. Tenenbaum, O. Shochet, I. Cohen, A. Czirók, and T. Vicsek. "Generic Modeling of Cooperative Growth Patterns in Bacterial Colonies." *Nature* 368 (1994): 46–49.

25. By far the most sophisticated studies of the signals bacterial cells use to tell each other of danger or discovery have been done by Eshel Ben-Jacob (see Ben-Jacob's papers cited above). It is on Ben-Jacob's work that this account is based.

26. Michael B. Yarmolinsky. "Programmed Cell Death in Bacterial Populations," *Science,* February 10, 1995: 836–837, 837.

27. Richard Lipkin. "Bacterial Chatter: How Patterns Reveal Clues about Bacteria's Chemical Communication." *Science News,* March 4, 1995: 137.

28. Eshel Ben-Jacob. Personal communication. April 23, 1996; Eshel Ben-Jacob. Personal communication. May 9, 1996.

29. John H. Holland. *Hidden Order: How Adaptation Builds Complexity.* New York: Addison-Wesley, 1995: 31.

30. John H. Holland. *Hidden Order: How Adaptation Builds Complexity.* New York: Addison-Wesley, 1995: 31.

31. See, for example, J. A. Shapiro. "Genome Organization, Natural Genetic Engineering and Adaptive Mutation." *Trends in Genetics,* March 1997: 98−104; J. A. Shapiro. "Natural Genetic Engineering in Evolution." *Genetica* 86:1−3 (1992): 99−111.

32. Sorin Sonea and Maurice Panisset. *A New Bacteriology.* Boston: James and Bartlett, 1983; Sorin Sonea. "The Global Organism: A New View of Bacteria." *The Sciences,* July/August 1988: 38−45; Lynn Margulis and Dorion Sagan. *Microcosmos: Four Billion Years of Microbial Evolution.* New York: Summit Books, 1986.

33. Didier Mazel, Broderick Dychinco, Vera A. Webb, and Julian Davies. "A Distinctive Class of Integron in the Vibrio Cholerae Genome." *Science* 24, April 1998: 605−608; Elizabeth Pennisi. "Versatile Gene Uptake System Found in Cholera Bacterium." *Science,* April 24, 1998: 521−522.

34. J. William Schopf. "Microfossils of the Early Archean Apex Chert: New Evidence of the Antiquity of Life." *Science,* April 30, 1993: 640−646.

2. NETWORKING IN PALEONTOLOGY'S "DARK AGES"

1. Karen Wright. "When Life Was Odd." *Discover,* March 1997: 53.

2. Jochen J. Brocks, Graham A. Logan, Roger Buick, and Roger E. Summons. "Archean Molecular Fossils and the Early Rise of Eukaryotes." *Science,* August 13, 1999: 1033−1036.

3. Graham Bell. "Model Metaorganism." *Science,* October 9, 1998: 248; Richard A. Kerr. "Early Life Thrived Despite Earthly Travails." *Science,* June 25, 1999: 2111−2113; Dr. Charles F. Delwiche. Personal communications. February 1997 and September 1999. Genetic research agrees with the fossil evidence from *Grypania,* placing the separation between archaebacteria and eukaryotes at more than 2 billion years ago. See: D. F. Feng, G. Cho, and R. F. Doolittle. "Determining Divergence Times with a Protein Clock: Update and Reevaluation." *Proceedings of the National Academy of Sciences of the United States of America,* November 25, 1997: 13028−13033.

4. Christian de Duve. "The Birth of Complex Cells." *Scientific American,* April 1996: 50.

5. Graham Bell. "Model Metaorganism." *Science:* 248; Ben Waggoner. "Introduction to the Proterozoic." Copyright 1994, 1995, 1996, 1997 by the Museum of Paleontology of the University of California at Berkeley and the Regents of the University of California. http://www.ucmp.berkeley.edu/. February 1997.

6. William Martin and Miklos Müller. "The Hydrogen Hypothesis for the First Eukaryote." *Nature,* March 5, 1998: 37−41. Pioneering biologist Lynn Margulis was the first to champion "the serial endosymbiotic theory" stating that organelles in eukaryotic cells began as invading or invited bacteria.

7. Lynn Margulis. *Symbiosis in Cell Evolution: Microbial Communities in the Archean and Proterozoic Eons,* 2nd ed. New York: W. H. Freeman, 1993.

8. Jeffrey D. Palmer. "Organelle Genomes: Going, Going, Gone!" *Science,* February 7, 1997: 790−791.

9. William Martin and Miklos Müller. "The Hydrogen Hypothesis for the First Eukaryote." *Nature;* Kathleen Sandman and John N. Reeve. "Origin of the Eukaryotic Nucleus." *Science,* April 24, 1998: 499d; Gretchen Vogel. "Did the First Complex Cell Eat Hydrogen?" *Science,* March 13,

1998: 1633–1634; Gretchen Vogel. "Searching for Living Relics of the Cell's Early Days." *Science,* September 12, 1997: 1604.

10. Lynn Margulis. *Symbiosis in Cell Evolution:* 5–335.

11. Margulis is supported by the theories Christian de Duve presents in "The Birth of Complex Cells." *Scientific American:* 56.

12. Roberto Kolter and Richard Losick. "One for All and All for One." *Science,* April 10, 1998: 226–227.

13. Ning Wang, James P. Butler, and Donald E. Ingber. "Mechanotransduction across the Cell Surface and through the Cytoskeleton." *Science,* May 21, 1993: 1124–1127; Steven R. Heidemann. "A New Twist on Integrins and the Cytoskeleton." *Science,* May 21, 1993: 1080–1081; A. J. Maniotis, C. S. Chen, and D. E. Ingber. "Demonstration of Mechanical Connections between Integrins, Cytoskeletal Filaments, and Nucleoplasm That Stabilize Nuclear Structure." *Proceedings of the National Academy of Sciences of the United States of America,* February 4, 1997: 849–54.

14. R. Lahoz-Beltra, S. R. Hameroff, and J. E. Dayhoff. "Cytoskeletal Logic: A Model for Molecular Computation via Boolean Operations in Microtubules and Microtubule-Associated Proteins." *Biosystems* 29:1, 1993: 1–23; M. Jibu, S. Hagan, S. R. Hameroff, K. H. Pribram, and K. Yasue. "Quantum Optical Coherence in Cytoskeletal Microtubules: Implications for Brain Function." *Biosystems* 32:3, 1994: 195–209.

15. Hiroaki Kawasaki, Gregory M. Springett, Naoki Mochizuki, Shinichiro Toki, Mie Nakaya, Michiyuki Matsuda, David E. Housman, and Ann M. Graybiel. "A Family of CAMP-Binding Proteins That Directly Activate Rap1." *Science,* December 18, 1998: 2275–2279; Robert D. Blitzer, John H. Connor, George P. Brown, Tony Wong, Shirish Shenolikar, Ravi Iyengar, and Emmanuel M. Landau. "Gating of CaMKII by CAMP-Regulated Protein Phosphatase Activity during LTP." *Science,* June 19, 1998: 1940–1943; Domenico Grieco, Antonio Porcellini, Enrico V. Avvedimento, and Max E. Gottesman. "Requirement for CAMP-PKA Pathway Activation by M Phase–Promoting Factor in the Transition from Mitosis to Interphase." *Science,* March 22, 1996: 1718–1723; Upinder S. Bhalla and Ravi Iyengar. "Emergent Properties of Networks of Biological Signaling Pathways." *Science,* January 15, 1999: 381–387.

16. Trisha Gura. "One Molecule Orchestrates Amoebae." *Science,* July 11, 1997: 182.

17. Ravi Iyengar. "Gating by Cyclic AMP: Expanded Role for an Old Signaling Pathway." *Science,* January 26, 1996: 461–463; Peter Satir and Ian Gibbons. "Cilia and Flagella." *McGraw-Hill Multimedia Encyclopedia of Science and Technology.* Ver. 2.0. New York: McGraw-Hill, 1995. CD-ROM.

18. N. Rieder. "Early Stages of Toxicyst Development in *Didinium Nasutum.*" *Zeitschrift für Naturforschung B,* April 1968: 569; Richard P. Hall. "Protozoa." *McGraw-Hill Multimedia Encyclopedia of Science and Technology.* Ver. 2.0. New York: McGraw-Hill, 1995. CD-ROM.

19. James A. Shapiro. "Thinking about Bacterial Populations as Multicellular Organisms." *Annual Review of Microbiology.* Palo Alto, Calif.: Annual Reviews, 1998: 88.

20. T. W. James, F. Crescitelli, E. R. Loew, and W. N. McFarland. "The Eyespot of *Euglena Gracilis:* A Microspectrophotometric Study." *Vision Research,* September 1992: 1583–1591; J. J. Wolken. "*Euglena:* The Photoreceptor System for Phototaxis." *Journal of Protozoology,* November 1977: 518–522; Richard P. Hall. "Protozoa." *McGraw-Hill Multimedia Encyclopedia of Science and Technology.*

21. Angus I. Lamond and William C. Earnshaw. "Structure and Function in the Nucleus." *Science,* April 24, 1998: 547–553; J. W. Bodnar. "A Domain Model for Eukaryotic DNA Organization: A Molecular Basis for Cell Differentiation and Chromosome Evolution." *Journal of Theoretical Biology,* June 22, 1988: 479–507; Experimental Study Group. Massachusetts Institute of Technology. "Characteristics of Prokaryotes and Eukaryotes." *MIT Biology Hypertextbook.* http://esg-www.mit.edu:8001/esgbio/cb/prok_euk.html, February 1999; Mark Dalton. *WWW Cell Biology Course.* http://www.cbc.umn.edu/~mwd/cell.html, February 1999.

22. Lynn Margulis. *Symbiosis in Cell Evolution:* 50.

23. Ibid.: 255–260.

24. David L. Kirk. *Volvox Molecular-Genetic Origins of Multicellularity and Cellular Differentiation.* Cambridge, U.K.: Cambridge University Press, 1998; Graham Bell. "Model Metaorganism." *Science,* 248.

25. Paul C. Silva and Richard L. Moe. "Volvocales." *McGraw-Hill Multimedia Encyclopedia of Science and Technology.* Ver. 2.0. New York: McGraw-Hill, 1995. CD-ROM.

26. J. Muñoz-Dorado and J. M. Arias. "The Social Behavior of Myxobacteria." *Microbiologia,* December 1995: 429–438; L. J. Shimkets. "Control of Morphogenesis in Myxobacteria." *Critical Reviews in Microbiology* 14:3 (1987): 195–227.

27. Helmut Sauer. "Cell Differentiation." *McGraw-Hill Multimedia Encyclopedia of Science and Technology.* Ver. 2.0. New York: McGraw-Hill, 1995. CD-ROM.

28. Gretchen Vogel. "Cell Biology: Parasites Shed Light on Cellular Evolution." *Science,* March 7, 1997: 1422–1430.

29. Lynn Margulis. *Symbiosis in Cell Evolution:* 233, 260; Lynn Margulis and Michael F. Dolan. "Swimming against the Current." *The Sciences,* January/February 1997: 20–25.

30. D. R. Smith and M. Dworkin. "Territorial Interactions between Two Myxococcus Species." *Journal of Bacteriology,* February 1994: 1201–1205.

31. Dr. Charles F. Delwiche. Personal communication. February 1997.

32. Helmut Sauer. "Cell Differentiation." *McGraw-Hill Multimedia Encyclopedia of Science and Technology.*

33. Current evidence fixes the date of evolution of multicellular animals before 1.2 billion years ago. However, each year new data rolls the rise of Animalia further back in time. See: Gregory A. Wray, Jeffrey S. Levinton, and Leo H. Shapiro. "Molecular Evidence for Deep Precambrian Divergences among Metazoan Phyla." *Science,* October 25, 1996: 568–573; Geerat J. Vermeij. "Animal Origins." *Science,* October 25, 1996: 525–526; Kenneth J. McNamara. "Dating the Origin of Animals." *Science,* December 20, 1996: 1993f–1997f.

34. The first multicellular animals we know of were wormlike creatures which dug tunnels under the thick sea mats in which some bacterial colonies gathered. (Adolf Seilacher, Pradip K. Bose, and Friedrich Pflüger. "Triploblastic Animals More Than One Billion Years Ago: Trace Fossil Evidence from India." *Science,* October 2, 1998: 80–83.)

35. The date of 720 million years ago for the first clam comes from Armand Kuris, professor of zoology, the University of California at Santa Barbara, an expert on marine ecology. The University of California at Berkeley's Ben Waggoner, a scholar who has been extremely helpful in providing information on early multicellular life for this book, has found evidence in Russia of a later date for the ancestors of clams: 550 million years ago. (Mikhail A. Fedonkin and Benjamin M. Waggoner. "The Late Precambrian Fossil Kimberella Is a Mollusc-like Bilaterian Organism." *Nature,* August 28, 1997: 868–871.) The two dates do not necessarily conflict.

36. Constance Holden. "Modeling Fossils in Jell-O." *Science,* July 24, 1998: 511.

37. Kevin Brett. Personal communication. March 4, 1997.

38. Kerry B. Clark. Personal communication. March 5, 1997.

39. See, for example, Ilya Prigogine and Isabelle Stengers. *Order Out of Chaos: Man's New Dialogue with Nature.* New York: Bantam, 1984.

3. THE EMBRYONIC MEME

1. D. A. Jackson and P. R. Cook. "The Structural Basis of Nuclear Function." *International Review of Cytology* 162A (1995): 125–149.

2. L. L. Wallrath, Q. Lu, H. Granok, and S. C. Elgin. "Architectural Variations of Inducible Eukaryotic Promoters: Preset and Remodeling Chromatin Structures." *Bioessays,* March 1994: 165–170.

3. R. M. Benbow. "Chromosome Structures." *Science Progress* 76, pt. 3–4 (1992): 301–302, 425–50.

4. Wallace Marshall. "Visualizing Nuclear Architecture: Specific Interactions between Chromatin and the Nuclear Envelope in Drosophila Embryos." http://util.ucsf.edu/sedat/marsh/interactions.html. February 1999; Wallace Marshall. "Individual Loci Occupy Defined Positions in the Nucleus." http://util.ucsf.edu/sedat/marsh/interactions_positioning.html. February 1999; Einar Hallberg, Henrik Sšderqvist, and Madeleine Kihlmark. "Proteins from the 'Pore Membrane' Domain of the Nuclear Envelope: Molecular Membrane Biogenesis and Post Mitotic Assembly." Stockholm University, Stockholm, Sweden. http://www.chem.su.se/Biochemfolder/Gallery/hallberg_e.html. February 1999; "Inner Surface of Nuclear Envelope from Electron Microscopy #13." Indigo Instruments, Tonawanda, N.Y. http://www.indigo.com/photocd/gphpcd/em13.html. February 1999.

5. K.S. Rózsa. "The Pharmacology of Molluscan Neurons." *Progress in Neurobiology,* 23:1-2 (1984): 79-150.

6. See, for example, John B. Connolly and Tim Tully. "You Must Remember This: Finding the Master Switch for Long-term Memory." *The Sciences,* May-June 1996: 42; I. P. Ashmarin. "Neurological Memory as a Probable Product of Evolution of Other Forms of Biological Memory." *Zhurnal Evoliutsionnoi Biokhimii I Fiziologii,* May-June 1973: 217-224.

7. T. Tully, T. Preat, S. C. Boynton, and M. Del Vecchio. "Genetic Dissection of Consolidated Memory in Drosophila." *Cell,* October 7, 1994: 35-47.

8. J. C. Yin, M. Del Vecchio, H. Zhou, and T. Tully. "CREB as a Memory Modulator: Induced Expression of a DCREB2 Activator Isoform Enhances Long-term Memory in Drosophila." *Cell,* April 7, 1995: 107-115; T. Tully, G. Bolwig, J. Christensen, J. Connolly, J. DeZazzo, J. Dubnau, C. Jones, S. Pinto, M. Regulski, F. Svedberg, and K. Velinzon. "Genetic Dissection of Memory in Drosophila." *Journal of Physiology, Paris* 90:5-6 (1996): 383; T. Tully, G. Bolwig, J. Christensen, J. Connolly, M. Del Vecchio, J. DeZazzo, J. Dubnau, C. Jones, S. Pinto, M. Regulski, B. Svedberg, and K. Velinzon. "A Return to Genetic Dissection of Memory in Drosophila." *Cold Spring Harbor Symposia on Quantitative Biology* 61 (1996): 207-218.

9. J. de Gunzburg. "Mode of Action of Cyclic AMP in Prokaryotes and Eukaryotes, CAP and CAMP-Dependent Protein Kinases." *Biochimie,* January 1985: 563-582.

10. J. C. Yin, M. Del Vecchio, H. Zhou, and T. Tully. "CREB as a Memory Modulator." *Cell.* CREB stands for cyclic AMP-responsive element-binding protein.

11. Richard Dawkins. *The Selfish Gene.* New York: Oxford University Press, 1976.

12. Aaron Lynch. *Thought Contagion: How Belief Spreads through Society.* New York: Basic Books, 1996; Francis Heylighen. *Principia Cybernetica Web.* http://pespmc1.vub.ac.be/Default.html. February 1999; Richard Brodie. *Virus of the Mind: The New Science of the Meme.* Seattle: Integral Press, 1996.

13. William Morton Wheeler. "The Ant Colony as an Organism." *Journal of Morphology* 22 (1911): 307-325; for more on the history and use of the term "superorganism," see: Howard Bloom. *The Lucifer Principle: A Scientific Expedition into the Forces of History.* New York: Atlantic Monthly Press, 1995.

14. S. J. Braddy and L. I. Anderson. "An Upper Carboniferous Eurypterid Trackway from Mostyn, Wales." *Proceedings of the Geologists' Association* 107 (1996): 51-56.

15. S. J. Braddy and J. A. Dunlop. "The Functional Morphology of Mating in the Silurian Eurypterid, *Baltoeurypterus Tetragonophthalmus* (Fischer, 1839)." *Zoological Journal of the Linnean Society,* August 1997: 435-461.

16. Kerry B. Clark. Personal communication. March 27, 1997.

17. Graziano Fiorito and Pietro Scotto. "Observational Learning in *Octopus Vulgaris.*" *Science,* April 24, 1992: 545-547. B. Moore. "The Evolution of Imitative Learning." In *Social Learning in Animals: The Roots of Culture,* ed. Cecilia M. Heyes and Bennett G. Galef Jr. San Diego: Academic Press, 1996: 245-265.

18. Kent G. Bailey. *Human Paleopsychology: Applications to Aggression and Pathological Processes.* Hillsdale, N.J.: Erlbaum Press, 1987: 293. Paul MacLean. "The Imitative-Creative Interplay of Our

Three Mentalities." In *Astride the Two Cultures: Arthur Koestler at 70,* ed. H. Harris. New York: Random House, 1976: 187–213.

19. Konrad Lorenz. *On Aggression.* New York: Harcourt Brace Jovanovich, 1974.

20. Lee Alan Dugatkin. "Interface between Culturally Based Preferences and Genetic Preferences: Female Mate Choice in *Poecilia Reticulata.*" *Proceedings of the National Academy of Sciences of the United States of America,* April 2, 1996: 2770–2773.

21. Keven N. Laland and Kerry Williams. "Social Transmission of Maladaptive Information in the Guppy." *Behavioral Ecology,* September 1998; Stephanie E. Briggs, Jean-Guy Godin, J. Dugatkin, and Lee Alan. "Mate-Choice Copying under Predation Risk in the Trinidadian Guppy *(Poecilia Reticulata).*" *Behavioral Ecology,* summer 1996; A. Kodric-Brown and P. F. Nicoletto. "Consensus among Females in Their Choice of Males in the Guppy *Poecilia Reticulata.*" *Behavioral Ecology and Sociobiology* 39:6 (1996); Constance Holden. "Nature v. Culture: A Lesson from the Guppy." *Science,* April 12, 1996: 203.

22. W. F. Herrnkind. "Evolution and Mechanisms of Mass Single-File Migration in Spiny Lobster." *Migration: Mechanisms and Adaptive Significance.* Contributions in Marine Science—Monographic Series, vol. 68. Austin: Marine Science Institute, University of Texas at Austin, 1985: 197–211.

23. P. Bushmann and J. Atema. "Aggression-Reducing Courtship Signals in the Lobster, *Homarus Americanus.*" *Biological Bulletin,* October 1994: 275–276; Christa Karavanich and Jelle Atema. "Individual Recognition and Memory in Lobster Dominance." *Animal Behaviour,* December 1998: 1553–1560; Paul J. Bushmann and Jelle Atema. "Shelter Sharing and Chemical Courtship Signals in the Lobster, *Homarus Americanus.*" *Canadian Journal of Fisheries and Aquatic Science,* March 1997; Christa Karavanich and Jelle Atema. "Olfactory Recognition of Urine Signals in Dominance Fights between Male Lobster, *Homarus Americanus.*" *Behaviour,* September 1998; Elisa B. Karnofsky, Jelle Atema, and Randall H. Elgin. "Field Observations of Social Behavior, Shelter Use, and Foraging in the Lobster, *Homarus Americanus.*" *The Biological Bulletin,* June 1989; Elisa B. Karnofsky, Jelle Atema, and Randall H. Elgin. "Natural Dynamics of Population Structure and Habitat Use of the Lobster, *Homarus Americanus,* in a Shallow Cove." *The Biological Bulletin,* June 1989; Wallace Ravven. "Lobster Lust: Don Juans of the Deep." *Discover,* December 1987: 34–40.

24. Marcia Barinaga. "Social Status Sculpts Activity of Crayfish Neurons." *Science,* January 19, 1996: 290–291; Shih-Rung Yeh, Russell Fricke, and Donald Edwards. "The Effect of Social Experience on Serotonergic Modulation of the Escape Circuit of Crayfish." *Science,* January 19, 1996: 366–369; Justine H. Lange. "Dominance in Crayfish." *Science,* April 5, 1996: 18; Shih-Rung Yeh, Barbara E. Musolf, and Donald H. Edwards. "Neuronal Adaptations to Changes in the Social Dominance Status of Crayfish." *Journal of Neuroscience,* January 1997: 697–708; T. Nagayama, H. Aonuma, and P. L. Newland. "Convergent Chemical and Electrical Synaptic Inputs from Proprioceptive Afferents onto an Identified Intersegmental Interneuron in the Crayfish." *Journal of Neurophysiology,* May 1997: 2826–2830.

25. Alisdair Daws. Personal communication. April 11, 1999.

26. Edward O. Wilson. *The Insect Societies.* Cambridge, Mass.: Harvard University Press, 1971: 4. Repeated in Bert Holldobler and Edward O. Wilson. *The Ants.* Cambridge, Mass.: Belknap Press, 1990: 27–28.

27. Kerry B. Clark. Personal correspondence. March 27, 1997.

28. Edward O. Wilson. *The Insect Societies:* 439.

29. Edward O. Wilson. *The Insect Societies:* 4. Repeated in Bert Holldobler and Edward O. Wilson. *The Ants:* 27–28.

30. Wilson makes it clear that key entomologist William Morton Wheeler already operated on the presumption that the solitary state was the evolutionary predecessor of sociality when Wheeler published *Social Life among the Insects* in 1923. (Edward O. Wilson. *The Insect Societies:* 120.)

31. Conrad C. Labandeira and Tom L. Phillips. "Trunk Borings and Rhachis Skulls of Tree Ferns from the Late Pennsylvanian (Kasimovian) of Illinois; Implications for the Origin of the Galling Functional Feeding Group and Holometabolous Insects." *Palaeontographica,* part A (in

press); Conrad C. Labandeira. Personal communication. February 16, 1999; Christine Nalepa. Personal communication. April 6, 1997.

32. Stephen T. Hasiotis, Russell F. Dubiel, Paul T. Kay, Timothy M. Demko, Krystyna Kowalska, and Douglas McDaniel. "Research Update on Hymenopteran Nests and Cocoons, Upper Triassic Chinle Formation, Petrified Forest National Park, Arizona." *National Park Service Paleontological Research,* January 1998: 116–121.

33. Edward O. Wilson. *The Insect Societies:* 273.

34. Karl von Frisch. *The Dance Language and Orientation of Bees,* trans. Leigh E. Chadwick. Cambridge, Mass.: Belknap Press, 1967: 17. This experiment has since been replicated informally and many of the phenomenon's additional details observed by Princeton University's James L. Gould. (James L. Gould. Personal communication. April 1997.)

35. The "arrangement" of the bee mind which turns an individual flier into a module in a calculating machine capable of mapping geography so precisely is given in L. A. Real. "Animal Choice Behavior and the Evolution of Cognitive Architecture." *Science,* August 30, 1991: 980–986. See also: P. K. Visscher. "Collective Decisions and Cognition in Bees." *Nature,* February 4, 1999: 400.

36. Edward O. Wilson. *The Insect Societies:* 265.

37. Karl von Frisch. *Bees: Their Vision, Chemical Senses, and Language.* Ithaca, N.Y.: Cornell University Press, 1950: 53–96; Edward O. Wilson. *The Insect Societies:* 267–271; Thomas D. Seeley. *Honeybee Ecology: A Study of Adaptation in Social Life.* Princeton, N.J.: Princeton University Press, 1985: 83–88; Thomas D. Seeley. *The Wisdom of the Hive: The Social Physiology of Honey Bee Colonies.* Cambridge, Mass.: Harvard University Press, 1995: 36–39.

38. Thomas D. Seeley. *Honeybee Ecology:* 41.

39. Ibid.: 40–43.

40. Ibid.: 101.

41. Ibid.: 83.

42. Ibid.: 30.

43. Edward O. Wilson. *The Insect Societies:* 297.

44. Thomas D. Seeley. *The Wisdom of the Hive:* 34–36.

45. Thomas D. Seeley. *Honeybee Ecology:* 71–75.

46. Bert Holldobler and Edward O. Wilson. *The Ants:* 27.

47. Edward O. Wilson. *The Insect Societies:* 452.

48. Ibid.: 250.

49. J. L. Deneubourg, S. Aron, G. Goss, J. M. Pasteels, and G. Duerinck. "Random Behaviour, Amplification Processes and Number of Participants: How They Contribute to the Foraging Properties of Ants." In *Evolution, Games and Learning: Models for Adaptation in Machines and Nature, Proceedings of the Fifth Annual International Conference of the Center for Nonlinear Studies,* ed. Doyne Farmer, Alan Lapedes, Norman Packard, and Burton Wendroff. Amsterdam: North-Holland Physics Publishing, 1985: 181.

50. Edward O. Wilson. *The Insect Societies:* 238.

4. FROM SOCIAL SYNAPSES TO SOCIAL GANGLIONS: COMPLEX ADAPTIVE SYSTEMS IN JURASSIC DAYS

1. R. T. Bakker. "The Dinosaur Renaissance." *Scientific American* 232 (1975): 58–72; R. T. Bakker. "Ecology of the Brontosaurs." *Nature,* January 15, 1971: 172–174.

2. Robert T. Bakker. *Raptor Red.* New York: Bantam, 1996: 7–217. Robert T. Bakker. *The Dinosaur Heresies: New Theories—Unlocking the Mystery of the Dinosaurs and Their Extinction.* New York: William Morrow, 1986.

3. Lianhai Hou, Larry D. Martin, Zhonghe Zhou, and Alan Feduccia. "Early Adaptive Radiation of Birds: Evidence from Fossils from Northeastern China." *Science,* November 15, 1996: 1164–1167.

4. University of North Carolina at Chapel Hill. "Discovery of New Bird Species in China, Old-est Beak Shows Evolution Complexity." http://www.sciencedaily.com/releases/1999/06/990617072348.html. Downloaded June 1999.

5. C. Feare. *The Starling.* Oxford: Oxford University Press, 1984: 24−67.

6. Fifty-four miles to food and back is the figure recorded for a roost of 836 ravens near Mal-heur Lake, Oregon. (Bernd Heinrich. *Ravens in Winter.* New York: Summit Books, 1989: 164−165.)

7. Robert Burton. *Bird Behavior.* New York: Knopf, 1985: 13.

8. E. Curio, V. Ernst, and W. Vieth. "Cultural Transmission of Enemy Recognition: One Func-tion of Mobbing." *Science* 202 (1978): 899−901. For similar behavior among rats, see: B. G. Galef Jr. and E. E. Whiskin. "Learning Socially to Eat More of One Food Than of Another." *Journal of Com-parative Psychology,* March 1995: 99−101.

9. P. Ward and A. Zahavi. "The Importance of Certain Assemblages of Birds as 'Information-centres' for Food Finding." *Ibis* 115:4 (1973): 517−534.

10. John Marzluff. Personal communication. March 4, 1996.

11. John M. Marzluff. Personal communication. 1997; John M. Marzluff, Bernd Heinrich, and Colleen S. Marzluff. "Raven Roosts Are Mobile Information Centres." *Animal Behaviour* 51 (1996): 89−103; see also Bernd Heinrich, "Why Ravens Share," *American Scientist,* July−August 1995: 342−350; B. Heinrich and J. M. Marzluff, "Do Common Ravens Yell Because They Want to Attract Others?" *Behavioral Ecology and Sociobiology* 28 (1991): 13−21; and P. de Groot. "Information Trans-fer in a Socially Roosting Weaver Bird *(Quelea Quelea; Ploceinae):* An Experimental Study." *Animal Behaviour* 28 (1980): 1249−1254.

12. Howard Bloom. "Beyond the Supercomputer: Social Groups as Self-invention Machines." In *Sociobiology and Biopolitics,* ed. Albert Somit and Steven A. Peterson. Research in Biopolitics, vol. 6. Greenwich, Conn.: JAI Press, 1998: 43−64.

13. Doyne Farmer, Alan Lapedes, Norman Packard, and Burton Wendroff, eds. *Evolution, Games and Learning: Models for Adaptation in Machines and Nature, Proceedings of the Fifth Annual Interna-tional Conference of the Center for Nonlinear Studies.* Amsterdam: North-Holland Physics Publishing, 1985: 188.

14. Inner-judges of many kinds show up in the brain and body. The work of Neil Greenberg indicates that one key location is the striatum. (Neil Greenberg, Enrique Font, and Robert C. Switzer III. "The Reptilian Striatum Revisited: Studies on *Anolis* Lizards." In *The Forebrain of Reptiles: Current Concepts of Structure and Function,* ed. Walter K. Schwerdtfeger and Willhelmus J. A. J. Smeets. Basel, Switz.: Karger, 1988: 162−177.) The striatum can wreak havoc on such basic things as our ability to translate thought into movement and our capacity to dampen the pain of stress. Greenberg also points to another inner-judge, one which lowers our sexuality. (Neil Greenberg. Personal communication. June 20, 1998.)

15. For one example, see: Myron F. Floyd. "Pleasure, Arousal, and Dominance: Exploring Affective Determinants of Recreation Satisfaction." *Leisure Sciences,* April−June 1997: 83−96.

16. In neural net terms, these would be referred to as mechanisms for "self-inhibition." (T. Fukai and S. Tanaka. "A Simple Neural Network Exhibiting Selective Activation of Neuronal Ensembles: From Winner-Take-All to Winners-Share-All." *Neural Computation,* January 1997: 77−97.)

17. B. S. McEwen. "Corticosteroids and Hippocampal Plasticity." *Brain Corticosteroid Receptors: Studies on the Mechanism, Function, and Neurotoxicity of Corticosteroid Action, Annals of the New York Acad-emy of Sciences,* 746, November 30, 1994: 142−144, 178−179; L. P. Reagan and B. S. McEwen. "Controversies Surrounding Glucocorticoid-Mediated Cell Death in the Hippocampus." *Journal of Chemical Neuroanatomy,* August 1997: 149−167.

18. Not only do the inner-judges dull our wits by decreasing brain circulation, they also give us headaches. (Roy J. Mathew and W. H. Wilson. "Intracranial and Extracranial Blood Flow during Acute Anxiety." *Psychiatry Research,* May 16, 1997: 93−107.)

19. R. J. Mathew, A. A. Swihart, and M. L. Weinman. "Vegetative Symptoms in Anxiety and Depression." *British Journal of Psychiatry,* August 1982: 162−165; Klaus Atzwanger and Alain

Schmitt. "Walking Speed and Depression: Are Sad Pedestrians Slow?" *Human Ethology Bulletin,* September 3, 1997.

20. *Bacillus subtilis* helps us out by producing the antibiotic bacitracin.

21. The wry use of the words "Central Dogma" to describe neo-Darwinism were coined by the authors of MIT's *Biology Hypertextbook* ("Central Dogma Directory." In Experimental Study Group, Massachusetts Institute of Technology. *Biology Hypertextbook.* Cambridge, Mass.: Massachusetts Institute of Technology. http://esg-www.mit.edu:8001/esgbio/dogma/dogmadir.html. April 1999.)

22. O. W. Godfrey. "Directed Mutation in *Streptomyces Lipmanii.*" *Canadian Journal of Microbiology,* November 1974: 1479–1485.

23. Tom Kiely. "Rethinking Darwin." *Technology Review,* May-June 1990: 19–20; W. Stolzenburg. "Hypermutation: Evolutionary Fast Track?" *Science News,* June 23, 1990: 391; R. Weiss. "Do-It-Yourself Evolution Appears Unlikely." *Science News,* March 10, 1990: 149; Richard Lipkin. "Bacterial Chatter: How Patterns Reveal Clues about Bacteria's Chemical Communication." *Science News,* March 4, 1995: 137; Richard Lipkin. "Stressed Bacteria Spawn Elegant Colonies." *Science News,* September 9, 1995: 167; G. Maenhaut-Michel and J. A. Shapiro. "The Roles of Starvation and Selective Substrates in the Emergence of araB-lacZ Fusion Clones." *EMBO Journal,* November 1, 1994: 5229–5239.

24. Eshel Ben-Jacob. "Bacterial Wisdom, Gödel's Theorem and Creative Genomic Webs." *Physica A* 248 (1998): 58–59.

25. Ibid.; J. A. Shapiro. "Natural Genetic Engineering in Evolution." *Genetica* 86:1–3 (1992): 99–111; J. A. Shapiro. "Natural Genetic Engineering of the Bacterial Genome." *Current Opinion in Genetics and Development,* December 1993: 845–848; J. A. Shapiro. "Genome Organization, Natural Genetic Engineering and Adaptive Mutation." *Trends in Genetics,* March 13, 1997: 98–104. Shapiro disagrees with Ben-Jacob's attribution of purposeful creativity and "smarts" to the collective intelligence of a bacterial colony. (James A. Shapiro. Personal communication. February 9, 1999.) Though this is a dispute between two giants in their field, the evidence from my thirty years of field-work in mass behavior leads me to side with Ben-Jacob.

26. Eshel Ben-Jacob. "Bacterial Wisdom, Gödel's Theorem and Creative Genomic Webs." *Physica A:* 57–59.

27. B. G. Hall. "Adaptive Evolution That Requires Multiple Spontaneous Mutations. I. Mutations Involving an Insertion Sequence." *Genetics,* December 1988: 887–897; L. L. Parker and B. G. Hall. "A Fourth *Escherichia Coli* Gene System with the Potential to Evolve Beta-glucoside Utilization." *Genetics,* July 1988: 485–490.

28. Normally the word "learning" would be scientifically unacceptable in the context of bacteria; however, it is the term Ben-Jacob employs. (Eshel Ben-Jacob. "Bacterial Wisdom, Gödel's Theorem and Creative Genomic Webs." *Physica A:* 70.)

29. Ibid.: 71.

30. James A. Shapiro. "Thinking about Bacterial Populations as Multicellular Organisms." *Annual Review of Microbiology.* Palo Alto, Calif.: Annual Reviews, 1998: 83–88.

31. Ibid.: 92–93; Eshel Ben-Jacob. Personal communications. 1996; J. M. Solomon and A. D. Grossman. "Who's Competent and When: Regulation of Natural Genetic Competence in Bacteria." *Trends in Genetics* 12 (1996): 150–155.

32. Eshel Ben-Jacob. Personal communications. 1996.

33. M. J. C. Waller. Personal communications. 1995–1997; M. J. C. Waller. "Darwinism and the Enemy Within." *Journal of Social and Evolutionary Systems* 18:3 (1995): 217–229.

34. Eshel Ben-Jacob. Personal communication. September 7, 1999.

35. J. A. Shapiro. "Genome Organization, Natural Genetic Engineering and Adaptive Mutation." *Trends in Genetics:* 98–104; J. A. Shapiro. "Natural Genetic Engineering in Evolution." *Genetica:* 99–111; J. A. Shapiro. "Natural Genetic Engineering of the Bacterial Genome." *Current Opinion in Genetics and Development:* 845–848.

36. Eshel Ben-Jacob. "Bacterial Wisdom, Gödel's Theorem and Creative Genomic Webs." *Physica A:* 70; Eshel Ben-Jacob. Personal communication. February 20, 1999.

37. Eshel Ben-Jacob. "Bacterial Wisdom, Gödel's Theorem and Creative Genomic Webs." *Physica A:* 79.

38. D. R. Smith and M. Dworkin. "Territorial Interactions between Two Myxococcus Species." *Journal of Bacteriology,* February 1994: 1201–1205.

39. F. Le Guyader, M. Pommepuy, and M. Cormier. "Implantation of *Escherichia Coli* in Pilot Experiments and the Influence of Competition on the Flora." *Canadian Journal of Microbiology,* February 1991: 116–121.

40. D. J. Bibel, R. Aly, C. Bayles, W. G. Strauss, H. R. Shinefield, and H. I. Maibach. "Competitive Adherence as a Mechanism of Bacterial Interference." *Canadian Journal of Microbiology,* June 1983: 700–703.

41. A. Onderdonk, B. Marshall, R. Cisneros, and S. B. Levy. "Competition between Congenic *Escherichia Coli* K-12 Strains In Vivo." *Infection and Immunity,* April 1981: 74–79.

42. N. Rikitomi, M. Akiyama, and K. Matsumoto. "Role of Normal Microflora in the Throat in Inhibition of Adherence of Pathogenic Bacteria to Host Cells: In Vitro Competitive Adherence between *Corynebacterium Pseudodiphtheriticum* and *Branhamella Catarrhalis.*" *Kansenshogaku Zasshi,* February 1989: 118–124.

43. C. S. Impey, G. C. Mead, and S. M. George. "Competitive Exclusion of Salmonellas from the Chick Caecum Using a Defined Mixture of Bacterial Isolates from the Caecal Microflora of an Adult Bird." *Journal of Hygiene,* December 1982: 479–490.

44. M. Aho, L. Nuotio, E. Nurmi, and T. Kiiskinen. "Competitive Exclusion of Campylobacters from Poultry with K-bacteria and Broilact." *International Journal of Food Microbiology,* March–April 1992: 265–275.

45. Comm Tech Lab and the Center for Microbial Ecology at Michigan State University. *The Microbe Zoo DLC-ME Project.* http://commtechlab.msu.edu/sites/dleme/zoo/. Downloaded September 1999.

46. Richard A. Kerr. "Life Goes to Extremes in the Deep Earth—and Elsewhere?" *Science,* May 2, 1997: 703–704.

47. Lianhai Hou, Larry D. Martin, Zhonghe Zhou, Alan Feduccia, and Fucheng Zhang. "A Diapsid Skull in a New Species of the Primitive Bird Confuciusornis." *Nature,* June 17, 1999: 679–682; Lianhai Hou, Larry D. Martin, Zhonghe Zhou, and Alan Feduccia. "Early Adaptive Radiation of Birds: Evidence from Fossils from Northeastern China." *Science;* Bernd Heinrich. *Ravens in Winter:* 139–140. For indications of the manner in which the five principles of the complex adaptive system work among modern birds, see: Robert Burton. *Bird Behavior:* 134; Robert Burton. Personal communication. April 15, 1996; Harold E. Burtt. *The Psychology of Birds: An Interpretation of Bird Behavior.* New York: Macmillan, 1967.

5. MAMMALS AND THE FURTHER RISE OF MIND

1. Jose Bonaparte. Personal communication. March 18, 1999; J. F. Bonaparte. "Sobre Messungulatum Houssayi y Nuevos Mammiferos Cretacicos de Patagonia, Argentina." *IV Cong. Argentino de Paleontologia y Bioestratigrafia, Mendoza, Argentina* 2 (1986): 48–61.

2. John F. Eisenberg. "Parental Behavior." In *The Oxford Companion to Animal Behavior,* ed. David McFarland. New York: Oxford University Press, 1982: 443–447.

3. John F. Eisenberg. Personal communication. November 13, 1997.

4. John F. Eisenberg. *The Mammalian Radiations: An Analysis of Trends in Evolution, Adaptation, and Behavior.* Chicago: University of Chicago Press, 1981: 70.

5. David Sloan Wilson. "Reply to Joseph Soltis, Robert Boyd, and Peter J. Richerson: Can Group Functional Behaviors Evolve by Cultural Group Selection? An Empirical Test." *Current Anthropology,* June 1995: 489.

6. Richard G. Coss and Donald H. Owings. "Rattler Battlers." *Natural History,* May 1989: 31−32.

7. Barry Holstun Lopez. *Of Wolves and Men.* New York: Charles Scribner's Sons, 1978: 16.

8. K. R. L. Hall. "Social Learning in Primates." In *Primates: Studies in Adaptation and Variability,* ed. Phyllis C. Jay. New York: Holt, Rinehart and Winston, 1968: 383−397.

9. Stuart A. Altmann. "The Structure of Primate Social Communication." In *Social Communication among Primates,* ed. Stuart A. Altmann. Chicago: University of Chicago Press, 1967: 346.

10. Shirley C. Strum. *Almost Human: A Journey into the World of Baboons.* New York: Random House, 1987: 132.

11. K. R. L. Hall. "Experiment and Quantification in the Study of Baboon Behavior in Its Natural Habitat." In *Primates: Studies in Adaptation and Variability,* ed. Phyllis C. Jay. New York: Holt, Rinehart and Winston, 1968: 120.

12. K. R. L. Hall. "Social Organization of the Old-World Monkeys and Apes." In *Primates: Studies in Adaptation and Variability,* ed. Phyllis C. Jay. New York: Holt, Rinehart and Winston, 1968: 30.

13. David McFarland, ed. *The Oxford Companion to Animal Behavior.* New York: Oxford University Press, 1982: 251; W. D. Hamilton. "Geometry for the Selfish Herd." *Journal of Theoretical Biology,* 31:3 (1971): 295−311.

14. Christopher Boehm. Personal communication. July 11, 1997.

15. K. R. L. Hall. "Tool-Using Performances as Indicators of Behavioral Adaptability." In *Primates: Studies in Adaptation and Variability,* ed. Phyllis C. Jay. New York: Holt, Rinehart and Winston, 1968: 143.

16. Stuart Altmann. "Baboon Behavior." *Wild Discovery Wired,* October 12, 1998. http://eagle.online.discovery.com/cgibin/conversations_view/dir/Wild%20Discovery%20Wired/Baboons2/Baboon%20behavior. April 1999.

17. A. Stolba. "Entscheidungsfindungin Verbanden von Papio Hamadryas." Ph.D. diss. Zurich: Universitat Zurich, 1979; H. Kummer. *Weisse Affen am Roten Meer: Das Soziale Leben der Wüstenpaviane.* Munich: R. Piper, 1992; Jan A. R. A. M. Van Hooff. "Understanding Chimpanzee Understanding." In *Chimpanzee Cultures,* ed. Richard W. Wrangham, W. C. McGrew, Frans B. M. de Waal, and Paul G. Heltne with assistance from Linda A. Marquardt. Cambridge, Mass.: Harvard University Press, 1994: 279.

18. Christopher Boehm. "Four Mechanical Routes to Altruism." Unpublished manuscript. March 1996: 23. See also: Christopher Boehm. "Rational Preselection from Hamadryas to *Homo sapiens:* The Place of Decisions in Adaptive Process." *American Anthropologist,* June 1978: 265−296; Hans Kummer. *Primate Societies: Group Techniques of Ecological Adaptation.* Chicago: Aldine Press, 1971.

19. Shirley C. Strum. *Almost Human:* 203−259.

20. K. R. L. Hall. "Experiment and Quantification in the Study of Baboon Behavior in Its Natural Habitat." In *Primates,* ed. Phyllis C. Jay: 122.

21. Cynthia Moss. *Elephant Memories: Thirteen Years in the Life of an Elephant Family.* New York: William Morrow, 1988.

22. I. Douglas-Hamilton and O. Douglas-Hamilton. *Among the Elephants.* New York: Viking, 1975; summarized in John Tyler Bonner. *The Evolution of Culture in Animals.* Princeton, N.J.: Princeton University Press, 1983: 177.

23. G. S.Wilkinson. "Reciprocal Food Sharing in the Vampire Bat." *Nature* 308 (1984): 181−184; G. S. Wilkinson. "The Social Organization of the Common Vampire Bat." *Behavioral Ecology and Sociobiology* 17 (1985): 111−121.

24. Robin Dunbar. *Grooming, Gossip and the Evolution of Language.* London: Faber and Faber, 1996: 65.

25. Dorothy L. Cheney and Robert M. Seyfarth. "The Representation of Social Relations by Monkeys." *Cognition* 37 (1990): 167−196; Dorothy L. Cheney and Robert M. Seyfarth. *How Monkeys See the World: Inside the Mind of Another Species.* Chicago: University of Chicago Press, 1992; Dorothy L. Cheney and Robert M. Seyfarth. "The Evolution of Social Cognition in Primates." In *Behavioral Mech-*

anisms in Evolutionary Ecology, ed. Leslie A. Real. Chicago: University of Chicago Press, 1994: 371–389; Robert M. Seyfarth and Dorothy L. Cheney. "Some General Features of Vocal Development in Nonhuman Primates." In *Social Influences on Vocal Development,* ed. Charles T. Snowdon and Martine Hausberger. Cambridge, U.K.: Cambridge University Press, 1996: 249–273.

6. THREADING A NEW TAPESTRY

1. R. A. Hinde and J. Fisher. "Further Observations on the Opening of Milk Bottles by Birds." *British Birds* 44 (1951): 393–396. See also: D. F. Sherry and B. G. Galef. "Social Learning without Imitation: More about Milk Bottle Opening by Birds." *Animal Behavior* 40 (1990): 987–989.

2. J. D. Goss-Custard and W. J. Sutherland. "Feeding Specializations in Oystercatchers *Haematopus Ostralegus." Animal Behavior,* February 1984: 299–301.

3. Douglass H. Morse. *Behavioral Mechanisms in Ecology.* Cambridge, Mass.: Harvard University Press, 1980: 107; P. B. Heppelston. "The Comparative Ecology of Oystercatchers *(Haematopus Ostralegus L.)* in Inland and Coastal Habitats." *Journal of Animal Ecology* 41 (1972): 23–51.

4. S. J. McNaughton, F. F. Banyikwa, and M. M. McNaughton. "Promotion of the Cycling of Diet-Enhancing Nutrients by African Grazers." *Science,* December 5, 1997: 1788–1800.

5. Gretchen Vogel. "Chimps in the Wild Show Stirrings of Culture." *Science,* June 25, 1999: 2070–2073; A. Whiten, J. Goodall, W. C. McGrew, T. Nishida, V. Reynolds, Y. Sugiyama, C. E. G. Tutin, R. W. Wrangham, and C. Boesch. "Cultures in Chimpanzees." *Nature,* June 17, 1999: 682–685.

6. Steven Mithen. *The Prehistory of the Mind: The Cognitive Origins of Art, Religion and Science.* London: Thames and Hudson, 1996: 76.

7. Richard W. Wrangham, Frans B. M. de Waal, and W. C. McGrew. "The Challenge of Behavioral Diversity." In *Chimpanzee Cultures,* ed. Richard W. Wrangham, W. C. McGrew, Frans B. M. de Waal, and Paul G. Heltne with assistance from Linda A. Marquardt. Cambridge, Mass.: Harvard University Press, 1994: 11.

8. N. Inoue-Nakamura and T. Matsuzawa. "Development of Stone Tool Use by Wild Chimpanzees *(Pan Troglodytes)." Journal of Comparative Psychology,* June 1997: 159–173.

9. Tetsuro Matsuzawa. "Field Experiments on Use of Stone Tools by Chimpanzees in the Wild." In *Chimpanzee Cultures,* ed. Richard W. Wrangham, W. C. McGrew, Frans B. M. de Waal, and Paul G. Heltne with assistance from Linda A. Marquardt. Cambridge, Mass.: Harvard University Press, 1994: 356.

10. K. R. L. Hall. "Tool-Using Performances as Indicators of Behavioral Adaptability." In *Primates: Studies in Adaptation and Variability,* ed. Phyllis C. Jay. New York: Holt, Rinehart and Winston, 1968: 134–138.

11. William H. Calvin. *The Throwing Madonna: Essays on the Brain.* New York: McGraw-Hill, 1983.

12. Steven Mithen. *The Prehistory of the Mind:* 24–26.

13. N. Toth, K. D. Schick, and E. S. Savage-Rumbaugh. "Pan the Tool-Maker: Investigations into the Stone Tool-Making and Tool-Using Capabilities of a Bonobo *(Pan paniscus)." Journal of Archaeological Science,* January 1993: 81–91.

14. M. W. Marzke, K. L. Wullstein, and S. F. Viegas. "Evolution of the Power ('Squeeze') Grip and Its Morphological Correlates in Hominids." *American Journal of Physical Anthropology,* November 1992: 283–298; Mary W. Marzke, N. Toth, and K. N. An. "EMG Study of Hand Muscle Recruitment during Hard Hammer Percussion Manufacture of Oldowan Tools." *American Journal of Physical Anthropology,* March 1998: 315–332.

15. Ann Gibbons. "Tracing the Identity of the First Toolmakers." *Science,* April 4, 1997: 32–40.

16. Ann Gibbons. "In China, a Handier *Homo Erectus." Science,* March 13, 1998: 1636.

17. Clive Gamble. *Timewalkers: The Prehistory of Global Colonization.* Cambridge, Mass.: Harvard University Press, 1994.

18. M. F. Hammer, T. Karafet, A. Rasanayagam, et al. "Out of Africa and Back Again: Nested Cladistic Analysis of Human Y Chromosome Variation." *Molecular Biology and Evolution* 15 (1998): 427–441; A. R. Templeton. "Out of Africa? What Do Genes Tell Us?" *Current Opinion in Genetics and Development* 7 (1997): 841–847; Kate Wong. "Is Out of Africa Going Out the Door?" *Scientific American,* August 1999. http://www.sciam.com/1999/0899issue/0899infocus.html. Downloaded July 1999.

19. Clive Gamble. *Timewalkers:* 121; Valerius Geist. *Life Strategies, Human Evolution, Environmental Design: Toward a Biological Theory of Health.* New York: Springer-Verlag, 1978.

20. Richard Potts. *Humanity's Descent.* New York: Avon Books, 1996; Richard Potts. "Environmental Hypotheses of Hominid Evolution." *Yearbook of Physical Anthropology.* New York: John Wiley and Sons, 1998; Bruce Bower. "Humanity's Imprecision Vision," *Science News,* July 12, 1997: 26–27.

21. Ann Gibbons. "In China, a Handier Homo erectus." *Science:* 1636; Clive Gamble. *Timewalkers:* 132; H. L. Movius. "The Lower Paleolithic Cultures of Southern and Eastern Asia." *Transactions of the American Philosophical Society* 38 (1948); G. G. Pope and J. E. Cronin. "The Asian *Hominidae.*" *Journal of Human Evolution* 13 (1984): 377–396.

22. P. Y. Sondaar. "Pleistocene Man and Extinctions of Island Endemics." *Societe Geologique de France. Memoires* 150 (1987): 159–165.

23. M. J. Morwood, P. B. O'Sullivan, F. Aziz, and A. Raza. "Fission-Track Ages of Stone Tools and Fossils on the East Indonesian Island of Flores." *Nature,* March 12, 1998: 173–176; Mark Rose. "First Mariners." *Archaeology,* May–June 1998. http://www.archaeology.org/9805/newsbriefs/mariners.html/. May 1999; Ann Gibbons. "Ancient Island Tools Suggest *Homo Erectus* Was a Seafarer." *Science,* March 13, 1998: 1635–1637.

24. Arguments have long been advanced that early hominids were scavengers, not hunters, and were not adept at spear use, but evidence unearthed in 1995 tends to indicate this is far from true. (Hartmut Thieme. "Lower Paleolithic Hunting Spears from Germany." *Nature,* February 27, 1997: 807–810; B. Bower. "German Mine Yields Ancient Hunting Spears." *Science News,* March 1, 1997: 134. B. Bower. "Early Humans Make Their Marks as Hunters." *Science News,* April 12, 1997: 222; "Stone Age Wood." *Discover,* June 1997: 14, 18.)

25. Nicholas Toth, Desmond Clark, and Giancarlo Ligabue. "The Last Stone Ax Makers." *Scientific American,* July 1992: 93.

26. Malcolm T. Smith and Robert Layton. "Still Human after All These Years." *The Sciences,* January–February 1989: 11.

27. Steven Mithen. *The Prehistory of the Mind:* 24–26; Clive Gamble. *Timewalkers:* 71–72.

28. Shirley C. Strum. *Almost Human: A Journey into the World of Baboons.* New York: Random House, 1987: 129–133.

29. Jane Goodall. *In the Shadow of Man.* 1971. Reprint, Boston: Houghton Mifflin, 1983: 29–30.

30. Olga Soffer. *The Upper Paleolithic of the Central Russian Plain.* New York: Academic Press, 1985.

31. James M. Adovasio, Olga Soffer, and Bohuslav Klima. "Upper Palaeolithic Fibre Technology: Interlaced Woven Finds from Pavlov I, Czech Republic, c. 26,000 Years Ago." *Antiquity; a Quarterly Review of Archaeology.* September 1996.

32. C. R. Harington. "Wooly Mammoth." In *Animals of Beringia.* Yukon Beringia Interpretive Centre. http://www.beringia.com/01student/mainb2.html#top. December 1995. For a picture of a reconstructed hut made of mammoth bones and skin, see: "Mammoth Bones Saved Ice Age Humans from Winter's Chill." Chicago: Field Museum. http://www.fmnh.org/exhibits/ttt/TTT4a.html. May 1999.

33. Randall White. "The Dawn of Adornment." *Natural History,* May 1993: 61–66.

34. Melvin Konner. *Why the Reckless Survive . . . and Other Secrets of Human Nature.* New York: Viking, 1990: 4.

35. Terrence Deacon. "The Neural Circuitry Underlying Primate Calls and Human Language." In *Language Origin: A Multidisciplinary Approach,* ed. J. Wind, B. Chiarelli, B. Bichakhian, and A. Nocentini. Dordrecht: Kluwer Academic Publishing, 1992: 121–162.

36. Robin Dunbar. *Grooming, Gossip and the Evolution of Language.* London: Faber and Faber, 1996: 110–115; Steven Mithen. *The Prehistory of the Mind:* 208–209; Leslie Aiello. "Terrestriality, Bipedalism and the Origin of Language." In *The Evolution of Social Behaviour Patterns in Primates and Man,* ed. J. Maynard Smith. London: Proceedings of the British Academy, 1996; Steven Pinker. *The Language Instinct.* New York: William Morrow, 1994: 353; Constance Holden. "No Last Word on Language Origins." *Science,* November 20, 1998: 1455.

37. Phillip Tobias. "The Brain of *Homo Habilis:* A New Level of Organisation in Cerebral Evolution." *Journal of Human Evolution* 16 (1987): 741–761; Steven Mithen. *The Prehistory of the Mind:* 110.

38. Richard Dawkins. *The Selfish Gene.* New York: Oxford University Press, 1976: 206.

39. Other experts on the meme—like Richard Brodie and Aaron Lynch—have followed in Dawkins's path. For example, all of the examples of memes Aaron Lynch gives in his book *Thought Contagion: How Belief Spreads through Society* (New York: Basic Books, 1996), or in papers such as Lynch's "Units, Events and Dynamics in Memetic Evolution" (*Journal of Memetics—Evolutionary Models of Information Transmission* 2 [1998]; http://www.cpm.mmu.ac.uk/jom-emit/1998/vol2/lynch_a.html) relate solely to human beings.

40. Antoine Bechara, Daniel Tranel, Hanna Damasio, Ralph Adolphs, Charles Rockland, and Antonio R. Damasio. "Double Dissociation of Conditioning and Declarative Knowledge Relative to the Amygdala and Hippocampus in Humans." *Science,* August 25, 1995: 1115–1118; Trevor W. Robbins. "Refining the Taxonomy of Memory." *Science,* September 6, 1996: 1353–1354; Felicia B. Gershberg. "Implicit and Explicit Conceptual Memory following Frontal Lobe Damage." *Journal of Cognitive Neuroscience,* January 1, 1997: 105; Brett K. Hayes and Ruth Hennessy. "The Nature and Development of Nonverbal Implicit Memory." *Journal of Experimental Child Psychology,* October 1, 1996: 22; Howard Eichenbaum. "How Does the Brain Organize Memories?" *Science,* July 18, 1997: 330–331; F. Vargha-Khadem, D. G. Gadian, K. E. Watkins, A. Connelly, W. Van Paesschen, and M. Mishkin. "Differential Effects of Early Pathology on Episodic and Semantic Memory." *Science,* July 18, 1997: 376–379; Bruce Schechter. "How the Brain Gets Rhythm." *Science,* October 18, 1996: 339–340; Richard M. Restak, M.D. *The Modular Brain: How New Discoveries in Neuroscience Are Answering Age-Old Questions about Memory, Free Will, Consciousness, and Personal Identity.* New York: Charles Scribner's Sons, 1994: 80; Minouche Kandel and Eric Kandel. "Flights of Memory." *Discover,* May 1994: 36–37.

41. Steven Pinker. *The Language Instinct:* 308–311, 353.

42. See the concept of "consensual validation" developed by interpersonal psychiatrist and social psychologist Harry Stack Sullivan. (F. Barton Evans III. *Harry Stack Sullivan: Interpersonal Theory and Psychotherapy.* London: Routledge, 1996.)

7. A TRIP THROUGH THE PERCEPTION FACTORY

1. "From a semiotic perspective . . . culture includes . . . constitutive meanings that serve to define and thereby create realities." Joan G. Miller. "Culture and the Self: Uncovering the Cultural Grounding of Psychological Theory." In *The Self across Psychology: Self-Recognition, Self-Awareness, and the Self Concept,* ed. Joan Gay Snodgrass and Robert L. Thompson. New York: New York Academy of Sciences: 1997: 219.

2. J. S. L. Gilmour. "Taxonomy and Philosophy." In *The New Systematics,* ed. Julian Huxley. Oxford: Oxford University Press, 1940: 467. The quotes I've used are paraphrases provided by David L. Hull in his *Science as a Process: An Evolutionary Account of the Social and Conceptual Development of Science* (Chicago: University of Chicago Press, 1988: 104).

3. Francisco J. Varela. "Autonomy and Autopoiesis." In *Self-organizing Systems: An Interdisciplinary Approach,* ed. Gerhard Roth and Helmut Schwegler. New York: Campus Verlag, 1981: 15.

4. Humberto R. Maturana. "The Organization of the Living: A Theory of the Living Organization." *International Journal of Man-Machine Studies* 7 (1975): 316.

5. "Introduction to Social Constructionism." Hausarbeiten.de.http://www.hausarbeiten.de/data/psychologie/psychosocial-constructionism.html. Downloaded June 1999.

6. Humberto R. Maturana. "Reality: The Search for Objectivity or the Quest for a Compelling Argument." *Irish Journal of Psychology* 9:1 (1988): 25–82.

7. Wilder Penfield. "Consciousness, Memory and Man's Conditioned Reflexes." In *On the Biology of Learning,* ed. Karl H. Pribram. New York: Harcourt, Brace and World, 1969.

8. Israel Rosenfield. *The Invention of Memory: A New View of the Brain.* New York: Basic Books, 1988: 202–203.

9. Bertram H. Raven and Jeffrey Z. Rubin. *Social Psychology.* New York: John Wiley and Sons, 1983: 96.

10. Stanley Coren, Clare Porac, and Lawrence M. Ward. *Sensation and Perception.* New York: Academic Press, 1979: 419.

11. Jerome S. Bruner. *Beyond the Information Given: Studies in the Psychology of Knowing.* New York: W. W. Norton, 1973: 10, 96.

12. Humberto R. Maturana. "Reality." *Irish Journal of Psychology:* 25–82.

13. Michael J. Berry II, Iman H. Brivanlou, Thomas A. Jordan, and Markus Meister. "Anticipation of Moving Stimuli by the Retina." *Nature,* March 25, 1999: 334–338; Stelios M. Smirnakis, Michael J. Berry, David K. Warland, William Bialek, and Markus Meister. "Adaptation of Retinal Processing to Image Contrast and Spatial Scale." *Nature,* March 6, 1997: 69–73; Sang-Hun Lee and Randolph Blake. "Visual Form Created Solely from Temporal Structure." *Science,* May 14, 1999: 1165–1168.

14. See, for example, D. H. Hubel and T. N. Wiesel. "Anatomical Demonstration of Columns in the Monkey Striate Cortex." *Nature,* February 22, 1969: 747–750.

15. J. J. Kulikowski and T. R. Vidyasagar. "Space and Spatial Frequency: Analysis and Representation in the Macaque Striate Cortex." *Experimental Brain Research* 64:1 (1986): 5–18.

16. John McCrone. *The Ape That Spoke: Language and the Evolution of the Human Mind.* New York: William Morrow, 1990: 52–58.

17. E. Mellet, L. Petit, B. Mazoyer, M. Denis, and N. Tzourio. "Reopening the Mental Imagery Debate: Lessons from Functional Anatomy." *Neuroimage,* August 1998: 129–139.

18. I. Fried, K. A. MacDonald, and C. L. Wilson. "Single Neuron Activity in Human Hippocampus and Amygdala during Recognition of Faces and Objects." *Neuron,* May 1997: 753–765.

19. For the location of the conscious "narrator" we call "self" in the left cortical hemisphere, see, for example, Michael S. Gazzaniga. "Organization of the Human Brain." *Science,* September 1, 1989: 947–952.

20. I. Kohler. "Experiments with Goggles." *Scientific American* 206 (1962): 62–86.

21. Plato. *Republic.* In *Library of the Future.* 4th ed., ver. 5.0. Irvine, Calif.: World Library, 1996. CD-ROM.

22. Michael S. Gazzaniga. *The Social Brain: Discovering the Networks of the Mind.* New York: Basic Books, 1985: 72.

23. Michael D. Gershon. *The Second Brain: The Scientific Basis of Gut Instinct.* New York: Harper-Collins, 1998.

24. P. V. Simonov. "The 'Bright Spot of Consciousness.'" *Zhurnal Vysshei Nervnoi Deiatelnosti Imeni I. P. Pavlova,* November–December 1990: 1040–1044.

25. J. Singh and R. T. Knight. "Effects of Posterior Association Cortex Lesions on Brain Potentials Preceding Self-Initiated Movements." *Journal of Neuroscience,* May 1993: 1820–1829.

26. Michael S. Gazzaniga. "Organization of the Human Brain." *Science:* 947–952.

27. Humberto R. Maturana. "Reality." *Irish Journal of Psychology:* 25–82.

28. Michael S. Gazzaniga. *Nature's Mind: The Biological Roots of Thinking, Emotions, Sexuality, Language, and Intelligence.* New York: Basic Books, 1992: 121–133.

29. Jerome Kagan. *Unstable Ideas: Temperament, Cognition and Self.* Cambridge, Mass.: Harvard University Press, 1989: 52.

30. "Introduction to Social Constructionism." Hausarbeiten.de.http://www.hausarbeiten. de/data/psychologie/psychosocial-constructionism.html. Downloaded June 1999.

8. REALITY IS A SHARED HALLUCINATION

1. M. J. C. Waller. Personal communication. May 1996; M. J. C. Waller. "Organization Theory and the Origins of Consciousness." *Journal of Social and Evolutionary Systems* 19:1 (1996): 17–30; T. Burns and G. M. Stalker. *The Management of Innovation.* London: Tavistock, 1961: 92–93, 233–234.

2. Bruce Bower. "Brain Faces Up to Fear, Social Signs." *Science News,* December 17, 1994: 406; Eric R. Kandel and Robert D. Hawkins. "The Biological Basis of Learning and Individuality." *Scientific American,* September 1992: 78–87; Joseph E. LeDoux. "Emotion, Memory and the Brain." *Scientific American,* June 1994: 50–57.

3. Elizabeth Loftus. *Memory: Surprising New Insights into How We Remember and Why We Forget.* Reading, Mass.: Addison-Wesley, 1980: 45–49.

4. Solomon E. Asch. "Studies of Independence and Conformity: I. A Minority of One against a Unanimous Majority." *Psychological Monographs* 70:9 (1956): (whole no. 416); Bertram H. Raven and Jeffrey Z. Rubin. *Social Psychology.* New York: John Wiley and Sons, 1983: 566–569, 575.

5. C. Faucheux and S. Moscovici. "Le Style de Comportement d'Une Minorité et Son Influence sur les Réponses d'Une Majorité." *Bulletin du Centre d'Études et Recherches Psychologiques* 16 (1967): 337–360; S. Moscovici, E. Lage, and M. Naffrechoux. "Influence of a Consistent Minority on the Responses of a Majority in a Color Perception Task." *Sociometry* 32 (1969): 365–380; S. Moscovici and B. Personnaz. "Studies in Social Influence. Part V: Minority Influence and Conversion Behavior in a Perceptual Task." *Journal of Experimental Social Psychology* 16 (1980): 270–282; Bertram H. Raven and Jeffrey Z. Rubin. *Social Psychology:* 584–585.

6. See, for example, L. Eisenberg. "The Social Construction of the Human Brain." *American Journal of Psychiatry* 152:11 (1995): 1563–1575.

7. John Skoyles. "Mirror Neurons and the Motor Theory of Speech." *Noetica* (1998). http://psy.uq.edu.au/CogPsych/Noetica/OpenForumIssue9/.

8. See, for example, Janet F. Werker and Renee N. Desjardins. "Listening to Speech in the First Year of Life: Experiential Influences on Phoneme Perception." *Current Directions in Psychological Science,* June 1995: 76–81; Janet F. Werker. "Becoming a Native Listener." *American Scientist,* January–February 1989: 54–59; and Janet F. Werker and Richard C. Tees. "The Organization and Reorganization of Human Speech Perception." *Annual Review of Neuroscience* 15 (1992): 377–402.

9. Erkki Ruoslahti. "Stretching Is Good for a Cell." *Science,* May 30, 1997: 1345–1346.

10. See, for example, C. Haanen and I. Vermes. "Apoptosis: Programmed Cell Death in Fetal Development." *European Journal of Obstetrics, Gynecology, and Reproductive Biology,* January 1996: 129–133; Steve Nadis. "Kids' Brainpower: Use It or Lose It." *Technology Review,* November-December 1993: 19–20.

11. See, for example, Christiana M. Leonard, Linda J. Lombardino, Laurie R. Mercado, Samuel R. Browd, Joshua I. Breier, and O. Frank Agee. "Cerebral Asymmetry and Cognitive Development in Children: A Magnetic Resonance Imaging Study." *Psychological Science,* March 1996: 93; S. Scarr. "Theoretical Issues in Investigating Intellectual Plasticity." In *Plasticity of Development,* ed. S. E. Brauth, W. S. Hall, and R. J. Dooling. Cambridge, Mass.: MIT Press, 1991: 57–71.

12. Without training, guidance, or positive reinforcement, newborns automatically begin to imitate their fellow humans during their first hours out of the womb. (W. Wyrwicka. "Imitative Behavior: A Theoretical View." *Pavlovian Journal of Biological Sciences,* July–September 1988: 125–131.)

13. R. L. Fantz. "Visual Perception from Birth as Shown by Pattern Selectivity." *Annals of the New York Academy of Sciences* 118 (1965): 793–814; Stanley Coren, Clare Porac, and Lawrence M. Ward. *Sensation and Perception.* New York: Academic Press, 1979: 379–380.

14. Linnda R. Caporael. "Sociality: Coordinating Bodies, Minds and Groups." Downloaded from http://www.ai.univie.ac.at/cgi-bin/mfs/31/wachau/www/archives/Psycoloquy/1995.V6/0043.html/?84#mfs,95/6/01. A baby begins imitating others when it is less than a week old. T. G. R. Bower. *A Primer of Infant Development.* New York: W. H. Freeman, 1977: 28.

15. Margaret Mead. *Sex and Temperament in Three Primitive Societies.* London: Routledge and Kegan Paul, 1977.

16. See, for example, Paul Ekman. "Facial Expressions of Emotion: An Old Controversy and New Findings." *Philosophical Transactions of the Royal Society of London. Series B: Biological Sciences,* January 29, 1992: 63–69.

17. M. L. Hoffman. "Is Altruism Part of Human Nature?" *Journal of Personality and Social Psychology* 40:1 (1981): 121–137; Bertram H. Raven and Jeffrey Z. Rubin. *Social Psychology:* 311–312.

18. M. L. Hoffman. "Is Altruism Part of Human Nature?" *Journal of Personality and Social Psychology:* 121–137; Bertram H. Raven and Jeffrey Z. Rubin. *Social Psychology:* 311–312.

19. D. Bischof-Köhler. "Self-object and Interpersonal Emotions: Identification of Own Mirror Image, Empathy and Prosocial Behavior in the Second Year of Life." *Zeitschrift für Psychologie mit Zeitschrift für Angewandte Psychologie* 202:4 (1994): 349–377.

20. Bruce M. Hood, J. Douglas Willen, and Jon Driver. "Adults' Eyes Trigger Shifts of Visual Attention in Human Infants." *Psychological Science,* March 1998: 131–133; Herbert Terrace. "Thoughts without Words." In *Mindwaves: Thoughts on Intelligence, Identity and Consciousness,* ed. Colin Blakemore and Susan Greenfield. Oxford: Basil Blackwell, 1989: 128–129.

21. Jerome Bruner. *Actual Minds, Possible Worlds.* Cambridge, Mass.: Harvard University Press, 1986: 60, 67–68; Uta Frith. "Autism." *Scientific American,* June 1993: 108–114.

22. Jerome Kagan. *Unstable Ideas: Temperament, Cognition and Self.* Cambridge, Mass.: Harvard University Press, 1989: 185–186. In the body of psychological literature, the effect we're discussing is called "social referencing." According to C. L. Russell, K. A. Bard, and L. B. Adamson, "It is a well-documented ability in human infants." (C. L. Russell, K. A. Bard, and L. B. Adamson. "Social Referencing by Young Chimpanzees *(Pan Troglodytes)*." *Journal of Comparative Psychology,* June 1997: 185–193.) For more on social referencing in infants as young as 8.5 months old, see: J. J. Campos. "A New Perspective on Emotions." *Child Abuse and Neglect* 8:2 (1984): 147–156.

23. But let's not get too homocentric. Rats flock just as madly to the imitative urge. Put them with others who love a beverage that they loathe and their tastes will also change dramatically. (B. G. Galef Jr, E. E. Whiskin, and E. Bielavska. "Interaction with Demonstrator Rats Changes Observer Rats' Affective Responses to Flavors." *Journal of Comparative Psychology,* December 1997: 393–398.)

24. Barbara Kantrowitz and Pat Wingert. "How Kids Learn." *Newsweek,* April 17, 1989: 53.

25. William S. Condon. "Communication: Rhythm and Structure." In *Rhythm in Psychological, Linguistic and Musical Processes,* ed. James R. Evans and Manfred Clynes. Springfield, Ill.: C. C. Thomas, 1986: 55–77; William S. Condon. "Method of Microanalysis of Sound Films of Behavior." *Behavior Research Methods, Instruments and Computers* 2:2 (1970): 51–54.

26. William S. Condon. Personal communication. June 10, 1999. For information indicating the probability of related forms of synchrony, see: M. Krams, M. F. Rushworth, M. P. Deiber, R. S. Frackowiak, and R. E. Passingham. "The Preparation, Execution and Suppression of Copied Movements in the Human Brain." *Experimental Brain Research,* June 1998: 386–398; and L. O. Lundqvist. "Facial EMG Reactions to Facial Expressions: A Case of Facial Emotional Contagion?" *Scandinavian Journal of Psychology,* June 1995: 130–141.

27. William S. Condon and Louis W. Sander. "Neonate Movement Is Synchronized with Adult Speech: Interactional Participation and Language Acquisition." *Science* 183:4120 (1974): 99–101.

28. Edward T. Hall. *Beyond Culture*. New York: Anchor Books, 1977: 72–77.

29. H. H. Kelley. "The Warm-Cold Variable in First Impressions of Persons." *Journal of Personality* 18 (1950): 431–439; Bertram H. Raven and Jeffrey Z. Rubin. *Social Psychology:* 88–89.

30. Donna Eder and Janet Lynne Enke. "The Structure of Gossip: Opportunities and Constraints on Collective Expression among Adolescents." *American Sociological Review,* August 1991: 494–508.

31. J. Z. Rubin, F. J. Provenzano, and Z. Luria. "The Eye of the Beholder: Parents' Views on Sex of Newborns." *American Journal of Orthopsychiatry* 44 (1974): 512–519; Bertram H. Raven and Jeffrey Z. Rubin. *Social Psychology:* 512.

32. P. A. Goldberg. "Are Women Prejudiced against Women?" *Transaction,* April 1968: 28–30; Bertram H. Raven and Jeffrey Z. Rubin. *Social Psychology:* 518.

33. *Webster's Revised Unabridged Dictionary,* ed. Noah Porter. G. & C. Merriam, 1913; DICT Development Group. http://www.dict.org. Downloaded June 1999.

34. Steven Morgan Friedman. "Etymologically Speaking." www.westegg.com/etymology/. Downloaded June 1999.

35. Merriam-Webster. WWWebster.com.http://www.m-w.com/netdict.html. Downloaded June 1999.

36. "Feminism/Terms." Ver. 1.5, last modified February 15, 1993. http://www.cis.ohio-state.edu/hypertext/faq/usenet/feminism/terms/faq.html. Downloaded June 11, 1999.

37. Jerome S. Bruner. *Beyond the Information Given: Studies in the Psychology of Knowing.* New York: W. W. Norton, 1973: 380–386; Paul van Geert. "Green, Red and Happiness: Towards a Framework for Understanding Emotion Universals." *Culture and Psychology,* June 1995: 264.

38. W. Bogoras. *The Chukchee.* New York: G. E. Stechert, 1904–1909; Jerome S. Bruner. *Beyond the Information Given:* 102–103.

39. Jared Diamond. "This Fellow Frog, Name Belong-Him Dakwo." *Natural History,* April 1989: 16–23.

40. Linnda R. Caporael. "Sociality: Coordinating Bodies, Minds and Groups." *Psycoloquy.* Downloaded from http://www.ai.univie.ac.at/cgi-bin/mfs/31/wachau/www/archives/Psycoloquy/1995.V6/0043.html?84#mfs, 95/6/01.

41. Peter N. Stearns. "The Rise of Sibling Jealousy in the Twentieth Century." In *Emotion and Social Change: Toward a New Psychohistory,* ed. Carol Z. Stearns and Peter N. Stearns. New York: Holmes and Meier, 1988: 197–209.

42. Robert B. Edgerton. *Sick Societies: Challenging the Myth of Primitive Harmony.* New York: Free Press, 1992: 49–50; R. M. Jones. "Why Did the Tasmanians Stop Eating Fish?" In *Explorations in Ethnoarchaeology,* ed. R. A. Gould. Albuquerque: University of New Mexico Press, 1978: 11–48.

43. Daniel J. Boorstin. *The Discoverers: A History of Man's Search to Know His World and Himself.* New York: Vintage Books, 1985: 344–357.

9. THE CONFORMITY POLICE

1. The quoted phrase is a summary of Eibl-Eibesfeldt's conclusions provided by Melvin Konner. Melvin Konner. *Why the Reckless Survive . . . and Other Secrets of Human Nature.* New York: Viking, 1990: 213; Irenaus Eibl-Eibesfeldt. *Human Ethology.* New York: Aldine de Gruyter, 1989.

2. Harry F. Harlow and Gary Griffin. "Induced Mental and Social Deficits in Rhesus Monkeys." In *Biological Basis of Mental Retardation,* ed. Sonia F. Osler and Robert E. Cooke. Baltimore: Johns Hopkins Press, 1965.

3. Shirley C. Strum. *Almost Human: A Journey into the World of Baboons.* New York: Random House, 1987: 64.

4. Niko Tinbergen. *The Herring Gull's World: A Study of the Social Behavior of Birds.* New York: Basic Books, 1961: 49–51.

5. Dian Fossey. *Gorillas in the Mist.* Boston: Houghton Mifflin, 1983: 82, 100–101.

6. Dennis C. Harper. "Children's Attitudes toward Physical Disability in Nepal: A Field Study." *Journal of Cross-Cultural Psychology,* November 1997: 710–729.

7. Daniel Chernin. "China: A Love Story." *World Monitor,* September 1989: 51.

8. Ruth Benedict. *Patterns of Culture.* 1934. Reprint, New York: New American Library, 1950: 199–200.

9. George Ostrogorsky. *History of the Byzantine State.* Trans. Joan Hussey. New Brunswick, N.J.: Rutgers University Press, 1969: 127, 140–145.

10. David P. Barash. *The Hare and the Tortoise: Culture, Biology, and Human Nature.* New York: Penguin Books, 1987: 144.

11. J. H. Langlois, L. A. Roggman, and L. S. Vaughn. "Facial Diversity and Infant Preferences for Attractive Faces." *Developmental Psychology* 27 (1991): 79–84; Judith H. Langlois, Lori A. Roggman, and Lisa Musselman. "What Is Average and What Is Not Average about Attractive Faces." *Psychological Science,* July 1994: 214; Gillian Rhodes and Tanya Tremewan. "Averageness, Exaggeration, and Facial Attractiveness." *Psychological Science,* March 1996: 105–110.

12. Constance Holden. "Ordinary Is Beautiful." *Science,* April 20, 1990: 306. Experiments indicating that we judge beauty by the symmetry of a face have been widely publicized. However, a careful piece of research in which symmetry was balanced against the extent to which a face represented a midpoint of the surrounding population showed that a face reflecting the group norm is far more appealing than one which is merely symmetrical. (Gillian Rhodes, Alex Sumich, and Graham Byatt. "Are Average Facial Configurations Attractive Only Because of Their Symmetry?" *Psychological Science,* January 1999: 52–58.)

13. Qi Dong, Glenn Weisfeld, Ronald H. Boardway, and Jiliang Shen. "Correlates of Social Status among Chinese Adolescents." *Journal of Cross-cultural Psychology,* July 1996: 476–477.

14. Jerome Kagan. *Unstable Ideas: Temperament, Cognition and Self.* Cambridge, Mass.: Harvard University Press, 1989: 223–228; Richard M. Restak, M.D. *The Mind.* New York: Bantam Books, 1988: 65.

15. M. J. Boulton and K. Underwood. "Bully/Victim Problems among Middle School Children." *British Journal of Educational Psychology* 62, pt. 1 (February 1992): 73–87.

16. M. Mushinski. "Violence in America's Public Schools." *Statistical Bulletin—Metropolitan Insurance Companies,* April–June 1994: 2–9.

17. Paul Chance. "Kids without Friends." *Psychology Today,* January–February 1989: 29–30.

18. Ibid.

19. Clyde V. Prestowitz Jr. *Trading Places: How We Allowed Japan to Take the Lead.* New York: Basic Books, 1988: 84; Suvendrini Kakuchi. "Japan—Education: Rampant Bullying Reflects Social Problems." Inter Press Service English News Wire, December 14, 1995.

20. Carol Gilligan. *In a Different Voice: Psychological Theory and Women's Development.* Cambridge, Mass.: Harvard University Press, 1982.

21. Daniel G. Freedman. *Human Sociobiology: A Holistic Approach.* New York: Free Press, 1979: 49.

22. Roberta L. Paikoff and Ritch C. Savin-Williams. "An Exploratory Study of Dominance Interactions among Adolescent Females at a Summer Camp." *Journal of Youth and Adolescence,* October 1983: 419–433; Daniel G. Freedman. *Human Sociobiology:* 47–49.

23. Tad Szulc. *Fidel: A Critical Portrait.* New York: William Morrow, 1986: 111–112.

24. Antonia Fraser. *Cromwell.* New York: Donald I. Fine, 1973.

25. Max Weber. *Charisma and Institution Building.* Chicago: University of Chicago Press, 1968: 178.

26. Allen W. Johnson and Timothy Earle. *The Evolution of Human Societies: From Foraging Group to Agrarian State.* Stanford, Calif.: Stanford University Press, 1987: 290.

27. Jean L. Briggs. *Never in Anger: Portrait of an Eskimo Family.* Cambridge, Mass.: Harvard University Press, 1970; John Klama. *Aggression: The Myth of the Beast Within.* New York: John Wiley and Sons, 1988: 64–66.

28. Daniel Bell, James S. Coleman, Alex Inkeles, et al. "Developments in Sociology: Discussion," ed. Andrei S. Markovits. In *Advances in the Social Sciences, 1900–1980: What, Who, Where, How?* ed. Karl W. Deutsch, Andrei S. Markovits, and John Platt. Lanham, Md.: University Press of America, 1986: 53.

29. Paul H. Weaver. *The Suicidal Corporation.* New York: Simon and Schuster, 1988: 98.

30. Charles Darwin. *The Expressions of the Emotions in Man and Animals.* 1872. Reprint, New York: Greenwood Press, 1969: 207.

31. Heinrich Harrer. *Seven Years in Tibet.* Trans. Richard Graves. Los Angeles: Jeremy P. Tarcher, 1982.

32. Norman H. Holland. *Laughing: A Psychology of Humor.* Ithaca, N.Y.: Cornell University Press, 1982: 43–46.

33. Ibid.: 77.

34. Dian Fossey. *Gorillas in the Mist:* 184–185.

35. Frans de Waal. *Peacemaking among Primates.* Cambridge, Mass.: Harvard University Press, 1989: 65.

36. I. M. Piliavin, J. A. Piliavin, and J. Rodin. "Costs, Diffusion, and the Stigmatized Victim." *Journal of Personality and Social Psychology* 32 (1975): 429–438; Bertram H. Raven and Jeffrey Z. Rubin. *Social Psychology.* New York: John Wiley and Sons, 1983: 318.

37. G. Stanley Hall. "A Study of Anger." *American Journal of Psychology* 10 (1899): 543; Carol Tavris. *Anger: The Misunderstood Emotion.* New York: Simon and Schuster, 1984: 68.

38. J. L. Freedman and A. N. Doob. *Deviancy.* New York: Academic Press, 1968; Bertram H. Raven and Jeffrey Z. Rubin. *Social Psychology:* 234.

39. Bertram H. Raven and Jeffrey Z. Rubin. *Social Psychology:* 574.

40. Donald R. Morris. *The Washing of the Spears: A History of the Rise of the Zulu Nation under Shaka and Its Fall in the Zulu War of 1879.* New York: Simon and Schuster, 1986: 36.

41. J. S. Gartlan and C. K. Brain. "Ecology and Social Variability in *Cercopithecus Aethiops* and *C. Mitis.*" In *Primates: Studies in Adaptation and Variability,* ed. Phyllis C. Jay. New York: Holt, Rinehart and Winston, 1968: 274–275; Frans de Waal. *Peacemaking among Primates:* 65.

42. Kaoru Yamamoto. "Children's Ratings of the Stressfulness of Experiences." *Developmental Psychology,* September 1979: 580–581.

43. K. Yamamoto, A. Soliman, J. Parsons, and O. L. Davies Jr. "Voices in Unison: Stressful Events in the Lives of Children in Six Countries." *Journal of Child Psychology and Psychiatry and Allied Disciplines,* November 1987: 855–864; L. L. Dibrell and K. Yamamoto. "In Their Own Words: Concerns of Young Children." *Child Psychiatry and Human Development,* fall 1988: 14–25; Daniel Goleman. "What Do Children Fear Most? Their Answers Are Surprising." *New York Times,* March 17, 1988.

44. Edwin O. Reischauer. *The Japanese.* Cambridge, Mass.: Harvard University Press, 1981: 141.

45. John Winthrop. "Experiencia"; quoted in John Demos. "Shame and Guilt in Early New England." In *Emotion and Social Change: Toward a New Psychohistory,* ed. Carol Z. Stearns and Peter N. Stearns. New York: Holmes and Meier, 1988: 78.

46. John Demos. "Shame and Guilt in Early New England." In *Emotion and Social Change,* ed. Carol Z. Stearns and Peter N. Stearns: 69–85; John Putnam Demos. *Entertaining Satan: Witchcraft and the Culture of Early New England.* New York: Oxford University Press, 1982.

47. Quoted in John Demos. "Shame and Guilt in Early New England." In *Emotion and Social Change,* ed. Carol Z. Stearns and Peter N. Stearns: 80.

10. DIVERSITY GENERATORS: THE HUDDLE AND THE SQUABBLE—GROUP FISSION

1. For an interesting but extremely slim discussion of diversity generation in the organic world, see: Jack Cohen and Ian Stewart. *The Collapse of Chaos: Discovering Simplicity in a Complex World.* New York: Viking, 1994: 129–133.

2. Paul Davies. *The Cosmic Blueprint: New Discoveries in Nature's Creative Ability to Order the Universe.* New York: Simon and Schuster, 1988: 55–56.

3. R. E. Michod, M. F. Wojciechowski, and M. A. Hoelzer. "DNA Repair and the Evolution of Transformation in the Bacterium *Bacillus Subtilis.*" *Genetics,* January 1988: 31–39; David W. Murray. "Toward a Science of Desire." *The Sciences,* July–August 1995: 45–46.

4. Richard A. Kerr. "Geomicrobiology: Life Goes to Extremes in the Deep Earth—and Elsewhere?" *Science,* May 2, 1997: 703–704.

5. William B. Whitman, David C. Coleman, and William J. Wiebe. "Prokaryotes: The Unseen Majority." *Proceedings of the National Academy of Sciences of the United States of America,* June 9, 1998: 6578–6583.

6. See, for example, John D. Berard, Peter Nurnberg, and Jorg T. Epplen. "Male Rank, Reproductive Behavior, and Reproductive Success in Free-ranging Rhesus Macaques." *Primates,* October 1993: 481.

7. B. Bower. "Human Origin Recedes in Australia." *Science News,* September 28, 1996: 196. The figure of 176,000 years ago was obtained by luminescence dating techniques and is highly controversial. The more usual estimate for Australian immigration is between 60,000 and 50,000 years ago. (Ann Gibbons. "Doubts over Spectacular Dates." *Science,* October 10, 1997: 220–222.)

8. Sigmund Freud. *Civilization and Its Discontents.* New York: W. W. Norton, 1989.

9. Emile Durkheim. *The Rules of Sociological Method.* 1895. Reprint, New York: Free Press, 1964: 69.

10. The definitive book on why animals close to each other tend to be foes is Konrad Lorenz's classic *On Aggression.* New York: Harcourt Brace Jovanovich, 1974.

11. Edward O. Wilson. *The Insect Societies.* Cambridge, Mass.: Harvard University Press, 1971: 360, 446–449.

12. W. L. Brown Jr. and E. O. Wilson. "Character Displacement." *Systematic Zoology* 5:2 (1956): 49–64.

13. D. Schluter. "Experimental Evidence That Competition Promotes Divergences in Adaptive Radiation." *Science,* November 4, 1994: 798; Ann Gibbons. "On the Many Origins of Species." *Science,* September 13, 1996: 1498.

14. The traditional view, promoted by Ernst Mayr, is that groups need to be separated by a considerable distance to develop the genetic alterations which lead to speciation—the inability to crossbreed. However, that model has proven to be incorrect, especially among fish. (Ernst Mayr. *Populations, Species, and Evolution.* Cambridge, Mass.: Harvard University Press, 1970; Tom Tregenza and Roger K. Butlin. "Speciation without Isolation." *Nature,* July 22, 1999: 311–312; Virginia Morell. "Ecology Returns to Speciation Studies." *Science,* June 25, 1999: 2106–2108.)

15. C. Sturmbauer and A. Meyer. "Genetic Divergence, Speciation and Morphological Stasis in a Lineage of African Cichlid Fishes." *Nature,* August 13, 1992: 578.

16. Numerous scientists feel, as evolutionary psychiatrists John Price and Anthony Stevens put it, that "human groups, like all groups of social animals . . . thrive and multiply until they reach a critical size." Then resources run low and "all the mechanisms which previously served to promote group solidarity are put into reverse so as to drive . . . subgroups apart." (Anthony Stevens and John Price. *Evolutionary Psychiatry: A New Beginning.* London: Routledge Press, 1996; prepublication draft: 172.)

17. David Smillie. "Human Nature and Evolution: Language, Culture, and Race." Paper given at the Biennial Meeting of the International Society of Human Ethology, Amsterdam, August 1992: 7; Napoleon A. Chagnon. "Male Competition, Forming Close Kin, and Village Fissioning among the Yanomamo Indians." In *Evolutionary Biology and Human Social Behavior: An Anthropological Perspective,* ed. N. A. Chagnon and W. Irons. North Scituate, Mass.: Duxbury, 1979.

18. For extremely good arguments on the limits of inclusive fitness, the theory that "the closer creatures are in the composition of their genes, the more they will help each other," see: Richard

Dawkins. *The Extended Phenotype: The Long Reach of the Gene.* Oxford: Oxford University Press, 1982: 28, 80.

19. Steven B. Johnson and Ronald C. Johnson. "Support and Conflict of Kinsmen: A Response to Hekala and Buell." *Ethology and Sociobiology,* January 1995: 83–89.

20. Napoleon Chagnon. *Yanomamo: The Fierce People.* New York: Holt, Rinehart and Winston, 1968: 126.

21. "Self-esteem Linked to Type-A Risk." *Brain/Mind Bulletin,* July 1990: 6.

22. Diane Swanbrow. "The Paradox of Happiness." *Psychology Today,* July–August 1989: 38. Tellegen's quote is supported by Cecilia Ridgeway. "Status in Groups: The Importance of Motivation." *American Sociological Review* 47 (1982): 76–88.

23. Eric Klinger. *Meaning and Void: Inner Experience and the Incentives in People's Lives.* Minneapolis: University of Minnesota Press, 1977: 161.

24. Theodore D. Kemper. "Power, Status, and Emotions: A Sociological Contribution to a Psychophysiological Domain." In *Approaches to Emotion,* ed. Klaus R. Scherer and Paul Ekman. Hillsdale, N.J.: Lawrence Erlbaum Associates, 1984: 375.

25. J. Rotton, T. Barry, J. Frey, and E. Soler. "Air Pollution and Interpersonal Attraction." *Journal of Applied Social Psychology* 8 (1978): 57–71; S. Cohen and N. Weinstein. "Non-auditory Effects of Noise on Behavior and Health." *Journal of Social Issues* 37 (1981): 36–70; C. J. Holahan. *Environmental Psychology.* New York: Random House, 1982; R. Veitch, R. DeWood, and K. Bosko. "Radio News Broadcasts: Their Effects on Interpersonal Helping." *Sociometry* 40 (1977): 383–386.

26. R. A. Baron and P. A. Bell. "Aggression and Heat: The Influence of Ambient Temperature, Negative Affect and a Cooling Drink on Aggression." *Journal of Personality and Social Psychology,* March 1976: 245–255.

27. C. A. Anderson. "Temperature and Aggression: Ubiquitous Effects of Heat on Occurrence of Human Violence." *Psychological Bulletin,* July 1989: 74–96.

28. C. Haertzen, K. Buxton, L. Covi, and H. Richards. "Seasonal Changes in Rule Infractions among Prisoners: A Preliminary Test of the Temperature-Aggression Hypothesis." *Psychological Reports,* February 1993: 195–200.

29. R. A. Baron and V. M. Ransberger. "Ambient Temperature and the Occurrence of Collective Violence: The 'Long, Hot Summer' Revisited." *Journal of Personality and Social Psychology* 36 (1978): 351–360.

30. Robert E. Thayer. *The Biopsychology of Mood and Arousal.* New York: Oxford University Press, 1989: 87.

31. J. B. Birdsell. "Some Population Problems Involving Pleistocene Man." *Cold Spring Harbor Symposium on Quantitative Biology* 22 (1957): 47–69; Ann Gibbons. "A New Face for Human Ancestors." *Science,* May 30, 1997: 1331–1333; J. M. Bermúdez de Castro, J. L. Arsuaga, E. Carbonell, A. Rosas, I. Martínez, and M. Mosquera. "A Hominid from the Lower Pleistocene of Atapuerca, Spain: Possible Ancestor to Neandertals and Modern Humans." *Science,* May 30, 1997: 1392–1395; Paul M. Dolukhanov. "The Neolithisation of Europe: A Chronological and Ecological Approach." In *The Explanation of Culture Change: Models in Prehistory,* ed. Colin Renfrew. Pittsburgh: University of Pittsburgh Press, 1973: 332. For hunger leading to irritability, the resulting aggression acting as a repulsion mechanism, and the consequence being dispersal, see Valerius Geist. *Life Strategies, Human Evolution, Environmental Design: Toward a Biological Theory of Health.* New York: Springer-Verlag, 1978: 82–83.

32. Steven Mithen. *The Prehistory of the Mind: The Cognitive Origins of Art, Religion and Science.* London: Thames and Hudson, 1996: 139.

33. Allen W. Johnson and Timothy Earle. *The Evolution of Human Societies: From Foraging Group to Agrarian State.* Stanford, Calif.: Stanford University Press, 1987: 84.

34. Heather Pringle. "Ice Age Communities May Be Earliest Known Net Hunters." *Science,* August 29, 1997: 1203–1204.

35. John Scarry. "World Prehistory: The Middle Paleolithic." Chapel Hill, N.C.: Research Laboratories of Archaeology, the University of North Carolina at Chapel Hill. http://www.unc.edu/courses/anth100/midpaleo.html. Downloaded June 1999.

36. Correspondence between Nikolas Lloyd, Bill Benzon, Valerius Geist, and Howard Bloom, August 1998. Elements of this concept also appear in Geist. *Life Strategies, Human Evolution, Environmental Design:* 343−344.

37. Allen W. Johnson and Timothy Earle. *The Evolution of Human Societies:* 56.

38. R. D. Alexander. *The Biology of Moral Systems.* New York: Aldine de Gruyter, 1987.

39. Randall White. "Substantial Acts: From Materials to Meaning in Upper Paleolithic Representation." In *Beyond Art: Upper Paleolithic Symbolism,* ed. D. Stratmann, M. Conkey, and O. Soffer. San Francisco: California Academy of Sciences, 1996; Randall White. "The Dawn of Adornment." *Natural History,* May 1993, 61−66.

40. A. Gilman. "Explaining the Upper Paleolithic Revolution." In *Marxist Perspectives in Archaeology,* ed. M. Spriggs. Cambridge, U.K.: Cambridge University Press, 1984: 115−126.

41. Bruce Bower. "Africa's Ancient Cultural Roots." *Science News,* December 2, 1995: 378.

42. Kathryn Coe. "Art: The Replicable Unit—an Inquiry into the Possible Origin of Art as a Social Behavior." *Journal of Social and Evolutionary Systems* 15:2 (1992): 224.

43. Ibid.

44. B. Bower. "Human Origin Recedes in Australia." *Science News:* 196.

45. C. Knight, C. Power, and I. Watts. "The Human Symbolic Revolution: A Darwinian Account." *Cambridge Archaeological Journal* 5:1 (1995): 75−114.

46. Alexander Marshack. "Evolution of the Human Capacity: The Symbolic Evidence." In *Yearbook of Physical Anthropology.* New York: Wiley-Liss, 1989; Alexander Marshack. "On 'Close Reading' and Decoration versus Notation." *Current Anthropology,* February 1997: 81; Alexander Marshack and Francesco d'Errico. "The La Marche Antler Revisited." *Cambridge Archaeological Journal,* April 1996: 99; Stephen Jay Gould. "Honorable Men and Women." *Natural History,* March 1988: 18.

47. David Smillie. "Group Processes and Human Evolution: Sex and Culture as Adaptive Strategies." Paper presented at the Nineteenth Annual Meeting of the European Sociobiological Society, Alfred, N.Y., July 25, 1996.

48. Steven Pinker. *The Language Instinct.* New York: William Morrow, 1994: 353; Robin Dunbar. *Grooming, Gossip and the Evolution of Language.* London: Faber and Faber, 1996: 132−151; Steven Mithen. *The Prehistory of the Mind:* 185−187.

49. Alexander Marshack and Francesco D'Errico. "On Wishful Thinking and Lunar 'Calendars.'" *Current Anthropology,* August 1989: 491; Alexander Marshack. "The Tai Plaque and Calendrical Notation in the Upper Palaeolithic." *Cambridge Archaeological Journal,* April 1991: 25.

50. Anthony Stevens and John Price. *Evolutionary Psychiatry:* 176.

51. David Smillie. "Human Nature and Evolution": 6, 11.

52. Ibid.: 7−8; Napoleon Chagnon. "Male Competition, Forming Close Kin, and Village Fissioning among the Yanomamo Indians." In *Evolutionary Biology and Human Social Behavior,* ed. N. A. Chagnon and W. Irons: 86−131.

53. David M. Buss. *The Evolution of Desire—Strategies of Human Mating.* New York: Basic Books, 1994.

54. H. T. Reis. "Physical Attractiveness in Social Interaction." *Journal of Personality and Social Psychology* 38 (1980): 604−617; E. Walster, V. Aronson, D. Abrahams, and L. Rottmann. "Importance of Physical Attractiveness in Dating Behavior." *Journal of Personality and Social Psychology* 4 (1966): 508−516.

55. Steven J. Schapiro, Pramod N. Nehete, Jaine E. Perlman, Mollie A. Bloomsmith, and Jagannadha K. Sastry. "Effects of Dominance Status and Environmental Enrichment on Cell-Mediated Immunity in Rhesus Macaques." *Applied Animal Behaviour Science,* March 1998: 319−332.

56. Napoleon Chagnon. "Life Histories, Blood Revenge, and Warfare in a Tribal Population." *Science,* February 1988: 988−989.

57. Charles H. Southwick. "Genetic and Environmental Variables Influencing Animal Aggression." In *Animal Aggression: Selected Readings*, ed. Charles H. Southwick. New York: Van Nostrand Reinhold, 1970; John Paul Scott and John L. Fuller. *Genetics and the Social Behavior of the Dog*. Chicago: University of Chicago Press: 1965.

58. M. J. West-Eberhard. "Sexual Selection, Social Competition, and Speciation." *Quarterly Review of Biology* 58 (1983): 155–183.

59. Norman A. Chance. *The Eskimo of North Alaska*. New York: Holt, Rinehart and Winston, 1966: 65–66, 78; Allen W. Johnson and Timothy Earle. *The Evolution of Human Societies:* 137–138.

60. Scott Antes, of Northern Arizona University and the University of Alaska at Fairbanks, reports that "precontact Eskimos—Alaskan and Siberian—conducted a *lot* of warfare. My Siberian Yupik friends speak often of the days when their ancestors warred with the Chukchi, for example. Yupik Eskimos fought with the Alutiit (Koniag and Chugach). (The Yupiks used to steal Alutiiq women.) Certainly the Unangan (Aleuts) conducted mucho warfare. Etc., etc. The examples are too numerous to list." (Scott Antes. Personal communication. January 25, 1999.) See also: Kaj Birket-Smith and Frederica de Laguna. *The Eyak Indians of the Copper River Delta, Alaska*. Copenhagen: Levin and Munksgaard, 1938: 146; Lydia T. Black. *Glory Remembered: Wooden Headgear of Alaska Sea Hunters*. Juneau: Alaska State Museums, 1991; Ales Hrdlicka. *The Aleutian and Commander Islands and Their Inhabitants*. Philadelphia: Wistar Institute of Anatomy and Biology, 1945: 144–146.

61. David Smillie. "Human Nature and Evolution."

62. C. Knight, C. Powers, and I. Watts. "The Human Symbolic Revolution: A Darwinian Account." *Cambridge Archaeological Journal* 5:1 (1995): 75–114; R. Grun and C. Stringer. "Electron Spin Resonance Dating and the Evolution of Modern Humans." *Archaeometry* 33:2 (1991): 153–199; P. A. Mellars. "The Character of the Middle-Upper Palaeolithic Transition in Southwest France." In *The Explanation of Culture Change: Models in Prehistory*, ed. Colin Renfrew, 267; Pittsburgh: University of Pittsburgh Press, 1973: 267; Steven Mithen. *The Prehistory of the Mind:* 182–183; Randall White. "Visual Thinking in the Ice Age." *Scientific American,* July 1989: 94; Bruce Bower. "Africa's Ancient Cultural Roots." *Science News:* 378.

63. Robert Boyd and Peter J. Richerson. "Life in the Fast Lane: Rapid Cultural Change and the Human Evolutionary Process." In *Origins of the Human Brain,* ed. Jean-Pierre Changeux and Jean Chavaillon. Oxford: Clarendon Press, 1995: 168.

11. THE END OF THE ICE AGE AND THE RISE OF URBAN FIRE

1. This date comes from the "sex-strike" theories of British anthropologist Chris Knight and his colleagues. (C. Knight. *Blood Relations: Menstruation and the Origins of Culture*. New Haven, Conn.: Yale University Press, 1990; Chris Knight. Personal communication. November 11, 1997.)

2. Claude Lévi-Strauss. *The Elementary Structures of Kinship*. Trans. James Harle Bell, John Richard von Sturmer, and Rodney Needham. Boston: Beacon Press, 1969: 69.

3. Christopher Boehm. Personal communication. December 6, 1997; Peter Frost. Personal communication. December 6, 1997; Allen Johnson. Personal communication. December 7, 1997; Lyle Steadman. Personal communication. December 7, 1997.

4. Ruth Benedict. *Patterns of Culture*. 1934. Reprint, New York: New American Library, 1950: 36, 38; Emile Durkheim. *The Elementary Forms of The Religious Life*. Trans. Joseph Ward Swain. 1915. Reprint, New York: Free Press, 1965: 133–159; Ki-baik Lee. *A New History of Korea*. Trans. Edward W. Wagner with Edward J. Shultz. Cambridge, Mass.: Harvard University Press, 1984: 5–7.

5. Emile Durkheim. *The Elementary Forms of the Religious Life:* 156–159.

6. For the dire consequences to those who strayed, see: Marcel Mauss. *Sociology and Psychology: Essays by Marcel Mauss*. Trans. Ben Brewster. London: Routledge and Kegan Paul, 1979: 37–43.

7. Sir George James Frazer. *The New Golden Bough*. Ed. Theodor H. Gaster. 1922. Reprint, Garden City, N.Y.: Doubleday, 1961: 9, 258, 272; Katharine Milton. "*Real Men Don't* Eat Deer." *Discover,* June 1997: 46–53.

8. Olga Soffer. Personal communication. December 7, 1997.

9. For the importance of the intergroup tournament in the enlargement of human social structures, see: Allen W. Johnson and Timothy Earle. *The Evolution of Human Societies: From Foraging Group to Agrarian State.* Stanford, Calif.: Stanford University Press, 1987: 57–60, 101–158.

10. Napoleon Chagnon. *Yanomamo: The Fierce People.* New York: Holt, Rinehart and Winston, 1968: 124–137.

11. Kathleen M. Kenyon. "Excavations at Jericho, 1957–58." *Palestine Excavation Quarterly* 92 (1960): 88–108.

12. Dora Jane Hamblin with C. C. Lamberg-Karlovsky and the editors of Time-Life Books. *The Emergence of Man: The First Cities.* New York: Time-Life Books, 1979: 910.

13. The Catalhoyuk Project. "Catalhoyuk Excavation Data." http://catal.arch.cam.ac.uk/ catal/database/scripts/excavation/feature.idc? March 1999; Dora Jane Hamblin with C. C. Lamberg-Karlovsky and the editors of Time-Life Books. *The Emergence of Man:* 46, 52–54.

14. Dora Jane Hamblin with C. C. Lamberg-Karlovsky and the editors of Time-Life Books. *The Emergence of Man:* 9–10, 46, 52–54.

15. Heather Pringle. "The Slow Birth of Agriculture." *Science,* November 20, 1998: 1446.

16. Michael Balter. "Why Settle Down? The Mystery of Communities." *Science,* November 20, 1998: 1442–1445; Heather Pringle. "The Slow Birth of Agriculture": 1446.

17. D. B. Grigg. *The Agricultural Systems of the World: An Evolutionary Approach.* Cambridge, U.K.: Cambridge University Press, 1974: 11.

18. Manfred Heun, Ralf Schafer-Pregl, Dieter Klawan, Renato Castagna, Monica Accerbi, Basilio Borghi, and Francesco Salamini. "Site of Einkorn Wheat Domestication Identified by DNA Fingerprinting." *Science,* November 14, 1997: 1312–1322; C. Mlot. "Wheat's DNA Points to First Farms." *Science News,* November 15, 1997: 308.

19. Jared Diamond. *Guns, Germs, and Steel: The Fates of Human Societies.* New York: W. W. Norton, 1997: 172–174.

20. For the critical significance of female sexual resistance in the rise of prehistoric culture, see Chris Knight's "sham menstruation/sex-strike theory of the origins of symbolic culture" in C. Knight. *Blood Relations: Menstruation and the Origins of Culture.* New Haven, Conn.: Yale University Press, 1990.

21. For the importance of religious ritual in orchestrating group synchrony, see: Mary Douglas. *Natural Symbols: Explorations in Cosmology.* New York: Pantheon, 1982.

22. James Mellaart. *Catal-Huyuk: A Neolithic Town in Anatolia.* New York: McGraw-Hill, 1967: 223.

23. William Benzon. "Culture as an Evolutionary Arena." *Journal of Social and Evolutionary Systems* 19:4 (1996): 321–362; Maurizio Tosi. "Early Urban Evolution and Settlement Patterns in the Indo-Iranian Borderland." In *The Explanation of Culture Change: Models in Prehistory,* ed. Colin Renfrew. Pittsburgh: University of Pittsburgh Press, 1973: 439–441.

24. Hans Helback. "First Impressions of the Catal Huyuk Plant Husbandry." *Anatolian Studies* 14 (1964): 121–123.

25. Maurizio Tosi. "Early Urban Evolution and Settlement Patterns in the Indo-Iranian Borderland." In *The Explanation of Culture Change,* ed. Colin Renfrew: 443.

26. For a treatment of primitive cities as conductors of flow, see: Manuel de Landa. *A Thousand Years of Nonlinear History.* New York: Zone Books, 1997: 27–28.

27. James Mellaart. *Catal-Huyuk:* 211–216, 224–225.

28. Ibid.: 211–213, 224–225.

29. McGuire Gibson. "Population Shift and the Rise of Mesopotamian Civilisation." In *The Explanation of Culture Change: Models in Prehistory,* ed. Colin Renfrew. Pittsburgh: University of Pittsburgh Press, 1973: 448–450.

30. Lewis Thomas used the word "hypothesis" to describe instinctive insect behaviors in his *Lives of a Cell: Notes of a Biology Watcher* (New York: Bantam Books, 1975).

31. Vernon Reynolds. "Ethology of Social Change." In *The Explanation of Culture Change: Models in Prehistory*, ed. Colin Renfrew. Pittsburgh: University of Pittsburgh Press, 1973: 474–479.

12. THE WEAVE OF CONQUEST AND THE GENES OF TRADE

1. Frans de Waal. *Peacemaking among Primates*. Cambridge, Mass.: Harvard University Press, 1989: 83.

2. Shirley C. Strum. *Almost Human: A Journey into the World of Baboons*. New York: Random House, 1987: 131–132.

3. Frans de Waal. *Peacemaking among Primates*: 12; Jane Goodall. *In the Shadow of Man*. 1971. Reprint, Boston: Houghton Mifflin, 1983: 113–116.

4. F. B. M. de Waal. "The Chimpanzee's Service Economy: Food for Grooming." *Evolution and Human Behavior* 18:6 (1997): 375–386.

5. Frans de Waal. *Peacemaking among Primates*: 82. For exchanges of food between monkeys in which each of the traders seems to keep a mental account of who owes what to whom, see: F. B. de Waal. "Food Transfers through Mesh in Brown Capuchins." *Journal of Comparative Psychology*, December 1997: 370–378.

6. Leda Cosmides and John Tooby. "Cognitive Adaptations for Social Exchange." In *The Adapted Mind: Evolutionary Psychology and the Generation of Culture*, ed. Jerome H. Barkow, Leda Cosmides, and John Tooby. New York: Oxford University Press, 1992: 212.

7. The multi-thousand-mile traditional trade routes over which these treks take place are called the "walkabout." Melville J. Herskovitz. *Economic Anthropology: The Economic Life of Primitive Peoples*. 1940. Reprint, New York: W. W. Norton, 1965: 195, 200–203.

8. Ibid.: 161.

9. Ibid.: 182.

10. Ibid.: 186.

11. Ibid.: 185. Herskovitz is referring to the accounts written by the great Islamic traveler Ibn Battutah, who covered enormous distances in Africa and Asia during the 1300s.

12. David B. Cohen. *Stranger in the Nest: Do Parents Really Shape Their Children's Personality, Intelligence, or Character?* New York: John Wiley and Sons, 1999.

13. R. Schropp. "Children's Use of Objects—Competitive or Interactive?" Paper presented at the Nineteenth International Ethological Conference, Toulouse, France, 1985; Frans de Waal. *Peacemaking among Primates*: 254.

14. Dennis T. Regan. "Effects of a Favor and Liking on Compliance." *Journal of Experimental Social Psychology* 7 (1971): 627–639; Robert B. Cialdini. *Influence: How and Why People Agree on Things*. New York: William Morrow, 1984: 31–33.

15. Virginia Morell. "Genes May Link Ancient Eurasians, Native Americans." *Science,* April 24, 1998: 520.

16. A. L. Kroeber. *The Nature of Culture*. Chicago: University of Chicago Press, 1952: 287.

17. Melville J. Herskovitz. *Economic Anthropology*.

18. Bronislaw Malinowski. *Argonauts of the Western Pacific*. 1922. Reprint, Prospect Heights, Ill.: Waveland Press, 1984.

19. Leda Cosmides and John Tooby. "Evolutionary Psychology: A Primer." Santa Barbara: Center for Evolutionary Psychology, University of California, Santa Barbara. http://www.clark.net/pub/ogas/evolution/EVPSYCH_primer.html. Downloaded June 26, 1999.

20. John N. Thompson. "The Evolution of Species Interactions." *Science,* June 25, 1999: 2116–2118.

21. Manuel de Landa. *A Thousand Years of Nonlinear History*. New York: Zone Books, 1997: 142; William H. Durham. *Coevolution: Genes, Culture, and Human Diversity*. Stanford, Calif.: Stanford University Press, 1991: 283.

22. For a review of many post-agricultural and post-urban genetic adaptations in humans, including those involving such basics as skull shape and the configuration of teeth, see: Valerius Geist. *Life Strategies, Human Evolution, Environmental Design: Toward a Biological Theory of Health*. New York: Springer-Verlag, 1978: 388–401.

23. S. L. Wiesenfeld. "Sickle-Cell Trait in Human Biological and Cultural Evolution: Development of Agriculture Causing Increased Malaria Is Bound to Gene-Pool Changes Causing Malaria Reduction." *Science*, September 8, 1967: 1134–1140. Several groups of genetic researchers have attempted to establish a far older date for the evolution of sickle-cell anemia. However even Stine et al., who champion an ancient origin for the sickle-cell gene, acknowledge that its appearance is "usually attributed to recent . . . mutations." (O. C. Stine, G. J. Dover, D. Zhu, and K. D. Smith. "The Evolution of Two West African Populations." *Journal of Molecular Evolution*, April 1992: 336–344.)

24. William H. McNeill. *Plagues and Peoples*. 1976. Reprint, New York: Anchor Books, 1998: 208–224; Jared Diamond. *Guns, Germs, and Steel: The Fates of Human Societies*. New York: W. W. Norton, 1997; Manuel de Landa. *A Thousand Years of Nonlinear History*: 132–133.

25. William Raymond Manchester. *The Arms of Krupp, 1587–1968*. Boston: Little, Brown, 1968.

26. H. W. Koch. *A History of Prussia*. New York: Dorset Press, 1978: 7–14; Geoffrey Barraclough. *The Origins of Modern Germany*. New York: W. W. Norton, 1984: 254; A. L. Rowse. *The Expansion of Elizabethan England*. London: MacMillan, 1955; Doris May Lessing. *Going Home*. London: Panther, 1968.

27. United Nations Population Division, Department for Economic and Social Information and Policy Analysis. "World Urbanization Prospects: The 1994 Revision." gopher://gopher.undp.org: 70/00/ungophers/popin/wdtrends/urban. Downloaded June 1999.

28. World Resources Institute. "Population and Human Well-Being: Urban Growth." http://www.wri.org/wr-98-99/citygrow.html. Downloaded June 1999.

29. Scott P. Carroll, Hugh Dingle, and Stephen P. Klassen. "Genetic Differentiation of Fitness-Associated Traits among Rapidly Evolving Populations of Soapberry Bug." *Evolution* 51:4 (1997): 1182–1188; Kelly Kissane. Personal communication. May 15, 1998.

30. David Sloan Wilson. "Adaptive Genetic Variation and Human Evolutionary Psychology." *Ethology and Sociobiology* 15 (1994): 219–235.

31. Ross E. Dunn. *The Adventures of Ibn Battuta: A Muslim Traveler of the Fourteenth Century*. Berkeley: University of California Press, 1986: 165–167.

32. David W. Anthony. "Horse, Wagon and Chariot: Indo-European Languages and Archaeology." *Antiquity: A Quarterly Review of Archaeology*, September 1995: 554–565.

33. John Travis. "Jomon Genes." *Science News*, February 15, 1997: 106–107; Edwin O. Reischauer. *Japan Past and Present*. 3rd ed. Tokyo: Charles E. Tuttle, 1964: 12; Edwin O. Reischauer. *The Japanese*. Cambridge, Mass.: Harvard University Press, 1981: 42; Jared Diamond. "Japanese Roots." *Discover*, June 1998: 86–94.

34. John Tyler Bonner. *The Evolution of Culture in Animals*. Princeton, N.J.: Princeton University Press, 1983: 83.

35. The bigger the group, the more the rate of offspring per adult falls off. (Edward O. Wilson. *The Insect Societies*. Cambridge, Mass.: Harvard University Press, 1971: 337.)

36. Thomas D. Seeley. *Honeybee Ecology: A Study of Adaptation in Social Life*. Princeton, N.J.: Princeton University Press, 1985: 108–118.

37. Edward O. Wilson. *The Insect Societies*.

38. For a vivid description of Mongol population decimations in East Asia, see: Ki-baik Lee. *A New History of Korea*. Trans. Edward W. Wagner with Edward J. Shultz. Cambridge, Mass.: Harvard University Press, 1984: 149.

39. Daniel J. Boorstin. *The Discoverers: A History of Man's Search to Know His World and Himself*. New York: Vintage Books, 1985: 124–143; Wolfram Eberhard. *A History of China*. London: Routledge and Kegan Paul, 1977; Morris Rossabi. *Khubilai Khan: His Life and Times*. Los Angeles: University of California Press, 1988.

40. This caught on primarily in the province of Yunnan. (E. N. Anderson. *The Food of China.* New Haven, Conn.: Yale University Press, 1988: 74.)

41. In roughly 300 B.C., Kautilya wrote the *Arthashastra*—a book summarizing the entire body of Indian political science and economics to his day. These quotes are either taken directly or paraphrased from the *Arthashastra* by Balkrishna Govind Gokhale in his *Asoka Maurya* (New York: Twayne Publishers, 1966: 56).

42. Balkrishna Govind Gokhale. *Asoka Maurya:* 79–80; Geoffrey Parrinder. *World Religions: From Ancient History to the Present.* New York: Facts on File, 1983: 281, 284.

13. GREECE, MILETUS, AND THALES: THE BIRTH OF THE BOUNDARY BREAKERS

1. Since new evidence for the Indo-European invasions and their origins is being unearthed and reinterpreted constantly, these dates are *extremely* approximate. Some experts give the date as 1900 B.C. And one of the most interesting dissidents on the subject, Robert Drews, feels the date of these civilizational takeovers is as late as 1600 B.C. (Robert Drews. *The Coming of the Greeks: Indo-European Conquests in the Aegean and the Near East.* Princeton, N.J.: Princeton University Press, 1988.)

2. Ibid.: 198.

3. Michael Grant. *The Rise of the Greeks.* New York: Charles Scribner's Sons, 1987: 1; John Boardman, Jasper Griffin, and Oswyn Murray, eds. *The Oxford History of the Classical World: Greece and the Hellenistic World.* New York: Oxford University Press, 1988: 4.

4. David Anthony. "The Origin of Horseback Riding." *Scientific American,* December 1991: 94–100.

5. Colin Renfrew. "The Origins of Indo-European Languages." *Scientific American,* October 1989: 106–114; Colin Renfrew. *Archaeology and Language: The Puzzle of Indo-European Origins.* New York: Cambridge University Press, 1988.

6. Eric P. Hamp. "On the Indo-European Origins of the Retroflexes in Sanskrit." *Journal of the American Oriental Society,* October 21, 1996: 719–724.

7. This is an inference derived from Herodotus's account of the Ionian capture of Miletus in roughly 1000 B.C., when the Ionian civilization was heavily Indo-European in character. (Herodotus. *The History of Herodotus.* In *Library of the Future,* 4th ed., ver. 5.0. Irvine, Calif.: World Library, 1996. CD-ROM.) The view I'm presenting is my own, but it is supported in its general outline by Robert Drews in his *Coming of the Greeks.*

8. The validity of applying information from India to Greece is supported by Robert Drews when he points out that "the Hellenization of Greece thus seems to parallel the Aryanization of northwest India." (Robert Drews. *The Coming of the Greeks:* 200.)

9. Hesiod. *The Homeric Hymns and Homerica (c. 8th–6th Century B.C.).* In *Library of the Future,* 4th ed., ver. 5.0. Irvine, Calif.: World Library, 1996. CD-ROM; Hesiod. *The Homeric Hymns and Homerica.* Trans. Hugh G. Evelyn-White. Cambridge, Mass.: Harvard University Press, 1982.

10. My reconstruction, combining Hesiod and other early works with archaeological data on the origins of the Indo-Europeans and of their invasion of India, is one of several possible extrapolations based on the evidence available at the moment.

11. There was even a major Mycenaean palace at Athens. (Geoffrey Barraclough, ed. *The Times Atlas of World History.* London: Times Books, 1984: 67.)

12. Michael Grant. *The Rise of the Greeks:* 52.

13. Oswyn Murray. "Life and Society in Classical Greece." In *The Oxford History of the Classical World: Greece and the Hellenistic World,* ed. John Boardman, Jasper Griffin, and Oswyn Murray. New York: Oxford University Press, 1988: 206.

14. Ibid.: 211–212.

15. Will Durant. *The Life of Greece.* Pt. 2 of *The Story of Civilization.* Irvine, Calif.: World Library, 1991–1994. CD-ROM.

16. Michael Grant. *The Rise of the Greeks:* 39–41.

17. Herodotus. *The Histories.* Trans. Aubrey de Selincourt. New York: Penguin Books, 1972.

18. Aristotle. *Politics.* In *Library of the Future,* 4th ed., ver. 5.0. Irvine, Calif.: World Library, 1996. CD-ROM.

19. Herodotus. *The History of Herodotus.* In *Library of the Future.*

20. Plutarch. *Agis.* In *Library of the Future,* 4th ed., ver. 5.0. Irvine, Calif.: World Library, 1996. CD-ROM.

21. Plutarch. *Solon.* In *Library of the Future,* 4th ed., ver. 5.0. Irvine, Calif.: World Library, 1996. CD-ROM.

22. Sun-tzu is reputed to have been an adviser living in the period of Thales, and is said to have been responsible for the insights in the book popularly attributed to him, *Ping-fa* (The Art of War).

23. George Forrest. "Greece: The History of the Archaic Period." In *The Oxford History of the Classical World: Greece and the Hellenistic World,* ed. John Boardman, Jasper Griffin, and Oswyn Murray. New York: Oxford University Press, 1988: 20.

24. Aristotle. *Soul.* In *Library of the Future,* 4th ed., ver. 5.0. Irvine, Calif.: World Library, 1996. CD-ROM.

25. Aristotle. *Nicomachaean Ethics.* In *Library of the Future,* 4th ed., ver. 5.0. Irvine, Calif.: World Library, 1996. CD-ROM; Aristotle. *Generation of Animals.* In *Library of the Future,* 4th ed., ver. 5.0. Irvine, Calif.: World Library, 1996. CD-ROM.

14. SPARTA AND BABOONERY: THE GUESSWORK OF COLLECTIVE MIND

1. See, for example, Janet F. Werker and Renee N. Desjardins. "Listening to Speech in the First Year of Life: Experiential Influences on Phoneme Perception." *Current Directions in Psychological Science,* June 1995: 76–81; and Janet F. Werker and Richard C. Tees. "The Organization and Reorganization of Human Speech Perception." *Annual Review of Neuroscience* 15 (1992): 377–402.

2. Hans Kummer. "Two Variations in the Social Organization of Baboons." In *Primates: Studies in Adaptation and Variability,* ed. Phyllis C. Jay. New York: Holt, Rinehart and Winston, 1968: 293–312.

3. Ibid.: 304.

4. Shirley C. Strum. *Almost Human: A Journey into the World of Baboons.* New York: Random House, 1987: 15.

5. This account is based on Shirley C. Strum. *Almost Human:* 131–133, 137, 182, 173–174, 177, 178, 179, 183, 203–204, 205, 249.

6. Historian W. G. Forrest feels that whether Lycurgus was real or not, "the chances are that Sparta . . . owed her look to one hand." (W. G. Forrest. *A History of Sparta: 950–192 B.C.* New York: W. W. Norton, 1968: 40.) Richard Talbert and Will Durant also lay out evidence which would indicate that the constitution of Sparta was largely shaped by a single man. (Richard J. A. Talbert, trans. *Plutarch on Sparta.* New York: Penguin Books, 1988: 2–3; Will Durant. *The Life of Greece.* Pt. 2 of *The Story of Civilization.* New York: Simon and Schuster, 1939: 77.)

7. W. G. Forrest. *A History of Sparta:* 38; Geoffrey Barraclough, ed. *The Times Atlas of World History.* London: Times Books, 1984: 67, 74.

8. Spartans were not even willing to spread out to live on the agricultural estates from which they earned their income. They stayed within the Spartan walls and simply collected their revenues. (W. G. Forrest. *A History of Sparta:* 44.)

9. For the substantial differences in attitude, outlook, and way of life between inland capitals and seacoast centers of trade, see: Manuel de Landa. *A Thousand Years of Nonlinear History.* New York: Zone Books, 1997: 50–53; and Anne Querrien. "The Metropolis and the Capital." In *Zone 1 / 2: The Contemporary City.* New York: Zone Books, 1987: 219–221.

10. Xenophon. *Spartan Society.* In *Plutarch on Sparta,* trans. Richard J. A. Talbert. New York: Penguin Books, 1988: 166.

11. W. G. Forrest. *A History of Sparta:* 25–27.

12. Ibid.: 30–31.

13. Ibid.: 33.

14. Ibid.: 37.

15. Plutarch. *Lycurgus.* In *Plutarch on Sparta,* trans. Richard J. A. Talbert. New York: Penguin Books, 1988: 7–13.

16. Herodotus. *The History of Herodotus.* In *Library of the Future,* 4th ed., ver. 5.0. Irvine, Calif.: World Library, 1996. CD-ROM.

17. W. G. Forrest. *A History of Sparta:* 61–68.

18. Ibid.: 67.

19. Ibid.: 38.

20. Ibid.: 39.

21. Ibid.: 42.

22. Ibid.: 28.

23. Ibid.: 30.

24. Ibid.: 38.

25. Plutarch. *Lycurgus.* In *Plutarch on Sparta,* trans. Richard J. A. Talbert: 36.

26. Xenophon. *Spartan Society.* In *Plutarch on Sparta,* trans. Richard J. A. Talbert: 167.

27. Plutarch. *Lycurgus. Aristotle and Xenophon on Democracy and Oligarchy,* trans. J. M. Moore. Berkeley: University of California Press, 1986: 97.

28. J. M. Moore, trans. *Aristotle and Xenophon on Democracy and Oligarchy:* 97.

29. Xenophon. *Spartan Society.* In *Plutarch on Sparta,* trans. Richard J. A. Talbert: 167.

30. Ibid.

31. J. M. Moore, trans. *Aristotle and Xenophon on Democracy and Oligarchy:* 94, 96. Moore derives this particular data from Plutarch and Euripides.

32. W. G. Forrest. *A History of Sparta:* 47.

33. Ibid.: 49–50.

34. Ibid.: 50.

35. The quote appears in J. M. Moore, trans. *Aristotle and Xenophon on Democracy and Oligarchy:* 94.

36. W. G. Forrest. *A History of Sparta:* 51–52.

37. This account of Spartan life is based on: Xenophon. *Spartan Society.* In *Plutarch on Sparta,* trans. Richard J. A. Talbert: 168–171.

38. W. G. Forrest. *A History of Sparta:* 30; J. M. Moore, trans. *Aristotle and Xenophon on Democracy and Oligarchy:* 97.

39. Richard J. A. Talbert, trans. *Plutarch on Sparta:* 37.

40. Ibid.: 53.

15. THE PLURALISM HYPOTHESIS: ATHENS' UNDERSIDE

1. Michael Grant. *The Rise of the Greeks.* New York: Charles Scribner's Sons, 1987: 35–36; George Forrest. "Greece: The History of the Archaic Period." In *The Oxford History of the Classical World: Greece and the Hellenistic World,* ed. John Boardman, Jasper Griffin, and Oswyn Murray. New York: Oxford University Press, 1988: 25.

2. See, for example, Michael Grant. *The Rise of the Greeks:* 41; and Will Durant. *The Life of Greece.* Pt. 2 of *The Story of Civilization.* New York: Simon and Schuster, 1939: 109f.

3. Oswyn Murray. "Life and Society in Classical Greece." In *The Oxford History of the Classical World: Greece and the Hellenistic World,* ed. John Boardman, Jasper Griffin, and Oswyn Murray. New York: Oxford University Press, 1988: 202–203.

4. David Tomei and Frederick O. Cope, eds. *Apoptosis: The Molecular Basis of Cell Death.* Plainview, N.Y.: Cold Spring Harbor Laboratory Press, 1991; C. M. Payne, C. Bernstein, and H. Bernstein. "Apoptosis Overview Emphasizing the Role of Oxidative Stress, DNA Damage and

Signal-Transduction Pathways." *Leukemia and Lymphoma,* September 1995; M. J. Staunton and E. F. Gaffney. "Apoptosis: Basic Concepts and Potential Significance in Human Cancer." *Archives of Pathology and Laboratory Medicine,* April 1998: 310–319.

5. N. H. Kalin, S. E. Shelton, M. Rickman, and R. J. Davidson. "Individual Differences in Freezing and Cortisol in Infant and Mother Rhesus Monkeys." *Behavioral Neuroscience,* February 1998: 251–254. Among the hormones produced by the stresses of social isolation and lack of control is melanocyte-stimulating hormone, MSH, which changes the color of a lizard or other animal. (S. E. Lindley, K. J. Lookingland, and K. E. Moore. "Dopaminergic and Beta-adrenergic Receptor Control of Alpha-melanocyte-stimulating Hormone Secretion during Stress." *Neuroendocrinology,* July 1990: 46–51; Neil Greenberg. Personal communication. June 20, 1998.)

6. Greenberg. Personal communication. June 20, 1998; James Vaughn Kohl. Personal communication. June 22, 1998.

7. True for lizards and mammals, which both have nervous systems as well as the chemical emitters involved in our neuroendocrinology. Neil Greenberg, Enrique Font, and Robert C. Switzer III. "The Reptilian Striatum Revisited: Studies on Anolis Lizards." In *The Forebrain of Reptiles: Current Concepts of Structure and Function,* ed. Walter K. Schwerdtfeger and Willhelmus J. A. J. Smeets. Basel: Karger, 1988: 162–177; Cliff H. Summers and Neil Greenberg. "Somatic Correlates of Adrenergic Activity during Aggression in the Lizards, *Anolis Carolinensis.*" *Hormones and Behavior* 28 (1994): 29–40; Neil Greenberg. "Behavioral Endocrinology of Physiological Stress in a Lizard." *Journal of Experimental Zoology* (Supplement 4), 1990: 170–173.

8. See, for example, R. E. Wheeler, R. J. Davidson, and A. J. Tomarken. "Frontal Brain Asymmetry and Emotional Reactivity: A Biological Substrate of Affective Style." *Psychophysiology,* January 1993: 82–89; and A. J. Tomarken, R. J. Davidson, R. Wheeler, and R. Doss. "Individual Differences in Anterior Brain Asymmetry and Fundamental Dimensions of Emotions." *Journal of Personality and Social Psychology* 62 (1992): 676–687.

9. For the manner in which natural selection mixes and matches our personalities on society's behalf, see: E. O. Wilson. "Epigenesis and the Evolution of Social Systems." *Journal of Heredity,* March–April 1981: 70–77.

10. Kagan's interest in Asian versus American babies was triggered by the work of Daniel Freedman. See: Daniel G. Freedman. "Cross Cultural Differences in Newborn Behavior." University Park, Pa.: Psychological Cinema, 1980. Motion picture; Daniel G. Freedman. *Human Sociobiology:A Holistic Approach.* New York: Free Press, 1979. For Kagan's speculations on the temperamental and genetic origins of Buddhism's emphasis on placidity versus Christianity's stress on guilt and anxiety, see: Jerome Kagan with the collaboration of Nancy Snidman, Doreen Arcus, and J. Steven Reznick. *Galen's Prophecy: Temperament in Human Nature.* New York: Basic Books, 1994.

11. Jerome Kagan. *Unstable Ideas: Temperament, Cognition and Self.* Cambridge, Mass.: Harvard University Press, 1989: 84.

12. Ibid.: 81–84.

13. J. Kagan, J. S. Reznick, and N. Snidman. "Biological Bases of Childhood Shyness." *Science,* April 8, 1988: 167–171; Jerome Kagan. *Unstable Ideas:* 145.

14. Jerome Kagan. *Unstable Ideas:* 146.

15. Ibid.: 167–168; J. Dunnette and R. Weinshilboum. "Family Studies of Plasma Dopamine–Beta Hydroxylase." *American Journal of Human Genetics* 34 (1982): 84–99; R. M. Weinshilboum. "Catecholamine Biochemical Genetics in Human Populations." In *Neurogenetics,* ed. X. O. Breakefield. New York: Elsevier, 1979: 257–282.

16. Jerome Kagan. *Unstable Ideas:* 170.

17. Alessandra Piontelli. *From Fetus to Child: An Observational and Psychoanalytic Study.* New York: Tavistock/Routledge, 1992.

18. For related information on how twins are forced to assume different personalities by their competition and by their intensely emotional relationship, see: R. Zazzo. "The Twin Condition and the Couple Effects on Personality Development." *Acta Geneticae Medicae et Gemellologiae* 25 (1976):

343–352; and R. Zazzo. "Genesis and Peculiarities of the Personality of Twins." *Progress in Clinical and Biological Research* 24A (1978): 1–11.

19. According to a study of 1,172 pregnancies, approximately 1 in 51 of us is conceived as twins. But only 1 in 106 of these conceptions results in the birth of 2 live twins. (E. Gerdts. "Ultrasonic Diagnosis in Abortion of One Twin: The Vanishing Twin Phenomenon." *Tidsskrift for den Norske Laegeforening*, November 20, 1989: 3328–3329.

20. This is one of the many mother-fetus conflicts reported in: Randolph M. Nesse, M.D., and George C. Williams, Ph.D. *Why We Get Sick: The New Science of Darwinian Medicine.* New York: Random House, 1995: 197–199.

21. See, for example, M. G. Forest. "Role of Androgens in Fetal and Pubertal Development." *Hormone Research* 18:1–3 (1983): 69–83; and N. McConaghy and R. Zamir. "Sissiness, Tomboyism, Sex-Role, Sex Identity and Orientation." *Australian and New Zealand Journal of Psychiatry,* June 1995: 278–283.

22. J. Kagan, J. S. Reznick, and N. Snidman. "The Physiology and Psychology of Behavioral Inhibition in Children." *Child Development,* December 1987: 1459–1473; J. Kagan, J. S. Reznick, and N. Snidman. "Biological Bases of Childhood Shyness." *Science:* 167–171; J. Kagan. "Behavior, Biology, and the Meanings of Temperamental Constructs." *Pediatrics* 90 (September 1992): 510–513.

23. J. F. Rosenbaum, J. Biederman, E. A. Bolduc-Murphy, S. V. Faraone, J. Chaloff, D. R. Hirshfeld, and J. Kagan. "Behavioral Inhibition in Childhood: A Risk Factor for Anxiety Disorders." *Harvard Review of Psychiatry,* May–June 1993: 2–16.

24. Jerome Kagan. "Temperament and the Reactions to Unfamiliarity." *Child Development,* February 1997: 139–143.

25. J. Kagan, J. S. Reznick, and J. Gibbons. "Inhibited and Uninhibited Types of Children." *Child Development,* August 1989: 838–845.

26. Jerome Kagan. *Unstable Ideas:* 87.

27. Randall White. "Substantial Acts: From Materials to Meaning in Upper Paleolithic Representation." In *Beyond Art: Upper Paleolithic Symbolism,* ed. D. Stratmann, M. Conkey, and O. Soffer. San Francisco: California Academy of Sciences, 1996; Randall White. "The Dawn of Adornment." *Natural History,* May 1993: 61–66; Randall White. "Beyond Art: Toward an Understanding of the Origins of Material Representation in Europe." *Annual Review of Anthropology* 21 (1992): 537–564; Randall White. "Technological and Social Dimensions of 'Aurignacian-Age' Body Ornaments across Europe." In *Before Lascaux: The Complex Record of the Early Upper Paleolithic,* ed. H. Knecht, A. Pike-Tay, and R. White. Boca Raton, Fla.: CRC Press: 1993: 247–299; Olga Soffer. *The Upper Paleolithic of the Central Russian Plain.* New York: Academic Press, 1985.

28. Jerome Kagan, *Unstable Ideas:* 55–56.

29. Daniel Goleman. "New Research Overturns a Milestone of Infancy." *New York Times,* June 6, 1989: C-1, 14. For the damage done by maternal rejection, see: A. Raine, P. Brennam, and S. A. Mednick. "Birth Complications Combined with Early Maternal Rejection at Age One Year Predispose to Violent Crime at Age Eighteen Years." *Archives of General Psychiatry* 51 (1994): 984–988. For the necessity of maternal involvement in the development of the human brain, see: M. A. Hofer. "Relationships as Regulators: A Psychobiologic Perspective on Bereavement." *Psychosomatic Medicine* 46 (1984): 183–196; M. A. Hofer. "Early Social Relationships: A Psychobiologist's View." *Child Development* 58 (1987): 633–647; and M. A. Hofer. "On the Nature and Consequences of Early Loss." *Psychosomatic Medicine* 58 (1996): 570–581.

30. For self-inhibition in neural nets, see: T. Fukai and S. Tanaka. "A Simple Neural Network Exhibiting Selective Activation of Neuronal Ensembles: From Winner-Take-All to Winners-Share-All." *Neural Computation,* January 1997: 77–97. For the manner in which the processors in neural nets adjust the strength of their own connections, see: Sunny Bains. "A Subtler Silicon Cell for Neural Networks." *Science,* September 26, 1997: 1935.

31. Jerome Kagan. *The Power and Limitations of Parents.* Austin: Hogg Foundation for Mental Health, University of Texas, 1986.

32. Jerome Kagan. *Unstable Ideas:* 173.

33. Ibid.: 126.

34. David B. Cohen. *Stranger in the Nest: Do Parents Really Shape Their Children's Personality, Intelligence, or Character?* New York: John Wiley and Sons, 1999.

35. Jerome Kagan. *Unstable Ideas:* 214−215. For similar conclusions, see: Kazimierz Dabrowski with Andrzej Kawczak and Michael M. Piechowski. *Mental Growth through Positive Disintegration.* London: Gryf Publications, 1970.

36. Jerome Kagan. *Unstable Ideas:* 208−209.

37. David Freeman Hawke. *John D.—the Founding Father of the Rockefellers.* New York: Harper and Row, 1980.

38. Xenophon. *Spartan Society.* In *Plutarch on Sparta,* trans. Richard J. A. Talbert. New York: Penguin Books, 1988: 170.

16. PYTHAGORAS, SUBCULTURES, AND PSYCHO-BIO-CIRCUITRY

1. A. L. Kroeber. "Psychosis or Social Sanction?" In *The Nature of Culture.* Chicago: University of Chicago Press, 1952: 313; Charles Winick. *Dictionary of Anthropology.* New York: Philosophical Library, 1956: 19, 67−68, 265; Torrey E. Fuller. *Witchdoctors and Psychiatrists.* New York: Harper and Row, 1986: 51; Ruth Benedict. *Patterns of Culture.* 1934. Reprint, New York: New American Library, 1950: 243.

2. Chris Bader and Alfred Demaris. "A Test of the Stark-Bainbridge Theory of Affiliation with Religious Cults and Sects." *Journal for the Scientific Study of Religion,* September 1996: 285−303.

3. Xenophon. *Memorabilia.* The Perseus Project, ed. Gregory R. Crane. http://hydra.perseus. tufts.edu. August 1998; Plato. *Protagoras.* In *Library of the Future,* 4th ed., ver. 5.0. Irvine, Calif.: World Library, 1996. CD-ROM.

4. Plato. *Protagoras.* In *Library of the Future,* 4th ed., ver. 5.0. Irvine, Calif.: World Library, 1996. CD-ROM.

5. The term "sophist" came from the Greek "sophia"—meaning skill, wisdom, or knowledge. (Diodorus. *Historical Library.* The Perseus Project, ed. Gregory R. Crane. http://hydra.perseus. tufts.edu. August 1998.) Plutarch was a bit more cynical in his definition. His take: "What was then called 'sophia' or wisdom . . . was really nothing more than cleverness in politics and practical sagacity." (Plutarch. *Themistocles.* The Perseus Project, ed. Gregory R. Crane. http://hydra. perseus.tufts.edu. August 1998.) Others saw sophia as mystic knowledge, so this was a word of many meanings.

6. The Sophists' opponents made much of their high fees. (Plato. *Apology.* In *Library of the Future,* 4th ed., ver. 5.0. Irvine, Calif.: World Library, 1996. CD-ROM.)

7. Patricia R. Bergquist. *Sponges.* Berkeley: University of California Press, 1978: 74; Lewis Thomas and Robin Bates. "Notes of a Biology Watcher." Produced and directed by Robin Bates. Nova program #818, TV script. Boston: WGBH, 1981: 3−4; Eric Jantsch. *The Self-Organizing Universe: Scientific and Human Implications of the Emerging Paradigm of Evolution.* Oxford: Pergamon Press, 1980: 128. Members of two myxobacterial colonies sieved together will also seek out their mates and reconstitute themselves as two separate and opposing fruiting bodies . . . without the telltale identifier of different cell coloring. (D. R. Smith and M. Dworkin. "Territorial Interactions between Two Myxococcus Species." *Journal of Bacteriology,* February 1994: 1201−1205.)

8. Eric Jantsch. *The Self-Organizing Universe:* 129.

9. Rodolfo Llinás. " 'Mindfulness' as a Functional State of the Brain." In *Mindwaves: Thoughts on Intelligence, Identity and Consciousness,* ed. Colin Blakemore and Susan Greenfield. Oxford: Basil Blackwell, 1989: 347−348.

10. G. W. Barlow and R. C. Francis. "Unmasking Affiliative Behavior among Juvenile Midas Cichlids *(Cichlasoma Citrinellum).*" *Journal of Comparative Psychology,* June 1988: 118−123.

11. Michael Patrick Ghiglieri. *The Chimpanzees of Kibale Forest: A Field Study of Ecology and Social Structure.* New York: Columbia University Press, 1984: 128–134; F. B. de Waal and L. M. Luttrell. "The Similarity Principle Underlying Social Bonding among Female Rhesus Monkeys." *Folia Primatologica* 46:4 (1986): 215–234.

12. E. O. Laumann. "Friends of Urban Men: An Assessment of Accuracy in Reporting Their Socioeconomic Attributes, Mutual Choice, and Attitude Agreement." *Sociometry* 32 (1969): 54–70.

13. Robert B. Cialdini. *Influence: How and Why People Agree on Things.* New York: William Morrow, 1984: 170; M. Claes and L. Poirier. "Characteristics and Functions of Friendship in Adolescence." *Psychiatrie de l'enfant et de l'adolescent* 36:1 (1993): 289–308. Human sociobiologist Daniel Freedman observes that San Francisco kids of different ethnic backgrounds play together until they're ten, then separate and cluster with their own kind. (Daniel G. Freedman. *Human Sociobiology: A Holistic Approach.* New York: Free Press, 1979: 138.)

14. Robert B. Cialdini. *Influence:* 169–170.

15. G. W. Evans and R. B. Howard. "Personal Space." *Psychological Bulletin,* October 1973: 334–344.

16. K. R. Truett, L. J. Eaves, J. M. Meyer, A. C. Heath, and M. G. Martin. "Religion and Education as Mediators of Attitudes: A Multivariate Analysis." *Behavior Genetics,* January 1992: 43–62; C. R. Cloninger, J. Rice, and T. Reich. "Multifactorial Inheritance with Cultural Transmission and Assortative Mating. II. A General Model of Combined Polygenic and Cultural Inheritance." *American Journal of Human Genetics,* March 1979: 176–198; C. T. Nagoshi, R. C. Johnson, and G. P. Danko. "Assortative Mating for Cultural Identification as Indicated by Language Use." *Behavior Genetics,* January 1990: 23–31; M. E. Procidano and L. H. Rogler. "Homogamous Assortative Mating among Puerto Rican Families: Intergenerational Processes and the Migration Experience." *Behavior Genetics,* May 1989: 343–354.

17. Stanley Schachter. *The Psychology of Affiliation.* Stanford, Calif.: Stanford University Press, 1959; I. Sarnoff and P. G. Zimbardo. "Anxiety, Fear, and Social Affiliation." *Journal of Abnormal and Social Psychology* 61 (1962): 356–363.

18. Harry F. Harlow. *Learning to Love.* New York: Jason Aronson, 1974: 85.

19. Ibid.: 142–143; S. J. Suomi, H. F. Harlow, and J. K. Lewis. "Effect of Bilateral Frontal Lobectomy on Social Preferences of Rhesus Monkeys." *Journal of Comparative Physiology,* March 1970: 448–453.

20. F. H. Farley and C. B. Mueller. "Arousal, Personality, and Assortative Mating in Marriage: Generalizability and Cross-Cultural Factors." *Journal of Sex and Marital Therapy,* spring 1978: 50–53.

21. T. A. Rizzo and W. A. Corsaro. "Social Support Processes in Early Childhood Friendship: A Comparative Study of Ecological Congruences in Enacted Support." *American Journal of Community Psychology,* June 1995: 389–417.

22. Jerome Bruner. *Actual Minds, Possible Worlds.* Cambridge, Mass.: Harvard University Press, 1986: 66.

23. T. M. Newcomb. "Stabilities Underlying Changes in Interpersonal Attraction." *Journal of Abnormal and Social Psychology* 66 (1963): 393–404; T. M. Newcomb. "Interpersonal Constancies: Psychological and Sociological Approaches." In *Perspectives in Social Psychology,* ed. O. Klineberg and R. Christie. New York: Holt, Rinehart and Winston, 1963: 38–49. Numerous groups concoct a system in which outsiders are somehow inferior. This is even true among subcultures of prostitutes and alcoholics. (Earl Rubington. "Deviant Subcultures." In *Sociology of Deviance,* ed. M. Michael Rosenberg, Robert A. Stebbins, and Allan Turowitz. New York: St. Martin's, 1982: 46–50.)

24. Howard Bloom. *The Lucifer Principle: A Scientific Expedition into the Forces of History.* New York: Atlantic Monthly Press, 1995: 97–106.

25. M. R. Gunnar, K. Tout, M. de Haan, S. Pierce, and K. Stansbury. "Temperament, Social Competence, and Adrenocortical Activity in Preschoolers." *Developmental Psychobiology,* July 1997: 65–85.

26. A. W. Harrist, A. F. Zaia, J. E. Bates, K. A. Dodge, and G. S. Pettit. "Subtypes of Social Withdrawal in Early Childhood: Sociometric Status and Social-Cognitive Differences across Four Years." *Child Development,* April 1997: 278–294; Jerome Kagan. *Unstable Ideas: Temperament, Cognition and Self.* Cambridge, Mass.: Harvard University Press, 1989: 208–209, 215.

27. Thomas J. Young. "Judged Political Extroversion-Introversion and Perceived Competence of U.S. Presidents." *Perceptual and Motor Skills,* October 1996: 578.

28. J. D. Higley, S. T. King Jr., M. F. Hasert, M. Champoux, S. J. Suomi, and M. Linnoila. "Stability of Interindividual Differences in Serotonin Function and Its Relationship to Severe Aggression and Competent Social Behavior in Rhesus Macaque Females." *Neuropsychopharmacology,* January 1996: 67–76.

29. A. L. Clair, T. P. Oei, and L. Evans. "Personality and Treatment Response in Agoraphobia with Panic Attacks." *Comprehensive Psychiatry,* September–October 1992: 310–318; Svend Erik Moller, E. L. Mortensen, L. Breum, C. Alling, O. G. Larsen, T. Boge-Rasmussen, C. Jensen, and K. Bennicke. "Aggression and Personality: Association with Amino Acids and Monoamine Metabolites." *Psychological Medicine,* March 1996: 323–331; Harry F. Harlow and Margaret Kuenne Harlow. "Social Deprivation in Monkeys." *Scientific American,* November 1962: 138; Harry F. Harlow and Gary Griffin. "Induced Mental and Social Deficits in Rhesus Monkeys." In *Biological Basis of Mental Retardation,* ed. Sonia F. Osler and Robert E. Cooke. Baltimore, Md.: Johns Hopkins Press, 1965: 99–105; Harry F. Harlow. *Learning to Love:* 113.

30. See: Chris Bader and Alfred Demaris. "A Test of the Stark-Bainbridge Theory of Affiliation with Religious Cults and Sects." *Journal for the Scientific Study of Religion:* 285–303.

31. Yvonne Walsh and Robert Bor. "Psychological Consequences of Involvement in a New Religious Movement or Cult." *Counselling Psychology Quarterly* 9:1 (1996): 47–60.

32. Jeff B. Bryson and Michael J. Driver. "Cognitive Complexity, Introversion, and Preference for Complexity." *Journal of Personality and Social Psychology,* September 1972: 320–327.

33. Joel Cooper and Charles J. Scalise. "Dissonance Produced by Deviations from Life Styles: The Interaction of Jungian Typology and Conformity." *Journal of Personality and Social Psychology,* April 1974: 566–571. Though it may seem otherwise on the surface, introverts secretly hanker for the kind of inclusion only extroverts seem to know. (Steven R. Brown and Clyde Hendrick. "Introversion, Extraversion and Social Perception." *British Journal of Social and Clinical Psychology,* December 1971: 313–319.)

34. C. P. Schmidt and J. W. McCutcheon. "Reexamination of Relations between the Myers-Briggs Type Indicator and Field Dependence-Independence." *Perceptual and Motor Skills,* December 1988: 691–695.

35. Tests done with the Strong Interest Inventory tend to indicate that high IQ introverts are explorers. A. S. Kaufman and S. E. McLean. "An Investigation into the Relationship between Interests and Intelligence." *Journal of Clinical Psychology* 54:2 (1998): 279–295. John Price and Anthony Stevens hypothesize that during our hunter-gatherer days, extroverts joined the crowd, taking on the common perceptual scheme and banding together for hunting and warfare. Meanwhile, according to Price and Stevens, introverts went off to homestead and start single-family units of their own; and schizotypal folks, those with extraordinarily vivid, inward-oriented imagination, started cult groups. (John S. Price and Anthony Stevens. "The Human Male Socialization Strategy Set: Cooperation, Defection, Individualism, and Schizotypy." *Evolution and Human Behavior,* January 1998: 57–70; Anthony Stevens and John Price. *Evolutionary Psychiatry: A New Beginning.* London: Routledge, 1996.) Other studies indicate that refuge-seeking introverts tend to settle into long and stable marriages, while the escape of adventure hunters from standard social boundaries may turn them into criminals or drug abusers. (Gerald D. Otis and John L. Louks. "Rebelliousness and Psychological Distress in a Sample of Introverted Veterans." *Journal of Psychological Type* 40 (1997): 20–30.)

36. For a synthesis of additional research on the solitary, stress-tinged origins of genius, see: Valerius Geist. *Life Strategies, Human Evolution, Environmental Design: Toward a Biological Theory of Health.* New York: Springer-Verlag, 1978: 374–376.

37. H. J. Eysenck. "Primary Mental Abilities." *British Journal of Educational Psychology* 9 (1939): 26–265.

38. Hans J. Eysenck and Michael W. Eysenck. *Personality and Individual Differences: A Natural Science Approach.* New York: Plenum Press, 1985: 327.

39. Ibid.

40. Jeff B. Bryson and Michael J. Driver. "Cognitive Complexity, Introversion, and Preference for Complexity." *Journal of Personality and Social Psychology:* 320–327.

41. Richang Zheng and Yijun Qiu. "A Study of the Personality Trend of the Players in the Chinese First-rate Women's Volleyball Teams." *Information on Psychological Sciences* 3 (1984): 22–27.

42. Hans J. Eysenck and Michael W. Eysenck. *Personality and Individual Differences:* 325.

43. Ibid.: 328.

44. M. M. Marinkovic. "Importance of Introversion for Science and Creativity." *Analytische Psychologie* 12:1 (1981): 1–35; Hans J. Eysenck and Michael W. Eysenck. *Personality and Individual Differences:* 325.

45. Among other things, introverts seem to process information in a manner all their own and are slow to knuckle under to the collective perception of the mainstream. Extroverts, on the other hand, are quicker to see things in the "normal" way. (Michael W. Eysenck and Christine Eysenck. "Memory Scanning, Introversion^Extraversion, and Levels of Processing." *Journal of Research in Personality,* September 1979: 305–315.) For more on the manner in which outsiders come up with vital solutions others fail to notice, see: Thomas Kuhn. *The Structure of Scientific Revolutions,* 2nd ed. Chicago: University of Chicago Press, 1970; and Dean Keith Simonton. *Greatness: Who Makes History and Why.* New York: Guilford Press, 1994.

46. Hakan Fischer, Gustav Wik, and Mats Fredrikson. "Extraversion, Neuroticism and Brain Function: A PET Study of Personality." *Personality and Individual Differences,* August 1997: 345–352.

47. D. L. Johnson, J. S. Wiebe, S. M. Gold, N. C. Andreasen, R. D. Hichwa, G. L. Watkins, and L. L. Boles Ponto. "Cerebral Blood Flow and Personality: A Positron Emission Tomography Study." *American Journal of Psychiatry,* February 1999: 252–257; R. J. Mathew, M. L. Weinman, and D. L. Barr. "Personality and Regional Cerebral Blood Flow." *British Journal of Psychiatry,* May 1984: 529–532; John R. Skoyles. Personal communication. July 20, 1999.

48. Sally J. Power and Lorman L. Lundsten. "Studies That Compare Type Theory and Left-Brain/Right-Brain Theory." *Journal of Psychological Type* 43 (1997): 23; Sally J. Power. Personal communication. July 28, 1998.

49. Darren R. Gitelman, Nathaniel M. Alpert, Stephen Kosslyn, and Kirk Daffner. "Functional Imaging of Human Right Hemispheric Activation for Exploratory Movements." *Annals of Neurology,* February 1996: 174–179.

50. J. C. Borod. "Interhemispheric and Intrahemispheric Control of Emotion: A Focus on Unilateral Brain Damage." *Journal of Consulting and Clinical Psychology,* June 1992: 339–348.

51. A. Bechara, H. Damasio, D. Tranel, and A. R. Damasio. "Deciding Advantageously before Knowing the Advantageous Strategy." *Science,* February 28, 1997: 1293–1295; Gretchen Vogel. "Scientists Probe Feelings Behind Decision-Making." *Science,* February 28, 1997: 1269; Bruce Bower. "Hunches Pack Decisive Punches." *Science News,* March 22, 1997: 183.

52. Sonia Ancoli and Kenneth F. Green. "Authoritarianism, Introspection, and Alpha Wave Biofeedback Training." *Psychophysiology,* January 1977: 40–44.

53. Ibid.

54. Hans J. Eysenck and Michael W. Eysenck. *Personality and Individual Differences:* 274.

55. Psychologist Dean Keith Simonton has examined the many factors behind monumental achievement in *Greatness: Who Makes History and Why.* His conclusion: persistence is the one quality without which all other endowments seem to end in nothing.

56. Jerome Kagan. *Unstable Ideas:* 214–215. For similar conclusions, see: Kazimierz Dabrowski with Andrzej Kawczak and Michael M. Piechowski. *Mental Growth through Positive Disintegration.* London: Gryf Publications, 1970.

57. When it came to digesting intellectual complexities, says Iamblichus, the Samians "lacked endurance." (Iamblichus. *The Life of Pythagoras; or, On the Pythagorean Life.* In *The Pythagorean Sourcebook and Library,* ed. Kenneth Sylvan Guthrie and David Fideler. Grand Rapids, Mich.: Phanes Press, 1987: 61.)

58. Ibid.: 59.

59. Thucydides. *History of the Peloponnesian War,* trans. Richard Crawley. In *Library of the Future,* 4th ed., ver. 5.0. Irvine, Calif.: World Library, 1996. CD-ROM.

60. Iamblichus. *The Life of Pythagoras.* In *The Pythagorean Sourcebook and Library,* ed. Kenneth Sylvan Guthrie and David Fideler: 59.

61. Diodorus. *Historical Library.* 10.10.1.

62. S. Franzoi. "Personality Characteristics of the Crosscountry Hitchhiker." *Adolescence,* fall 1985: 655–668.

63. Historian Michael Grant feels that Pythagoras picked up knowledge from the shamans of Scythia and Thrace, then popularized them within the Greek world. (Michael Grant. *The Rise of the Greeks.* New York: Charles Scribner's Sons, 1987: 229.)

64. Will Durant. *The Life of Greece.* Pt. 2 of *The Story of Civilization.* New York: Simon and Schuster, 1939: 161.

65. The *Brihadaranyaka* and the *Chandogya.*

66. Thomas Bulfinch. *The Age of Fable.* In *Library of the Future,* 4th ed., ver. 5.0. Irvine, Calif.: World Library, 1996. CD-ROM. Even Michael Grant, a skeptic about all things Pythagorean, is convinced that Pythagoras was influenced by the Hindu Upanishads. (Michael Grant. *The Rise of the Greeks:* 229.)

67. This account is based on: Iamblichus. *The Life of Pythagoras.* In *The Pythagorean Sourcebook and Library,* ed. Kenneth Sylvan Guthrie and David Fideler: 63.

68. Valerius Geist. *Life Strategies, Human Evolution, Environmental Design:* 24–35.

69. A. G. Smithers and D. M. Lobley. "Dogmatism, Social Attitudes and Personality." *British Journal of Social and Clinical Psychology,* June 1978: 135–142. Which dogma flockers dive into for refuge doesn't seem to matter. When it comes to beliefs on economics, aesthetics, politics, and religion, both introverts and extroverts can go either left or right. (Km. Kamlesh. "A Study of the Effect of Personality on Value Pattern." *Indian Psychological Review,* January 1981: 13–17.)

70. Will Durant. *The Life of Greece:* 162. Durant appears to be quoting Diogenes Laertius.

71. Diogenes Laertius. *The Life of Pythagoras.* In *The Pythagorean Sourcebook and Library,* ed. Kenneth Sylvan Guthrie and David Fideler. Grand Rapids, Mich.: Phanes Press, 1987: 145–147; Will Durant. *The Life of Greece:* 162.

72. Will Durant. *The Life of Greece:* 163.

73. Ibid.

74. Yvonne Walsh and Robert Bor. "Psychological Consequences of Involvement in a New Religious Movement or Cult." *Counselling Psychology Quarterly* 9:1 (1996): 47–60.

75. Bulfinch, for example, titles one chapter of his *Age of Fable* "Pythagoras–Egyptian Deities-Oracles."

76. Michael Grant. *The Rise of the Greeks:* 228.

77. There is no sense in removing the sexual titillation from our scholarship when it is historically legitimate. Speculation that Alcibiades and Socrates shared more than just their dialogues was rife in antiquity and remains a subject still up for grabs today.

78. Will Durant. *The Life of Greece:* 445.

79. Thucydides. *History of the Peloponnesian War,* trans. Richard Crawley.

80. Xenophon. *Memorabilia:* i. 2. 46.

81. Thucydides. *History of the Peloponnesian War,* trans. Richard Crawley.

82. Ibid.

83. The specific charge on which Socrates was "impeached" was that of corrupting Athens' youth. (Plato. *Euthyphro.* In *Library of the Future,* 4th ed., ver. 5.0. Irvine, Calif.: World Library, 1996. CD-ROM.) And Alcibiades was the most famous of those the philosopher had "corrupted."

84. Saint Augustine. *The City of God.* In *Library of the Future,* 4th ed., ver. 5.0. Irvine, Calif.: World Library, 1996. CD-ROM.

85. M. Hassler. "Testosterone and Musical Talent." *Experimental and Clinical Endocrinology and Diabetes* 98:2 (1991): 89–98. See also: Barbara M. Yarnold. "Steroid Use among Miami's Public School Students, 1992: Alternative Subcultures: Religion and Music versus Peers and the 'Body Cult.'" *Psychological Reports,* February 1998: 19–24.

86. Chris Bader and Alfred Demaris. "A Test of the Stark-Bainbridge Theory of Affiliation with Religious Cults and Sects." *Journal for the Scientific Study of Religion:* 285–303; A. Holland and T. Andre. "The Relationship of Self-Esteem to Selected Personal and Environmental Resources of Adolescents." *Adolescence,* summer 1994: 345–360.

87. The word "shyness" is one Jerome Kagan, the king of this field of study, has used frequently to sum up the temperamental trait which both he and other researchers also call "withdrawal," "limbic sensitivity," "overexcitability," and "introversion." (J. Kagan, J. S. Reznick, and N. Snidman. "Biological Bases of Childhood Shyness." *Science,* April 8, 1988: 167–171; I. R. Bell, M. L. Jasnoski, J. Kagan, and D. S. King. "Is Allergic Rhinitis More Frequent in Young Adults with Extreme Shyness? A Preliminary Survey." *Psychosomatic Medicine,* September–October 1990: 517–525.)

88. Iamblichus. *The Life of Pythagoras.* In *The Pythagorean Sourcebook and Library,* ed. Kenneth Sylvan Guthrie and David Fideler: 81.

89. Diogenes Laertius. *Lives of Eminent Philosophers.* The Perseus Project, ed. Gregory R. Crane: 8.1.46. See also: Diogenes Laertius. *Lives of Eminent Philosophers,* trans. R. D. Hicks. 1925. Reprint, Cambridge, Mass.: Harvard University Press, 1972. For the use of the phrase *"Autos epha ipse dixit"* by slaves and another group disciplined with the lash—pupils—as a sign of absolute subservience, see: Herbert Weir Smyth. *Greek Grammar,* 1st ed. The Perseus Project. ed., Gregory R. Crane. http://hydra.perseus.tufts.edu. August 1998.

90. Iamblichus. *The Life of Pythagoras.* In *The Pythagorean Sourcebook and Library,* ed. Kenneth Sylvan Guthrie and David Fideler: 64. Diogenes Laertius, less prone to hyperbole than Iamblichus, agrees that the Pythagorean dominion over Croton was handled "in a most excellent manner, so that the constitution was very nearly an aristocracy." (Diogenes Laertius. *The Life of Pythagoras.* In *The Pythagorean Sourcebook and Library,* ed. Kenneth Sylvan Guthrie and David Fideler: 142.)

91. David Fideler. Personal communication. July 15, 1998.

92. Aristotle. *Metaphysics.* In *Library of the Future,* 4th ed., ver. 5.0. Irvine, Calif.: World Library, 1996. CD-ROM; Aristotle. *Heavens.* In *Library of the Future,* 4th ed., ver. 5.0. Irvine, Calif.: World Library, 1996. CD-ROM.

93. Pausanias. *Description of Greece.* The Perseus Project, ed. Gregory R. Crane. http://hydra.perseus.tufts.edu. August 1998.

94. Herodotus. *Histories:* 2.80.1. The Perseus Project, ed. Gregory R. Crane. http://hydra.perseus.tufts.edu. July 1998.

95. Diodorus. *Historical Library:* 10.4.1.

96. Euclid. *The Thirteen Books of Euclid's Elements,* trans. Sir Thomas L. Heath. The Perseus Project, ed. Gregory R. Crane. http://hydra.perseus.tufts.edu. July 1998. Heath's commentaries cite a swarm of Euclid's Pythagoreanisms.

97. Plutarch. *Numa Pompilius.* In *Library of the Future,* 4th ed., ver. 5.0. Irvine, Calif.: World Library, 1996. CD-ROM.

98. Demosthenes calls Archytas an administrator of remarkable ability. (Demosthenes. *Erotic Essay 46.* The Perseus Project, ed. Gregory R. Crane. http://hydra.perseus.tufts.edu. August 1998: 61.46.)

99. Diodorus. *Historical Library:* 10.12.3.

100. Already in the first century B.C., Diodorus, writing from Roman Sicily, just below the Italian boot, says of the Pythagoreans that "mankind speaks of them as if they were alive today." Diodorus. *Historical Library.*

101. Montaigne, writing in the 1500s, refers to "the Pythagoreans" as a living presence. (Michel Eyquem de Montaigne. *Essays.* In *Library of the Future,* 4th ed., ver. 5.0. Irvine, Calif.: World Library, 1996. CD-ROM.)

17. SWIVELING EYES AND PIVOTING MINDS: THE PULL OF INFLUENCE ATTRACTORS

1. Herodotus. *The History of Herodotus.* In *Library of the Future,* 4th ed., ver. 5.0. Irvine, Calif.: World Library, 1996. CD-ROM.

2. Manuel de Landa. *A Thousand Years of Nonlinear History.* New York: Zone Books, 1997: 40, 50.

3. Will Durant. *The Life of Greece.* Pt. 2 of *The Story of Civilization.* New York: Simon and Schuster, 1939: 85.

4. Ibid.: 245–250.

5. Thucydides. *History of the Peloponnesian War,* trans. Richard Crawley. In *Library of the Future,* 4th ed., ver. 5.0. Irvine, Calif.: World Library, 1996. CD-ROM.

6. Histiaea was one city-state whose inhabitants were turned to refugees. (Thucydides. *History of the Peloponnesian War,* trans. Richard Crawley.)

7. Ibid.

8. Will Durant. *The Life of Greece:* 251.

9. The final series of contests which ended the long-running Peloponnesian War pitted a politics-torn Athens against Cyrus, gifted son of the Persian emperor, and Lysander, a remarkable Spartan general. (Thucydides. *History of the Peloponnesian War,* trans. Richard Crawley.)

10. Will Durant. *The Life of Greece:* 463.

11. D. Bray. "Protein Molecules as Computational Elements in Living Cells." *Nature,* July 27, 1995: 307–312.

12. G. J. Schütz, W. Trabesinger, and T. Schmidt. "Direct Observation of Ligand Colocalization on Individual Receptor Molecules." *Biophysical Journal,* May 1998: 2223–2226; T. J. Graddis, K. Brasel, D. Friend, S. Srinivasan, S. Wee, S. D. Lyman, C. J. March, and J. T. McGrew. "Structure-Function Analysis of FLT3 Ligand–FLT3 Receptor Interactions Using a Rapid Functional Screen." *Journal of Biological Chemistry,* July 1998: 17626–17633; A. Moore, J. P. Basilion, E. A. Chiocca, and R. Weissleder. "Measuring Transferrin Receptor Gene Expression by NMR Imaging." *Biochimica et Biophysica Acta,* April 1998: 239–249; B. J. Willett, K. Adema, N. Heveker, A. Brelot, L. Picard, M. Alizon, J. D. Turner, J. A. Hoxie, S. Peiper, J. C. Neil, and M. J. Hosie. "The Second Extracellular Loop of CXCR4 Determines Its Function as a Receptor for Feline Immunodeficiency Virus." *Journal of Virology,* August 1998: 6475–6481. The molecules described in these studies are as complex in their architecture and as riddled with information-processing devices as a computer. Hence my use of the otherwise coy and anthropomorphic-sounding term "smart molecule." Its application to receptors is not metaphoric but literal.

13. D. Bray, M. D. Levin, and C. J. Morton-Firth. "Receptor Clustering as a Cellular Mechanism to Control Sensitivity." *Nature,* May 7, 1998: 85–88; Martin Brookes. "Get the Message." *New Scientist,* August 15, 1998: 40–43. Bray has compared the web of molecules in a cell to nodes in a neural net or in a parallel distributed processor. (Dennis Bray. "Intracellular Signalling as a Parallel Distributed Process." *Journal of Theoretical Biology,* March 22, 1990: 215–231.) In other words, he sees the mesh of molecules which operate a cell as a computer or learning machine operating along complex adaptive system lines.

14. ". . . hierarchical structures are a pervasive feature of all *cas* [complex adaptive systems]." (John H. Holland. *Hidden Order: How Adaptation Builds Complexity.* New York: Addison-Wesley, 1995: 107.)

15. Yukimaru Sugiyama. "Social Characteristics and Socialization of Wild Chimpanzees." In *Primate Socialization,* ed. Frank E. Poirier. New York: Random House, 1972: 151.

16. John H. Kaufmann. "Social Relations in Hamadryas Baboons." In *Social Communication among Primates,* ed. Stuart A. Altmann. Chicago: University of Chicago Press, 1967: 82, 84, 96.

17. Frans de Waal calls the ability of the dominant male to focus the attention of the group "the 'control role' of the alpha male." (F. B. de Waal. "The Organization of Agonistic Relations within Two Captive Groups of Java-Monkeys [*Macaca Fascicularis*]." *Zeitschrift für Tierpsychologie,* July 1977: 225–282.) De Waal has also demonstrated one of the points I'm making here: that a dominance hierarchy is not primarily a structure for the exercise of aggression; rather, it is one which generates social integration. (F. B. de Waal. "The Integration of Dominance and Social Bonding in Primates." *Quarterly Review of Biology,* December 1986: 459–479.)

18. Frans de Waal. *Peacemaking among Primates.* Cambridge, Mass.: Harvard University Press, 1989; Dian Fossey. *Gorillas in the Mist.* Boston: Houghton Mifflin, 1983; Lionel Tiger and Robin Fox. *The Imperial Animal.* New York: Holt, Rinehart and Winston, 1971; Daniel G. Freedman. *Human Sociobiology: A Holistic Approach.* New York: Free Press, 1979: 38.

19. Dian Fossey. *Gorillas in the Mist:* 74.

20. Hans Kummer. "Tripartite Relations in Hamadryas Baboons." In *Social Communication among Primates,* ed. Stuart A. Altmann. Chicago: University of Chicago Press, 1967: 64.

21. Dian Fossey. *Gorillas in the Mist:* 82.

22. R. Noe, F. B. de Waal, and J. A. van Hooff. "Types of Dominance in a Chimpanzee Colony." *Folia Primatologica* 34:1–2 (1980): 90–110; D. L. Cheney, R. M. Seyfarth, and J. B. Silk. "The Responses of Female Baboons *(Papio Cynocephalus Ursinus)* to Anomalous Social Interactions: Evidence for Causal Reasoning?" *Journal of Comparative Psychology,* June 1995: 134–141; K. R. L. Hall. "Social Interactions of the Adult Males and Adult Females of a Patas Monkey Troop." In *Social Communication among Primates,* ed. Stuart A. Altmann. Chicago: University of Chicago Press, 1967: 270.

23. S. L. Washburn and D. A. Hamburg. "Aggressive Behavior in Old World Monkeys and Apes." In *Primates: Studies in Adaptation and Variability,* ed. Phyllis C. Jay. New York: Holt, Rinehart and Winston, 1968: 471.

24. Phyllis C. Jay. "The Social Behavior of the Langur Monkey." Ph.D. diss., University of Chicago, 1962; Stuart A. Altmann. "The Structure of Primate Social Communication." In *Social Communication among Primates,* ed. Stuart A. Altmann. Chicago: University of Chicago Press, 1967: 349.

25. Jane Van Lawick-Goodall. "A Preliminary Report on Expressive Movements and Communication in the Gombe Stream Chimpanzees." In *Primates: Studies in Adaptation and Variability,* ed. Phyllis C. Jay. New York: Holt, Rinehart and Winston, 1968: 323; Daniel G. Freedman. *Human Sociobiology:* 37.

26. John E. Frisch, S.J. "Individual Behavior and Intertroop Variability in Japanese Macaques." In *Primates: Studies in Adaptation and Variability,* ed. Phyllis C. Jay. New York: Holt, Rinehart and Winston, 1968: 246–247; Daniel G. Freedman. *Human Sociobiology:* 38; I. DeVore, ed. *Behavior: Field Studies of Monkeys and Apes.* New York: Holt, Rinehart and Winston, 1965.

27. Toshisada Nishida. "Review of Recent Findings on Mahale Chimpanzees: Implications and Future Research Directions." In *Chimpanzee Cultures,* ed. Richard W. Wrangham, W. C. McGrew, Frans B. M. de Waal, and Paul G. Heltne with assistance from Linda A. Marquardt. Cambridge, Mass.: Harvard University Press, 1994: 377; Konrad Lorenz. *On Aggression.* New York: Harcourt Brace Jovanovich, 1974: 46; Daniel G. Freedman. *Human Sociobiology:* 37–38.

28. For a brief roundup of information on attention structures, see: Daniel G. Freedman. *Human Sociobiology:* 36–39.

29. D. R. Omark, M. Omark, and M. S. Edelman. "Formation of Dominance Hierarchies in Young Children: Action and Perception." In *Psychological Anthropology,* ed. T. Williams. The Hague: Mouton, 1975; D. R. Omark and M. S. Edelman. "The Development of Attention Structures in Young Children." In *Attention Structures in Primates and Man,* ed. M. R. A. Chance and R. Larson. New York: John Wiley and Sons, 1977.

30. Linda Schele and David Freidel. *A Forest of Kings: The Untold Story of the Ancient Maya*. New York: William Morrow, 1990: 264–265.

31. Terry Williams. *The Cocaine Kids*. Reading, Mass.: Addison-Wesley, 1989; Daniel G. Freedman. *Human Sociobiology*: 96.

32. By calming things down, serotonin lowers rates of aggression. To keep combativeness at bay, serotonin has to make its mark on a particular variety of receptor, that called 5-HT1B. (F. Saudou, D. A. Amara, A. Dierich, M. LeMeur, S. Ramboz, L. Segu, M. C. Buhot, and R. Hen. "Enhanced Aggressive Behavior in Mice Lacking 5-HT1B Receptor." *Science*, September 23, 1994: 1875–1878; Marcia Barinaga. "This Is Your Brain on Stress." *Science*, November 19, 1993: 1210.)

33. Michael A. Goldberg and Barry Katz. "The Effect of Nonreciprocated and Reciprocated Touch on Power/Dominance Perception." *Journal of Social Behavior and Personality* 5:5 (1990): 379–386; Stephen Thayer. "Close Encounters." *Psychology Today*, March 1988: 34.

34. Michael Argyle. "Innate and Cultural Aspects of Human Non-verbal Communication." In *Mindwaves: Thoughts on Intelligence, Identity and Consciousness*, ed. Colin Blakemore and Susan Greenfield. Oxford: Basil Blackwell, 1989: 60; M. Argyle and M. Cook. *Gaze and Mutual Gaze*. Cambridge, U.K.: Cambridge University Press, 1976.

35. Nancy M. Henley. *Body Politics: Power, Sex, and Nonverbal Communication*. Englewood Cliffs, N.J.: Prentice-Hall, 1977.

36. B. Erickson, E. A. Lind, B. C. Johnson, and W. M. O'Barr. "Speech Style and Impression Formation in a Court Setting: The Effects of 'Powerful' and 'Powerless' Speech." *Journal of Experimental Social Psychology* 14 (1978): 266–279; R. T. Lakoff. *Language and Woman's Place*. New York: Harper and Row, 1975.

37. J. I. Hurwitz, A. F. Zander, and B. Hymovitch. "Some Effects of Power on the Relations among Group Members." In *Group Dynamics: Research and Theory*, ed. Dorwin Cartwright and Alvin Zander. New York: Harper and Row, 1953: 483–492.

38. Adam Smith described stored labor as "a certain quantity of labour stocked and stored up to be employed, if necessary, upon some other occasion." (Adam Smith. *An Inquiry into the Nature and Causes of the Wealth of Nations*. Dublin: Whitestone, 1776.) Gerald Marwell and Pamela Oliver. *The Critical Mass and Collective Action*. Cambridge, U.K.: Cambridge University Press, 1993.

39. Eshel Ben-Jacob and Herbert Levine. "The Artistry of Microbes." *Scientific American*, October 1988: 87; Trisha Gura. "One Molecule Orchestrates Amoebae." *Science*, July 11, 1997: 182; Dictyostelium WWW Server. http://dicty.cmb.nwu.edu/dicty/dicty.html. September 1998.

40. Thomas D. Seeley. *Honeybee Ecology: A Study of Adaptation in Social Life*. Princeton, N.J.: Princeton University Press, 1985; Thomas D. Seeley. *The Wisdom of the Hive: The Social Physiology of Honey Bee Colonies*. Cambridge, Mass.: Harvard University Press, 1995; Thomas D. Seeley and Royce A. Levien. "A Colony of Mind: The Beehive as Thinking Machine." *The Sciences*, July–August 1987: 38–42.

41. Edward O. Wilson. *The Insect Societies*: Cambridge, Mass.: Harvard University Press, 1971: 247–271.

42. In the 1930s, Thorstein Veblen showed how prestige drives culture into many of its most intriguing peculiarities via his theory of conspicuous consumption. A few years later Melville J. Herskovitz, in his 1940 book *Economic Anthropology: The Economic Life of Primitive Peoples*, went a step further and demonstrated the manner in which anthropology reveals a basic of human affairs: that the lust for prestige is more powerful than material greed. (Thorstein Veblen. *The Theory of the Leisure Class; an Economic Study of Institutions*. New York: Modern Library, 1934; Melville J. Herskovitz. *Economic Anthropology: The Economic Life of Primitive Peoples*. 1940. Reprint, New York: W. W. Norton, 1965.)

43. J. W. Thibaut and H. W. Riecken. "Some Determinants and Consequences of the Perception of Social Causality." *Journal of Personality* 24 (1955): 113–133; J. I. Hurwitz, A. F. Zander, and B. Hymovitch. "Some Effects of Power on the Relations among Group Members." In *Group Dynamics: Research and Theory*, ed. Dorwin Cartwright and Alvin Zander. New York: Harper and Row, 1953: 483–492. For the effects of this irrational impulse in one of the most scrupulously rational of

arenas, science, see: David L. Hull. *Science as a Process: An Evolutionary Account of the Social and Conceptual Development of Science.* Chicago: University of Chicago Press, 1988.

44. Irving Lorge. "Prestige, Suggestion, and Attitudes." *Journal of Social Psychology* 7 (1936): 386–402; Harry F. Harlow. *Learning to Love.* New York: Jason Aronson, 1974: 152; Irving Janis, Peter Defares, and Paul Grossman. "Hypervigilant Reactions to Threat." In *Selye's Guide to Stress Research,* vol. 3, ed. Hans Selye. New York: Scientific and Academic Editions, Van Nostrand Reinhold, 1983: 19; H. Sigall and R. Helmreich. "Opinion Change as a Function of Stress and Communicator Credibility." *Journal of Experimental and Social Psychology* 5 (1969): 70–78; Roland Radloff. "Opinion Evaluation and Affiliation." *Journal of Abnormal and Social Psychology* 62 (1961): 578–585; E. P. Torrance. "Some Consequences of Power Differences on Decision Making in Permanent and Temporary Three-Man Groups." *Research Studies* (Washington State College) 22 (1954): 130–140; K. Dion, E. Berscheid, and E. Walster. "What Is Beautiful Is Good." *Journal of Personality and Social Psychology* 24 (1972): 285–290.

45. C. Hovland and W. Weiss. "The Influence of Source Credibility on Communication Effectiveness." *Public Opinion Quarterly* 15 (1952): 635–650.

46. Ellen Langer. *Mindfulness.* Reading, Mass.: Addison-Wesley, 1989: 155–158.

47. Paul R. Wilson. "Perceptual Distortion of Height as a Function of Ascribed Academic Status." *Journal of Social Psychology,* February 1968: 97–102; Robert B. Cialdini. *Influence: How and Why People Agree on Things.* New York: William Morrow, 1984: 216.

48. This observation by Robert Decker, director of the Palo Alto Center for Stress Related Disorders, was reported in: Larry Reibstein with Nadine Joseph. "Mimic Your Way to the Top." *Newsweek,* August 8, 1988: 50.

49. Elias Canetti. *Crowds and Power,* trans. Carol Stewart. New York: Farrar, Straus, Giroux, 1984: 417.

50. Ibid.: 417.

51. Ruth Benedict. *Patterns of Culture.* 1934. Reprint, New York: New American Library, 1950: 176.

52. Thomas Carlyle. *Voltaire.* 1829. Newport Beach, Calif.: Books on Tape, c. 1983.

53. F. W. Jones, A. J. Wills, and I. P. McLaren. "Perceptual Categorization: Connectionist Modelling and Decision Rules." *Quarterly Journal of Experimental Psychology. B. Comparative and Physiological Psychology,* February 1998: 33–58; R. Desimone. "Neural Mechanisms for Visual Memory and Their Role in Attention." *Proceedings of the National Academy of Sciences of the United States of America,* November 26, 1996: 13494–13499.

54. Anna Freud called this "identification with the aggressor," a term she coined in 1946 at the end of World War II, a conflict which had demonstrated the phenomenon all too persuasively. (Anna Freud. *The Ego and Mechanisms of Defense.* New York: International Universities Press, 1946; H. P. Blum. "The Role of Identification in the Resolution of Trauma: The Anna Freud Memorial Lecture." *Psychoanalytic Quarterly,* October 1987: 609–627; M. Hirsch. "Two Forms of Identification with the Aggressor—according to Ferenczi and Anna Freud." *Praxis der Kinderpsychologie und Kinderpsychiatrie,* July–August 1996: 198–205.)

55. Suzanne Ripley. "Intertroop Encounters among Ceylon Gray Langurs *(Presbytis Entellus)."* In *Social Communication among Primates,* ed. Stuart A. Altmann. Chicago: University of Chicago Press, 1967: 237–254.

56. David P. Barash. *The Hare and the Tortoise: Culture, Biology, and Human Nature.* New York: Penguin Books, 1987: 264.

57. Per Bruno Bettelheim, psychologist and former concentration camp victim, who had the questionable privilege of witnessing this weirdness at first hand. (Bruno Bettelheim. "Individual and Mass Behavior in Extreme Situations." *Journal of Abnormal and Social Psychology* 38 (1943): 417–452.)

58. Fernand Braudel. *The Structures of Everyday Life: The Limits of the Possible.* Vol. 1 of *Civilization and Capitalism, 15th–18th Century,* trans. Siân Reynolds. New York: Harper and Row, 1982: 316.

59. For an instance in which an entire nation was reshaped by identification with the aggressor, see: K. Michio. "Japanese Responses to the Defeat in World War II." *International Journal of Social Psychiatry,* autumn 1984: 178–187.

60. Daniel Burstein. *Yen: Japan's New Financial Empire and Its Threat to America.* New York: Simon and Schuster, 1988: 37.

61. William Van Dusen Wishard. "The Twenty-first Century Economy." In *The 1990s and Beyond,* ed. Edward Cornish. Bethesda, Md.: World Future Society, 1990: 133.

62. Paul Kennedy. *The Rise and Fall of the Great Powers: Economic Change and Military Conflict from 1500 to 2000.* New York: Random House, 1987: 475–476.

63. Victor Hao Li. "The New Orient Express." *World Monitor,* November 1988: 24–35.

64. Daniel Burstein. *Yen:* 38.

65. Robert Whiting. *Ya Gotta Have Wa: When Two Cultures Collide on the Baseball Diamond.* New York: MacMillan, 1989: 5. Do not be misled by the subtitle of this book. When it emerged, critics said it gave more of a feel for the Japanese culture than 90 percent of the more erudite-sounding books on the subject. The critics were right.

66. John Naisbitt and Patricia Aburdene. *Megatrends 2000: Ten New Directions for the 1990's.* New York: William Morrow, 1990: 180.

67. James Fallows. *More Like Us: Making America Great Again.* Boston: Houghton Mifflin, 1989: 42.

68. Victor Hao Li. "The New Orient Express." *World Monitor:* 32.

69. Ezra Vogel. "Pax Nipponica?" *Foreign Affairs,* spring 1986: 752–767; Leon Hollerman. *Japan's Economic Strategy in Brazil: Challenge for the United States.* Lexington, Mass.: Lexington Books, 1988: 13.

70. Doug Henwood. "Japan in Latin America." *Left Business Observer,* May 15, 1989: 2; Victor Hao Li. "The New Orient Express": 24–35; *Today's Japan.* Tokyo: NHK-TV, June 22, 1990. Television news broadcast; Mike Mansfield. "The U.S. and Japan: Sharing Our Destinies." *Foreign Affairs,* spring 1989: 11.

71. Leon Hollerman. *Japan's Economic Strategy in Brazil:* 15.

72. Bruno Thomas. N.t. article from *Le Monde,* reprinted in *World Press Review,* April 1988: 20. For additional statistics on Japan's position in the global economy toward the end of the 1980s, see: Paul Kennedy. *The Rise and Fall of the Great Powers:* 458–468.

73. Daniel Burstein. *Yen:* 288–289.

74. Leon Hollerman. *Japan's Economic Strategy in Brazil:* 254–257.

75. Hassan Ziady. "An African View of Debt." *Jeune Afrique Economie,* reprinted in *World Press Review,* August 1989: 50.

76. Barry Shelby. "Japan and Africa." *World Press Review,* January 1989: 44.

77. Daniel Burstein. *Yen:* 281.

78. Ibid.

79. Vladimir Voinovich. "An Exile's Dilemma." *Wilson Quarterly,* August 1990: 114–120.

80. Daniel Burstein. *Yen:* 281.

81. Ian Buruma. *God's Dust: A Modern Asian Journey.* New York: Farrar, Straus, and Giroux, 1989: 7.

82. Ibid.: 230–231.

83. The trendsetting impact of Japanese fashion and architecture hit particularly hard in Asia. ("So You Want to Live in a World Capital." *Economist,* reprinted in *World Press Review,* October 1988: 35.)

84. John Naisbitt and Patricia Aburdene. *Megatrends 2000:* 181.

85. Deyan Sudjic. "Tokyo's 'Spectacular' Stores." *Times* (London), reprinted in *World Press Review,* October 1989: 72.

86. Clyde Prestowitz. "Japanese vs. Western Economies: Why Each Side Is a Mystery to the Other." *Technology Review,* May–June 1988: 34.

87. Clyde V. Prestowitz Jr. *Trading Places: How We Allowed Japan to Take the Lead.* New York: Basic Books, 1988: 60–61, 76; Pat Choate. *Agents of Influence.* New York: Knopf, 1990.

88. *Today's Japan.* Tokyo: NHK-TV, July 2, 1990. Television news broadcast.

89. Ryuji Katayama. "Can Japan Rescue the Philippines." *Business Tokyo,* August 1990: 31.

90. Ian Buruma. *God's Dust:* 139.

91. Ira Magaziner and Mark Patinkin. *The Silent War: Inside the Global Business Battles Shaping America's Future.* New York: Random House, 1989: 40.

92. Peter McGill. "Anxiety on the Road to Foreign Leadership." *Weekly Observer of London,* reprinted in *World Press Review,* August 1992: 16.

93. *Today's Japan.* Tokyo: NHK-TV, January 22, 1990. Television news broadcast.

94. Andrew Clark. "Japan Goes to Europe." *World Monitor,* April 1990: 40.

95. Robert Graham. "Latin America's Reawakening." *Financial Times of London,* reprinted in *World Press Review,* November 1990: 60–61.

96. Sterett Pope. "Japan in the Gulf." *World Press Review,* June 1989: 42.

97. Glen S. Fukushima. "Affirmative Action, Japanese Style." *Tokyo Business,* February 1994: 58. See also *Beyond Capitalism: The Japanese Model of Market Economics,* by Eisuke Sakakibara (Lanham, Md.: University Press of America, 1993); Sakakibara was deputy general of Japan's all-powerful Ministry of Finance.

98. Socrates was an acquaintance of Plato's uncle Charmides, subject of one of the Platonic dialogues. (Plato. *Charmides.* In *Library of the Future,* 4th ed., ver. 5.0. Irvine, Calif.: World Library, 1996. CD-ROM.)

99. Diogenes Laertius. *Lives of Eminent Philosophers,* trans. R. D. Hicks. 1925. Reprint, Cambridge, Mass.: Harvard University Press, 1972: 281–282.

100. R. M. Doty, B. E. Peterson, and D. G. Winter. "Threat and Authoritarianism in the United States, 1978–1987." *Journal of Personality and Social Psychology,* October 1991: 629–640.

101. Diodorus. *Historical Library.* The Perseus Project, ed. Gregory R. Crane. http://hydra.perseus.tufts.edu. August 1998.

102. Manuel de Landa. *A Thousand Years of Nonlinear History:* 168.

103. Plato. *Laws.* In *Library of the Future,* 4th ed., ver. 5.0. Irvine, Calif.: World Library, 1996. CD-ROM.

104. N. G. L. Hammond. *Alexander the Great, King, Commander, and Statesman.* Park Ridge, N.J.: Noyes Press, 1980; Mary Renault. *Fire from Heaven.* New York: Vintage Books, 1977; Mary Renault, *The Nature of Alexander.* London: Allen Lane, 1975.

105. Diodorus. *Historical Library:* 15.29.4.

18. OUTSTRETCH, UPGRADE, AND IRRATIONALITY: SCIENCE AND THE WARPS OF MASS PSYCHOLOGY

1. Melvin Konner. *Why the Reckless Survive . . . and Other Secrets of Human Nature.* New York: Viking, 1990: 4.

2. Since the wind blew reliably against the river's current, the Egyptians learned they could sail upstream, then drift or row downstream again, turning the Nile into a people-mover par excellence. (Anne Millard. *The Egyptians.* Morristown, N.J.: Silver Burdett, 1975: 12–13.)

3. Herodotus. *The History of Herodotus.* In *Library of the Future,* 4th ed., ver. 5.0. Irvine, Calif.: World Library, 1996. CD-ROM; Ernle Dusgate Selby Bradford. *Hannibal.* New York: McGraw-Hill, 1981; Will Durant. *Our Oriental Heritage.* Pt. 1 of *The Story of Civilization.* New York: Simon and Schuster, 1935: 291–295; Philip D. Curtin. *Cross-Cultural Trade in World History.* New York: Cambridge University Press, 1984: 76.

4. Claudius Ptolemy's *Sexta Asiae Tabula* credits this discovery to the Greek Hippalus in 50 B.C. Ptolemy, in turn, derived his information from a sailing manual called "The Periplus of the

Erythraean Sea." (Susan Ludmer-Gliebe. "Sinbads of the Sea." *Mercator's World.* http://www.mercatormag.com/306_sinbad.html. December 1998.) William H. McNeill. *Plagues and Peoples.* 1976. Reprint, New York: Anchor Books, 1998; Will Durant. *Our Oriental Heritage.* Pt. 1 of *The Story of Civilization.* Irvine, Calif.: World Library, 1991–1994. CD-ROM.)

5. Daniel J. Boorstin. *The Discoverers:A History of Man's Search to Know His World and Himself.* New York: Vintage Books, 1985: 179–181.

6. Albert A. Trever. *History of Ancient Civilization.* Vol. 2, *The Roman World.* New York: Harcourt, Brace, 1939: 546.

7. Suetonius. *C. Suetonius Tranquillus. Divi Augusti Vita,* ed. Michael Adams. London: Macmillan, 1939; Henry G. Pflaum. *Essai sur la Cursus Publicus sous le Haut-Empire Romaine* (Study of the Cursus-Publicus in the High Roman Empire). Paris: Imprimerie Nationale, 1950; PTA AG, Telekom Austria. "Kommunikation und Informationsaustausch im Wandel der Jahrhunderte—Vor 2000 Jahren: Der römische 'Cursus Publicus.'" *Geschichte des Post- und Fernmeldewesens.* http://www.pta.at/fr/ag/geschichte/index-fr.html. December 1998.

8. Albert A. Trever. *History of Ancient Civilization.* Vol. 2, *The Roman World:* 274; J. G. Landels. *Engineering in the Ancient World.* Berkeley: University of California Press, 1978: 132–166; Paul Veyne, ed. *A History of Private Life.* Vol. 1, *From Pagan Rome to Byzantium.* Cambridge, Mass.: Harvard University Press, 1987: 143, 154.

9. Geoffrey Barraclough, ed. *The Times Atlas of World History.* London: Times Books, 1984: 92–93.

10. Edward Gibbon. *The Decline and Fall of the Roman Empire.* Vol. 1, A.D. 180–A.D. 395. New York: Modern Library, n.d.: 345.

11. Robert M. Hartwell. "Fifteen Centuries of Chinese Environmental History: Creating a Retroactive Decision-support System." CITAS (The China in Time and Space project) and the University of Pennsylvania, CSS94, June 2, 1994. http://citas.csde.washington.edu/org/report_c. Downloaded September 1999.

12. Wolfram Eberhard. *A History of China.* London: Routledge and Kegan Paul, 1977: 172; Dennis Bloodworth and Ching Ping Bloodworth. *The Chinese Machiavelli: 3,000 Years of Chinese Statecraft.* New York: Farrar, Straus and Giroux, 1976: 201.

13. Allen W. Johnson and Timothy Earle. *The Evolution of Human Societies: From Foraging Group to Agrarian State.* Stanford, Calif.: Stanford University Press, 1987: 262; Philip D. Curtin. *Cross-Cultural Trade in World History:* 88.

14. Allen W. Johnson and Timothy Earle. *The Evolution of Human Societies:* 265.

15. A common language, uniform weights and measures, and the regulation of axle widths were instituted under the Ch'in, from 250 to 200 B.C. in China. (Wolfram Eberhard. *A History of China:* 63.)

16. Philip D. Curtin. *Cross-Cultural Trade in World History:* 93–96.

17. Will Durant. *Caesar and Christ.* Pt. 3 of *The Story of Civilization.* Irvine, Calif.: World Library, 1991–1994. CD-ROM.

18. Andrea Schulte-Peevers. "The Brothers Grimm and the Evolution of the Fairy Tale." *German Life,* March 31, 1996; Robert Darnton. *The Great Cat Massacre and Other Episodes in French Cultural History.* New York: Random House, 1985: 21.

19. Will Durant. *The Age of Faith.* Pt. 4 of *The Story of Civilization.* Irvine, Calif.: World Library, 1991–1994. CD-ROM.

20. E. N. Anderson. *The Food of China.* New Haven, Conn.: Yale University Press, 1988: 79; D. B. Grigg. *The Agricultural Systems of the World: An Evolutionary Approach.* Cambridge, U.K.: Cambridge University Press, 1974: 28–29.

21. D. B. Grigg. *The Agricultural Systems of the World:* 21.

22. E. N. Anderson. *The Food of China:* 80.

23. This information comes from a historical project carried out by percussionist Ralph MacDonald in which I was a participant. See also: John Storm Roberts. *Black Music of Two Worlds:African, Caribbean, Latin, and African-American Traditions.* New York: Schirmer, 1998.

24. Desiderius Erasmus. *The Correspondence of Erasmus,* trans. R. A. B. Mynors and D. F. S. Thomson. Toronto: University of Toronto Press, 1974. The collected version of Erasmus's surviving letters fills two hefty volumes.

25. William H. McNeill. *Plagues and Peoples:* 120, 134−135, 141, 188, 212−214, 215.

26. H. Philip Spratt. "The Marine Steam-Engine." In *A History of Technology.* Vol. 5, *The Late Nineteenth Century, c. 1850 to c. 1900,* ed. Charles Singer, E.J. Holmyard, A. R. Hall, and Trevor I. Williams. Oxford: Oxford University Press, 1958; Robert L. O'Connell. *Of Arms and Men: A History of War, Weapons, and Aggression.* New York: Oxford University Press, 1989: 191−192, 233; Paul Kennedy. *The Rise and Fall of the Great Powers: Economic Change and Military Conflict from 1500 to 2000.* New York: Random House, 1987: 150.

27. Wolfram Eberhard. *A History of China:* 298.

28. "Hong Kong—from Opium War to 1997 and Beyond." http://www.interlog.com/~yuan/hk.html. December 1998.

29. Aristotle. *Posterior Analytics:* 1.31. The Perseus Project, ed. Gregory R. Crane. http://hydra.perseus.tufts.edu. December 1998; Aristophanes. *Clouds.* In *Library of the Future,* 4th ed., ver. 5.0. Irvine, Calif.: World Library, 1996. CD-ROM; Cicero. *For Aulus Caecina:* 52. The Perseus Project, ed. Gregory R. Crane. http://hydra.perseus.tufts.edu. December 1998; Ovid. *Metamorphoses:* 15.352. The Perseus Project, ed. Gregory R. Crane. http://hydra.perseus.tufts.edu. December 1998; Plutarch. *Octavius:* Para 2. The Perseus Project, ed. Gregory R. Crane. http://hydra.perseus.tufts.edu. December 1998; Thomas E. Jones. "History of the Light Microscope." 1997. http://www.utmem.edu/personal/thjones/hist/c1.html. December 1998.

30. Thomas E. Jones. "History of the Light Microscope"; Daniel J. Boorstin. *The Discoverers:* 312.

31. Lewis Wolpert. "The Evolution of 'the Cell Theory'." *Current Biology* 6 (1996): 225−228. http://www.biomednet.com/library/fulltext/JCUB.bb63m2?render. December 1998; J. H. Scharf. "Turning Points in Cytology." *Acta Histochemica: Supplementband* 39 (1990): 11−47.

32. Thomas S. Hall. *Ideas on Life and Matter,* vol. 2. Chicago: University of Chicago Press, 1969: 194.

33. H. Franke. "Moritz Traube (1826−1894)—Life and Work of an Universal Private Scholar and a Pioneer of the Physiological Chemistry." http://home.t-online.de/home/henrik.franke/english.html. December 1998; Dr. Michael Engel. "Studien und Quellen zur Geschichte der Chemie." Verlag für Wissenschafts- und Regionalgeschichte. http://www.gnt-verlag.com/Engel/vp/chemie.html#chemie. December 1998.

34. Gilbert Ling. "The So-called 'Sodium-' and 'Potassium Channels.'" http://www.gilbertling.org/lp16.html,lp16. November 1998.

35. Ling claims that over two hundred journal articles have been published presenting his Association-Induction Hypothesis, disproving the sodium-potassium pump hypothesis, dealing with the criticisms of his detractors, and attempting to demonstrate the broader ramifications of his theories. In addition, he has laid out his work in three books spanning three decades: *A Physical Theory of the Living State: The Association-Induction Hypothesis; with Considerations of the Mechanics Involved in Ionic Specificity.* New York: Blaisdell. 1962; *In Search of the Physical Basis of Life.* New York: Plenum Press, 1984; *A Revolution in the Physiology of the Living Cell.* Malabar, Fla.: Krieger, 1992. See also: G. N. Ling. "Maintenance of Low Sodium and High Potassium Levels in Resting Muscle Cells." *Journal of Physiology* (Cambridge), July 1978: 105−123; G. N. Ling. "Oxidative Phosphorylation and Mitochondrial Physiology: A Critical Review of Chemiosmotic Theory and Reinterpretation by the Association-Induction Hypothesis." *Physiological Chemistry and Physics* 13:1 (1981): 29−96; G. N. Ling. "Debunking the Alleged Resurrection of the Sodium Pump Hypothesis." *Physiological Chemistry and Physics and Medical NMR* 29:2 (1997): 123−198.

36. Gilbert Ling. "Why Science Cannot Cure Cancer and AIDS without Your Help?" http://www.gilbertling.org/. November 1998.

37. G. N. Ling. "The New Cell Physiology: An Outline, Presented against Its Full Historical Background, Beginning from the Beginning." *Physiological Chemistry and Physics and Medical*

NMR 26:2 (1994): 121–203; Gilbert Ling. "Why Science Cannot Cure Cancer and AIDS without Your Help?"

38. Gilbert Ling. "Why Science Cannot Cure Cancer and AIDS without Your Help?"

39. Ibid. For more on Troshin, see: A. D. Braun, A. A. Vereninov, and A. B. Kaulin. "Afanasii Semenovich Troshin (on His Seventieth Birthday)." *Tsitologiia,* June 1983: 726–732.

40. James Mattson and Merrill Simon. *The Pioneers of NMR and Magnetic Resonance in Medicine: The Story of MRI.* Ramat Gan, Israel: Bar-Ilan University Press, 1996; *The Pioneers of NMR and Magnetic Resonance in Medicine . . . The Story of MRI.* http://www.mribook.com/index.html. December 1998; Gilbert Ling. "Raymond V. Damadian." http://www.gilbertling.org/lp30a.html. November 1998.

41. The number of researchers besides Ling who have applied the concept of thermodynamic cooperative states to biology is small, but their results are intriguing. For examples see: T. Jobe, R. Vimal, A. Kovilparambil, J. Port, and M. Gaviria. "A Theory of Cooperativity Modulation in Neural Networks as an Important Parameter of CNS Catecholamine Function and Induction of Psychopathology." *Neurological Research,* October 1994: 330–341; Y. Omura. "Inhibitory Effect of NaCl on Hog Kidney Mitochondrial Membrane-Bound Monoamine Oxidase: pH and Temperature Dependences." *Japanese Journal of Pharmacology,* December 1995: 293–302.

42. E. D. Isaacs, A. Shukla, P. M. Platzman, D. R. Hamann, B. Barbiellini, and C. A. Tulk. "Covalency of the Hydrogen Bond in Ice: A Direct X-ray Measurement." *Physical Review Letters,* January 18, 1999: 600–603; P. Weiss. "Electron Mix Binds Water Molecules." *Science News,* January 23, 1999: 52.

43. G. N. Ling. "A Physical Theory of the Living State: Application to Water and Solute Distribution." *Scanning Microscopy,* June 1988: 899–913. Explains Ling, "The self diffusion coefficient of water in living cells is one half of that in normal liquid water. The self diffusion coefficient of water molecules in ice is one one millionth (0.000001) of that in liquid water." (Gilbert Ling. Personal communication. December 9, 1998.)

44. G. N. Ling, M. M. Ochsenfeld, C. Walton, and T. J. Bersinger. "Experimental Confirmation, from Model Studies, of a Key Prediction of the Polarized Multilayer Theory of Cell Water." *Physiological Chemistry and Physics* 10:1 (1978): 87–88; G. N. Ling. "The Functions of Polarized Water and Membrane Lipids: A Rebuttal." *Physiological Chemistry and Physics* 9:4–5 (1977): 301–311; G. N. Ling and W. Hu. "Studies on the Physical State of Water in Living Cells and Model Systems. X. The Dependence of the Equilibrium Distribution Coefficient of a Solute in Polarized Water on the Molecular Weights of the Solute: Experimental Confirmation of the 'Size Rule' in Model Studies." *Physiological Chemistry and Physics and Medical NMR* 20:4 (1988): 293–307.

45. G. N. Ling. "Solute Exclusion by Polymer and Protein-Dominated Water: Correlation with Results of Nuclear Magnetic Resonance (NMR) and Calorimetric Studies and Their Significance for the Understanding of the Physical State of Water in Living Cells." *Scanning Microscopy,* June 1988: 871–884.

46. G. N. Ling. "Can We See Living Structure in a Cell?" *Scanning Microscopy,* June 1992: 405–439.

47. Ibid.; Gilbert Ling. "Some High Lights of the Association-Induction Hypothesis." http://www.gilbertling.org/lp6c.html. November 1998.

48. J. Gulati, M. M. Ochsenfeld, and G. N. Ling. "Metabolic Cooperative Control of Electrolyte Levels by Adenosine Triphosphate in the Frog Muscle." *Biophysical Journal,* December 1971: 973–980.

49. Gilbert N. Ling. *A Revolution in the Physiology of the Living Cell;* Gilbert N. Ling. Personal communication. July 12–14, 1999.

50. G. N. Ling and A. Fisher. "Cooperative Interaction among Cell Surface Sites: Evidence in Support of the Surface Adsorption Theory of Cellular Electrical Potentials." *Physiological Chemistry and Physics and Medical NMR* 15:5 (1983): 369–378.

51. Charles S. Zuker and Rama Ranganathan. "The Path to Specificity." *Science,* January 29, 1999: 650–651.

52. L. Yamasaki, L. P. Kanda, and R. E. Lanford. "Identification of Four Nuclear Transport Signal-Binding Proteins That Interact with Diverse Transport Signals." *Molecular and Cellular Biology,* July 1989: 3028–3036; D. R. Finlay, D. D. Newmeyer, P. M. Hartl, J. Horecka, and D. J. Forbes. "Nuclear Transport In Vitro." *Journal of Cell Science* (Supplement 11), 1989: 225–42; J. M. Gerrard, S. P. Saxena, and A. McNicol. "Histamine as an Intracellular Messenger in Human Platelets." *Advances in Experimental Medicine and Biology* 344 (1993): 209–219; H. Rasmussen, P. Barrett, J. Smallwood, W. Bollag, and C. Isales. "Calcium Ion as Intracellular Messenger and Cellular Toxin." *Environmental Health Perspectives,* March 1990: 17–25; R. H. Kramer. "Patch Cramming: Monitoring Intracellular Messengers in Intact Cells with Membrane Patches Containing Detector Ion Channels." *Neuron,* March 1990: 335–341; Carol Featherstone. "Coming to Grips with the Golgi." *Science,* December 18, 1998: 3172–3174; John Travis. "Outbound Traffic." *Science News,* November 15, 1997: 316–317.

53. R. Lahoz-Beltra, S. R. Hameroff, and J. E. Dayhoff. "Cytoskeletal Logic: A Model for Molecular Computation via Boolean Operations in Microtubules and Microtubule-Associated Proteins." *Biosystems* 29:1 (1993): 1–23; M. Jibu, S. Hagan, S. R. Hameroff, K. H. Pribram, and K. Yasue. "Quantum Optical Coherence in Cytoskeletal Microtubules: Implications for Brain Function." *Biosystems* 32:3 (1994): 195–209.

54. Marc Kirschner and John Gerhart. "Evolvability." *Proceedings of the National Academy of Sciences, USA,* July 1998: 8420–8427.

55. I. B. Heath. "The Cytoskeleton in Hyphal Growth, Organelle Movements and Mitosis." In *The Mycota,* vol. 1, ed. J. G. H. Wessels and F. Mienhardt. Berlin: Springer-Verlag, 1994: 43; R. Lahoz-Beltra, S. R. Hameroff, and J. E. Dayhoff. "Cytoskeletal Logic." *Biosystems:* 1–23; Evelyn Strauss. "Society for Developmental Biology Meeting: How Embryos Shape Up." *Science,* July 10, 1998: 166–167; Stella M. Hurtley. "Cell Biology of the Cytoskeleton." *Science,* March 19, 1999: 1931–1934; Yasushi Okada and Nobutaka Hirokawa. "A Processive Single-Headed Motor: Kinesin Superfamily Protein KIF1A." *Science,* February 19, 1999: 1152–1157.

56. Upinder S. Bhalla and Ravi Iyengar. "Emergent Properties of Networks of Biological Signaling Pathways." *Science,* January 15, 1999: 381–387.

57. Marc Kirschner and John Gerhart. "Evolvability." *Proceedings of the National Academy of Sciences, USA.*

58. Erkki Ruoslahti. "Stretching Is Good for a Cell." *Science,* May 30, 1997: 1345–1346.

59. K.S. Vogel. "Development of Trophic Interactions in the Vertebrate Peripheral Nervous System." *Molecular Neurobiology,* fall–winter 1993: 363–382; C. Haanen and I. Vermes. "Apoptosis: Programmed Cell Death in Fetal Development." *European Journal of Obstetrics, Gynecology, and Reproductive Biology,* January 1996: 129–133; Daniel S. Levine. "Survival of the Synapses." *The Sciences,* November–December 1988:51.

60. G. N. Ling. "Oxidative Phosphorylation and Mitochondrial Physiology." *Physiological Chemistry and Physics.*

61. Edelmann had read Ling's 1962 book, *A Physical Theory of the Living State: The Association-Induction Hypothesis.*

62. Gilbert Ling. "Corralled in an Overarching Net." http://www.gilbertling.org/lp12.html. February 1998.

63. Edelmann has demonstrated his underlying ability in his new area, publishing forty-three articles in such peer-reviewed journals as the *Journal of Biological Chemistry,* the *Journal of Neuroscience,* the *Journal of Microscopy,* and the *American Journal of Clinical Pathology.*

64. Gilbert Ling. "Affidavit." http://www.gilbertling.org/lp18.html#job_loss. November 1998.

65. Ling is not the only one to feel that the peer review process suppresses originality. E. I. Cantekin, T. W. McGuire, and R. L. Potter, writing in the *Journal of the American Medical Association,*

concluded that "the current peer review system . . . is unable to embrace dissent within the peer review process and to use dissent to serve scientific truth and the public interest." (E. I. Cantekin, T. W. McGuire, R. L. Potter. "Biomedical Information, Peer Review, and Conflict of Interest as They Influence Public Health." *Journal of the American Medical Association,* March 9, 1990: 1427–1430. See also: D. F. Horrobin. "The Philosophical Basis of Peer Review and the Suppression of Innovation." *Journal of the American Medical Association,* March 9, 1990: 1438–1441.)

66. Dr. George N. Eaves. "The Project-grant Application of the National Institutes of Health." Bethesda, Md.: National Institutes of Health, 1972. Quoted in Gilbert Ling. N.t. http://www.gilbertling.org/lp11.html#sci_gro. November 1998.

67. Ling's testimony was given before the Congressional Investigation of the Peer Review System of the National Science Foundation in 1975. (Gilbert Ling. "Peer Review System Suppresses Innovation and Progress." http://www.gilbertling.org/lp21.html, November 1998.)

68. Gilbert Ling. "Why Science Cannot Cure Cancer and AIDS without Your Help?"

69. According to *Medline,* the leading medical database, there were 1,567 articles published in peer-reviewed journals on just one of these pumps—the sodium pump—between 1966 and 1998. Each of the studies on which these articles was based had been the beneficiary of one or more funding grants.

70. I. M. Glynn and S. J. Karlish. "The Sodium Pump." *Annual Review of Physiology* 37 (1975): 13–55.

71. Gilbert Ling. "Corruption." http://www.gilbertling.org/lp15.html#Na_plp15.

72. When the two National Institutes of Health officials who had saved Ling from the funding guillotine retired, the blade finally fell and his last trickles of NIH and ONR (Office of Naval Research) financing disappeared. (Gilbert Ling. "Why Science Cannot Cure Cancer and AIDS without Your Help?")

73. Damadian began the quest which led him to develop whole-body NMR scanning with a search for Ling's nemesis, the sodium pump. His work, performed at Washington University's School of Medicine in Saint Louis, led him to Ling's conclusion—that the sodium-pump model is erroneous. Damadian credits Ling's contribution to nuclear magnetic resonance imaging in the following words: "On the morning of July 3, 1977, at 4:45 A.M. . . . we achieved with great jubilation the world's first MRI image of the live human body. The achievement originated in the modern concepts of salt water biophysics which you are the grand pioneer with your classic treatise, the association-induction hypothesis." (Gilbert Ling. "Why Science Cannot Cure Cancer and AIDS without Your Help?") Ling also "played a central role in the development of the Ling-Gerard micro-electrode. The microelectrode has subsequently proven to be one of the most important devices applied to the study of cellular physiology." (NIH Summary Statement 1 R) 11 HL 39249-01, April 30, 1987. Quoted in Gilbert Ling. "Why Science Cannot Cure Cancer and AIDS without Your Help?" For further information on Damadian, see: James Mattson and Merrill Simon. "Raymond V. Damadian (1936–)." *The Pioneers of NMR and Magnetic Resonance in Medicine . . . The Story of MRI.* http://www.mribook.com/Damad.html. December 1998.

74. The National Inventors Hall of Fame. "Raymond V. Damadian." *Inventure Place: The National Inventors Hall of Fame!* http://www.nforce.com/projects/inventure/book/book-text/28.html. December 1998; "Welcome to the FONAR Home Page—a Leading Manufacturer of MRI Scanners." http://www.fonar.com/. February 1998.

75. For some of Ling's work on cancer and drug activity, see: G. N. Ling and R. C. Murphy. "Apparent Similarity in Protein Compositions of Maximally Deviated Cancer Cells." *Physiological Chemistry and Physics* 14:3 (1982): 213; G. N. Ling and Y. Z. Fu. "An Electronic Mechanism in the Action of Drugs, ATP, Transmitters and Other Cardinal Adsorbents. II. Effect of Ouabain on the Relative Affinities for Li^+, Na^+, K^+, and Rb^+ of Surface Anionic Sites That Mediate the Entry of Cs^+ into Frog Ovarian Eggs." *Physiological Chemistry and Physics and Medical NMR* 20:1 (1988): 61–77; and G. N. Ling and M. M. Ochsenfeld. "Control of Cooperative Adsorption of Solutes and Water in Living Cells by Hormones, Drugs, and Metabolic Products." *Annals of the New York Academy*

of Sciences 204 (March 30, 1973): 325–336. Ling's lab is now known officially as the Damadian Foundation for Basic and Cancer Research.

76. Gilbert Ling. "Why Science Cannot Cure Cancer and AIDS without Your Help?"

77. Ibid.

78. F. Baquero. "Gram-Positive Resistance: Challenge for the Development of New Antibiotics." *Journal of Antimicrobial Chemotherapy* (Supplement A). May 1997: 1–6; O. Cars. "Colonisation and Infection with Resistant Gram-positive Cocci: Epidemiology and Risk Factors." *Drugs* 54 (Supplement 6), 1997: 4–10; G. G. Rao. "Risk Factors for the Spread of Antibiotic-Resistant Bacteria." *Drugs,* March 1998: 323–330; R. N. Jones. "Impact of Changing Pathogens and Antimicrobial Susceptibility Patterns in the Treatment of Serious Infections in Hospitalized Patients." *American Journal of Medicine,* June 24, 1996: 3S–12S; Valentina Stosor, Gary A. Noskin, and Lance R. Peterson. "The Management and Prevention of Vancomycin-Resistant Enterococci." *Infections in Medicine* 13:6 (1996): 487–488, 493–498.

79. The 1990s saw a spate of books predicting microbial catastrophe. See: Laurie Garrett. *The Coming Plague: Newly Emerging Diseases in a World Out of Balance.* New York: Farrar, Straus and Giroux, 1994; Richard Rhodes. *Deadly Feasts: Tracking the Secrets of a Terrifying New Plague.* New York: Simon and Schuster, 1997; Frank Ryan. *Virus-X: Tracking the New Killer Plagues: Out of the Present and into the Future.* Boston: Little, Brown, 1997; C. J. Peters and Mark Olshaker. *Virus Hunter: Thirty Years of Battling Hot Viruses around the World.* New York: Anchor Books, 1997; Rodney Barker. *And the Waters Turned to Blood: The Ultimate Biological Threat.* New York: Simon and Schuster, 1997; Richard Preston, *The Hot Zone.* New York: Random House, 1994; Edward Regis. *Virus Ground Zero: Stalking the Killer Viruses with the Centers for Disease Control.* New York: Pocket Books, 1996; and Peter Radetsky. *The Invisible Invaders: The Story of the Emerging Age of Viruses.* Boston: Little, Brown, 1991.

19. THE KIDNAP OF MASS MIND: FUNDAMENTALISM, SPARTANISM, AND THE GAMES SUBCULTURES PLAY

1. Neal Horsley. "Secession via Nuclear Weapons." The Creator's Rights Party (web page). http://www.christiangallery.com/strategy.html. November 1998.

2. *Meet the Press.* NBC-TV, October 24, 1998.

3. For an article which claims that common elements unite Christian fundamentalism and fascism, see: Charles B. Strozier. "Christian Fundamentalism, Nazism, and the Millennium." *Psychohistory Review,* winter 1990: 207–217.

4. Richard Louis Schanck. "A Study of a Community and Its Groups and Institutions Conceived of as Behaviors of Individuals." *Psychological Monographs* 43:2 (1933): 15–16.

5. Attitudes toward alcohol and tobacco don't appear in Schanck's paper, but are reported by Bertram H. Raven in Bertram H. Raven and Jeffrey Z. Rubin. *Social Psychology.* New York: John Wiley and Sons, 1983: 403–404.

6. Schanck's 133-page scholarly article reporting on his research hints at the story which appears here, but fails to tell it—a frequent lapse in scientific monographs. Schanck mentions what he calls Mrs. Salt's "personality tyranny" (126), then returns to commenting on tables of statistics. The tale behind the manner in which (quoting Schanck again) "she dominated" the community was recorded forty-four years later by UCLA's Bertram H. Raven in his book *Social Psychology.* Raven obtained the details from a colleague of Schanck's, Dan Katz. Since those who performed the Elm Hollow study are now dead, Raven has rescued a gem of social scientific information which would otherwise have been lost to posterity. (Bertram H. Raven and Jeffrey Z. Rubin. *Social Psychology:* 403–404; Bertram H. Raven. Personal communication, November 6, 1998.)

7. Richard Louis Schanck. "A Study of a Community and Its Groups and Institutions Conceived of as Behaviors of Individuals." *Psychological Monographs:* 126.

8. Ibid.: 74.

9. Ibid.: 126.

278 NOTES TO PAGES 194–195

10. See, for example, Jamie Arndt, Jeff Greenberg, Tom Pyszczynski, and Sheldon Solomon. "Subliminal Exposure to Death-Related Stimuli Increases Defense of the Cultural Worldview." *Psychological Science,* September 1997: 379–385; J. Krause. "Differential Fitness Returns in Relation to Spatial Position in Groups." *Biological Reviews of the Cambridge Philosophical Society,* May 1994: 187–206.

11. Eshel Ben-Jacob and Herbert Levine. "The Artistry of Microorganisms." *Scientific American,* October 1998: 82–87. http://www.sciam.com/1998/1098issue/1098levine.html. November 1998 (web version); E. Ben-Jacob. "From Snowflake Formation to the Growth of Bacterial Colonies. Part II. Cooperative Formation of Complex Colonial Patterns." *Contemporary Physics* 38 (1997): 205–241.

12. H. J. Rothgerber. "External Intergroup Threat as an Antecedent to Perceptions of In-group and Out-group Homogeneity." *Journal of Personality and Social Psychology,* December 1997: 1206–1212; W. N. Morris, S. Worchel, J. L. Bios, J. A. Pearson, C. A. Rountree, G. M. Samaha, J. Wachtler, and S. L. Wright. "Collective Coping with Stress: Group Reactions to Fear, Anxiety, and Ambiguity." *Journal of Personality and Social Psychology,* June 1976: 674–679.

13. I. L. Janis. *Victims of Groupthink: A Psychological Study of Foreign-policy Decisions and Fiascos.* Boston: Houghton Mifflin, 1972; M. R. Callaway, R. G. Marriott, and J. K. Esser. "Effects of Dominance on Group Decision Making: Toward a Stress-Reduction Explanation of Groupthink." *Journal of Personality and Social Psychology,* October 1985: 949–952.

14. S. E. Taylor and M. Lobel. "Social Comparison Activity under Threat: Downward Evaluation and Upward Contacts." *Psychological Review,* October 1989: 569–575.

15. L. S. Newman, K. J. Duff, R. F. Baumeister. "A New Look at Defensive Projection: Thought Suppression, Accessibility, and Biased Person Perception." *Journal of Personality and Social Psychology,* May 1997: 980–1001; P. Laudedale, P. Smith-Cunnien, J. Parker, and J. Inverarity. "External Threat and the Definition of Deviance." *Journal of Personality and Social Psychology,* May 1984: 1058–1068; Howard Bloom. *The Lucifer Principle: A Scientific Expedition into the Forces of History.* New York: Atlantic Monthly Press, 1995: 73–96.

16. The tendency of short (acute) bursts of energizers to produce benefits, and of those same energizers to do damage in lengthy (chronic) doses, shows up particularly among stress hormones, which prime the body for reaction to a crisis but, if not packed away quickly, do enormous damage. (Neil Greenberg. "Behavioral Endocrinology of Physiological Stress in a Lizard." *Journal of Experimental Zoology* (Supplement 4), 1990: 171–172; Neil Greenberg. "Ethologically Informed Design in Husbandry and Research." In *Health and Welfare of Captive Reptiles,* ed. Clifford Warwick, Fredric L. Frye, and James B. Murphy. London: Chapman and Hall, 1995: 253.

17. *ITN World News.* Independent Television Network, September 15, 1998.

18. Renee Schoof. "China Cult Could Be Formidable Foe." Associated Press, April 27, 1999.

19. Joseph Kahn. "Notoriety Now for Movement's Leader." *New York Times,* April 27, 1999.

20. Andrew Browne. "Cultists Besiege China's Seat of Power." Reuters, April 25, 1999.

21. Patrick Smith. "You Have a Job, but How About a Life?" *Business Week,* November 16, 1998; Richard Sennett. *The Corrosion of Character: The Personal Consequences of Work in the New Capitalism.* New York: W. W. Norton, 1998.

22. Aaron Bernstein. "A Floor under Foreign Factories?" *Business Week,* November 2, 1998.

23. *Pursuing Excellence. A Study of U.S. Eighth Grade Mathematics and Science Teaching, Learning, Curriculum and Achievement in International Context.* U.S. Department of Education. Washington, D.C. http://nces.ed.gov/timss/97198-4.html. November 1998.

24. Joseph Needham. *Science and Civilisation in China.* Cambridge, U.K.: Cambridge University Press, 1954–1998; Dennis Bloodworth and Ching Ping Bloodworth. *The Chinese Machiavelli: 3,000 Years of Chinese Statecraft.* New York: Farrar, Straus and Giroux, 1976.

25. Anita Chan and Robert A. Senser. "China's Troubled Workers." *Foreign Affairs,* March–April 1997: 104–117; John I. Fialka. "While America Sleeps." *Wilson Quarterly,* winter 1997: 48–63; Ethan B. Kapstein. "Workers and the World Economy." *Foreign Affairs,* May–June 1996: 16–37.

26. "Lean Company versus Job Security: Sides Battle over GM's Future." ABC news.com— Business, June 13, 1998. http://www.abcnews.com/sections/business/DailyNews/gmstrike2_ 980613/index.html. November 1998.

27. George J. Church. "Are We Better Off?" *Time*, February 26, 1998.

28. This figure was cited during October 1998 reports on *Asia Market Wrap*, a nightly television news show broadcast from Singapore. (*Asia Market Wrap*. Singapore: CNBC Asia Business News—a Service of NBC and Dow Jones.)

29. Associated Press. "Congress Approves Final 2000 Budget." April 15, 1999.

30. Edward J. Lincoln. "Japan's Financial Mess." *Foreign Affairs*, May–June 1998: 57–66; "The Japan Crisis." Institute for International Economics. Washington, D.C. http://www.iie.com/ hotjapan.html. November 1998; Adam Posen. "Some Background Q&A on Japanese Economic Stagnation. Briefing Memo Submitted to the Trade Subcommittee, Committee on Ways and Means, U.S. House of Representatives, July 13, 1998." Institute for International Economics. Washington, D.C. http://www.iie.com/japanwm.html. November 1998. The Japanese economy actually shrank by 2.8 percent in 1998. (Rich Miller, John Larkin, Peter Hadfield, Paul Wiseman, and Traci Romine. "Global Economies on Rebound? Hard Times Still Loom, So Don't Celebrate Just Yet, Experts Say." *USA Today*, April 22, 1999.)

31. "Impact of Asian Crisis on Individual States, by State." Washington, D.C.: International Trade Administration of the Department of Congress. http://www.ita.doc.gov/media/ !325asia.html. November 1998.

32. "Economic Survey of Latin America and the Caribbean, 1997–1998." ECLAC, Impact of the Asian Crisis on Latin America (LC/G.2026), May 1998. United Nations Economic Commission for Latin America and the Caribbean. http://www.cepal.org/english/Publications/survey98/ contents.html. November 1998.

33. For the manner in which peripheral movements attract the lonely and confused, see: Kelton Rhoads. "Social Influence: The Science of Persuasion and Compliance." Ca. 1997. http:// www.public.asu.edu/~kelton/Cult.html. July 1998; and Margaret Thaler Singer and Janja Lalich. *Cults in Our Midst*. San Francisco: Jossey-Bass Publishers, 1995.

34. V. Spruiell. "Crowd Psychology and Ideology: A Psychoanalytic View of the Reciprocal Effects of Folk Philosophies and Personal Actions." *International Journal of Psychoanalysis* 69:2 (1988): 171–178; I. G. Sarason and B. R. Sarason. "Concomitants of Social Support: Attitudes, Personality Characteristics, and Life Experiences." *Journal of Personality*, September 1982: 331–344.

35. Br. Abu Ruqaiyah. "The Islamic Legitimacy of the 'Martyrdom Operations,'" trans. Br. Hussein El-Chamy. *Nida'ul Islam Magazine (The Call of Islam: A Comprehensive Intellectual Magazine)*, December–January 1996–1997. http://www.islam.org.au/articles/16/martyrdom.html. August 1998.

36. A. Dean and N. Lin. "The Stress-Buffering Role of Social Support: Problems and Prospects for Systematic Investigation." *Journal of Nervous and Mental Disease*, December 1977: 403–417.

37. Peter Beinart. "Battle for the 'Burbs: Nostalgia and the Christian Right." *New Republic*. October 19, 1998: 25–29.

38. Muzafer Sherif and Hadley Cantril. *The Psychology of Ego-Involvements: Social Attitudes and Identifications*. New York: John Wiley and Sons, 1947: 83.

39. In what may be the most famous experiment on this subject, Muzafer Sherif triggered groups to chafe against each other, then worked to unify them again. To his dismay, the easiest way to "reduce intergroup tensions" turned out to be what he called "the common enemies approach." (Muzafer Sherif, O. H. Harvey, B. Jack White, William R. Hood, and Carolyn W. Sherif. *The Robbers Cave Experiment: Intergroup Conflict and Cooperation*. Middletown, Conn.: Wesleyan University Press, 1988: 45–49; see also: Howard Bloom. *The Lucifer Principle*: 77–97.)

40. Muzafer Sherif, O. H. Harvey, B. Jack White, William R. Hood, and Carolyn W. Sherif. *The Robbers Cave Experiment*.

41. Jamie Arndt, Jeff Greenberg, Tom Pyszczynski, and Sheldon Solomon. "Subliminal Exposure

to Death-Related Stimuli Increases Defense of the Cultural Worldview." *Psychological Science,* September 1997: 379–385.

42. H. A. McGregor, J. D. Lieberman, J. Greenberg, S. Solomon, J. Arndt, L. Simon, and T. Pyszczynski. "Terror Management and Aggression: Evidence that Mortality Salience Motivates Aggression against Worldview-Threatening Others." *Journal of Personality and Social Psychology,* March 1998: 590–605.

43. F. Leichsenring and H. A. Meyer. "Reducing Ambiguity: Semantic Statistical Studies of 'Normal' Probands, Neurotic Patients, Borderline Patients and Schizophrenic Patients." *Zeitschrift für klinische Psychologie, Psychopathologie und Psychotherapie* 42:4 (1994): 355–372.

44. Anne Harrington. *Reenchanted Science: Holism in German Culture from Wilhelm II to Hitler.* Princeton, N.J.: Princeton University Press, 1996: 30.

45. W. N. Morris, S. Worchel, J. L. Bios, J. A. Pearson, C. A. Rountree, G. M. Samaha, J. Wachtler, and S. L. Wright. "Collective Coping with Stress." *Journal of Personality and Social Psychology:* 674–679.

46. F. W. Wicker and E. L. Young. "Instance-Based Clusters of Fear and Anxiety: Is Ambiguity an Essential Dimension?" *Perceptual and Motor Skills,* February 1990: 115–121.

47. R. M. Doty, B. E. Peterson, and D. G. Winter. "Threat and Authoritarianism in the United States, 1978–1987." *Journal of Personality and Social Psychology,* October 1991: 629–640; Robert S. Robins. "Paranoid Ideation and Charismatic Leadership." *Psychohistory Review,* fall 1986: 15–55.

48. Charles B. Strozier. "Christian Fundamentalism, Nazism, and the Millennium." *Psychohistory Review:* 207–217.

49. The age of extremist outsiders armed with atomic, biological, and chemical weapons had already come upon us as early as 1998, declared no less an authority than *Foreign Affairs,* the journal of America's statesmen. (Ashton Carter, John Deutch, and Philip Zelikow. "Catastrophic Terrorism." *Foreign Affairs,* November–December 1998: 80–94.)

50. James Walsh. "Shoko Asahara: The Making of a Messiah." *Time,* April 3, 1995. http://cgi.pathfinder.com/time/. September 1998.

51. "Aum Shinrikyo, which is an extraordinary and extraordinarily disturbing example of the worldwide epidemic of fundamentalism and the particular danger of apocalyptic violence, . . . had in mind nothing short of worldwide genocide. They wanted to murder everybody in the world. Such fantasies are widespread." (Robert Jay Lifton. "Beyond Armageddon: New patterns of ultimate violence." *Modern Psychoanalysis* 22:1 (1997): 17–29; James Walsh. "Shoko Asahara." *Time.*)

52. Ian MacKenzie. "Philippine Negotiators Agree on Peace Accord." Reuters, August 29, 1996; "Manila Extends Truce in Southern Philippines." Xinhua News Agency, China: July 10, 1997.

53. Office of the United Nations High Commissioner for Refugees. "Background Paper on Refugees and Asylum Seekers from Afghanistan." UNHCR Centre for Documentation and Research, Geneva, June 1997. http://www.unhcr.ch/refworld/country/cdr/cdrafg.html. November 1998.

54. United Nations Office for the Coordination of Humanitarian Affairs. "Humanitarian Report 1997: Gender Issues in Afghanistan." http://www.reliefweb.int/dha_ol/pub/humrep97/gender.html. November 1998; Shadaba Islam. "International Campaign for Afghan Women." *Dawn* (Pakistan Herald Publications). http://dawn.com/daily/19980204/top12.html. November 1998; Feminist Majority Foundation. "Stop Gender Apartheid in Afghanistan." *Feminist Majority Foundation Online.* http://www.feminist.org/afghan/facts.html. November 1998.

55. In addition to chronicling the incidents cited here, Amnesty International reported that the Taliban had forced a father and brothers to shoot their own family member before an audience of thirty thousand gathered in a stadium. Toppling walls was a means of execution for those deemed guilty of sodomy. In the opinion of Amnesty International, none of the accused were given access to a true court proceeding, hence their alleged crimes were highly questionable. (Amnesty International News Service. "Afghanistan: Public Executions and Amputations on Increase." AI INDEX: ASA 11/05/98, May 21, 1998. http://www2.amnesty.se/isext98.nsf/

eb4dea1ac25099c4c12565b7005b76ae/debc87e79452a698c125660a0063e0a8?OpenDocument. November 1998.)

56. Agence France Presse. "Taleban: 'Film and Music Leads to Moral Corruption,' " July 8, 1998.

57. Amnesty International News Service. "Afghanistan: Thousands of Civilians Killed Following Taleban Takeover of Mazar-e Sharif." AI INDEX: ASA 11/07/98, September 3, 1998. http://www2.amnesty.se/isext98.nsf/eb4dea1ac25099c4c12565b7005b76ae/598fd5236b885a2fc125667b003a63d5?OpenDocument. November 1998.

58. *ITN World News.* Independent Television Network News, September 4, 1998.

59. The Commonwealth Business Forum. "Pakistan—Society." http://www.tcol.co.uk/pakistan/pak2.html#society. November 1998; Government Statistical Service—the Office for National Statistics, United Kingdom. http://www.statistics.gov.uk/stats/ukinfigs/pop.html. November 1998; Information Please. "World's Twenty Most Populous Countries: 1998 and 2028." infoplease.com. Information Please LLC. http://www.infoplease.com/ipa/A0004391.html.

60. Robert Fisk. "Saudis Secretly Funding Taleban." *Independent* (London), September 2, 1998: 9.

61. Robert Fisk. "Anti-Americans of Saudi Arabia." *Independent on Sunday,* reprinted in *World Press Review,* October 1998: 16.

62. John Stackhouse. "Will Karachi Be the Next Kabul?" *Toronto Globe and Mail,* reprinted in *World Press Review,* October 1998: 17.

63. "The Friday Speech from Alaqsa Mosque in Occupied Muslim Land." January 28, 1988. *Ramadhan/Alaqsa Islamic Website.* http://www.ramadhan.org/hadith/hbody12.txt. August 1998.

64. It was Pakistani prime minister Zulfikar Ali Bhutto who first popularized the slogan "the Islamic Bomb." (Kathy Gannon. "New Capability Meant to Bring Pride to Islamic World—Pakistani Nuclear Testing." Associated Press, May 31, 1998.)

65. Greg Sheridan. Untitled. *Australian,* reprinted in *World Press Review,* October 1998: 15.

66. "Fear Along the Silk Road: The Taleban Virus." *Middle East,* October 1998, reprinted in *World Press Review,* December 1998: 6–7.

67. Carlotta Gall. "Dagestan Skirmish Is Big Russian Risk." *New York Times,* August 13, 1999.

68. Gareth Jones. "Russian Minister Briefs FBI Boss on Dagestan." Moscow: Reuters, September 8, 1999.

69. Martin Nesirky. "FOCUS-CIS Faces International Terror Menace—Putin." Moscow: Reuters, September 15, 1999.

70. abusulaymaan@my-dejanews.com.Posting to misc.activism.militia,3 Sep 1998. When I asked abusulaymaan for information on his (or her) identity and background, I received no reply.

71. Greg Sheridan. Untitled. *Australian,* reprinted in *World Press Review;* C. Raja Mohan. Untitled. *Hindu,* reprinted in *World Press Review,* October 1998: 15.

72. For coverage of Germany's reborn fascisms, see: Jacob Heilbrunn. "Germany's New Right: Liberating Nationalism from History." *Foreign Affairs,* November–December 1996: 80–98.

73. Boguslaw Zajac. " 'Death to the Jews and Masons.' " *Rzeczpospolita,* reprinted in *World Press Review,* October 1998: 11–12; Rod Usher, Yuri Zarakhovich. "The New Russian Nazis." *Time,* June 9, 1997. http://cgi.pathfinder.com/time, downloaded November 1998.

74. "Teetering toward Fascism." *Economist,* reprinted in *World Press Review,* October 1998: 8–9.

75. *ITN World News.* "Death Penalty for McVeigh: Jury Must Decide." Independent Television Network News, June 2, 1997. http://www.itn.co.uk/World/world0602/060206w.html. November 1998.

76. For a chilling enunciation of the Christian Identity movement's message of hatred in the name of Yahweh, see: Meggie. "Can a Christian Hate?" *Christian Identity* (online magazine). http://www4.stormfront.org/posterity/ci/hate.html. For the full, hate-charged, racist, neo-Nazi, allegedly biblical worldview of the movement, see the Christian Identity Homepage—"Our Legacy of Truth." http://www4.stormfront.org/posterity/index.html.

77. America's Promise Ministries. "America's Promise Ministries Homepage." http://www.amprom.org/index.html. November 1998.

78. Frederick Clarkson. "Christian Reconstructionism: Theocratic Dominionism Gains Influence." *Public Eye Magazine*. http://www.publiceye.org/pra/magazine/chrisrel.html. November 1998; Frederick Clarkson. *Eternal Hostility: The Struggle between Theocracy and Democracy.* Monroe, Maine: Common Courage Press, 1997.

79. Aryan Nations (web site). http://www.nidlink.com/~aryanvic/. November 1988.

80. "The old Jerusalem [the Jews] now has the role of Canaan and is to be destroyed (Matt. 24). The whole world is the new Canaan, to be judged and conquered: 'Go ye into all the world . . .'" David Chilton. *The Days of Vengeance: An Exposition of the Book of Revelation.* Ft. Worth, Tex.: Dominion Press, 1984: 217; quoted in Paul Thibodeau. "The New Canaan." *Christian Reconstruction: God's Glorious Millennium? Anthology of Quotations.* http://www.serve.com/thibodep/cr/canaan.html. November 1998.

81. James Ridgeway. "The Far Right's Bomb Squad." *Freedom Writer Magazine,* March–April 1997: 7–8; James Ridgeway. "Identity Politics." *Freedom Writer Magazine,* July–August 1997: 3–5.

82. "Witness Says Michigan Officer Passed Information to Militia." Associated Press, November 4, 1998. *Nando Times.* http://www.nandotimes.com. November 1998.

83. Anonymous. "Letters from the Underground—Part II." *No Compromise: The Militant, Direct Action Magazine of Grassroots Animal Liberationists and Their Supporters.* http://www.nocompromise. org/. November 1998.

84. Federal Bureau of Investigation. "Top Ten Fugitive: Eric Robert Rudolph." *The FBI's Ten Most Wanted Fugitives.* http://www.fbi.gov/mostwant/rudolph.html. November 1998; "Search Intensifies after Bombing Suspect Sighted." Associated Press, July 14, 1998. http://interactive.cfra.com/1998/07/14/47646.html. November 1998.

85. Twenty-one-year-old Matthew Shepard was the gay student. His death was cheered by the Christian fundamentalist group God Hates Fags, run by Reverend Fred Phelps. (God Hates Fags [web site]. http://www.godhatesfags.com/. November 1998; God Hates Fags. "Picket of Matthew Shepard's Funeral, Wyoming." God Hates Fags [web site]. http://www.godhatesfags.com/shepard_funeral.jpg. November 1998; Fred Phelps Resource Page. http://members.tripod.com/~fredphelps/. November 1998.) Featured at the top of a key God Hates Fags webpage was the following quote: "'If a man also lie with mankind, as he lieth with a woman, both of them have committed an abomination: they shall surely be put to death; their blood shall be upon them.'—Leviticus 20:13." (God Hates Fags. "Facts and Statistics about Fags." www.godhatesfags.com/fagfacts.html. November 1998.)

86. Christian Gallery. "The Nuremberg Files: Alleged Abortionists and Their Accomplices." http://www.bestchoice.com/atrocity/aborts.html. November 1998.

87. Kenneth S. Stern. "Militias and the Religious Right." *Freedom Writer,* October 1996. http://apocalypse.berkshire.net/~ifas/fw/9610/militias.html. November 1998.

88. For a compilation of quotations on the militant Christian Fundamentalist mission of world conquest in the name of a biblical God and his chosen people, see: Paul Thibodeau. *Christian Reconstruction: God's Glorious Millennium?—Anthology of Quotations.* http://www.serve.com/thibodep/cr/worldcnq.html. November 1998. Thibodeau features the following statement on a focal webpage: "Christian politics has as its primary intent the conquest of the land—of men, families, institutions, bureaucracies, courts, and governments for the Kingdom of Christ." Other quotations make clear that this conquest has been ordained by God to include all of the globe and of humanity.

89. Plato. *Laws.* In *Library of the Future,* 4th ed., ver. 5.0. Irvine, Calif.: World Library, 1996. CD-ROM.

90. Paranoia sometimes proves an accurate mode of perception. The reason: leaders of the movements which oppose you may very well use paranoid delusions about *you* to fuel their movements. See: Robert S. Robins. "Paranoid Ideation and Charismatic Leadership." *Psychohistory Review,* fall 1986: 15–55.

91. Karl Dietrich Bracher. *The German Dictatorship: The Origins, Structure, and Effects of National Socialism.* New York: Praeger, 1970; William L. Shirer. *Twentieth Century Journey: The Nightmare Years, 1930–1940.* New York: Bantam Books, 1985.

92. The secret to productivity, according to *Business Week* magazine's Gene Koretz, is "learning growth that is facilitated by close contact between diverse individuals." (Gene Koretz. "Bright Lights, Midsize City—What Towns Are Best for Business?" *Business Week,* November 2, 1998.) The mathematical models of Stuart Kauffman produce the same conclusion, as does the economic analysis of history provided by Jane Jacobs. (Stuart Kauffman. *At Home in the Universe: The Search for the Laws of Self-Organization and Complexity.* New York: Oxford University Press, 1995: 295; Jane Jacobs. *Cities and the Wealth of Nations: Principles of Economic Life.* New York: Random House, 1984.)

93. Alfred Darnell and Darren E. Sherkat. "The Impact of Protestant Fundamentalism on Educational Attainment." *American Sociological Review,* April 1997: 306–315.

94. Quoted in Anne Harrington. *Reenchanted Science:* 85. Harrison Salisbury, in his history of the Russian Revolution, noted a similar retreat from thought to rigid emotional reflex among the intellectuals of Russia upon the outbreak of World War I. (Harrison Evans Salisbury. *Black Night, White Snow: Russia's Revolutions, 1905–1917.* New York: Da Capo, 1981.)

95. R. L. Montgomery, S. W. Hinkle, and R. F. Enzie. "Arbitrary Norms and Social Change in High- and Low-Authoritarian Societies." *Journal of Personality and Social Psychology,* June 1976: 698–708.

96. T. Fukai and S. Tanaka. "A Simple Neural Network Exhibiting Selective Activation of Neuronal Ensembles: From Winner-Take-All to Winners-Share-All." *Neural Computation,* January 1997: 77–97.

97. Stuart Kauffman. *At Home in the Universe:* 255–257. Kauffman regards a social group as a search device whose individual members can be freed to pursue their own interests or be yoked to just one dictatorial purpose. When only one search approach is allowed, Kauffman shows how easily the system can be frozen in a dead-end strategy. Giving heft to Kauffman's conclusions, businesses at the end of the twentieth century learned that an authoritarian corporate structure kills collective intelligence, and that to revive, a company must tap its strains of internal diversity. (Russell Lincoln Ackoff. *The Democratic Corporation: A Radical Prescription for Recreating Corporate America and Rediscovering Success.* New York: Oxford University Press, 1994; Michael Ray Rinzler and Alan Rinzler, eds. *The New Paradigm in Business: Emerging Strategies for Leadership and Organizational Change.* New York: J. P. Tarcher/Perigee, 1993; Peter Corning. "The Sociobiology of Democracy." *Social Science Information,* in press.)

98. D. Cohen, R. E. Nisbett, B. F. Bowdle, and N. Schwarz. "Insult, Aggression, and the Southern Culture of Honor: An 'Experimental Ethnography.'" *Journal of Personality and Social Psychology,* May 1996: 945–959.

99. *ABC-TV Weekend News,* October 25, 1997.

100. For historical background on southern cultural attitudes toward violence, see Samuel C. Hyde, Jr. *Pistols and Politics: The Dilemma of Democracy in Louisiana's Florida Parishes, 1810–1899.* Baton Rouge: Louisiana State University Press, 1996. See also: Dov Cohen. "Culture, Social Organization and Patterns of Violence." *Journal of Personality and Social Psychology,* August 1998: 408–419.

101. Shihadeh, of the Louisiana Population Data Center, is a reviewer for sociology's leading journal, *American Sociological Review.* His research focuses on the structural determinants of violent crime in black and white urban communities. For more on his background, see: Louisiana Population Data Center. "Dr. Edward S. Shihadeh." http://www.lapop.lsu.edu/shihadeh.html. November 1998.

102. C. Russell and W. M. Russell. "The Natural History of Violence." *Journal of Medical Ethics,* September 1979: 108–116.

103. T. Fukai and S. Tanaka. "A Simple Neural Network Exhibiting Selective Activation of Neuronal Ensembles." *Neuronal Computation:* 77–97.

104. Deborah M. Gordon. "The Expandable Network of Ant Exploration." *Animal Behavior* 50

(1995): 995–1007; Deborah M. Gordon. "How Do Ants Find the Cake Crumb You Drop?" *Natural History,* September 1997. Gordon's work on ants is mirrored by findings on search patterns in bee colonies and among dictyostelium (cellular slime mold). It is also reflected in Valerius Geist's work on maintenance versus dispersal phenotypes among large mammals. (Valerius Geist. *Life Strategies, Human Evolution, Environmental Design: Toward a Biological Theory of Health.* New York: Springer-Verlag, 1978: 262.)

105. Peter Beinart. "Battle for the 'Burbs." *New Republic.*

106. William L. Shirer. *Twentieth Century Journey: The Nightmare Years, 1930–1940.*

107. Ibn Khaldun. *The Muqaddimah: An Introduction to History,* ed. and abr. N. J. Dawood, trans. Franz Rosenthal. Princeton, N.J.: Princeton University Press, 1967: 91–115.

108. Writes historian Willard B. Gatewood, "Since 'one can be a "ruralist" by adoption as well as inheritance,' fundamentalism gained popularity among people who moved to the city without becoming 'urban-minded.' " Willard B. Gatewood Jr. *Controversy in the Twenties: Fundamentalism, Modernism, and Evolution.* Nashville: Vanderbilt University Press, 1969: 37–38.

20. INTERSPECIES GLOBAL MIND

1. David McFarland, ed. *The Oxford Companion to Animal Behavior.* New York: Oxford University Press, 1982: 319.

2. Christine R. Hurd. "Interspecific Attraction to the Mobbing Calls of Black-Capped Chickadees *(Parus Antricapillus)*." *Behavioral Ecology and Sociobiology,* April 1996: 287–292. For numerous additional examples of these interspecies meshes, see: Peter A. Corning. *The Synergism Hypothesis: A Theory of Progressive Evolution.* New York: McGraw-Hill, 1983; and John N. Thompson. "The Evolution of Species Interactions." *Science,* June 25, 1999: 2116–2118.

3. Thirteen/WNET and BBC-TV, coproducers. "Hunters of the Sea Wind." New York: Thirteen/WNET, 1996.

4. Christine A. Nalepa. "Nourishment and the Origin of Termite Eusociality." In *Nourishment and Evolution in Insect Societies,* ed. James H. Hunt and Christine A. Nalepa. Boulder, Colo.: Westview Press, 1994: 57–96.

5. Some cooperative bacteria go further than simply exchanging services. They actually remodel the tissue of their host to provide advantages for both species. (J. S. Foster and M. J. McFall-Ngai. "Induction of Apoptosis by Cooperative Bacteria in the Morphogenesis of Host Epithelial Tissues." *Development Genes and Evolution,* August 1998: 295–303.)

6. "Bugs That Energize Termites." *Science,* January 29, 1999: 601j; J. R. Leadbetter, T. M. Schmidt, J. R. Graber, and J. A. Breznak. "Acetogenesis from H_2 Plus CO_2 by Spirochetes from Termite Guts." *Science,* January 29, 1999: 686–689.

7. The interspecies web sometimes turns vicious, as when the malarial plasmodium manipulates the nervous system of a mosquito and makes it bite recklessly—a great way to spread malarial plasmodia to new victims and also to shorten the mosquito's life. (Virginia Morell. "How the Malaria Parasite Manipulates Its Hosts." *Science,* October 10, 1997: 223.)

8. Nancy E. Beckage. "The Parasitic Wasp's Secret Weapon." *Scientific American,* November 1997: 82.

9. The Comm Tech Lab and the Center for Microbial Ecology at Michigan State University. *The Microbe Zoo DLC-ME Project.* http://commtechlab.msu.edu/sites/dleme/zoo/. Downloaded September 1999.

10. James Vaughn Kohl and Robert T. Francoeur. *The Scent of Eros: Mysteries of Odor in Human Sexuality.* New York: Continuum, 1995; James Vaughn Kohl. Personal communication, May 9, 1996; James Vaughn Kohl. Poster presentation: Eighteenth Annual Meeting of the Association for Chemoreception Sciences, Sarasota, Fla., April 17–21, 1996; K. Grammer and A. Jütte. "Battle of

Odors: Significance of Pheromones for Human Reproduction." *Gynakologisch-Geburtshilfliche Rundschau* 37:3 (1997): 150–153.

11. K. Stern and M. K. McClintock. "Regulation of Ovulation by Human Pheromones." *Nature,* March 12, 1998: 177–179.

12. William B. Whitman, David C. Coleman, and William J. Wiebe. "Prokaryotes: The Unseen Majority." *Proceedings of the National Academy of Sciences of the United States of America,* June 9, 1998: 6578–6583; Richard Gallagher. "Monie a Mickle Maks a Muckle." *Science,* July 10, 1998: 186.

13. Katrin Pütsep, Carl-Ivar Brändén, Hans G. Boman, and Staffan Normark. "Antibacterial Peptide from *H. Pylori.*" *Nature,* April 22, 1999: 671–672.

14. These bits of repetitive DNA, though a part of the genetic code, are called microsatellites. (E. Richard Moxon and Christopher Wills. "DNA Microsatellites: Agents of Evolution." *Scientific American,* January 1999: 94–99; M. Radman, I. Matic, J. A. Halliday, and F. Taddei. "Editing DNA Replication and Recombination by Mismatch Repair: From Bacterial Genetics to Mechanisms of Predisposition to Cancer in Humans." *Philosophical Transactions of the Royal Society of London. Series B: Biological Sciences,* January 30, 1995: 97–103; E. R. Moxon, P. B. Rainey, M. A. Nowak, and R. E. Lenski. "Adaptive Evolution of Highly Mutable Loci in Pathogenic Bacteria." *Current Biology,* January 1, 1994: 24–33.)

15. Hidde L. Ploegh. "Viral Strategies of Immune Evasion." *Science,* April 10, 1998: 248–253.

16. Michael Balter. "Viruses Have Many Ways to Be Unwelcome Guests." *Science,* April 10, 1998: 204–205.

17. T. Hatch. "Chlamydia: Old Ideas Crushed, New Mysteries Bared." *Science,* October 23, 1998: 638; Richard S. Stephens, Sue Kalman, Claudia Lammel, Jun Fan, Rekha Marathe, L. Aravind, Wayne Mitchell, Lynn Olinger, Roman L. Tatusov, Qixun Zhao, Eugene V. Koonin, and Ronald W. Davis. "Genome Sequence of an Obligate Intracellular Pathogen of Humans: *Chlamydia trachomatis.*" *Science,* October 23, 1998: 754–759; J. Travis. "Chlamydia Bacterium Yields Surprise Genome." *Science News,* October 24, 1998: 261.

18. D. McKenney, K. E. Brown, and D. G. Allison. "Influence of *Pseudomonas Aeruginosa* Exoproducts on Virulence Factor Production in *Burkholderia Cepacia:* Evidence of Interspecies Communication." *Journal of Bacteriology,* December 1995: 6989–6992; Evelyn Strauss. "Mob Action: Peer Pressure in the Bacterial World." *Science News,* August 23, 1997: 124–125. For more on the manner in which bacteria converse chemically with other bacterial species, plants, and animals, see: Richard Losick and Dale Kaiser. "Why and How Bacteria Communicate." *Scientific American,* February 1997: 68–73.

19. S. Iyobe. "Mechanism of Acquiring Drug-Resistance Genes in Pathogenic Bacteria." *Nippon Rinsho: Japanese Journal of Clinical Medicine.* May 1997: 1179–1184; D. G. Guiney Jr. "Promiscuous Transfer of Drug Resistance in Gram-Negative Bacteria." *Journal of Infectious Diseases,* March 1984: 320–329; Stuart B. Levy. "The Challenge of Antibiotic Resistance." *Scientific American,* March 1998: 46–53.

20. Julian Davies. "Inactivation of Antibiotics and the Dissemination of Resistance Genes." *Science,* April 15, 1994: 375–382.

21. A colony of *E. coli* doubles in size every twenty minutes.

22. Inon Cohen. "Re: Mechanisms of Genes Transfer." Unpublished manuscript. Tel Aviv, 1999.

23. Biotechnology Industry Organization. "Cloning Genes." In *Biology Hypertextbook.* Cambridge, Mass.: Experimental Study Group, Massachusetts Institute of Technology. Massachusetts Institute of Technology. http://esg-www.mit.edu:8001/esgbio/rdna/cloning.html. January 1999.

24. The bacterially derived enzyme used in the polymerase chain reaction is Taq DNA Polymerase. For a simple explanation of PCR, see: Genentech. "Polymerase Chain Reaction—From Simple Ideas." Access Excellence. http://www.gene.com/ae/AB/IE/PCR_From_Simple_Ideas. html. January 1999.

25. Philip H. Abelson. "Pharmaceuticals Based on Biotechnology." *Science,* August 9, 1996: 719.

26. G. Chen, Z. Guan, C. T. Chen, L. Fu, V. Sundaresan, and F. H. Arnold. "A Glucose-Sensing Polymer." *Nature Biotechnology,* April 14, 1997: 354–357; G. W. Wilson. "Making an Imprint on Blood-Glucose Monitoring." *Nature Biotechnology,* April 14, 1997: 322; J. Raloff. "New Glucose Test on the Way for Diabetes." *Science News,* March 29, 1997: 190.

27. Wendy A. Petka, James L. Harden, Kevin P. McGrath, Denis Wirtz, and David A. Tirrell. "Reversible Hydrogels from Self-Assembling Artificial Proteins." *Science,* July 17, 1998: 389–392.

28. J. Schmidt and B. K. Ahring. "Granular Sludge Formation in Upflow Anaerobic Sludge Blanket (UASB) Reactors." *Biotechnology and Bioengineering* 49 (1996): 229–246; James A. Shapiro. "Thinking about Bacterial Populations as Multicellular Organisms." *Annual Review of Microbiology.* Palo Alto, Calif.: Annual Reviews, 1998: 81–104, 91–92.

29. Robert F. Service. "Microbiologists Explore Life's Rich, Hidden Kingdoms." *Science,* March 21, 1997: 1740–1742.

30. T. Abee, L. Krockel, and C. Hill. "Bacteriocins: Modes of Action and Potentials in Food Preservation and Control of Food Poisoning." *International Journal of Food Microbiology,* December 1995: 169–185; Janet Raloff. "Staging Germ Warfare in Foods." *Science News,* February 7, 1998: 89–90.

31. David Schneider. "Where No Brush Can Reach." *Scientific American,* December 1998: 40, 44.

32. J. Travis. "Novel Bacteria Have a Taste for Aluminum." *Science News,* May 30, 1998: 341.

33. U.S. Department of Energy. DOE Microbial Genome Program. http://www.er.doe.gov/ production/ober/EPR/mig_top.html. January 1999.

34. M. D. Smith, C. I. Masters, E. Lennon, L. B. McNeil, and K. W. Minton. "Gene Expression in *Deinococcus Radiodurans.*" *Gene,* February 1991: 45–52; Michael J. Daly and Kenneth W. Minton. "Contribution to Sequencing of the *Deinococcus Radiodurans* Genome." DOE Human Genome Program Contractor-Grantee Workshop VI, November 9–13, 1997, Santa Fe, N.Mex. http:// www.ornl.gov/hgmis/publicat/97santa/sequence.html#164. January 1999.

35. C. C. Lange, L. P. Wackett, K. W. Minton, and M. J. Daly. "Engineering a Recombinant *Deinococcus Radiodurans* for Organopollutant Degradation in Radioactive Mixed Waste Environments." *Nature Biotechnology,* October 1998: 929; John Travis. "Meet the Superbug." *Science News,* December 12, 1998: 376–378.

36. Anne Simon Moffat. "Toting Up the Early Harvest of Transgenic Plants." *Science,* December 18, 1998: 2176–2178.

37. David A. O'Brochta and Peter W. Atkinson. "Building the Better Bug." *Scientific American,* December 1998: 90–95.

38. David E. Kerr, Fengxia Liang, Kenneth R. Bondioli, Haiping Zhao, Gert Kreibich, Robert J. Wall, and Tung-Tien Sun. "The Bladder as a Bioreactor: Urothelium Production and Secretion of Growth Hormone into Urine." *Nature Biotechnology,* January 1998: 75; H. Meade and C. Ziomek. "Urine as a Substitute for Milk?" *Nature Biotechnology,* January 1998: 21; S. Milius. "Future Farmers May Collect Urine, Not Milk." *Science News,* January 10, 1998, http://www.sciencenews.org/ sn_arc98/1_10_98/fob3.html. January 1998.

39. Anne Simon Moffat. "Improving Gene Transfer into Livestock." *Science,* November 27, 1998: 1619a.

40. Anne Simon Moffat. "Down on the Animal Pharm." *Science,* December 18, 1998: 2177.

41. Wade Roush. "Biotech Finds a Growth Industry." *Science,* July 19, 1996: 300–301.

42. G. Filice, L. Soldini, P. Orsolini, E. Razzini, R. Gulminetti, D. Campisi, L. Chiapparoli, E. Cattaneo, and G. Achilli. "Sensitivity and Specificity of Anti-HIV ELISA Employing Recombinant (p24, p66, gp120) and Synthetic (gp41) Viral Antigenic Peptides." *Microbiologica,* July 1991: 185–194.

43. B. Weber, E. W. Fall, A. Berger, and H. W. Doerr. "Reduction of Diagnostic Window by New Fourth-Generation Human Immunodeficiency Virus Screening Assays." *Journal of Clinical Microbiology,* August 1998: 2235–2239; A. Tanabe-Tochikura, M. T. Ang Singh, H. Tsuchie, J. Zhang, F. J. Paladin, and T. Kurimura. "A Newly Developed Immunofluorescence Assay for Simultaneous Detection of Antibodies to Human Immunodeficiency Virus Type 1 and Type 2." *Journal of Virological Methods,* April 1995: 239–246.

44. Q. J. Sattentau, J. P. Moore, F. Vignaux, F. Traincard, and P. Poignard. "Conformational Changes Induced in the Envelope Glycoproteins of the Human and Simian Immunodeficiency Viruses by Soluble Receptor Binding." *Journal of Virology,* December 1993: 7383–7393; Michael Emerman and Michael H. Malim. "HIV-1 Regulatory/Accessory Genes: Keys to Unraveling Viral and Host Cell Biology." *Science,* June 19, 1998: 1880–1884.

45. M. Grez, U. Dietrich, P. Balfe, H. von Briesen, J. K. Maniar, G. Mahambre, E. L. Delwart, J. I. Mullins, and H. Rübsamen-Waigmann. "Genetic Analysis of Human Immunodeficiency Virus Type 1 and 2 (HIV-1 and HIV-2) Mixed Infections in India Reveals a Recent Spread of HIV-1 and HIV-2 from a Single Ancestor for Each of These Viruses." *Journal of Virology,* April 1994: 2161–2168.

46. David Baltimore and Carole Heilman. "HIV Vaccines: Prospects and Challenges." *Scientific American,* July 1998: 98.

47. Ibid.

48. John G. Bartlett and Richard D. Moore. "Improving HIV Therapy." *Scientific American,* July 1998: 93. For vigorous agreement that the research effort against AIDS has advanced medical science substantially, see: Harold Varmus and Neal Nathanson. "Science and the Control of AIDS." *Science,* June 19, 1998: 1815a.

49. C. Scholtissek. "Source for Influenza Pandemics." *European Journal of Epidemiology,* August 1994: 455–488.

50. V. S. Hinshaw, R. G. Webster, and B. Turner. "Water-borne Transmission of Influenza A Viruses?" *Intervirology* 11:1 (1979): 66–68.

51. V. F. Nettles, J. M. Wood, and R. G. Webster. "Wildlife Surveillance Associated with an Outbreak of Lethal H5N2 Avian Influenza in Domestic Poultry." *Avian Diseases,* July–September 1985: 733–741; R. G. Webster. "Influenza Virus: Transmission between Species and Relevance to Emergence of the Next Human Pandemic." *Archives of Virology Supplementum* 13 (1997): 105–113; Patricia Gadsby. "Fear of Flu." *Discover,* January 1999: 82–89.

52. J. Süss, J. Schäfer, H. Sinnecker, and R. G. Webster. "Influenza Virus Subtypes in Aquatic Birds of Eastern Germany." *Archives of Virology* 135:1–2 (1994): 101–114.

53. R. G. Webster. "Influenza Virus." *Archives of Virology Supplementum.*

54. L. P. Shu, G. B. Sharp, Y. P. Lin, E. C. Claas, S. L. Krauss, K. F. Shortridge, and R. G. Webster. "Genetic Reassortment in Pandemic and Interpandemic Influenza Viruses: A Study of 122 Viruses Infecting Humans." *European Journal of Epidemiology,* February 1996: 63–70; B. R. Murphy, J. Harper, D. L. Sly, W. T. London, N. T. Miller, and R. G. Webster. "Evaluation of the A/Seal/ Mass/1/80 Virus in Squirrel Monkeys." *Infection and Immunity,* October 1983: 424–426.

55. J. R. Schäfer, Y. Kawaoka, W. J. Bean, J. Süss, D. Senne, and R. G. Webster. "Origin of the Pandemic 1957 H2 Influenza A Virus and the Persistence of Its Possible Progenitors in the Avian Reservoir." *Virology,* June 1993: 781–788.

56. R. G. Webster. "Influenza Virus." *Archives of Virology Supplementum.*

57. Patricia Gadsby. "Fear of Flu." *Discover.*

58. W. Graeme Laver, Norbert Bischofberger, and Robert G. Webster. "Disarming Flu Viruses." *Scientific American,* January 1999: 78–87; Jon Cohen. "The Flu Pandemic That Might Have Been." *Science,* September 12, 1997: 1600–1601; C. Scholtissek. "Source for Influenza Pandemics." *European Journal of Epidemiology.*

59. W. Graeme Laver, Norbert Bischofberger, and Robert G. Webster. "Disarming Flu Viruses." *Scientific American:* 86.

60. E. C. Claas, A. D. Osterhaus, R. van Beek, J. C. De Jong, G. F. Rimmelzwaan, D. A. Senne, S. Krauss, K. F. Shortridge, and R. G. Webster. "Human Influenza A H5N1 Virus Related to a Highly Pathogenic Avian Influenza." *Lancet,* February 1998: 472−477; R. G. Webster. "Isolation of the Potentially Pandemic H5N1 Influenza Virus." In *St. Jude Children's Research Hospital 1997 Research Report.* http://www.stjude.org/sr/vmb.html. December 1998.

61. The Dutch team was centered at the Erasmus University Department of Virology and WHO National Influenza Centre in Rotterdam.

62. Leaders of the Dutch team included J. C. de Jong, W. L. Lim, E. C. J. Claas, and A. D. M. E. Osterhaus. (*St. Jude Children's Research Hospital 1997 Research Report.* http://www.stjude.org/sr/vmb.html. December 1998.)

63. R. G. Webster. "Isolation of the Potentially Pandemic H5N1 Influenza Virus." In *St. Jude Children's Research Hospital 1997 Research Report.*

64. Jon Cohen. "The Flu Pandemic That Might Have Been." *Science;* Patricia Gadsby. "Fear of Flu." *Discover;* J. C. de Jong, E. C. Claas, A. D. Osterhaus, R. G. Webster, and W. L. Lim. "A Pandemic Warning?" *Nature,* October 9, 1997: 554; R. G. Webster. "Predictions for Future Human Influenza Pandemics." *Journal of Infectious Diseases* (Supplement 1), August 1997: S14−19.

65. R. G. Webster. "Predictions for Future Human Influenza Pandemics." *Journal of Infectious Diseases* (Supplement 1).

66. *St. Jude Children's Research Hospital 1997 Research Report.*

67. Webster has done research on AIDS in West African monkeys, so was very familiar with the knowledge gained from medical science's HIV wars. See: P. A. Marx, Y. Li, N. W. Lerche, S. Sutjipto, A. Gettie, J. A. Yee, B. H. Brotman, A. M. Prince, A. Hanson, R. G. Webster, et al. "Isolation of a Simian Immunodeficiency Virus Related to Human Immunodeficiency Virus Type 2 from a West African Pet Sooty Mangabey." *Journal of Virology,* August 1991: 4480−4485.

68. D. A. Senne, B. Panigrahy, Y. Kawaoka, J. E. Pearson, J. Süss, M. Lipkind, H. Kida, and R. G. Webster. "Survey of the Hemagglutinin (HA) Cleavage Site Sequence of H5 and H7 Avian Influenza Viruses: Amino Acid Sequence at the HA Cleavage Site as a Marker of Pathogenicity Potential." *Avian Diseases,* April−June 1996: 425−437; E. C. Claas, A. D. Osterhaus, R. van Beek, J. C. De Jong, G. F. Rimmelzwaan, D. A. Senne, S. Krauss, K. F. Shortridge, and R. G. Webster. "Human Influenza A H5N1 Virus Related to a Highly Pathogenic Avian Influenza." *Lancet.*

69. D. M. Feltquate, S. Heaney, R. G. Webster, and H. L. Robinson. "Different T Helper Cell Types and Antibody Isotypes Generated by Saline and Gene Gun DNA Immunization." *Journal of Immunology,* March 1, 1997: 2278−2284; C. M. Boyle, M. Morin, R. G. Webster, and H. L. Robinson. "Role of Different Lymphoid Tissues in the Initiation and Maintenance of DNA-Raised Antibody Responses to the Influenza Virus H1 Glycoprotein." *Journal of Virology,* December 1996: 9074−9078.

70. E. A. Govorkova, G. Murti, B. Meignier, C. de Taisne, and R. G. Webster. "African Green Monkey Kidney (Vero) Cells Provide an Alternative Host Cell System for Influenza A and B Viruses." *Journal of Virology,* August 1996: 5519−5524; J. M. Katz and R. G. Webster. "Amino Acid Sequence Identity between the HA1 of Influenza A (H3N2) Viruses Grown in Mammalian and Primary Chick Kidney Cells." *Journal of General Virology,* May 1992: 1159−1165.

71. W. Graeme Laver, Norbert Bischofberger and Robert G. Webster. "Disarming Flu Viruses." *Scientific American.*

72. National Center for Infectious Diseases. "About the NCID." http://www.cdc.gov/ncidod/about.html. December 1998.

73. World Health Organization. "Division of Emerging and Other Communicable Diseases Surveillance and Control: Partnerships to Meet the Challenge." http://www.who.int/emc/aboutemc.html. December 1998.

74. R. G. Webster. "Influenza Virus." *Archives of Virology Supplementum.*

75. W. Graeme Laver, Norbert Bischofberger, and Robert G. Webster. "Disarming Flu Viruses." *Scientific American;* Maggie Fox. "Fight Germs before It's Too Late, WHO Warns." Reuters, June 17, 1999.

21. CONCLUSION: THE REALITY OF THE MASS MIND'S DREAMS: TERRAFORMING THE COSMOS

1. Robert F. Service. "Quantum Computing Makes Solid Progress." *Science,* April 30, 1999: 722–723; Yu Y. Nakamura, A. Pashkin, and J. S. Tsai. "Coherent Control of Macroscopic Quantum States in a Single-Cooper-Pair Box." *Nature,* April 29, 1999: 786–788; S. J. van Enk, J. I. Cirac, and P. Zoller. "Photonic Channels for Quantum Communication." *Science,* January 9, 1998: 205–208.

2. Richard Borshay Lee. "The Hunters: Scarce Resources in the Kalahari." In *Conformity and Conflict: Readings in Cultural Anthropology,* ed. James P. Spradley and David W. McCurdy. Boston: Little, Brown, 1987: 192–207.

3. Melvin Konner. *The Tangled Wing: Biological Constraints on the Human Spirit.* New York: Holt, Rinehart and Winston, 1982: 371.

4. M. H. Bornstein, O. M. Haynes, L. Pascual, K. M. Painter, and C. Galperín. "Play in Two Societies: Pervasiveness of Process, Specificity of Structure." *Child Development,* March–April 1999: 317–331; L. E. Crandell and R. P. Hobson. "Individual Differences in Young Children's IQ: A Social-Developmental Perspective." *Journal of Child Psychology and Psychiatry and Allied Disciplines,* March 1999: 455–464; L. Vedeler. "Dramatic Play: A Format for 'Literate' Language?" *British Journal of Educational Psychology* 67:2 (June 1997): 153–167; Gordon M. Burghardt. *Play Behavior in Animals.* New York: Chapman and Hall, 1998; Marc Bekoff and John Byers, eds. *Animal Play.* Cambridge, U.K.: Cambridge University Press, 1998.

5. Caleb Project. "The Gabbra of Kenya." http://www.grmi.org/~jhanna/obj15.html. Downloaded August 1999.

6. John Reader. *Man on Earth.* Austin: University of Texas Press, 1988: 93.

7. Paul Somerson. "I've Got a Server in My Pants." *ZD Equip Magazine.* http://www.zdnet.co.uk/athome/feature/serverinpants/welcome.html. Downloaded August 1999; IBM. "The Personal Area Network: Communicating Body-to-Body." http://www.ibm.com/stories/1996/11/pan_main.html. Downloaded August 1999; IBM. "The Personal Area Network: Are You Ready for Prosthetic Memory?" http://www.ibm.com/stories/1996/11/pan_side4.html. Downloaded August 1999; Steve Mann. " 'Smart' Clothing." http://wearcam.org. Downloaded August 1999.

8. Alexander Chislenko. Personal communication. December 1996.

9. Richard Dawkins. *The Selfish Gene.* New York: Oxford University Press, 1976.

10. Fernand Braudel. *The Structures of Everyday Life: The Limits of the Possible.* Vol. 1 of *Civilization and Capitalism, 15th–18th Century,* trans. Siân Reynolds. New York: Harper and Row, 1982: 397.

11. Fred Howard. *Wilbur and Orville: A Biography of the Wright Brothers.* New York: Dover Publications, 1998.

12. Computer modelers using simulations of complex adaptive systems have come to the same conclusion as that presented here. The computer modelers' terms for the properties which emerge from their simulations are "polarization" (squabbling) and "clustering" (rallying around common threads). See Robert Axelrod. *The Complexity of Cooperation: Agent-Based Models of Competition and Collaboration.* Princeton, N.J.: Princeton University Press, 1997; Bibb Latane, Andrzej Nowak and James H. Liu. "Measuring Emergent Social Phenomena: Dynamism, Polarization and Clustering as Order Parameters of Social Systems." *Behavioral Science* 39 (1994): 1–24.

13. C. P. Collier, E. W. Wong, M. Belohradsky, F. M. Raymo, J. F. Stoddart, P. J. Kuekes, R. S. Williams, and J. R. Heath. "Electronically Configurable Molecular-Based Logic Gates." *Science,* July 16, 1999: 391–394; Robert F. Service. "Organic Molecule Rewires Chip Design." *Science,* July 16,

1999: 313–315; Maggie Fox. "Crystal Computer Chip Uses Chemistry for Speed." Reuters, July 16, 1999.

14. Dr. Edward L. Wright. *Ned Wright's Cosmology Tutorial.* http://www.astro.ucla.edu/~wright/glossary.html#VED. Downloaded August 1999.

15. Gary C. Smith. "Some Biological Considerations for a Permanent, Manned Lunar Base." *American Biology Teacher,* February 1995: 92; Comm Tech Lab and the Center for Microbial Ecology at Michigan State University. *The Microbe Zoo DLC-ME Project.* http://commtechlab.msu.edu/sites/dleme/zoo/. Downloaded September 1999.

16. Julian Krolik. *Active Galactic Nuclei: From Central Black Hole to the Galactic Environment.* Princeton, N.J.: Princeton University Press, 1999; James Glanz. "Gamma Rays Open a View Down a Cosmic Gun Barrel." *Science,* November 14, 1997: 1225; Dennis Normile. "New Ground-Based Arrays to Probe Cosmic Powerhouses." *Science,* April 30, 1999: 734–735; Daniel W. Weedman. "Making Sense of Active Galaxies." *Science,* October 16, 1998: 423–424; Ethan J. Schreier, Alessandro Marconi, and Chris Packham. "Evidence for a 20 Parsec Disk at the Nucleus of Centaurus A." *Astrophysical Journal,* June 1, 1998: 2; Space Telescope Science Institute. "Hubble Provides Multiple Views of How to Feed a Black Hole." Press Release No. STScI-PR98-14, posted May 15, 1998. http://oposite.stsci.edu/pubinfo/pr/1998/14/. Downloaded August 1998; National Aeronautics and Space Administration. " 'Old Faithful' Black Hole in Our Galaxy Ejects Mass Equal to an Asteroid at Fantastic Speeds." Posted January 9, 1998. http://www.sciencedaily.com/releases/1998/01/980109074724.html. Downloaded August 1999.

17. Louisiana State University. "Physicist Finds Out Why 'We Are Stardust . . .' " Posted June 25, 1999. http://www.sciencedaily.com/releases/1999/06/990625080416.html. Downloaded August 1999.

18. David Arnett and Grant Bazan. "Nucleosynthesis in Stars: Recent Developments." *Science,* May 30, 1997: 1359–1362; Linda Rowan. "Stellar Birth and Death." *Science,* May 30, 1997: 1315; University of Illinois at Urbana-Champaign. "Simulation Reveals Very First Stars That Formed in the Universe." Posted March 25, 1999. http://www.sciencedaily.com/releases/1999/03/990325054316.html. Downloaded August 1999.

19. Werner R. Loewenstein. *The Touchstone of Life: Molecular Information, Cell Communication, and the Foundations of Life.* New York: Oxford University Press, 1999: 24.

BIBLIOGRAPHY

Note the ethnic derivations of the names in this bibliography. Some are Japanese, Russian, Chinese, Arabic, Greek, African, Italian, Hispanic, Anglo-Saxon, Jewish, etc. All indicate the extent to which this volume is a product of a transnational synaptic weave—a hypothesis brought to fruition by the operation of the global brain.

T. Abee, L. Krockel, and C. Hill. "Bacteriocins: Modes of Action and Potentials in Food Preservation and Control of Food Poisoning." *International Journal of Food Microbiology,* December 1995: 169–185.

Robert Ader. "Behavioral Conditioning and the Immune System." In *Emotions in Health and Illness: Theoretical and Research Foundations,* edited by Lydia Temoshok, Craig Van Dyke, and Leonard S. Zegans. Orlando, Fla.: Grune and Stratton, 1983: 137–151.

R. Adolphs, D. Tranel, and A. R. Damasio. "The Human Amygdala in Social Judgment." *Nature,* June 4, 1998: 470–474.

James M. Adovasio, Olga Soffer, and Bohuslav Klima. "Upper Palaeolithic Fibre Technology: Interlaced Woven Finds from Pavlov I, Czech Republic, c. 26,000 Years Ago." *Antiquity: A Quarterly Review of Archaeology,* September 1996. http://intarch.ac.uk/antiquity/adovasio1.html. Downloaded February 2000.

M. Aho, L. Nuotio, E. Nurmi, and T. Kiiskinen. "Competitive Exclusion of Campylobacters from Poultry with K-bacteria and Broilact." *International Journal of Food Microbiology,* March–April 1992: 265–275.

Leslie Aiello. "Terrestriality, Bipedalism and the Origin of Language." In *The Evolution of Social Behaviour Patterns in Primates and Man,* edited by J. Maynard Smith. London: Proceedings of the British Academy, 1996.

C. K. Akins and T. R. Zentall. "Imitative Learning in Male Japanese Quail *(Coturnix japonica)* Using the Two-Action Method." *Journal of Comparative Psychology,* September 1996: 316–320.

T. G. Aldsworth, R. L. Sharman, C. E. Dodd, and G. S. Stewart. "A Competitive Microflora Increases the Resistance of *Salmonella Typhimurium* to Inimical Processes: Evidence for a Suicide Response." *Applied and Environmental Microbiology,* April 1998: 1323–1327.

C. Allison and C. Hughes. "Bacterial Swarming: An Example of Prokaryotic Differentiation and Multicellular Behaviour." *Science Progress* 75:298, pt. 3–4 (1991): 403–422.

C. Allison, H. C. Lai, D. Gygi, and C. Hughes. "Cell Differentiation of *Proteus Mirabilis* Is Initiated by Glutamine, a Specific Chemoattractant for Swarming Cells." *Molecular Microbiology,* April 1993: 53–60.

Stuart A. Altmann. "The Structure of Primate Social Communication." In *Social Communication among Primates,* edited by Stuart A. Altmann. Chicago: University of Chicago Press, 1967: 325–362.

V. E. Amassian, K. Henry, H. Durkin, S. Chice, J. B. Cracco, M. Somasundaram, N. Hassan, R. Q. Cracco, P. J. Maccabee, and L. Eberle. "The Human Immune System Is Differentially Influenced by Magnetic Stimulation of Left vs. Right Cerebral Cortex." *Society for Nueroscience Abstracts* 20 (1994): 104.

Jean Claude Ameisen. "The Origin of Programmed Cell Death." *Science,* May 31, 1996: 1278–1279.

America's Promise Ministries. "America's Promise Ministries Homepage." http://www.amprom.org/index.html. November 1998.

Amnesty International News Service. "Afghanistan: Public Executions and Amputations on Increase." AI INDEX: ASA 11/05/98, May 21, 1998. http://www2.amnesty.se/isext98.nsf/

eb4dea1ac25099c4c12565b7005b76ae/debc87e79452a698c125660a0063e0a8?OpenDocument. November 1998.

Sonia Ancoli and Kenneth F. Green. "Authoritarianism, Introspection, and Alpha Wave Biofeedback Training." *Psychophysiology*, January 1977: 40–44.

C. A. Anderson. "Temperature and Aggression: Ubiquitous Effects of Heat on Occurrence of Human Violence." *Psychological Bulletin*, July 1989: 74–96.

E. N. Anderson. *The Food of China*. New Haven, Conn.: Yale University Press, 1988.

R. J. Andrew. "The Displays of the Primates." In *Evolutionary and Genetic Biology of the Primates*, edited by J. Buettner-Janusch. New York: Academic Press, 1963.

Anonymous. "Letters from the Underground—Part II." *No Compromise: The Militant, Direct Action Magazine of Grassroots Animal Liberationists and Their Supporters*. http://www.nocompromise. org/. November 1998.

Scott Antes. Personal communication. January 25, 1999.

David W. Anthony. "The Archaeology of Indo-European Origins." *The Journal of Indo-European Studies*. Fall 1991: 193–222.

———. "Horse, Wagon and Chariot: Indo-European Languages and Archaeology." *Antiquity; a Quarterly Review of Archaeology*. September 1995: 554–565.

———. Personal communication. April 8, 1998.

M. Argyle and M. Cook. *Gaze and Mutual Gaze*. Cambridge, U.K.: Cambridge University Press, 1976.

Aristophanes. *Clouds*. In *Library of the Future*. 4th ed., ver. 5.0. Irvine, Calif.: World Library, 1996. CD-ROM.

Aristotle. *Heavens*. In *Library of the Future*. 4th ed., ver. 5.0. Irvine, Calif.: World Library, 1996. CD-ROM.

———. *Metaphysics*. In *Library of the Future*. 4th ed., ver. 5.0. Irvine, Calif.: World Library, 1996. CD-ROM.

———. *Politics*. In *Library of the Future*. 4th ed., ver. 5.0. Irvine, Calif.: World Library, 1996. CD-ROM.

———. *Nicomachaean Ethics*. The Perseus Project, edited by Gregory R. Crane. http://hydra. perseus.tufts.edu. August 1998.

———. *Physics*. The Perseus Project, edited by Gregory R. Crane. http://hydra.perseus. tufts.edu. August 1998.

———. *Posterior Analytics*. The Perseus Project, edited by Gregory R. Crane. http://hydra. perseus.tufts.edu. August 1998.

———. *Rhetoric*. The Perseus Project, edited by Gregory R. Crane. http://hydra.perseus. tufts.edu. August 1998.

Jamie Arndt, Jeff Greenberg, Tom Pyszczynski, and Sheldon Solomon. "Subliminal Exposure to Death-Related Stimuli Increases Defense of the Cultural Worldview." *Psychological Science*, September 1997: 379–385.

David Arnett and Grant Bazan. "Nucleosynthesis in Stars: Recent Developments." *Science*, May 30, 1997: 1359–1362.

B. B. Arnetz, T. Theorell, L. Levi, A. Kallner, and P. Eneroth. "An Experimental Study of Social Isolation of Elderly People." *Psychosomatic Medicine* 45 (1983): 395–406.

Eliot Aronson and Darwyn Linder. "Gain and Loss of Esteem as Determinants of Interpersonal Attractiveness." *Journal of Experimental Social Psychology* 1 (1965): 156–171.

Aryan Nations (web site). http://www.nidlink.com/~aryanvic/. November 1988.

Solomon E. Asch. "Studies of Independence and Conformity: I. A Minority of One against a Unanimous Majority." *Psychological Monographs* 70:9 (1956): (Whole no. 416).

Berhane Asfaw, Tim White, Owen Lovejoy, Bruce Latimer, Scott Simpson, and Gen Suwa. "*Australopithecus Garhi*: A New Species of Early Hominid from Ethiopia." *Science*, April 23, 1999: 629–635.

I. P. Ashmarin. "Neurological Memory as a Probable Product of Evolution of Other Forms of Biological Memory." *Zhurnal Evoliutsionnoi Biokhimii I Fiziologii,* May–June 1973: 217–224.

Klaus Atzwanger and Alain Schmitt. "Walking Speed and Depression: Are Sad Pedestrians Slow?" *Human Ethology Bulletin,* September 3, 1997.

Saint Augustine. *The City of God.* In *Library of the Future.* 4th ed., ver. 5.0. Irvine, Calif.: World Library, 1996. CD-ROM.

Nathalie Auphan, Joseph A. DiDonato, Caridad Rosette, Arno Helmsberg, and Michael Karin. "Immunosuppression by Glucocorticoids: Inhibition of NF-B Activity through Induction of I B Synthesis." *Science,* October 13, 1995: 286–290.

Robert Axelrod. *The Complexity of Cooperation: Agent-Based Models of Competition and Collaboration.* Princeton, N.J.: Princeton University Press, 1997.

Loren E. Babcock, Molly F. Miller, and Stephen T. Hasiotis. "Paleozoic-Mesozoic Crayfish from Antarctica: Earliest Evidence of Freshwater Decapod Crustaceans." *Geology,* June 1998: 539–543.

J. Backon. "Etiology of Alcoholism: Relevance of Prenatal Hormonal Influences on the Brain, Anomalous Dominance, and Neurochemical and Pharmacological Brain Asymmetry." *Medical Hypothesis,* May 1989: 59–63.

Chris Bader and Alfred Demaris. "A Test of the Stark-Bainbridge Theory of Affiliation with Religious Cults and Sects." *Journal for the Scientific Study of Religion,* September 1996: 285–303.

Kent G. Bailey. *Human Paleopsychology: Applications to Aggression and Pathological Processes.* Hillsdale, N.J.: Erlbaum Press, 1987.

R. T. Bakker. "Ecology of the Brontosaurs." *Nature,* January 15, 1971: 172–174.

———. *The Dinosaur Heresies: New Theories—Unlocking the Mystery of the Dinosaurs and Their Extinction.* New York: William Morrow, 1986.

Guillaume Balavoine and André Adoutte. "One or Three Cambrian Radiations?" *Science,* April 17, 1998: 397–398.

J. M. Baldwin. "A New Factor in Evolution." *American Naturalist* 30 (1896): 441–451.

Michael Balter. "Why Settle Down? The Mystery of Communities." *Science,* November 20, 1998: 1442–1445.

Martin S. Banks, Krishna V. Shenoy, Richard A. Andersen, and James A. Crowell. "Visual Self-Motion Perception during Head Turns." *Nature Neuroscience,* December 1998: 732–737.

F. Baquero. "Gram-positive Resistance: Challenge for the Development of New Antibiotics." *Journal of Antimicrobial Chemotherapy.* (Supplement A) May 1997: 1–6.

David P. Barash. *Sociobiology and Behavior.* New York: Elsevier Scientific Publishing, 1977.

———. *The Hare and the Tortoise: Culture, Biology, and Human Nature.* New York: Penguin Books, 1987.

Marcia Barinaga. "Watching the Brain Remake Itself." *Science,* December 2, 1994: 1475.

———. "Neurobiology: Social Status Sculpts Activity of Crayfish Neurons." *Science,* January 19, 1996: 290–291.

Jerome H. Barkow, Leda Cosmides, and John Tooby, eds. *The Adapted Mind: Evolutionary Psychology and the Generation of Culture.* New York: Oxford University Press, 1992.

G. W. Barlow and R. C. Francis. "Unmasking Affiliative Behavior among Juvenile Midas Cichlids (Cichlasoma Citrinellum)." *Journal of Comparative Psychology,* June 1988: 118–123.

Horace Barlow. "The Biological Role of Consciousness." In *Mindwaves: Thoughts on Intelligence, Identity and Consciousness,* edited by Colin Blakemore and Susan Greenfield. Oxford: Basil Blackwell, 1989: 361–374.

R. A. Baron and P. A. Bell. "Aggression and Heat: The Influence of Ambient Temperature, Negative Affect and a Cooling Drink on Aggression." *Journal of Personality and Social Psychology,* March 1976: 245–255.

R. A. Baron and V. M. Ransberger. "Ambient Temperature and the Occurrence of Collective Violence: The 'Long, Hot Summer' Revisited." *Journal of Personality and Social Psychology* 36 (1978): 351–360.

Geoffrey Barraclough. *The Origins of Modern Germany.* New York: W. W. Norton, 1984.

J. Baruch. "The Diffusion of Medical Technology." *Medical Instrumentation,* January–February 1979: 11–13.

Abd El Baset and Mahmoud El Aziz. "Impact of Advance Organizers of Interaction and Extraversion/Introversion on Scholastic Achievement for Middle College Female Students." *Derasat Nafseyah,* January 1994: 119–151.

W. G. Beasley. *The Meiji Restoration.* Stanford, Calif.: Stanford University Press, 1972.

A. Bechara, H. Damasio, D. Tranel, and A. R. Damasio. "Deciding Advantageously before Knowing the Advantageous Strategy." *Science,* February 28, 1997: 1293–1295.

Antoine Bechara, Daniel Tranel, Hanna Damasio, Ralph Adolphs, Charles Rockland, and Antonio R. Damasio. "Double Dissociation of Conditioning and Declarative Knowledge Relative to the Amygdala and Hippocampus in Humans." *Science,* August 25, 1995: 1115–1118.

Nancy E. Beckage. "The Parasitic Wasp's Secret Weapon." *Scientific American,* November 1997: 82.

N. Behera and V. Nanjundiah. "An Investigation into the Role of Phenotypic Plasticity in Evolution." *Journal of Theoretical Biology* 172 (1995): 225–234.

R. K. Belew and M. Mitchell, eds. *Adaptive Individuals in Evolving Populations: Models and Algorithms.* Reading, Mass.: Addison-Wesley, 1996.

Graham Bell. "Model Metaorganism." *Science,* October 9, 1998: 248.

I. R. Bell, M. L. Jasnoski, J. Kagan, and D. S. King. "Is Allergic Rhinitis More Frequent in Young Adults with Extreme Shyness? A Preliminary Survey." *Psychosomatic Medicine,* September–October 1990: 517–525.

Jay Belsky, Becky Spritz, and Keith Crnic. "Infant Attachment Security and Affective-Cognitive Information Processing at Age 3." *Psychological Science,* March 1996: 111–114.

R. M. Benbow. "Chromosome Structures." *Science Progress* 76, pt. 3–4 (1992): 301–302; 425–450.

Ruth Benedict. *Patterns of Culture.* 1934. Reprint, New York: New American Library, 1950.

S. Ben-Eliyahu, R. Yirmiya, Y. Shavit, and J. C. Liebeskind. "Stress-Induced Suppression of Natural Killer Cell Cytotoxicity in the Rat: A Naltrexone-Insensitive Paradigm." *Behavioral Neuroscience,* February 1990: 235–238.

E. Ben-Jacob. "From Snowflake Formation to the Growth of Bacterial Colonies. Part I. Diffusive Patterning in Azoic Systems." *Contemporary Physics,* September 1993: 247.

———. "From Snowflake Formation to the Growth of Bacterial Colonies. Part II. Cooperative Formation of Complex Colonial Patterns." *Contemporary Physics,* May 1997: 205–241.

E. Ben-Jacob, I. Cohen, and D. L. Gutnick. "Chemomodulation of Cellular Movement, Collective Formation of Vortices by Swarming Bacteria, and Colonial Development." *Physica A,* April 15, 1997: 181–197.

E. Ben-Jacob, H. Shmueli, O. Shochet, and A. Tenenbaum. "Adaptive Self-organization during Growth of Bacterial Colonies." *Physica A,* September 15, 1992: 378.

E. Ben-Jacob, A. Tenenbaum, and O. Shochet. "Holotransformations of Bacterial Colonies and Genome Cybernetics." *Physica A,* January 1994: 1–47.

E. Ben-Jacob, A. Tenenbaum, O. Shochet, I. Cohen, A. Czirók, and T. Vicsek. "Generic Modeling of Cooperative Growth Patterns in Bacterial Colonies." *Nature* 368 (1994): 46–49.

E. Ben-Jacob, A. Tenenbaum, O. Shochet, I. Cohen, A. Czirók, and T. Vicsek. "Communication, Regulation and Control during Complex Patterning of Bacterial Colonies." *Fractals* 2:1 (1994): 14–44.

E. Ben-Jacob, A. Tenenbaum, O. Shochet, I. Cohen, A. Czirók, and T. Vicsek. "Cooperative Strategies in Formation of Complex Bacterial Patterns." *Fractals* 3:4 (1995): 849–868.

Eshel Ben-Jacob. Personal communications. April 1996–December 1999.

————. "Bacterial Wisdom, Gödel's Theorem and Creative Genomic Webs." *Physica A* 248 (1998): 57–76.

William Benzon. "Culture as an Evolutionary Arena." *Journal of Social and Evolutionary Systems* 19:4 (1996): 321–362.

Peter L. Berger and Thomas Luckmann. *The Social Construction of Reality: A Treatise in the Sociology of Knowledge.* New York: Doubleday, 1966.

Patricia R. Bergquist. *Sponges.* Berkeley: University of California Press, 1978.

L. F. Berkman, "Assessing the Physical Health Effects of Social Networks and Social Support." *Annual Review of Public Health* 5 (1984): 413–432.

Edward L. Bernays. *Propaganda: The Public Mind in the Making.* New York: Horace Liveright, 1928.

Michael J. Berry II, Iman H. Brivanlou, Thomas A. Jordan, and Markus Meister. "Anticipation of Moving Stimuli by the Retina." *Nature,* March 25, 1999: 334–338.

Bruno Bettelheim. "Individual and Mass Behavior in Extreme Situations." *Journal of Abnormal and Social Psychology* 38 (1943): 417–452.

Upinder S. Bhalla and Ravi Iyengar. "Emergent Properties of Networks of Biological Signaling Pathways." *Science,* January 15, 1999: 381–387.

K. L. Bierman. "The Clinical Significance and Assessment of Poor Peer Relations: Peer Neglect versus Peer Rejection." *Journal of Developmental and Behavioral Pediatrics,* August 1987: 233–240.

James H. Billington. *The Icon and the Axe: An Interpretive History of Russian Culture.* New York: Vintage Books, 1970.

Alfred Binet. *The Psychic Life of Micro-Organisms: A Study in Experimental Psychology.* 1888. Reprint, Philadelphia: Albert Saifer, 1970.

J. B. Birdsell. "Some Population Problems Involving Pleistocene Man." *Cold Spring Harbor Symposium on Quantitative Biology* 22 (1957): 47–69.

Kaj Birket-Smith and Frederica de Laguna. *The Eyak Indians of the Copper River Delta, Alaska.* Copenhagen: Levin and Munksgaard, 1938.

I. B. Black. "Trophic Molecules and Evolution of the Nervous System." *Proceedings of the National Academy of Sciences of the United States of America,* November 1986: 8249–8252.

Jonah Blank. *Arrow of the Blue Skinned God: Retracing the Ramayana through India.* Boston: Houghton Mifflin, 1992.

Robert D. Blitzer, John H. Connor, George P. Brown, Tony Wong, Shirish Shenolikar, Ravi Iyengar, and Emmanuel M. Landau. "Gating of CaMKII by cAMP-Regulated Protein Phosphatase Activity during LTP." *Science,* June 19, 1998: 1940–1943.

R. A. Block and J. V. McConnell. "Classically Conditioned Discrimination in the Planarian, *Dugesia Dorotocephala.*" *Nature,* September 30, 1967: 1465–1466.

Dennis Bloodworth and Ching Ping Bloodworth. *The Chinese Machiavelli: 3,000 Years of Chinese Statecraft.* New York: Farrar, Straus and Giroux, 1976.

Howard Bloom. *The Lucifer Principle: A Scientific Expedition into the Forces of History.* New York: Atlantic Monthly, 1995.

————. "A History of the Global Brain—Creative Nets in the Pre-Cambrian Age." *ASCAP— Across-Species Comparisons and Psychopathology Society* 10:3 (March 1997): 7–11.

————. "Group Selection and the Social Sciences: A New Evolutionary Synthesis." In *Research in Biopolitics.* Vol. 6. Greenwich, Conn.: JAI Press, 1998: 43–63.

————. *Global Brain: Die Evolution Sozialer Intelligenz.* Stuttgart: Deutsche Verlags-Anstalt, 1999.

Howard Bloom and Michael J. Waller. "The Group Mind: Groups as Complex Adaptive Systems." Human Behavior and Evolution Society Annual Meeting, 1996.

John Boardman, Jasper Griffin, and Oswyn Murray, eds. *The Oxford History of the Classical World: Greece and the Hellenistic World.* New York: Oxford University Press, 1988.

J. W. Bodnar. "A Domain Model for Eukaryotic DNA Organization: A Molecular Basis for Cell Differentiation and Chromosome Evolution." *Journal of Theoretical Biology,* June 22, 1988: 479–507.

Christopher Boehm. "Rational Preselection from Hamadryas to *Homo Sapiens:* The Place of Decisions in Adaptive Process." *American Anthropologist,* June 1978: 265–296.

———. "Four Mechanical Routes to Altruism." Unpublished manuscript. March 1996.

W. Bogoras. *The Chukchee.* New York: G. E. Stechert, 1904–1909.

Johan Bollen and Francis Heylighen. "Algorithms for the Self-Organization of Distributed, Multi-User Networks: Possible Application to the Future World Wide Web." In *Cybernetics and Systems '96,* edited by R. Trappl. Vienna: Austrian Society for Cybernetics, 1996: 911–916.

G. M. Bolwig, M. Del Vecchio, G. Hannon, and T. Tully. "Molecular Cloning of Linotte in Drosophila: A Novel Gene That Functions in Adults during Associative Learning." *Neuron,* October 1995: 829–842.

M. G. Bolyard and S. T. Lord. "High-level Expression of a Functional Human Fibrinogen Gamma Chain in *Escherichia Coli.*" *Gene,* June 30, 1988: 183–192.

J. F. Bonaparte. "Sobre messungulatum houssayi y nuevos mammiferos cretacicos de Patagonia, Argentina." *IV Cong. Argentino de Paleontologia y Bioestratigrafia, Mendoza, Argentina* 2 (1986): 48–61.

Jose Bonaparte. Personal communication. March 18, 1999.

John Tyler Bonner. *The Evolution of Culture in Animals.* Princeton, N.J.: Princeton University Press, 1983.

Daniel J. Boorstin. "Our Cultural Hypochondria and How to Cure It." In *The Genius of American Politics.* Chicago: University of Chicago Press, 1953: 161–189.

S. P. Borriello and F. E. Barclay. "Protection of Hamsters against *Clostridium Difficile Ileocaecitis* by Prior Colonisation with Non-pathogenic Strains." *Journal of Medical Microbiology,* June 1985: 339–350.

T. J. Bouchard, D. T. Lykken, M. McGue, N. L. Segal, and A. Tellegen. "Sources of Human Psychological Differences: The Minnesota Study of Twins Reared Apart." *Science* 250 (1990): 223–228.

M. J. Boulton and K. Underwood. "Bully/Victim Problems among Middle School Children." *British Journal of Educational Psychology* 62, pt. 1 (February 1992): 73–87.

B. Bower. "Human Origin Recedes in Australia." *Science News,* September 28, 1996: 196.

Bruce Bower. "Brain Faces Up to Fear, Social Signs." *Science News,* December 17, 1994: 406.

B. B. Boycott. "Learning in *Octopus Vulgaris* and Other Cephalopods." *Pubblicazioni. Stazione Zoologica di Napoli* 25 (1954): 67–93.

Robert Boyd and Peter J. Richerson. *Culture and Evolutionary Process.* Chicago: University of Chicago Press, 1985.

———. "Life in the Fast Lane: Rapid Cultural Change and the Human Evolutionary Process." In *Origins of the Human Brain,* edited by Jean-Pierre Changeux and Jean Chavaillon. Oxford: Clarendon Press, 1995: 155–169.

S. Boynton and T. Tully. "Iatheo, a New Game Involved in Associative Learning and Memory in *Drosophila Melanogaster,* Identified from P Element Mutagenesis." *Genetics,* July 1992: 655–672.

Karl Dietrich Bracher. *The German Dictatorship: The Origins, Structure, and Effects of National Socialism.* New York: Praeger, 1970.

S. J. Braddy and L. I. Anderson. "An Upper Carboniferous Eurypterid Trackway from Mostyn, Wales." *Proceedings of the Geologists' Association* 107 (1996): 51–56.

Ernle Bradford. *The Battle for the West: Thermopylae.* New York: McGraw-Hill, 1980.

Ernle Dusgate Selby Bradford. *Hannibal.* New York: McGraw-Hill, 1981.

David C. Bradley, Marsha Maxwell, Richard A. Andersen, Martin S. Banks, and Krishna V. Shenoy. "Mechanisms of Heading Perception in Primate Visual Cortex." *Science,* September 13, 1996: 1544–1547.

Joseph V. Brady. "Ulcers in Executive Monkeys." *Scientific American,* October 1958: 95–100.

V. Braitenberg and C. Braitenberg. "Geometry of Orientation Columns in the Visual Cortex." *Biological Cybernetics,* August 1979: 179–186.

Fernand Braudel. *The Structures of Everyday Life: The Limits of the Possible.* Vol. 1 of *Civilization and Capitalism, 15th–18th Century.* Translated by Siân Reynolds. New York: Harper and Row, 1982.

A. D. Braun, A. A. Vereninov, and A. B. Kaulin. "Afanasii Semenovich Troshin (on His Seventieth Birthday)." *Tsitologiia,* June 1983: 726–732.

D. Bray. "Protein Molecules as Computational Elements in Living Cells." *Nature,* July 27, 1995: 307–312.

Dennis Bray. "Intracellular Signalling as a Parallel Distributed Process." *Journal of Theoretical Biology,* March 22, 1990: 215–231.

M. P. Brenner, L. S. Levitov, and E. O. Budrene. "Physical Mechanisms for Chemotactic Pattern Formation by Bacteria." *Biophysical Journal,* April 1998: 1677–1693.

Kevin Brett. Personal communications. February 25–26, 1997; March 4, 1997.

Jean L. Briggs. *Never in Anger: Portrait of an Eskimo Family.* Cambridge, Mass.: Harvard University Press, 1970.

Stephanie E. Briggs, Jean-Guy Godin, J. Dugatkin, and Lee Alan. "Mate-Choice Copying under Predation Risk in the Trinidadian Guppy *(Poecilia reticulata)."* *Behavioral Ecology,* Summer 1996: 151–157.

W. E. Broadhead, B. H. Kaplan, S. A. James, E. H. Wagner, V. J. Schoenbach, R. Grimson, A. Heyden, G. Tibblin, and S. H. Gehlbach. "The Epidemiological Evidence for a Relationship between Social Support and Health." *American Journal of Epidemiology* 117 (1983): 521–537.

Jochen J. Brocks, Graham A. Logan, Roger Buick, and Roger E. Summons. "Archean Molecular Fossils and the Early Rise of Eukaryotes." *Science,* August 13, 1999: 1033–1036.

Richard Brodie. *Virus of the Mind: The New Science of the Meme.* Seattle: Integral Press, 1996.

A. S. Brown, E. S. Susser, S. P. Lin, R. Neugebauer, and J. M. Gorman. "Increased Risk of Affective Disorders in Males after Second Trimester Prenatal Exposure to the Dutch Hunger Winter of 1944–45." *British Journal of Psychiatry,* May 1995: 601–606.

G. W. Brown, B. Andrews, T. Harris, Z. Adler, and L. Bridge. "Social Support, Self-esteem and Depression." *Psychological Medicine* 16 (1986): 813–831.

P. S. Brown, R. D. Humm, and R. B. Fischer. "The Influence of a Male's Dominance Status on Female Choice in Syrian Hamsters." *Hormones and Behavior,* June 1988: 143–149.

Steven R. Brown and Clyde Hendrick. "Introversion, Extraversion and Social Perception." *British Journal of Social and Clinical Psychology,* December 1971: 313–319.

W. L. Brown Jr. and E. O. Wilson. "Character Displacement." *Systematic Zoology* 5:2 (1956): 49–64.

W. M. Brown, M. George Jr, and A. C. Wilson. "Rapid Evolution of Animal Mitochondrial DNA." *Proceedings of the National Academy of Sciences of the United States of America,* April 1979: 1967–1971.

Jerome Bruner. *Actual Minds, Possible Worlds.* Cambridge, Mass.: Harvard University Press, 1986.

Jerome S. Bruner. *Beyond the Information Given: Studies in the Psychology of Knowing.* New York: W. W. Norton, 1973.

Jeff B. Bryson and Michael J. Driver. "Cognitive Complexity, Introversion, and Preference for Complexity." *Journal of Personality and Social Psychology,* September 1972: 320–327.

E. O. Budrene and H. C. Berg. "Dynamics of Formation of Symmetrical Patterns by Chemotactic Bacteria." *Nature,* July 6, 1995: 49–53.

Thomas Bulfinch. *Age of Fable.* In *Library of the Future.* 4th ed., ver. 5.0. Irvine, Calif.: World Library, 1996. CD-ROM.

Carol J. Bult, Owen White, Gary J. Olsen, Lixin Zhou, Robert D. Fleischmann, Granger G. Sutton, Judith A. Blake, Lisa M. FitzGerald, Rebecca A. Clayton, Jeannine D. Gocayne, Anthony R. Kerlavage, Brian A. Dougherty, Jean-Francois Tomb, Mark D. Adams, Claudia I. Reich, Ross Overbeek, Ewen F. Kirkness, Keith G. Weinstock, Joseph M. Merrick, Anna Glodek, John L. Scott, Neil S. M. Geoghagen, Janice F. Weidman, Joyce L. Fuhrmann, Dave Nguyen, Teresa R. Utterback, Jenny M. Kelley, Jeremy D. Peterson, Paul W. Sadow, Michael C. Hanna, Matthew

D. Cotton, Kevin M. Roberts, Margaret A. Hurst, Brian P. Kaine, Mark Borodovsky, Hans-Peter Klenk, Claire M. Fraser, Hamilton O. Smith, Carl R. Woese, and J. Craig Venter. "Complete Genome Sequence of the Methanogenic Archaeon, *Methanococcus jannaschii.*" *Science,* August 23, 1996: 1058–1073.

Gordon M. Burghardt. *Play Behavior in Animals.* New York: Chapman and Hall, 1998.

D. C. Burr, M. C. Morrone, and D. Spinelli. "Evidence for Edge and Bar Detectors in Human Vision." *Vision Research* 29:4 (1989): 419–431.

Daniel Burstein. *Yen: Japan's New Financial Empire and Its Threat to America.* New York: Simon and Schuster, 1988.

Robert Burton. *Bird Behavior.* New York: Alfred Knopf, 1985.

————. Personal communication. April 15, 1996.

Harold E. Burtt. *The Psychology of Birds: An Interpretation of Bird Behavior.* New York: Macmillan, 1967.

Ian Buruma. *God's Dust: A Modern Asian Journey.* New York: Farrar, Straus and Giroux, 1989.

P. Bushmann and J. Atema. "Aggression-Reducing Courtship Signals in the Lobster, *Homarus Americanus.*" *Biological Bulletin,* October 1994: 275–276.

David M. Buss. *The Evolution of Desire—Strategies of Human Mating.* New York: Basic Books, 1994.

————. *Evolutionary Psychology: The New Science of the Mind.* Boston: Allyn and Bacon, 1999.

Leo W. Buss. *The Evolution of Individuality.* Princeton, N.J.: Princeton University Press, 1987.

Donn Byrne. *The Attraction Paradigm.* New York: Academic Press, 1971.

M. R. Callaway, R. G. Marriott, and J. K. Esser. "Effects of Dominance on Group Decision Making: Toward a Stress-Reduction Explanation of Groupthink." *Journal of Personality and Social Psychology,* October 1985: 949–952.

L. Campitelli, I. Donatelli, E. Foni, M. R. Castrucci, C. Fabiani, Y. Kawaoka, S. Krauss, and R. G. Webster. "Continued Evolution of H1N1 and H3N2 Influenza Viruses in Pigs in Italy." *Virology,* June 9, 1997: 310–318.

Elias Canetti. *Crowds and Power.* Translated by Carol Stewart. New York: Farrar, Straus and Giroux, 1984.

Walter B. Cannon. *Bodily Changes in Pain, Hunger, Fear and Rage.* New York: Appleton, 1923.

Linnda R. Caporael. "Sociality: Coordinating Bodies, Minds and Groups." *Psycoloquy.* Downloaded from http://www.ai.univie.ac.at/cgi-bin/mfs/31/wachau/www/archives/Psycoloquy/1995. V6/0043.html?84#mfs,95/6/01.

Thomas Carlyle. *Voltaire.* 1829. Newport Beach, Calif.: Books on Tape, c. 1983.

L. A. Caron-Leslie, R. A. Schwartzman, M. L. Gaido, M. M. Compton, and J. A. Cidlowski. "Identification and Characterization of Glucocorticoid-Regulated Nuclease(s) in Lymphoid Cells Undergoing Apoptosis." *Journal of Steroid Biochemistry and Molecular Biology* 40:4–6 (1991): 661–671.

Scott P. Carroll, Hugh Dingle, and Stephen P. Klassen. "Genetic Differentiation of Fitness-Associated Traits among Rapidly Evolving Populations of Soapberry Bug." *Evolution* 51:4 (1997): 1182–1188.

Ashton Carter, John Deutch, and Philip Zelikow. "Catastrophic Terrorism." *Foreign Affairs,* November–December 1998: 80–94.

Denys de Catanzaro. "Reproductive Status, Family Interactions, and Suicidal Ideation: Surveys of the General Public and High-Risk Groups." *Ethology and Sociobiology* 16 (1995): 393.

Luigi Cavalli Sforza. "Diffusion of Culture and Genes." In *Issues in Biological Anthropology,* edited by B. J. Williams. Malibu, Calif.: Undena, 1986.

Napoleon Chagnon. *Yanomamo: The Fierce People.* New York: Holt, Rinehart and Winston, 1968.

————. "Life Histories, Blood Revenge, and Warfare in a Tribal Population." *Science,* February 1988: 988–989.

————. "Male Competition, Forming Close Kin, and Village Fissioning among the Yanomamo Indians." In *Evolutionary Biology and Human Social Behavior: An Anthropological Perspective,* edited by N. A. Chagnon and W. Irons. North Scituate, Mass.: Duxbury, 1979.

Anita Chan and Robert A. Senser. "China's Troubled Workers." *Foreign Affairs,* March–April 1997: 104–117.

M. R. A. Chance and R. Larson, eds. *The Structure of Social Attention.* New York: John Wiley and Sons, 1976.

Michael Chance. "Attention Structure as the Basis of Primate Rank Orders." *Man* 2 (n.d.): 503–518.

————. "Attention Structure as the Basis of Primate Rank Orders." *J.R.A.I.* 2: 4 (1967): 503–518.

Michael R. A. Chance, ed., assisted by Donald R. Omark. *Social Fabrics of the Mind.* Sussex, U.K.: Psychology Press, 1988.

Norman A. Chance. *The Eskimo of North Alaska.* New York: Holt, Rinehart and Winston, 1966.

G. Chapouthier. "Contribution of Studies on Animal Memory to Understanding Human Memory." *Encephale,* September–October 1997: 327–331.

D. L. Cheney, R. M. Seyfarth, and J. B. Silk. "The Responses of Female Baboons *(Papio Cynocephalus Ursinus)* to Anomalous Social Interactions: Evidence for Causal Reasoning?" *Journal of Comparative Psychology,* June 1995: 134–141.

Dorothy L. Cheney and Robert M. Seyfarth. "The Representation of Social Relations by Monkeys." *Cognition* 37 (1990): 167–196.

————. *How Monkeys See the World: Inside the Mind of Another Species.* Chicago: University of Chicago Press, 1992.

————. "The Evolution of Social Cognition in Primates." In *Behavioral Mechanisms in Evolutionary Ecology,* edited by Leslie A. Real. Chicago: University of Chicago Press, 1994: 371–389.

David Chilton. *The Days of Vengeance: An Exposition of the Book of Revelation.* Ft. Worth, Tex.: Dominion Press, 1984.

Alexander Chislenko. "Collaborative Information Filtering and Semantic Transports." December 1996 draft of "Information Institutions: The Technological Imperative." *Computers and Software* (Kiev, Ukraine, root@cplusp.kiev.ua—this version appears in Russian) 6:39 (1997). English translation downloadable from http://www.lucifer.com/~sasha/articles/ACF.html.

S. T. Christensen, V. Leick, L. Rasmussen, and D. N. Wheatley. "Signaling in Unicellular Eukaryotes." *International Review of Cytology* 177 (1998): 181–253.

Christian Gallery. "The Nuremberg Files: Alleged Abortionists and Their Accomplices." http://www.bestchoice.com/atrocity/aborts.html. November 1998.

Christian Identity Homepage. "Our Legacy of Truth." http://www4.stormfront.org/posterity/index.html. November 1998.

Robert B. Cialdini. *Influence: How and Why People Agree on Things.* New York: William Morrow, 1984.

E. C. Claas, A. D. Osterhaus, R. van Beek, J. C. De Jong, G. F. Rimmelzwaan, D. A. Senne, S. Krauss, K. F. Shortridge, and R. G. Webster. "Human Influenza A H5N1 Virus Related to a Highly Pathogenic Avian Influenza." *Lancet,* February 1998: 472–477.

M. Claes and L. Poirier. "Characteristics and Functions of Friendship in Adolescence." *Psychiatrie de l'enfant et de l'adolescent* 36:1 (1993): 289–308.

Anne B. Clark, David Sloan Wilson, and Ted Dearstyne. "Shyness and Boldness in Humans and Other Animals." *Trends in Ecology and Evolution,* November 1994: 442.

P. G. Clarke and S. Clarke. "Nineteenth Century Research on Naturally Occurring Cell Death and Related Phenomena." *Anatomy and Embryology,* February 1996: 81–99.

C. R. Cloninger, J. Rice, and T. Reich. "Multifactorial Inheritance with Cultural Transmission and Assortative Mating. II. A General Model of Combined Polygenic and Cultural Inheritance." *American Journal of Human Genetics,* March 1979: 176–198.

Kathryn Coe. "Art: The Replicable Unit—an Inquiry into the Possible Origin of Art as a Social Behavior." *Journal of Social and Evolutionary Systems* 15:2 (1992): 217–234.

D. Cohen, R. E. Nisbett, B. F. Bowdle, and N. Schwarz. "Insult, Aggression, and the Southern Culture of Honor: An 'Experimental Ethnography.' " *Journal of Personality and Social Psychology,* May 1996: 945–959.

David B. Cohen. *Stranger in the Nest: Do Parents Really Shape Their Children's Personality, Intelligence, or Character?* New York: John Wiley and Sons, 1999.

I. Cohen, A. Czirok, and E. Ben-Jacob. "Chemotactic-Based Adaptive Self-Organization during Colonial Development." *Physica A,* December 1996: 678–698.

Inon Cohen. "Re: Mechanisms of Genes Transfer." Unpublished manuscript. Tel Aviv, 1999.

S. Cohen and S. L. Syme, eds. *Social Support and Health.* New York: Academic Press, 1985.

S. Cohen and N. Weinstein. "Non-auditory Effects of Noise on Behavior and Health." *Journal of Social Issues* 37 (1981): 36–70.

Sheldon Cohen, J. R. Kaplan, Joan E. Kunick, Steven E. Manuck, and Bruce S. Rabin. "Chronic Social Stress Affiliation and Cellular Immune Response in Non-Human Primates." *Psychological Science,* September 1992: 301–304.

C. P. Collier, E. W. Wong, M. Belohradsky, F. M. Raymo, J. F. Stoddart, P. J. Kuekes, R. S. Williams, and J. R. Heath. "Electronically Configurable Molecular-Based Logic Gates." *Science,* July 16, 1999: 391–394.

M. M. Compton, L. A. Caron, and J. A. Cidlowski. "Glucocorticoid Action on the Immune System." *Journal of Steroid Biochemistry* 27:1–3 (1987): 201–208.

William S. Condon. "Communication: Rhythm and Structure." In *Rhythm in Psychological, Linguistic and Musical Processes,* edited by James R. Evans and Manfred Clynes. Springfield, Ill.: C. C. Thomas, 1986: 55–77.

———. Personal communication. June 10, 1999.

William S. Condon and Louis W. Sander. "Neonate Movement Is Synchronized with Adult Speech: Interactional Participation and Language Acquisition." *Science* 183:4120 (1974): 99–101.

John B. Connolly and Tim Tully. "You Must Remember This: Finding the Master Switch for Long-term Memory." *The Sciences,* May–June 1996: 42.

M. Conrad. "Bootstrapping Model of the Origin of Life." *Biosystems* 15:3 (1982): 209–219.

Peter A. Corning. *The Synergism Hypothesis: A Theory of Progressive Evolution.* New York: McGraw-Hill, 1983.

———. "On the Concept of Synergy and Its Role in the Evolution of Complex Systems." Prepublication manuscript prepared for the *Journal of Social and Evolutionary Systems,* 1996.

T. N. Cornsweet. *Visual Perception.* New York: Academic Press, 1970.

Leda Cosmides and John Tooby. "Cognitive Adaptations for Social Exchange." In *The Adapted Mind: Evolutionary Psychology and the Generation of Culture,* edited by Jerome H. Barkow, Leda Cosmides, and John Tooby. New York: Oxford University Press, 1992: 163–228.

Kenneth Courtis. Speech to the 1994 Pacific Rim Forum. Reprinted in *Asia, Inc.,* Hong Kong, and reprinted from *Asia, Inc.* in *The National Times,* August-September 1995: 22–25.

D. Crews. "Species Diversity and the Evolution of Behavioral Controlling Mechanisms." In *The Integrative Neurobiology of Affiliation,* edited by C. Sue Carter, I. Izja Lederhendler, and Brian Kirkpatrick. New York: New York Academy of Sciences, 1997: 1–21.

Helena Cronin. *The Ant and the Peacock.* New York: Oxford University Press, 1991.

P. H. Crowley, L. Provencher, S. Sloane, L. A. Dugatkin, B. Spohn, L. Rogers, and M. Alfieri. "Evolving Cooperation: The Role of Individual Recognition." *Biosystems* 37:1–2 (1996): 49–66.

Elizabeth Culotta. "A New Human Ancestor?" *Science,* April 23, 1999: 572–573.

E. Curio, V. Ernst, and W. Vieth. "Cultural Transmission of Enemy Recognition: One Function of Mobbing." *Science* 202 (1978): 899–901.

Philip D. Curtin. *Cross-Cultural Trade in World History.* New York: Cambridge University Press, 1984.

Kazimierz Dabrowski. *Psychoneurosis Is Not an Illness.* London: Gryf Publications, 1972.

Kazimierz Dabrowski with Andrzej Kawczak and Michael M. Piechowski. *Mental Growth through Positive Disintegration.* London: Gryf Publications, 1970.

Michael J. Daly and Kenneth W. Minton. "Contribution to Sequencing of the *Deinococcus Radiodurans* Genome." DOE Human Genome Program Contractor-Grantee Workshop VI, November 9–13, 1997, Santa Fe, N. Mex. http://www.ornl.gov/hgmis/publicat/97santa/sequence. html#164. January 1999.

F. R. D'Amato. "Neurobiological and Behavioral Aspects of Recognition in Female Mice." *Physiology and Behavior,* December 1997: 1311–1317.

Alfred Darnell and Darren E. Sherkat. "The Impact of Protestant Fundamentalism on Educational Attainment." *American Sociological Review,* April 1997: 306–315.

Robert Darnton. *The Kiss of Lamourette: Reflections in Cultural History.* New York: W. W. Norton, 1990.

Charles Darwin. *The Descent of Man, and Selection in Relation to Sex.* New York: D. Appleton, 1871.

———. *The Expressions of the Emotions in Man and Animals.* 1872. Reprint, New York: Greenwood Press, 1969.

———. *Origin of Species.* In *Library of the Future.* 4th ed., ver. 5.0. Irvine, Calif.: World Library, 1996. CD-ROM.

Aniruddha Das and Charles D. Gilbert. "Topography of Contextual Modulations Mediated by Short-range Interactions in Primary Visual Cortex." *Nature* 17 (June 1999): 655–661.

Richard J. Davidson. "Emotion and Affective Style, Hemispheric Substrates." *Psychological Science,* January 1992: 39–43.

D. G. Davies, M. R. Parsek, J. P. Pearson, B. H. Iglewski, J. W. Costerton, and E. P. Greenberg. "The Involvement of Cell-to-Cell Signals in the Development of a Bacterial Biofilm." *Science,* April 10, 1998: 295–298.

Julian Davies. "Inactivation of Antibiotics and the Dissemination of Resistance Genes." *Science,* April 15, 1994: 375–382.

Richard Dawkins. *The Selfish Gene.* New York: Oxford University Press, 1976.

———. *The Extended Phenotype: The Long Reach of the Gene.* Oxford: Oxford University Press, 1982.

Alisdair Daws. Personal communication. April 11, 1999.

Terrence Deacon. "The Neural Circuitry Underlying Primate Calls and Human Language." In *Language Origin: A Multidisciplinary Approach,* edited by J. Wind, B. Chiarelli, B. Bichakhian, and A. Nocentini. Dordrecht: Kluwer Academic Publishing, 1992: 121–162.

A. Dean and N. Lin. "The Stress-Buffering Role of Social Support: Problems and Prospects for Systematic Investigation." *Journal of Nervous and Mental Disease,* December 1977: 403–417.

J. M. Bermúdez de Castro, J. L. Arsuaga, E. Carbonell, A. Rosas, I. Martínez, and M. Mosquera. "A Hominid from the Lower Pleistocene of Atapuerca, Spain: Possible Ancestor to Neandertals and Modern Humans." *Science,* May 30, 1997: 1392–1395.

Denys de Catanzaro. "Reproductive Status, Family Interactions, and Suicidal Ideation: Surveys of the General Public and High-Risk Groups." *Ethology and Sociobiology* 16 (1995): 385–394.

Christian de Duve. "RNA without Protein or Protein without RNA?" In *What Is Life? The Next Fifty Years—Speculations on the Future of Biology,* edited by Michael P. Murphy and Luke A. J. O'Neill. Cambridge, U.K.: Cambridge University Press, 1995: 79–82.

———. "The Birth of Complex Cells." *Scientific American,* April 1996: 50–57.

P. de Groot. "Information Transfer in a Socially Roosting Weaver Bird *(Quelea Quelea; Ploceinae):* An Experimental Study." *Animal Behaviour* 28 (1980): 1249–1254.

J. de Gunzburg. "Mode of Action of Cyclic AMP in Prokaryotes and Eukaryotes, CAP and CAMP-Dependent Protein Kinases." *Biochimie,* January 1985: 563–582.

Jean de Heinzelin, J. Desmond Clark, Tim White, William Hart, Paul Renne, Giday WoldeGabriel, Yonas Beyene, and Elisabeth Vrba. "Environment and Behavior of 2.5-Million-Year-Old Bouri Hominids." *Science,* April 23, 1999: 625–629.

J. C. de Jong, E. C. Claas, A. D. Osterhaus, R. G. Webster, and W. L. Lim. "A Pandemic Warning?" *Nature,* October 9, 1997: 554.

Manuel de Landa. *A Thousand Years of Nonlinear History.* New York: Zone Books, 1997.

Edward DeLong. "Archaeal Means and Extremes." *Science,* April 24, 1998: 542–543.

Dr. Charles F. Delwiche. Personal communication. February 1997.

S. Demeter, J. Ringo, and R. Dorty. "Morphometric Analysis of the Human Corpus Callosum and Anterior Commissure." *Human Neurobiology* 6 (1988): 219–226.

John D'Emilio and Estelle B. Freedman. *Intimate Matters: A History of Sexuality in America.* New York: Harper and Row, 1988.

John Demos. "Shame and Guilt in Early New England." In *Emotion and Social Change: Toward a New Psychohistory,* edited by Carol Z. Stearns and Peter N. Stearns. New York: Holmes and Meier, 1988: 69–85.

John Putnam Demos. *Entertaining Satan: Witchcraft and the Culture of Early New England.* New York: Oxford University Press, 1982.

Demosthenes. *Erotic Essay 46.* The Perseus Project, edited by Gregory R. Crane. http://hydra.perseus.tufts.edu. August 1998.

———. *Philippic.* The Perseus Project, edited by Gregory R. Crane. http://hydra.perseus.tufts.edu. September 1998.

J. L. Deneubourg, S. Aron, G. Goss, J. M. Pasteels, and G. Duerinck. "Random Behaviour, Amplification Processes and Number of Participants: How They Contribute to the Foraging Properties of Ants." In *Evolution, Games and Learning: Models for Adaptation in Machines and Nature, Proceedings of the Fifth Annual International Conference of the Center for Nonlinear Studies, Los Alamos, NM 87545, USA, May 20–24, 1985,* edited by Doyne Farmer, Alan Lapedes, Norman Packard, and Burton Wendroff. Amsterdam: North-Holland Physics Publishing, 1985: 176–186.

John S. Denker. "Neural Network Models of Learning and Adaptation." In *Evolution, Games and Learning: Models for Adaptation in Machines and Nature, Proceedings of the Fifth Annual International Conference of the Center for Nonlinear Studies, Los Alamos, NM 87545, USA, May 20–24, 1985,* edited by Doyne Farmer, Alan Lapedes, Norman Packard, and Burton Wendroff. Amsterdam: North-Holland Physics Publishing, 1985: 216–232.

Koen DePryck. *Knowledge, Evolution, and Paradox: The Ontology of Language.* Albany: State University of New York Press, 1993.

Charles D. Derby. Personal communication. March 2, 1999.

R. Desimone. "Neural Mechanisms for Visual Memory and Their Role in Attention." *Proceedings of the National Academy of Sciences of the United States of America,* November 26, 1996: 13494–13499.

I. DeVore, ed. *Behavior: Field Studies of Monkeys and Apes.* New York: Holt, Rinehart and Winston, 1965.

Henk de Vos and Evelien Zeggelink. "The Emergence of Reciprocal Altruism and Group-Living: An Object-Oriented Simulation Model of Human Social Evolution." *Social Sciences: Information sur les Sciences Sociales,* September 1994: 496–499.

———. "Reciprocal Altruism in Human Social Evolution: The Viability of Reciprocal Altruism with a Preference for 'Old-Helping-Partners.'" *Evolution and Human Behavior,* July 1997: 261–278.

F. B. de Waal. "The Organization of Agonistic Relations within Two Captive Groups of Java-Monkeys *(Macaca fascicularis).*" *Zeitschrift für Tierpsychologie,* July 1977: 225–282.

———. "The Integration of Dominance and Social Bonding in Primates." *Quarterly Review of Biology,* December 1986: 459–479.

———. "Food Transfers through Mesh in Brown Capuchins." *Journal of Comparative Psychology,* December 1997: 370–378.

F. B. de Waal and L. M. Luttrell. "The Similarity Principle Underlying Social Bonding among Female Rhesus Monkeys." *Folia Primatologica* 46:4 (1986): 215–234.

F. B. M. de Waal. "The Chimpanzee's Service Economy: Food for Grooming." *Evolution and Human Behavior* 18:6 (1997): 375–386.

Frans B. M. de Waal. "Chimpanzees' Adaptive Potential: A Comparison of Social Life under Captive and Wild Conditions." In *Chimpanzee Cultures,* edited by Richard W. Wrangham, W. C. McGrew,

Frans B. M. de Waal, and Paul G. Heltne with assistance from Linda A. Marquardt. Cambridge, Mass.: Harvard University Press, 1994: 243–260.

Frans B. M. de Waal and Filippo Aureli. "Conflict Resolution and Distress Alleviation in Monkeys and Apes." In *The Integrative Neurobiology of Affiliation,* edited by C. Sue Carter, I. Izja Lederhendler, and Brian Kirkpatrick. New York: New York Academy of Sciences, 1997: 317–328.

Frans de Waal. *Peacemaking among Primates.* Cambridge, Mass.: Harvard University Press, 1989.

J. DeZazzo and T. Tully. "Dissection of Memory Formation: From Behavioral Pharmacology to Molecular Genetics." *Trends in Neurosciences,* May 1995: 212–218.

Jared Diamond. "This Fellow Frog, Name Belong-Him Dakwo." *Natural History,* April 1989: 16–23.

———. *Guns, Germs, and Steel: The Fates of Human Societies.* New York: W. W. Norton, 1997.

———. "Japanese Roots." *Discover,* June 1998: 86–94.

L. L. Dibrell and K. Yamamoto. "In Their Own Words: Concerns of Young Children." *Child Psychiatry and Human Development,* fall 1988: 14–25.

Diodorus. *Historical Library.* The Perseus Project, edited by Gregory R. Crane. http:// hydra.perseus.tufts.edu. August 1998.

K. Dion, E. Berscheid, and E. Walster. "What Is Beautiful Is Good." *Journal of Personality and Social Psychology* 24 (1972): 285–290.

K. K. Dion. "Physical Attractiveness and Evaluation of Children's Transgressions." *Journal of Personality and Social Psychology* 24 (1972): 207–213.

A. F. Dixson, T. Bossi, and E. J. Wickings. "Male Dominance and Genetically Determined Reproductive Success in the Mandrill *(Mandrillus sphinx)*." *Primates,* October 1993: 525–532.

Alan F. Dixson. Personal correspondence. June 15–17, 1998.

Paul M. Dolukhanov. "The Neolithisation of Europe: A Chronological and Ecological Approach." In *The Explanation of Culture Change: Models in Prehistory,* edited by Colin Renfrew. Pittsburgh: University of Pittsburgh Press, 1973: 329–342.

Qi Dong, Glenn Weisfeld, Ronald H. Boardway, and Jiliang Shen. "Correlates of Social Status among Chinese Adolescents." *Journal of Cross-cultural Psychology,* July 1996: 476–493.

Anthony N. Doob and Alan E. Gross. "Status of Frustrator as an Inhibitor of Horn-Honking Responses." *Journal of Social Psychology* 76:2 (1968): 213–218.

Russell F. Doolittle. "Microbial Genomes Opened Up." *Nature,* March 26, 1998: 339–342.

F. Doricchi, C. Armati, G. Martorano, and C. Violani. "Generation of Skeletal and Multipart Mental Visual Images in the Cerebral Hemispheres: A Study in Normal Subjects." *Neuropsychologia,* February 1995: 181–201.

R. M. Doty, B. E. Peterson, and D. G. Winter. "Threat and Authoritarianism in the United States, 1978–1987." *Journal of Personality and Social Psychology,* October 1991: 629–640.

C. W. Douglas and K. A. Bisset. "Development of Concentric Zones in the Proteus Swarm Colony." *Journal of Medical Microbiology,* November 1976: 497–500.

Mary Douglas. *Natural Symbols: Explorations in Cosmology.* New York: Pantheon, 1982.

C. M. Drea. "Social Context Affects How Rhesus Monkeys Explore Their Environment." *American Journal of Primatology* 44:3 (1998): 205–214.

Robert Drews. *The Coming of the Greeks: Indo-European Conquests in the Aegean and the Near East.* Princeton, N.J.: Princeton University Press, 1988.

Josh Dubnau and Tim Tully. "Gene Discovery in Drosophila: New Insights for Learning and Memory." *Annual Review of Neuroscience* 21 (1998): 407–444.

S. Ducrocq, Y. Chaimanee, and J. J. Jaeger. "Mammalian Faunas and the Ages of Continental Tertiary Fossiliferous Localities from Thailand." *Journal of Southeast Asian Earth Sciences,* July 1995: 65–78.

Lee Alan Dugatkin. "Interface between Culturally Based Preferences and Genetic Preferences: Female Mate Choice in *Poecilia Reticulata.*" *Proceedings of the National Academy of Sciences of the United States of America,* April 2, 1996: 2770–2773.

William J. Duiker. *Cultures in Collision: The Boxer Rebellion.* San Rafael, Calif.: Presidio Press, 1978.

Robin Dunbar. *Grooming, Gossip and the Evolution of Language*. London: Faber and Faber, 1996.

———. "The Monkeys' Defence Alliance." *Nature* 386 (1997): 555–557.

J. Duncan. "Converging Levels of Analysis in the Cognitive Neuroscience of Visual Attention." *Philosophical Transactions of the Royal Society of London. Series B: Biological Sciences,* August 29, 1998: 1307–1317.

Ross E. Dunn. *The Adventures of Ibn Battuta: A Muslim Traveler of the Fourteenth Century.* Berkeley: University of California Press, 1986.

William H. Durham. *Coevolution: Genes, Culture, and Human Diversity.* Stanford, Calif.: Stanford University Press, 1991.

Emile Durkheim. *The Elementary Forms of the Religious Life.* Translated by Joseph Ward Swain. 1915. Reprint, New York: Free Press, 1965.

"Early Buddhist Art." Hiindia. http://www.hiindia.com/arts/docs/buddhist.html. December 1998.

Wolfram Eberhard. *A History of China.* London: Routledge and Kegan Paul, 1977.

"Economic Survey of Latin America and the Caribbean, 1997–1998. ECLAC, Impact of the Asian Crisis on Latin America (LC/G.2026), May 1998." United Nations Economic Commission for Latin America and the Caribbean. http://www.cepal.org/english/Publications/survey98/contents.html. November 1998.

"Eco-Terror Suspected in Fires." *Oregon Lands Coalition Network News,* November 4, 1996. http://www.speaksoftly.com/olc/olc_pages/o0042.phtml. October 1998.

F. W. Edens, C. R. Parkhurst, I. A. Casas, and W. J. Dobrogosz. "Principles of Ex Ovo Competitive Exclusion and In Ovo Administration of *Lactobacillus Reuteri.*" *Poultry Science,* January 1997: 179–196.

Donna Eder and Janet Lynne Enke. "The Structure of Gossip: Opportunities and Constraints on Collective Expression among Adolescents." *American Sociological Review,* August 1991: 494–508.

Robert B. Edgerton. *Sick Societies: Challenging the Myth of Primitive Harmony.* New York: Free Press, 1992.

Don Edwards. Personal communication. February 24, 1999.

C. L. Ehardt. "Absence of Strongly Kin-Preferential Behavior by Adult Female Sooty Mangabeys (*Cercocebus Atys*)." *American Journal of Physical Anthropology,* June 1988: 233–243.

Irenaus Eibl-Eibesfeldt. *Human Ethology.* New York: Aldine de Gruyter, 1989.

Howard Eichenbaum. "How Does the Brain Organize Memories?" *Science,* July 18, 1997: 330–331.

John F. Eisenberg. *The Mammalian Radiations: An Analysis of Trends in Evolution, Adaptation, and Behavior.* Chicago: University of Chicago Press, 1981.

———. "Parental Behavior." In *The Oxford Companion to Animal Behavior,* edited by David McFarland. New York: Oxford University Press, 1982: 443–447.

———. Personal communication. November 13, 1997.

L. Eisenberg. "The Social Construction of the Human Brain." *American Journal of Psychiatry* 152:11 (1995): 1563–1575.

Paul Ekman. "Facial Expressions of Emotion: An Old Controversy and New Findings." *Philosophical Transactions of the Royal Society of London. Series B: Biological Sciences,* January 29, 1992: 63–69.

———. "Facial Expression and Emotion." *American Psychologist,* April 1993: 384–392.

Thomas Elbert, Christo Pantev, Christian Wienbruch, Brigitte Rockstroh, and Edward Taub. "Increased Cortical Representation of the Fingers of the Left Hand in Stringed Players." *Science,* October 13, 1995: 305–307.

Dr. Alexander Elder. *Rubles to Dollars: Making Money on Russia's Exploding Financial Frontier.* New York: New York Institute of Finance, 1999.

Niles Eldredge. *The Pattern of Evolution.* New York: W. H. Freeman, 1998.

Robert N. Emde. "Levels of Meaning for Infant Emotions: A Biosocial View." In *Approaches to Emotion,* edited by Klaus R. Scherer and Paul Ekman. Hillsdale, N.J.: Lawrence Erlbaum Associates, 1984: 77–107.

Michael Emerman and Michael H. Malim. "HIV-1 Regulatory/Accessory Genes: Keys to Unraveling Viral and Host Cell Biology." *Science,* June 19, 1998: 1880–1884.

G. R. Emory. "Comparison of Spatial and Orientational Relationships as Manifestations of Divergent Modes of Social Organization in Captive Groups of *Mandrillus Sphinx* and *Theropithecus Gelada.*" *Folia Primatologica* 24:4 (1975): 293–314.

Enviroweb. "Don't XXXX Genetic Experiment with Our Food." http://www.enviroweb.org/ shag/info/leaflets/vs.html. November 1998.

S. Epstein. "The Self-Concept Revisited: Or a Theory of a Theory." *American Psychologist* 28 (1973): 404–416.

Desiderius Erasmus. *The Correspondence of Erasmus.* Translated by R. A. B. Mynors and D. F. S. Thomson. Toronto: University of Toronto Press, 1974.

B. Erickson, E. A. Lind, B. C. Johnson, and W. M. O'Barr. "Speech Style and Impression Formation in a Court Setting: The Effects of 'Powerful' and 'Powerless' Speech." *Journal of Experimental Social Psychology* 14 (1978): 266–279.

S. Erisman, M. Carnes, L. K. Takahashi, and S. J. Lent. "The Effects of Stress on Plasma ACTH and Corticosterone in Young and Aging Pregnant Rats and Their Fetuses." *Life Sciences* 47:17 (1990): 1527–1533.

"Ernst von Glasersfeld: Biography." Amherst, Mass.: Scientific Reasoning Research Institute, University of Amherst. http://www.umass.edu/srri/vonGlasersfeld/biography.html. Downloaded June 1999.

S. Esipov and J. A. Shapiro. "Kinetic Model of *Proteus Mirabilis* Swarm Colony Development." *Journal of Mathematical Biology* 36 (1998): 249–263.

Richard D. Estes, Daniel Otte (illustrator), and Kathryn Fuller. *The Safari Companion: A Guide to Watching African Mammals.* Post Mills, Vt.: Chelsea Green, 1993.

Euclid. *The Thirteen Books of Euclid's Elements.* Translated by Sir Thomas L. Heath. The Perseus Project, edited by Gregory R. Crane. http://hydra.perseus.tufts.edu. July 1998.

F. Barton Evans III. *Harry Stack Sullivan: Interpersonal Theory and Psychotherapy.* London: Routledge, 1996.

H. J. Eysenck. "Primary Mental Abilities." *British Journal of Educational Psychology* 9 (1939): 26–265.

Hans J. Eysenck and Michael W. Eysenck. *Personality and Individual Differences: A Natural Science Approach.* New York: Plenum Press, 1985.

Michael W. Eysenck and M. Christine Eysenck. "Memory Scanning, Introversion^Extraversion, and Levels of Processing." *Journal of Research in Personality,* September 1979: 305–315.

James Fallows. *More Like Us: Making America Great Again.* Boston: Houghton Mifflin, 1989.

R. L. Fantz. "Visual Perception from Birth as Shown by Pattern Selectivity." *Annals of the New York Academy of Sciences* 118 (1965): 793–814.

F. H. Farley and C. B. Mueller. "Arousal, Personality, and Assortative Mating in Marriage: Generalizability and Cross-cultural Factors." *Journal of Sex and Marital Therapy,* spring 1978: 50–53.

J. D. Farmer, S. A. Kauffman, N. H. Packard, and A. S. Perelson. "Adaptive Dynamic Networks as Models for the Immune System and Autocatalytic Sets." In *Perspectives in Biological Dynamics and Theoretical Medicine,* edited by S. H. Koslow, A. J. Mandell, and M. F. Shlesinger. New York: New York Academy of Sciences, 1987: 118–131.

Doyne Farmer, Alan Lapedes, Norman Packard, and Burton Wendroff, eds. *Evolution, Games and Learning: Models for Adaptation in Machines and Nature, Proceedings of the Fifth Annual International Conference of the Center for Nonlinear Studies.* Amsterdam: North-Holland Physics Publishing, 1985.

J. Doyne Farmer, Norman H. Packard, and Alan S. Perelson. "The Immune System, Adaptations, and Machine Learning." In *Evolution, Games and Learning: Models for Adaptation in Machines and Nature, Proceedings of the Fifth Annual International Conference of the Center for Nonlinear Studies,* edited by Doyne Farmer, Alan Lapedes, Norman Packard, and Burton Wendroff. Amsterdam: North-Holland Physics Publishing, 1985: 187–204.

C. Faucheux and S. Moscovici. "Le style de comportement d'une minorité et son influence sur les résponses d'une majorité." *Bulletin du Centre d'Études et Recherches Psychologiques* 16 (1967): 337–360.

"Fear along the Silk Road: The Taleban Virus." *Middle East,* October 1998, reprinted in *World Press Review,* December 1998: 6–7.

Carol Featherstone. "Coming to Grips with the Golgi." *Science,* December 18, 1998: 3172–3174.

Federal Bureau of Investigation. "Top Ten Fugitive: Eric Robert Rudolph." *The FBI's Ten Most Wanted Fugitives.* http://www.fbi.gov/mostwant/rudolph.html. November 1998.

Mikhail A. Fedonkin and Benjamin M. Waggoner. "The Late Precambrian Fossil Kimberella Is a Mollusc-like Bilaterian Organism." *Nature,* August 28, 1997: 868–871.

D. M. Feltquate, S. Heaney, R. G. Webster, and H. L. Robinson. "Different T Helper Cell Types and Antibody Isotypes Generated by Saline and Gene Gun DNA Immunization." *Journal of Immunology,* March 1, 1997: 2278–2284.

"Feminism/Terms." Ver. 1.5, last modified February 15, 1993. http://www.cis.ohio-state.edu/hypertext/faq/usenet/feminism/terms/faq.html. Downloaded June 1999.

D. F. Feng, G. Cho, and R. F. Doolittle. "Determining Divergence Times with a Protein Clock: Update and Reevaluation." *Proceedings of the National Academy of Sciences of the United States of America,* November 25, 1997: 13028–13033.

David Fideler. Personal communication. July 15, 1998.

David Fideler, ed. *The Pythagorean Sourcebook and Library.* Compiled and translated by Kenneth Sylvan Guthrie. Grand Rapids, Mich.: Phanes Press, 1987.

G. Filice, L. Soldini, P. Orsolini, E. Razzini, R. Gulminetti, D. Campisi, L. Chiapparoli, E. Cattaneo, and G. Achilli. "Sensitivity and Specificity of Anti-HIV ELISA Employing Recombinant (p24, p66, gp120) and Synthetic (gp41) Viral Antigenic Peptides." *Microbiologica,* July 1991: 185–194.

D. R. Finlay, D. D. Newmeyer, P. M. Hartl, J. Horecka, and D. J. Forbes. "Nuclear Transport In Vitro." *Journal of Cell Science* (Supplement 11), 1989: 225–242.

G. Fiorito, C. von Planta, and P. Scotto. "Problem Solving Ability of *Octopus Vulgaris* Lamarck (Mollusca, Cephalopoda)." *Behavioral and Neural Biology,* March 1990: 217–230.

Graziano Fiorito and Pietro Scotto. "Observational Learning in *Octopus vulgaris.*" *Science* 256:5056 (1992): 545–547.

Joshua Fischman. "Evidence Mounts for Our African Origins—and Alternatives." *Science,* March 8, 1996: 1364.

Hakan Fisher, Gustav Wik, and Mats Fredrikson. "Extraversion, Neuroticism and Brain Function: A PET Study of Personality." *Personality and Individual Differences,* August 1997: 345–352.

M. G. Forest. "Role of Androgens in Fetal and Pubertal Development." *Hormone Research* 18:1–3 (1983): 69–83.

George Forrest. "Greece: The History of the Archaic Period." In *The Oxford History of the Classical World: Greece and the Hellenistic World,* edited by John Boardman, Jasper Griffin, and Oswyn Murray. New York: Oxford University Press, 1988.

W. G. Forrest. *A History of Sparta: 950–192 B.C.* New York: W. W. Norton, 1968.

Dian Fossey. *Gorillas in the Mist.* Boston: Houghton Mifflin, 1983.

J. S. Foster and M. J. McFall-Ngai. "Induction of Apoptosis by Cooperative Bacteria in the Morphogenesis of Host Epithelial Tissues." *Development Genes and Evolution,* August 1998: 295–303.

A. L. Fowden and M. J. Silver. "The Effects of Thyroid Hormones on Oxygen and Glucose Metabolism in the Sheep Fetus during Late Gestation." *Journal of Physiology* 482, pt. 1, January 1, 1995: 203–213.

Don C. Fowles, Richard Roberts, and Kim E. Nagel. "The Influence of Introversion/Extraversion on the Skin Conductance Response to Stress and Stimulus Intensity." *Journal of Research in Personality,* June 1977: 129–146.

N. A. Fox, M. A. Bell, and N. A. Jones. "Individual Differences in Response to Stress and Cerebral Asymmetry." *Developmental Neuropsychology* 8 (1992): 161–184.

S. Franzoi. "Personality Characteristics of the Crosscountry Hitchhiker." *Adolescence,* fall 1985: 655–668.

Antonia Fraser. *Cromwell.* New York: Donald I. Fine, 1973.

————. *The Warrior Queens.* New York: Alfred A. Knopf, 1989.

Sir George James Frazer. *The New Golden Bough.* Edited by Theodor H. Gaster. 1922. Reprint, Garden City, N.Y.: Doubleday, 1961.

J. B. Free. "The Social Organization of the Bumble-bee Colony: A Lecture to the Central Association of Bee-Keepers on 18th Jan 1961." Fleet, Hants, U.K.: North Hants Printing and Publishing Co., Ltd.

Daniel G. Freedman. "A Case for the Evolutionary Function of Ethnocentrism." Unpublished manuscript. N.d.

————. "Cross Cultural Differences in Newborn Behavior." University Park, Pa.: Psychological Cinema, 1980. Motion picture.

————. *Human Sociobiology: A Holistic Approach.* New York: Free Press, 1979.

J. L. Freedman and A. N. Doob. *Deviancy.* New York: Academic Press, 1968.

R. French and A. Messinger. "Genes, Phenes and the Baldwin Effect." In *Artificial Life IV,* edited by Rodney Brooks and Patricia Maes. Cambridge, Mass.: MIT Press, 1994.

Anna Freud. *The Ego and Mechanisms of Defense.* New York: International Universities Press, 1946.

Sigmund Freud. *Civilization and Its Discontents.* New York: W. W. Norton, 1989.

"The Friday Speech from Alaqsa Mosque in Occupied Muslim Land." January 28, 1988. Ramadhan/Alaqsa Islamic Web Site. http://www.ramadhan.org/hadith/hbody12.txt. August 1998.

I. Fried, K. A. MacDonald, and C. L. Wilson. "Single Neuron Activity in Human Hippocampus and Amygdala during Recognition of Faces and Objects." *Neuron,* May 1997: 753–765.

Ariella Friedman, Judith Todd, and Priscilla Wanjiru Kariuki. "Comparing Cooperative and Competitive Behavior in Rural and Urban Children in Kenya." *Cross Cultural Psychology,* July 1995: 374.

J. Y. Frigon and L. Granger. "Extraversion-Introversion and Arousal in a Visual Vigilance Task." *Bulletin de Psychologie,* November–December 1978: 33–40.

John E. Frisch, S.J. "Individual Behavior and Intertroop Variability in Japanese Macaques." In *Primates: Studies in Adaptation and Variability,* edited by Phyllis C. Jay. New York: Holt, Rinehart and Winston, 1968: 243–252.

Karl von Frisch. *The Dance Language and Orientation of Bees.* Translated by Leigh E. Chadwick. Cambridge, Mass.: Belknap Press, 1967.

Honor Frost. "How Carthage Lost the Sea." *Natural History,* December 1987: 58–67.

A. W. Frye, B. F. Richards, E. W. Bradley, and J. R. Philp. "The Consistency of Students' Self-Assessments in Short-Essay Subject Matter Examinations." *Medical Education,* July 1992: 310–316.

T. Fukai and S. Tanaka. "A Simple Neural Network Exhibiting Selective Activation of Neuronal Ensembles: From Winner-Take-All to Winners-Share-All." *Neural Computation,* January 1997: 77–97.

Glen S. Fukushima. "Affirmative Action, Japanese Style." *Tokyo Business,* February 1994: 58.

Francis Fukuyama. *The End of History and the Last Man.* New York: Avon Books, 1993.

Torrey E. Fuller. *Witchdoctors and Psychiatrists.* New York: Harper and Row, 1986.

John Kenneth Galbraith. *The Great Crash: 1929.* 1954. Reprint, Boston: Houghton Mifflin, 1988.

B. G. Galef Jr and E. E. Whiskin. "Learning Socially to Eat More of One Food Than of Another." *Journal of Comparative Psychology,* March 1995: 99–101.

B. G. Galef Jr, E. E. Whiskin, and E. Bielavska. "Interaction with Demonstrator Rats Changes Observer Rats' Affective Responses to Flavors." *Journal of Comparative Psychology,* December 1997: 393–398.

Winifred Gallagher. *The Power of Place: How Our Surroundings Shape Our Thoughts, Emotions and Actions.* New York: Poseidon Press, 1993.

Clive Gamble. *Timewalkers: The Prehistory of Global Colonization.* Cambridge, Mass.: Harvard University Press, 1994.

K. Gardiner. "Clonability and Gene Distribution on Human Chromosome 21: Reflections of Junk DNA Content?" *Gene,* December 31, 1997: 39–46.

J. S. Gartlan and C. K. Brain. "Ecology and Social Variability in *Cercopithecus Aethiops* and *C. Mitis.*" In *Primates: Studies in Adaptation and Variability,* edited by Phyllis C. Jay. New York: Holt, Rinehart and Winston, 1968: 253–292.

Edward Gaten. Personal communication. April 2, 1997.

Willard B. Gatewood Jr. *Controversy in the Twenties: Fundamentalism, Modernism, and Evolution.* Nashville, Tenn.: Vanderbilt University Press, 1969.

M. S. Gazzaniga. "Brain and Conscious Experience." *Advances in Neurology* 77 (1998): 181–192.

M. S. Gazzaniga, J. C. Eliassen, L. Nisenson, C. M. Wessinger, R. Fendrich, and K. Baynes. "Collaboration between the Hemispheres of a Callosotomy Patient: Emerging Right Hemisphere Speech and the Left Hemisphere Interpreter." *Brain* (4). August 1996: 1255–1262.

Michael S. Gazzaniga. *The Social Brain: Discovering the Networks of the Mind.* New York: Basic Books, 1985.

———. "Organization of the Human Brain." *Science,* September 1, 1989: 947–952.

———. *Nature's Mind: The Biological Roots of Thinking, Emotions, Sexuality, Language, and Intelligence.* New York: Basic Books, 1992.

Valerius Geist. *Life Strategies, Human Evolution, Environmental Design: Toward a Biological Theory of Health.* New York: Springer-Verlag, 1978.

———. Personal communications. 1997–1999.

Genetic Engineering Network. Genetic Engineering—Direct Action. http://www.k2net.co.uk/ef/toxmut/action.html. November 1998.

Genetix. "Superheroes against Genetix." http://www.enviroweb.org/shag/genetix.html. November 1998.

E. Gerdts. "Ultrasonic Diagnosis in Abortion of One Twin: The Vanishing Twin Phenomenon." *Tidsskrift for den Norske Laegeforening,* November 20, 1989: 3328–3329.

J. M. Gerrard, S. P. Saxena, and A. McNicol. "Histamine as an Intracellular Messenger in Human Platelets." *Advances in Experimental Medicine and Biology* 344 (1993): 209–219.

C. J. Gerrish and J. R. Alberts. "Differential Influence of Adult and Juvenile Conspecifics on Feeding by Weanling Rats *(Rattus norvegicus):* A Size-Related Explanation." *Journal of Comparative Psychology,* March 1995: 61–67.

Felicia B. Gershberg. "Implicit and Explicit Conceptual Memory following Frontal Lobe Damage." *Journal of Cognitive Neuroscience,* January 1, 1997: 105.

Raymond F. Gesteland, Thomas R. Cech, and John F. Atkins. *The RNA World.* 2nd ed. Monograph 37. Plainview, N.Y.: Cold Spring Harbor Laboratory Press, 1999.

Michael Patrick Ghiglieri. *The Chimpanzees of Kibale Forest: A Field Study of Ecology and Social Structure.* New York: Columbia University Press, 1984.

Edward Gibbon. *The Decline and Fall of the Roman Empire.* Vol. 1, A.D. 180–A.D. 395. New York: Modern Library, n.d.

———. *The Decline and Fall of the Roman Empire.* Vol. 2, A.D. 395–A.D. 1185. New York: Modern Library, n.d.

Ann Gibons. "On the Many Origins of Species." *Science,* September 13, 1996: 1496–1499.

———. "Tracing the Identity of the First Toolmakers." *Science,* April 4, 1997: 32–40.

———. "A New Face for Human Ancestors." *Science,* May 30, 1997: 1331–1333.

———. "Archaeologists Rediscover Cannibals." *Science,* August 1, 1997: 635–637.

———. "Doubts over Spectacular Dates." *Science,* October 10, 1997: 220–222.

————. "Ancient Island Tools Suggest *Homo erectus* Was a Seafarer." *Science,* March 13, 1998: 1635–1637.

————. "In China, a Handier Homo erectus." *Science,* March 13, 1998: 1636.

————. "Which of Our Genes Make Us Human?" *Science,* September 4, 1998: 1432–1434.

McGuire Gibson. "Population Shift and the Rise of Mesopotamian Civilisation." In *The Explanation of Culture Change: Models in Prehistory,* edited by Colin Renfrew. Pittsburgh: University of Pittsburgh Press, 1973: 447–463.

J. N. Giedd, C. Vaituzis, S. D. Hamburger, N. Lange, J. Rajapakse, D. Kaysen, Y. C. Vauss, and J. L. Rapoport. "Quantitative MRI of the Temporal Lobe, Amygdala, and Hippocampus in Normal Human Development: Ages 4–18 Years." *Journal of Comparative Neurology* 366 (1996): 223–230.

P. Gilbert and S. Allan. "Assertiveness, Submissive Behaviour and Social Comparison." *British Journal of Clinical Psychology* 33 (1994): 295–306.

P. Gilbert, J. Price, and S. Allan. "Social Comparison, Social Attractiveness and Evolution: How Might They Be Related?" *New Ideas in Psychology* 13:2 (1994): 149–165.

Carol Gilligan. *In a Different Voice: Psychological Theory and Women's Development.* Cambridge, Mass.: Harvard University Press, 1982.

A. Gilman. "Explaining the Upper Paleolithic Revolution." In *Marxist Perspectives in Archaeology,* edited by M. Spriggs. Cambridge, U.K.: Cambridge University Press, 1984: 115–126.

J. S. L. Gilmour. "Taxonomy and Philosophy." In *The New Systematics,* edited by Julian Huxley. Oxford: Oxford University Press, 1940: 401–474.

M. Gimbutas. "Wall Paintings of Catal Huyuk." *Review of Archaeology,* fall 1990.

Darren R. Gitelman, Nathaniel M. Alpert, Stephen Kosslyn, and Kirk Daffner. "Functional Imaging of Human Right Hemispheric Activation for Exploratory Movements." *Annals of Neurology,* February 1996: 174–179.

A. G. Gitter, H. Black, and A. Goldman. "Role of Nonverbal Communication in the Perception of Leadership." *Perceptual and Motor Skills,* April 1975: 463–466.

James Glanz. "No Backing Off from the Accelerating Universe." *Science,* November 13, 1998: 1249–1251.

————. "Exploding Stars Point to a Universal Repulsive Force." *Science,* January 30, 1998: 651–652.

I. M. Glynn and S. J. Karlish. "The Sodium Pump." *Annual Review of Physiology* 37 (1975): 13–55.

O. W. Godfrey. "Directed Mutation in *Streptomyces Lipmanii.*" *Candian Journal of Microbiology,* November 1974: 1479–1485.

God Hates Fags (web site). http://www.godhatesfags.com/. November 1998.

Erving Goffman. *The Presentation of Self in Everyday Life.* New York: Anchor Books, 1959.

M. Gogarten-Boekels, E. Hilario, and J. P. Gogarten. "The Effects of Heavy Meteorite Bombardment on the Early Evolution—the Emergence of the Three Domains of Life." *Origins of Life and Evolution of the Biosphere,* June 1995: 251–264.

Balkrishna Govind Gokhale. *Asoka Maurya.* New York: Twayne Publishers, 1966.

Michael A. Goldberg and Barry Katz. "The Effect of Nonreciprocated and Reciprocated Touch on Power/Dominance Perception." *Journal of Social Behavior and Personality* 5:5 (1990): 379–386.

P. A. Goldberg. "Are Women Prejudiced against Women?" *Transaction,* April 1968: 28–30.

P. Goldman-Rakic and P. Rakic. "Experimental Modification of Gyral Patterns." In *Cerebral Dominance: The Biological Foundation,* edited by N. Geschwind and A. M. Galaburda. Cambridge, Mass.: Harvard University Press, 1984: 179–192.

Donald Goldsmith. "Supernovae Offer a First Glimpse of the Universe's Fate." *Science,* April 4, 1997: 37–40.

Alvin G. Goldstein and Judy Jeffords. "Status and Touching Behavior." *Bulletin of the Psychonomic Society,* February 1981: 79–81.

S. Golubic, L. Seong-Joo, and K. M. Browne. "Cyanobacteria: Architects of Sedimentary Structures." In *Microbial Sediments,* edited by R. Riding and S. M. Awramik. Berlin: Springer-Verlag, in press.

J. C. S. Gonzalez. "El Efecto Baldwin. La propuesta funcionalista para una søntesis psicobiologica" (The Baldwin Effect. The Functionalistic Proposal for a Psychobiological Synthesis). Ph.D. diss. Oviedo, Spain: University of Oviedo, 1994.

Jane Goodall. *In the Shadow of Man.* 1971. Reprint, Boston: Houghton Mifflin, 1983.

F. K. Goodwin and R. M. Post. "5-Hydroxytryptamine and Depression: A Model for the Interaction of Normal Variance with Pathology." *British Journal of Clinical Pharmacology* 15 (1983): 393S–405S.

Deborah M. Gordon. "The Expandable Network of Ant Exploration." *Animal Behavior* 50 (1995): 995–1007.

M. J. Gordon. "A Review of the Validity and Accuracy of Self-Assessments in Health Professions Training." *Academic Medicine,* December 1991: 762–769.

Naama Goren-Inbar and Idit Saragusti. "An Acheulian Biface Assemblage from Gesher Benot Ya'aqov, Israel: Indications of African Affinities." *Journal of Field Archaeology* 23:1 (spring 1996): 15–30.

J. Gormly, A. Gormly, and C. Johnson. "Interpersonal Attraction. Competence Motivation and Reinforcement Theory." *Journal of Personality and Social Psychology* 19 (1971): 375–380.

J. D. Goss-Custard and W. J. Sutherland. "Feeding Specializations in Oystercatchers *Haematopus Ostralegus.*" *Animal Behavior,* February 1984: 299–301.

Ian H. Gotlib. "Interpersonal and Cognitive Aspects of Depression." *Current Directions in Psychological Science,* October 1992: 149–154.

Elizabeth Gould. "The Effects of Adrenal Steroids and Excitatory Input on Neuronal Birth and Survival." In *Hormonal Restructuring of the Adult Brain: Basic and Clinical Perspectives,* edited by Victoria N. Luine and Cheryl F. Harding. Annals of the New York Academy of Sciences, vol. 743. New York: New York Academy of Sciences, 1994: 73–93.

James L. Gould. Personal communication. April 10, 1997.

Stephen Jay Gould. *Wonderful Life: The Burgess Shale and the Nature of History.* New York: W. W. Norton, 1989.

Government Statistical Service—the Office for National Statistics, United Kingdom. http://www.statistics.gov.uk/stats/ukinfigs/pop.html. November 1998.

E. A. Govorkova, G. Murti, B. Meignier, C. de Taisne, and R. G. Webster. "African Green Monkey Kidney (Vero) Cells Provide an Alternative Host Cell System for Influenza A and B Viruses." *Journal of Virology,* August 1996: 5519–5524.

T. J. Graddis, K. Brasel, D. Friend, S. Srinivasan, S. Wee, S. D. Lyman, C. J. March, and J. T. McGrew. "Structure-Function Analysis of FLT3 Ligand-FLT3 Receptor Interactions Using a Rapid Functional Screen." *Journal of Biological Chemistry,* July 1998: 17626–17633.

J. Grafman, S. C. Vance, H. Weingartner, A. M. Salazar, and D. Amin. "The Effects of Lateralised Frontal Lesions on Mood Regulation." *Brain* 109 (1986): 1127–1148.

K. Grammer and A. Jütte. "Battle of Odors: Significance of Pheromones for Human Reproduction." *Gynakologisch-Geburtshilfliche Rundschau* 37:3 (1997): 150–153.

Michael Grant. *The Rise of the Greeks.* New York: Charles Scribner's Sons, 1987.

Peter R. Grant. "Ecological Character Displacement." *Science,* November 4, 1994: 746.

K. M. Gray. "Intercellular Communication and Group Behavior in Bacteria." *Trends in Microbiology,* May 1997: 184–188.

Neil Greenberg. "Observations of Social Feeding in Lizards." *Herpetologica,* September 1976: 348–352.

———. "Behavioral Endocrinology of Physiological Stress in a Lizard." *Journal of Experimental Zoology* (Supplement 4), 1990: 170–173.

———. "Central and Endocrine Aspects of Tongue-Flicking and Exploratory Behavior in *Anolis carolinensis.*" *Brain, Behavior, and Evolution* 41 (1993): 210–218.

————. "Ethologically Informed Design in Husbandry and Research." In *Health and Welfare of Captive Reptiles,* edited by Clifford Warwick, Fredric L. Frye, and James B. Murphy. London: Chapman and Hall, 1995: 239–261.

————. Personal communications. June 17–20, 1998.

Neil Greenberg, Enrique Font, and Robert C. Switzer III. "The Reptilian Striatum Revisited: Studies on *Anolis* Lizards." In *The Forebrain of Reptiles: Current Concepts of Structure and Function,* edited by Walter K. Schwerdtfeger and Willhelmus J. A. J. Smeets. Basel, Switz.: Karger, 1988: 162–177.

Neil Greenberg, Paul D. MacLean, and John L. Ferguson. "Role of the Paleostriatum in Species-Typical Display Behavior of the Lizard *(Anolis Carolinensis)."* *Brain Research,* August 1979: 229–241.

M. Grez, U. Dietrich, P. Balfe, H. von Briesen, J. K. Maniar, G. Mahambre, E. L. Delwart, J. I. Mullins, and H. Rübsamen-Waigmann. "Genetic Analysis of Human Immunodeficiency Virus Type 1 and 2 (HIV-1 and HIV-2) Mixed Infections in India Reveals a Recent Spread of HIV-1 and HIV-2 from a Single Ancestor for Each of These Viruses." *Journal of Virology,* April 1994: 2161–2168.

Domenico Grieco, Antonio Porcellini, Enrico V. Avvedimento, and Max E. Gottesman. "Requirement for cAMP-PKA Pathway Activation by M Phase–Promoting Factor in the Transition from Mitosis to Interphase." *Science,* March 22, 1996: 1718–1723.

James W. Grier. *Biology of Animal Behavior.* Saint Louis: Times Mirror, 1984.

William Griffitt. "Environmental Effects on Interpersonal Affective Behavior: Ambient Effective Temperature and Attraction." *Journal of Personality and Social Psychology* 15 (1970): 240–244.

W. Griffitt and R. Veitch. "Hot and Crowded: Influences of Population Density and Temperature on Interpersonal Affective Behavior." *Journal of Personality and Social Psychology* 17 (1971): 92–98.

D. B. Grigg. *The Agricultural Systems of the World: An Evolutionary Approach.* Cambridge, U.K.: Cambridge University Press, 1974.

John P. Grotzinger, Samuel A. Bowring, Beverly Z. Saylor, and Alan J. Kaufman. "Biostratigraphic and Geochronologic Constraints on Early Animal Evolution." *Science,* October 27, 1995: 598–604.

John P. Grotzinger and Daniel H. Rothman. "An Abiotic Model for Stromatolite Morphogenesis." *Nature,* October 3, 1996: 423–425.

R. Grun and C. Stringer. "Electron Spin Resonance Dating and the Evolution of Modern Humans." *Archaeometry* 33:2 (1991): 153–199.

D. G. Guiney Jr. "Promiscuous Transfer of Drug Resistance in Gram-Negative Bacteria." *Journal of Infectious Diseases,* March 1984: 320–329.

Haldun Gulalp. "Islamism and Post-Modernism." *Contention,* winter 1995: 59–73.

B. B. Gump and J. A. Kulik. "Stress, Affiliation, and Emotional Contagion." *Journal of Personality and Social Psychology,* February 1997: 305–319.

M. R. Gunnar, K. Tout, M. de Haan, S. Pierce, and K. Stansbury. "Temperament, Social Competence, and Adrenocortical Activity in Preschoolers." *Developmental Psychobiology,* July 1997: 65–85.

Trisha Gura. "One Molecule Orchestrates Amoebae." *Science,* July 11, 1997: 182.

Alan H. Guth. "Inflation." *Proceedings of the National Academy of Sciences of the United States of America,* June 1, 1993: 4871–4877.

————. *The Inflationary Universe: The Quest for a New Theory of Cosmic Origins.* Reading, Mass.: Addison-Wesley, 1997.

C. Haanen and I. Vermes. "Apoptosis: Programmed Cell Death in Fetal Development." *European Journal of Obstetrics, Gynecology, and Reproductive Biology,* January 1996: 129–133.

C. Haertzen, K. Buxton, L. Covi, and H. Richards. "Seasonal Changes in Rule Infractions among Prisoners: A Preliminary Test of the Temperature-Aggression Hypothesis." *Psychological Reports,* February 1993: 195–200.

B. M. Hagerty and R. A. Williams. "The Effects of Sense of Belonging, Social Support, Conflict, and Loneliness on Depression." *Nursing Research,* July–August 1999: 215–219.

Jason Hagey. "Flocking Birds and Schooling Fish: Flocking and Schooling as Complex Systems." http://www.susqu.edu/facstaff/b/brakke/complexity/hagey/flock.html. February 1999.

B. G. Hall. "Adaptive Evolution That Requires Multiple Spontaneous Mutations. I. Mutations Involving an Insertion Sequence." *Genetics,* December 1988: 887–897.

Edward T. Hall. *Beyond Culture.* New York: Anchor Books, 1977.

G. Stanley Hall. "A Study of Anger." *American Journal of Psychology* 10 (1899): 516–591.

K. R. L. Hall. "Experiment and Quantification in the Study of Baboon Behavior in Its Natural Habitat." In *Primates: Studies in Adaptation and Variability,* edited by Phyllis C. Jay. New York: Holt, Rinehart and Winston, 1968: 120–130.

————. "Tool-Using Performances as Indicators of Behavioral Adaptability." In *Primates: Studies in Adaptation and Variability,* edited by Phyllis C. Jay. New York: Holt, Rinehart and Winston, 1968: 131–148.

————. "Social Learning in Primates." In *Primates: Studies in Adaptation and Variability,* edited by Phyllis C. Jay. New York: Holt, Rinehart and Winston, 1968: 383–397.

Thomas S. Hall. *Ideas on Life and Matter.* Vol. 2. Chicago: University of Chicago Press, 1969.

W. D. Hamilton. "The Genetical Theory of Social Behaviour." *Journal of Theoretical Biology* 7:1 (1964): 1–52.

————. "Geometry for the Selfish Herd." *Journal of Theoretical Biology* 31:3 (1971): 295–311.

————. "Selection of Selfish and Altruistic Behavior in Some Extreme Models." In *Man and Beast: Comparative Social Behavior,* edited by J. F. Eisenberg and W. S. Dillon. Washington, D.C.: Smithsonian Institution Press, 1971: 57–91.

————. "Altruism and Related Phenomena, Mainly in Social Insects." *Annual Review of Ecology and Systematics* 3 (1972): 193–232.

M. F. Hammer, T. Karafet, A. Rasanayagam, et al. "Out of Africa and Back Again: Nested Cladistic Analysis of Human Y Chromosome Variation." *Molecular Biology and Evolution* 15 (1998): 427–441.

N. G. L. Hammond. *Alexander the Great, King, Commander, and Statesman.* Park Ridge, N.J.: Noyes Press, 1980.

Eric P. Hamp. "On the Indo-European Origins of the Retroflexes in Sanskrit." *Journal of the American Oriental Society,* October 21, 1996: 719–724.

Terence P. Hannigan. "Body Odor: The International Student and Cross-cultural Communication." *Culture and Psychology,* December 1995: 407–503.

D. Hansen, H. C. Lou, M. Nordentoft, O. A. Pryds, J. R. Jensen, J. Nim, and R. P. Hemmingsen. "The Significance of Psychosocial Stress for Pregnancy Course and Fetal Development." *Ugeskrift for Laeger,* April 1996: 2369–2372.

R. Hardison. "Hemoglobins from Bacteria to Man: Evolution of Different Patterns of Gene Expression." *Journal of Experimental Biology* (pt. 8). April 1998: 1099–1117.

S. Harkins and R. G. Geen. "Discriminability and Criterion Differences between Extraverts and Introverts during Vigilance." *Journal of Research in Personality* 9 (December 1975): 335–340.

Stevan Harnad. "Consciousness: An Afterthought." *Cognition and Brain Theory* 5 (1982): 29–47.

Harry F. Harlow. *Learning to Love.* New York: Jason Aronson, 1974.

Harry F. Harlow and Gary Griffin. "Induced Mental and Social Deficits in Rhesus Monkeys." In *Biological Basis of Mental Retardation,* edited by Sonia F. Osler and Robert E. Cooke. Baltimore: Johns Hopkins Press, 1965: 87–106.

Harry F. Harlow and Margaret Kuenne Harlow. "Social Deprivation in Monkeys." *Scientific American,* November 1962: 136–146.

Charles L. Harper Jr. and Stein B. Jacobsen. "Noble Gases and Earth's Accretion." *Science,* September 27, 1996: 1814–1818.

Dennis C. Harper. "Children's Attitudes toward Physical Disability in Nepal: A Field Study." *Journal of Cross-Cultural Psychology,* November 1997: 710–729.

Anne Harrington. *Reenchanted Science: Holism in German Culture from Wilhelm II to Hitler.* Princeton, N.J.: Princeton University Press, 1996.

Eugene E. Harris and Jody Hey. "X Chromosome Evidence for Ancient Human Histories." *Proceedings of the National Academy of Sciences of the United States of America,* March 16, 1999: 3320–3324.

Lawrence E. Harrison. *Underdevelopment Is a State of Mind: The Latin American Case.* Center for International Affairs, Harvard University. Lanham, Md.: Madison Books, 1988.

A. W. Harrist, A. F. Zaia, J. E. Bates, K. A. Dodge, and G. S. Pettit. "Subtypes of Social Withdrawal in Early Childhood: Sociometric Status and Social-Cognitive Differences across Four Years." *Child Development,* April 1997: 278–294.

R. M. Harshey. "Bees Aren't the Only Ones: Swarming in Gram-negative Bacteria." *Molecular Microbiology,* May 1994: 389–394.

Steve E. Hartman, Jaime Hylton, and Ronda F. Sanders. "The Influence of Hemispheric Dominance on Scores of the Myers-Briggs Type Indicator." *Educational and Psychological Measurement,* June 1997: 440–449.

Allen Harvey. Personal communication. March 31, 1997.

Stephen T. Hasiotis, Russell F. Dubiel, Paul T. Kay, Timothy M. Demko, Krystyna Kowalska, and Douglas McDaniel. "Research Update on Hymenopteran Nests and Cocoons, Upper Triassic Chinle Formation, Petrified Forest National Park, Arizona." *National Park Service Paleontological Research,* January 1998: 116–121.

M. Hassler. "Testosterone and Musical Talent." *Experimental and Clinical Endocrinology and Diabetes* 98:2 (1991): 89–98.

T. Hatch. "Chlamydia: Old Ideas Crushed, New Mysteries Bared." *Science,* October 23, 1998: 638.

David Freeman Hawke. *John D.—the Founding Father of the Rockefellers.* New York: Harper and Row, 1980.

Brett K. Hayes and Ruth Hennessy. "The Nature and Development of Nonverbal Implicit Memory." *Journal of Experimental Child Psychology,* October 1, 1996: 22.

C. Vance Haynes Jr., T. A. Maxwell, A. El-Hawary, K. A. Nicoll, and S. Stokes. "An Acheulian Site Near Bir Kiseiba in the Darb el Arba' in Desert, Egypt." *Geoarchaeology* 12, December 1997: 819–832.

I. B. Heath. "The Cytoskeleton in Hyphal Growth, Organelle Movements and Mitosis." In *The Mycota,* vol. 1, edited by J. G. H. Wessels and F. Mienhardt. Berlin: Springer-Verlag, 1994.

Steven R. Heidemann. "A New Twist on Integrins and the Cytoskeleton." *Science,* May 21, 1993: 1080–1081.

Jacob Heilbrunn. "Germany's New Right: Liberating Nationalism from History." *Foreign Affairs,* November-December 1996: 80–98.

Bernd Heinrich. *Ravens in Winter.* New York: Summit Books, 1989.

———. "Why Ravens Share." *American Scientist,* July–August 1995: 342–350.

B. Heinrich and J. M. Marzluff. "Do Common Ravens Yell Because They Want to Attract Others?" *Behavioral Ecology and Sociobiology* 28 (1991): 13–21.

Hans Helback. "First Impressions of the Catal Huyuk Plant Husbandry." *Anatolian Studies* 14 (1964): 121–123.

Nancy M. Henley. *Body Politics: Power, Sex, and Nonverbal Communication.* Englewood Cliffs, N.J.: Prentice-Hall, 1977.

Doug Henwood. "Japan in Latin America." *Left Business Observer,* May 15, 1989: 1–3.

———. "Japan in Three Countries." *Left Business Observer,* May 15, 1989: 3.

P. B. Heppleston. "The Comparative Breeding Ecology of Oystercatchers *(Haematopus ostralegus)* in Inland and Coastal Habitats." *Journal of Animal Ecology* 41 (1972): 23–51.

Herodotus. *The Histories.* Translated by Aubrey de Selincourt. New York: Penguin Books, 1972.

W. F. Herrnkind. "Evolution and Mechanisms of Mass Single-file Migration in Spiny Lobster." *Migration: Mechanisms and Adaptive Significance.* Contributions in Marine Science—monographic ser., vol. 68. Austin: Marine Science Institute, University of Texas at Austin, 1985: 197–211.

William Herrnkind. Personal communication. February 26, 1999; April 5, 2000.

Melville J. Herskovitz. *Economic Anthropology: The Economic Life of Primitive Peoples.* 1940. Reprint, New York: W. W. Norton, 1965.

Hesiod. *The Homeric Hymns:And Homerica.* Translated by Hugh G. Evelyn-White. Cambridge, Mass.: Harvard University Press, 1982.

Manfred Heun, Ralf Schafer-Pregl, Dieter Klawan, Renato Castagna, Monica Accerbi, Basilio Borghi, and Francesco Salamini. "Site of Einkorn Wheat Domestication Identified by DNA Fingerprinting." *Science,* November 14, 1997: 1312–1322.

Francis Heylighen. "Towards a Global Brain: Integrating Individuals into the World-Wide Electronic Network." In *Der Sinn der Sinne,* edited by Uta Brandes and Claudia Neumann. Göttingen: Steidl Verlag, 1997.

Francis Heylighen and Johan Bollen. "The World-Wide Web as a Super-Brain: From Metaphor to Model." In *Cybernetics and Systems '96,* edited by R. Trappl. Vienna: Austrian Society for Cybernetic Studies, 1996.

F. Heylighen, C. Joslyn, and V. Turchin, eds. *Principia Cybernetica Web.* Brussels: Principia Cybernetica. http://pespmc1.vub.ac.be/Default.html. February 1999.

J. D. Higley, S. T. King Jr, M. F. Hasert, M. Champoux, S. J. Suomi, and M. Linnoila. "Stability of Interindividual Differences in Serotonin Function and Its Relationship to Severe Aggression and Competent Social Behavior in Rhesus Macaque Females." *Neuropsychopharmacology,* January 1996: 67–76.

C. T. Hill and D. E. Stull. "Sex Differences in Effects of Social and Value Similarity in Same-sex Friendship." *Journal of Personality and Social Psychology* 41 (1981): 488–502.

R. A. Hinde and J. Fisher. "Further Observations on the Opening of Milk Bottles by Birds." *British Birds* 44 (1951): 393–396.

V. S. Hinshaw, R. G. Webster, and B. Turner. "Water-borne Transmission of Influenza A Viruses?" *Intervirology* 11:1 (1979): 66–68.

M. Hinton, G. C. Mead, and C. S. Impey. "Protection of Chicks against Environmental Challenge with *Salmonella Enteritidis* by 'Competitive Exclusion' and Acid-Treated Feed." *Letters in Applied Microbiology,* March 1991.

M. Hirsch. "Two Forms of Identification with the Aggressor—according to Ferenczi and Anna Freud." *Praxis der Kinderpsychologie und Kinderpsychiatrie,* July–August 1996: 198–205.

M. A. Hofer. "Early Social Relationships: A Psychobiologist's View." *Child Development* 58 (1987): 633–647.

———. "On the Nature and Consequences of Early Loss." *Psychosomatic Medicine* 58 (1996): 570–581.

Joseph F. Hoffman and Bliss Forbush III, eds. *Structure, Mechanism, and Function of the Na/K Pump.* New York: Academic Press, 1983.

M. L. Hoffman. "Is Altruism Part of Human Nature?" *Journal of Personality and Social Psychology* 40:1 (1981): 121–137.

Constance Holden. "Ordinary Is Beautiful." *Science,* April 20, 1990: 306.

———. "Sensing Music." *Science,* October 13, 1995: 237.

———. "Nature v. Culture: A Lesson from the Guppy." *Science,* April 12, 1996: 203.

———. "The First Tool Kit." *Science,* January 31, 1997: 623.

———. "A Billion Years of Starvation." *Science,* July 3, 1998: 39a.

———. "Modeling Fossils in Jell-O." *Science,* July 24, 1998: 511.

———. "No Last Word on Language Origins." *Science,* November 20, 1998: 1455.

K. E. Holekamp, L. Smale, and M. Szykman. "Rank and Reproduction in the Female Spotted Hyaena." *Journal of Reproduction and Fertility,* November 1996: 229–237.

A. Holland and T. Andre. "The Relationship of Self-esteem to Selected Personal and Environmental Resources of Adolescents." *Adolescence,* summer 1994: 345–360.

Heinrich D. Holland. "Evidence for Life on Earth More Than 3850 Million Years Ago." *Science,* January 3, 1997: 38–39.

John H. Holland. *Hidden Order: How Adaptation Builds Complexity.* New York: Addison-Wesley, 1995.

Norman H. Holland. *Laughing: A Psychology of Humor.* Ithaca, N.Y.: Cornell University Press, 1982.

E. P. Hollander. "Conformity, Status, and Idiosyncrasy Credit." *Psychological Review* 65 (1958): 117–127.

Laura J. Hollar and Mark S. Springer. "Old World Fruitbat Phylogeny: Evidence for Convergent Evolution and an Endemic African Clade." *Proceedings of the National Academy of Sciences of the United States of America* 94 (1997): 5716–5721.

Bert Holldobler and Edward O. Wilson. *The Ants.* Cambridge, Mass.: Belknap Press, 1990.

Leon Hollerman. *Japan's Economic Strategy in Brazil: Challenge for the United States.* Lexington, Mass.: Lexington Books, 1988.

Dr. Cornelius Holtorf. "Knowing without Metaphysics and Pretentiousness: A Radical Constructivist Proposal." Goteborg, Sweden: Arkeologiska institutionen Goteborgs universitet. http://www.lamp.ac.uk/~cjh. Downloaded May 1999.

Bruce M. Hood, J. Douglas Willen, and Jon Driver. "Adults' Eyes Trigger Shifts of Visual Attention in Human Infants." *Psychological Science,* March 1998: 131–133.

Robert Hooke. *Micrographia.* 1665. Octavo Posters on Mediaflex.com. http://www.mediaflex.com/octavo/hkemic.html. December 1998.

J. J. Hopfield. "Neural Networks and Physical Systems with Emergent Collective Computational Abilities." *Proceedings of the National Academy of Sciences of the United States of America,* April 1982: 2554–2558.

————. Personal communication. November 1988.

Horace. "Odes." 2.10.5. The Perseus Project, edited by Gregory R. Crane. http://hydra.perseus.tufts.edu. September 1998.

Simon Hornblower. "Greece: The History of the Classical Period." In John Boardman, *The Oxford History of the Classical World: Greece and the Hellenistic World,* edited by John Boardman, Jasper Griffin, and Oswyn Murray. New York: Oxford University Press, 1988: 118–149.

Harvey A. Hornstein. *Cruelty and Kindness: A New Look at Aggression and Altruism.* Englewood Cliffs, N.J.: Prentice-Hall, 1976.

D. F. Horrobin. "The Philosophical Basis of Peer Review and the Suppression of Innovation." *Journal of the American Medical Association,* March 9, 1990: 1438–1441.

Lianhai Hou, Larry D. Martin, Zhonghe Zhou, and Alan Feduccia. "Early Adaptive Radiation of Birds: Evidence from Fossils from Northeastern China." *Science,* November 15, 1996: 1164–1167.

Lianhai Hou, Larry D. Martin, Zhonghe Zhou, Alan Feduccia, and Fucheng Zhang. "A Diapsid Skull in a New Species of the Primitive Bird Confuciusornis." *Nature,* June 17, 1999: 679–682.

X. G. Hou and J. Bergström. "Cambrian Lobopodians—Ancestors of Extant Onychophorans?" *Zoological Journal of the Linnean Society* 114 (1995): 3–19.

James S. House, Karl R. Landis, and Debra Umberson. "Social Relationships and Health." *Science,* July 29, 1988: 540–545.

C. Hovland and W. Weiss. "The Influence of Source Credibility on Communication Effectiveness." *Public Opinion Quarterly* 15 (1952): 635–650.

Fred Howard. *Wilbur and Orville: A Biography of the Wright Brothers.* New York: Dover Publications, 1998.

D. H. Hubel and T. N. Wiesel. "Receptive Fields and Functional Architecture of Monkey Striate Cortex." *Journal of Physiology,* March 1968: 215–243.

————. "Anatomical Demonstration of Columns in the Monkey Striate Cortex." *Nature,* February 22, 1969: 747–750.

David Hubel. *Eye, Brain and Vision.* New York: Methuen, 1986.

Robert Huber. Personal communication. February 22, 1999.

David L. Hull. *Science as a Process: An Evolutionary Account of the Social and Conceptual Development of Science.* Chicago: University of Chicago Press, 1988.

M. E. Hume, J. A. Byrd, L. H. Stanker, and H. L. Ziprin. "Reduction of Caecal Listeria Monocytogenes in Leghorn Chicks Following Treatment with a Competitive Exclusion Culture (PREEMPT)." *Letters in Applied Microbiology,* June 1998: 432–436.

Christine R. Hurd. "Interspecific Attraction to the Mobbing Calls of Black-Capped Chickadees *(Parus antricapillus)." Behavioral Ecology and Sociobiology,* April 1996: 287–292.

Stella M. Hurtley. "Cell Biology of the Cytoskeleton." *Science,* March 19, 1999: 1931–1934.

J. I. Hurwitz, A. F. Zander, and B. Hymovitch. "Some Effects of Power on the Relations among Group Members." In *Group Dynamics: Research and Theory,* edited by Dorwin Cartwright and Alvin Zander. New York: Harper and Row, 1953: 483–492.

N. H. Hussain, M. Goodson, and R. J. Rowbury. "Recent Advances in Biology: Intercellular Communication and Quorum Sensing in Micro-organisms." *Science Progress* 81, pt. 1 (1998): 69–80.

Samuel C. Hyde Jr. *Pistols and Politics: The Dilemma of Democracy in Louisiana's Florida Parishes, 1810–1899.* Baton Rouge: Louisiana State University Press, 1996.

Iamblichus. *The Life of Pythagoras; or, On the Pythagorean Life.* In *The Pythagorean Sourcebook and Library,* edited by David Fideler and compiled and translated by Kenneth Sylvan Guthrie. Grand Rapids, Mich.: Phanes Press, 1987: 57–123.

IBM. "The Personal Area Network: Are You Ready for Prosthetic Memory?" http://www.ibm.com/stories/1996/11/pan_side4.html. Downloaded August 1999.

Ibn Khaldun. *The Muqaddimah: An Introduction to History.* Edited and abridged by N. J. Dawood. Translated by Franz Rosenthal. Princeton, N.J.: Princeton University Press, 1967.

"Impact of Asian Crisis on Individual States, by State." Washington, D.C.: International Trade Administration of the Department of Congress. http://www.ita.doc.gov/media/!325asia.html. November 1998.

C. S. Impey, G. C. Mead, and S. M. George. "Competitive Exclusion of Salmonellas from the Chick Caecum Using a Defined Mixture of Bacterial Isolates from the Caecal Microflora of an Adult Bird." *Journal of Hygiene,* December 1982: 479–490.

N. Inoue-Nakamura and T. Matsuzawa. "Development of Stone Tool Use by Wild Chimpanzees *(Pan troglodytes)." Journal of Comparative Psychology,* June 1997: 159–173.

"Introduction to Social Constructionism." Hausarbeiten.de.http://www.hausarbeiten.de/data/psychologie/psychosocial-constructionism.html. Downloaded June 1999.

E. D. Isaacs, A. Shukla, P. M. Platzman, D. R. Hamann, B. Barbiellini, and C. A. Tulk. "Covalency of the Hydrogen Bond in Ice: A Direct X-ray Measurement." *Physical Review Letters,* January 18, 1999: 600–603.

J. Itani. "Social Structures of African Great Apes." *Journal of Reproduction and Fertility* (Supplement 28), 1980: 33–41.

T. Ito, K. Okazaki, Y. Kawaoka, A. Takada, R. G. Webster, and H. Kida. "Perpetuation of Influenza A Viruses in Alaskan Waterfowl Reservoirs." *Archives of Virology* 140:7 (1995): 1163–1172.

Ravi Iyengar. "Gating by Cyclic AMP: Expanded Role for an Old Signaling Pathway." *Science,* January 26, 1996: 461–463.

S. Iyobe. "Mechanism of Acquiring Drug-resistance Genes in Pathogenic Bacteria." *Nippon Rinsho: Japanese Journal of Clinical Medicine,* May 1997: 1179–1184.

D. A. Jackson and P. R. Cook. "The Structural Basis of Nuclear Function." *International Review of Cytology* 162A (1995): 125–149.

Jane Jacobs. *Cities and the Wealth of Nations: Principles of Economic Life.* New York: Random House, 1984.

T. W. James, F. Crescitelli, E. R. Loew, and W. N. McFarland. "The Eyespot of *Euglena Gracilis:* A Microspectrophotometric Study." *Vision Research,* September 1992: 1583–1591.

William James. *On a Certain Blindness in Human Beings—What Makes a Life Significant.* New York: Henry Holt, 1899.

I. L. Janis. *Victims of Groupthink: A Psychological Study of Foreign-policy Decisions and Fiascos.* Boston: Houghton Mifflin, 1972.

"The Japan Crisis." Institute for International Economics. Washington, D.C. http://www.iie.com/hotjapan.html. November 1998.

K. Jarvinen. "Can Ward Rounds Be a Danger to Patients with Myocardial Infarction?" *British Medical Journal* 1:4909 (1955): 318–320.

S. Jasanoff. "Innovation and Integrity in Biomedical Research." *Academic Medicine* (Supplement S), September 1993: 91–95.

Phyllis C. Jay. "The Social Behavior of the Langur Monkey." Ph.D. diss. University of Chicago, 1962.

C. M. Jenkinson, R. J. Madeley, J. R. Mitchell, and I. D. Turner. "The Influence of Psychosocial Factors on Survival after Myocardial Infarction." *Public Health,* September 1993: 305–317.

M. Jibu, S. Hagan, S. R. Hameroff, K. H. Pribram, and K. Yasue. "Quantum Optical Coherence in Cytoskeletal Microtubules: Implications for Brain Function." *Biosystems* 32:3 (1994): 195–209.

T. Jobe, R. Vimal, A. Kovilparambil, J. Port, and M. Gaviria. "A Theory of Cooperativity Modulation in Neural Networks as an Important Parameter of CNS Catecholamine Function and Induction of Psychopathology." *Neurological Research,* October 1994: 330–341.

Allen Johnson. Personal communication. December 7, 1997.

Allen W. Johnson and Timothy Earle. *The Evolution of Human Societies: From Foraging Group to Agrarian State.* Stanford, Calif.: Stanford University Press, 1987.

B. J. Johnson, G. H. Miller, M. L. Fogel, J. W. Magee, M. K. Gagan, and A. R. Chivas. "65,000 Years of Vegetation Change in Central Australia and the Australian Summer Monsoon." *Science,* May 14, 1999: 1150–1152.

D. L. Johnson, J. S. Wiebe, S. M. Gold, N. C. Andreasen, R. D. Hichwa, G. L. Watkins, and L. L. Boles Ponto. "Cerebral Blood Flow and Personality: A Positron Emission Tomography Study." *American Journal of Psychiatry,* February 1999: 252–257.

Steven B. Johnson and Ronald C. Johnson. "Support and Conflict of Kinsmen: A Response to Hekala and Buell." *Ethology and Sociobiology,* January 1995: 83–89.

F. W. Jones, A. J. Wills, and I. P. McLaren. "Perceptual Categorization: Connectionist Modelling and Decision Rules." *Quarterly Journal of Experimental Psychology. Section B: Comparative and Physiological Psychology,* February 1998: 33–58.

I. H. Jones, D. M. Stoddart, and J. Mallick. "Towards a Sociobiological Model of Depression." *British Journal of Psychiatry* 166 (1995): 475–479.

R. N. Jones. "Impact of Changing Pathogens and Antimicrobial Susceptibility Patterns in the Treatment of Serious Infections in Hospitalized Patients." *American Journal of Medicine,* June 24, 1996: 3S–12S.

W. H. Jones, S. A. Hobbs, and D. Hockenbury. "Loneliness and Social Skill Deficits." *Journal of Personality and Social Psychology* 42 (1981): 682–689.

P. G. Judge and F. B. de Waal. "Intergroup Grooming Relations between Alpha Females in a Population of Free-Ranging Rhesus Macaques." *Folia Primatologica* 63:2 (1994): 63–70.

J. Kagan, J. S. Reznick, and J. Gibbons. "Inhibited and Uninhibited Types of Children." *Child Development,* August 1989: 838–845.

J. Kagan, J. S. Reznick, and N. Snidman. "The Physiology and Psychology of Behavioral Inhibition in Children." *Child Development,* December 1987: 1459–1473.

———. "Biological Bases of Childhood Shyness." *Science,* April 8, 1988: 167–171.

Jerome Kagan. *The Power and Limitations of Parents.* Austin: Hogg Foundation for Mental Health, University of Texas, 1986.

———. *Unstable Ideas: Temperament, Cognition and Self.* Cambridge, Mass.: Harvard University Press, 1989.

————. "Behavior, Biology, and the Meanings of Temperamental Constructs." *Pediatrics* 90 (September 1992): 510–513.

————. "Temperament and the Reactions to Unfamiliarity." *Child Development,* February 1997: 139–143.

Jerome Kagan and Howard A. Moss. *Birth to Maturity: A Study in Psychological Development.* New York: John Wiley and Sons, 1962.

Jerome Kagan, Nancy Snidman, and Doreen M. Arcus. "Initial Reactions to Unfamiliarity." *Current Directions in Psychological Science,* December 1992: 171–174.

Jerome Kagan with the collaboration of Nancy Snidman, Doreen Arcus, and J. Steven Reznick. *Galen's Prophecy: Temperament in Human Nature.* New York: Basic Books, 1994.

Dale Kaiser. "Bacteria Also Vote." *Science,* June 14, 1996: 1598–1599.

Suvendrini Kakuchi. "Japan—Education: Rampant Bullying Reflects Social Problems." Inter Press Service English News Wire, December 14, 1995.

N. H. Kalin, S. E. Shelton, M. Rickman, and R. J. Davidson. "Individual Differences in Freezing and Cortisol in Infant and Mother Rhesus Monkeys." *Behavioral Neuroscience,* February 1998: 251–254.

Km. Kamlesh. "A Study of the Effect of Personality on Value Pattern." *Indian Psychological Review,* January 1981: 13–17.

Eric R. Kandel and Robert D. Hawkins. "The Biological Basis of Learning and Individuality." *Scientific American,* September 1992: 78–87.

Minouche Kandel and Eric Kandel. "Flights of Memory." *Discover,* May 1994: 32–38.

D. Kang, R. J. Davidson, C. L. Coe, R. E. Wheeler, A. J. Tomarken, and W. B. Ershler. "Frontal Brain Asymmetry and Immune Function." *Behavioral Neuroscience* 105 (1991): 860–869.

Krzysztof Kaniasty and Fran H. Norris. "Mobilization and Deterioration of Social Support Following Natural Disasters." *Current Directions in Psychological Science,* June 1995: 94.

H. Kaplan and K. Hill. "Hunting Ability and Reproductive Success among Male Ache Foragers: Preliminary Results." *Current Anthropology* 26 (1985): 131–133.

Jay R. Kaplan, Stephen B. Manuck, Thomas B. Clarkson, Frances M. Lusso, David M. Taub, and Eric W. Miller. "Social Stress and Atherosclerosis in Normocholesterolemic Monkeys." *Science,* May 13, 1983: 733–735.

Ethan B. Kapstein. "Workers and the World Economy." *Foreign Affairs,* May–June 1996: 16–37.

Christa Karavanich and Jelle Atema. "Individual Recognition and Memory in Lobster Dominance." *Animal Behaviour,* December 1998: 1553–1560.

Elisa B. Karnofsky, Jelle Atema, and Randall H. Elgin. "Field Observations of Social Behavior, Shelter Use, and Foraging in the Lobster, *Homarus americanus.*" *The Biological Bulletin,* June 1989: 239–246.

James F. Kasting. "Planetary Atmospheres: Warming Early Earth and Mars." *Science,* May 23, 1997: 1213–1215.

John H. Kaufmann. "Social Relations in Hamadryas Baboons." In *Social Communication among Primates,* edited by Stuart A. Altmann. Chicago: University of Chicago Press, 1967: 73–98.

Stuart Kauffman. *At Home in the Universe: The Search for the Laws of Self-Organization and Complexity.* New York: Oxford University Press, 1995.

————. "What Is Life: Was Schrödinger Right." In *What Is Life? The Next Fifty Years—Speculations on the Future of Biology,* edited by Michael P. Murphy and Luke A. J. O'Neill. Cambridge, U.K.: Cambridge University Press, 1995: 83–114.

Stuart A. Kauffman. *The Origins of Order: Self-Organization and Selection in Evolution.* New York: Oxford University Press, 1993.

Hiroaki Kawasaki, Gregory M. Springett, Naoki Mochizuki, Shinichiro Toki, Mie Nakaya, Michiyuki Matsuda, David E. Housman, and Ann M. Graybiel. "A Family of cAMP-Binding Proteins That Directly Activate Rap1." *Science,* December 18, 1998: 2275–2279.

John Keegan. *A History of Warfare.* New York: Alfred A. Knopf, 1993.

H. H. Kelley. "The Warm-Cold Variable in First Impressions of Persons." *Journal of Personality* 18 (1950): 431–439.

Theodore D. Kemper. "Power, Status, and Emotions: A Sociological Contribution to a Psychophysiological Domain." In *Approaches to Emotion,* edited by Klaus R. Scherer and Paul Ekman. Hillsdale, N.J.: Lawrence Erlbaum Associates, 1984: 369–383.

———. *Social Structure and Testosterone.* New Brunswick, N.J.: Rutgers University Press, 1990.

Kathleen M. Kenyon. "Excavations at Jericho, 1957–58." *Palestine Excavation Quarterly* 92 (1960): 88–108.

Michael Keon. *Korean Phoenix: A Nation from the Ashes.* Englewood Cliffs, N.J.: Prentice-Hall, 1977.

Derrick De Kerckhove. *Connected Intelligence: The Arrival of the Web Society.* Toronto: Somerville House Books, 1997.

D. M. Kermack and K. A. Kermack. *The Evolution of Mammalian Characters.* London: Croom Helm Kapitan Szabo Publishers, 1984.

Richard A. Kerr. "Cores Document Ancient Catastrophe." *Science,* February 28, 1997: 1265–1270.

———. "Geomicrobiology: Life Goes to Extremes in the Deep Earth—and Elsewhere?" *Science,* May 2, 1997: 703–704.

———. "Early Life Thrived Despite Earthly Travails." *Science,* June 25, 1999: 2111–2113.

"Kingdom Divinity Ministries: Doctrinal Statement of Beliefs." http://www.kingidentity.com/doctrine.html. October 1998.

S. A. Kiritz. "Hand Movements and Clinical Ratings at Admission and Discharge for Hospitalized Psychiatric Patients." Ph.D. diss. San Francisco: University of California, San Francisco, 1971. Cited in P. Ekman and W. V. Friesen. "Non-Verbal Behaviour and Psychopathology." In *The Psychology of Depression: Contemporary Theory and Research,* edited by R. J. Friedman and M. M. Katz. New York: John Wiley and Sons, 1974: 203–232.

David L. Kirk. *Volvox Molecular-Genetic Origins of Multicellularity and Cellular Differentiation.* Cambridge, U.K.: Cambridge University Press, 1998.

Marc Kirschner and John Gerhart. "Evolvability." *Proceedings of the National Academy of Sciences of the United States of America,* July 1998: 8420–8427.

M. Kleerebezem, L. E. Quadri, O. P. Kuipers, and W. M. de Vos. "Quorum Sensing by Peptide Pheromones and Two-Component Signal-Transduction Systems in Gram-positive Bacteria." *Molecular Microbiology,* June 1997: 895–904.

Eric Klinger. *Meaning and Void: Inner Experience and the Incentives in People's Lives.* Minneapolis: University of Minnesota Press, 1977: 161.

C. Knight. *Blood Relations: Menstruation and the Origins of Culture.* New Haven, Conn.: Yale University Press, 1990.

Chris Knight. Personal communications. November 11–November 29, 1997.

Ferdinand Knobloch. "Individual, Group, or Meta-Selection." *ASCAP* (Galveston, Tex.: Across Species Comparison and Psychopathology Society), June 1996: 15–16.

———. "Integrated Psychotherapy and Evolutionary Psychology." *ASCAP* (Galveston, Tex.: Across Species Comparison and Psychopathology Society), September 1997: n.p.

A. L. Koch. "Speculations on the Growth Strategy of Prosthecate Bacteria." *Canadian Journal of Microbiology,* April 1988: 390–394.

A. Kodric-Brown and P. F. Nicoletto. "Consensus among Females in Their Choice of Males in the Guppy *Poecilia reticulata.*" *Behavioral Ecology and Sociobiology* 39:6 (1996): 395–400.

O. Koenig, L. P. Reiss, and S. M. Kosslyn. "The Development of Spatial Relation Representations: Evidence from Studies of Cerebral Lateralization." *Journal of Experimental Child Psychology,* August 1990: 119–130.

Robert Koenig. "European Researchers Grapple with Animal Rights." *Science,* June 4, 1999: 1604–1606.

James Vaughn Kohl. Poster presentation: Eighteenth Annual Meeting of the Association for Chemoreception Sciences, Sarasota, Fla., April 17–21, 1996.

————. Personal communication. May 9, 1996.

James Vaughn Kohl and Robert T. Francoeur. *The Scent of Eros: Mysteries of Odor in Human Sexuality.* New York: Continuum, 1995.

I. Kohler. "Experiments with Goggles." *Scientific American* 206 (1962): 62–86.

————. "The Formation and Transformation of the Perceptual World." *Psychological Issues* 3 (1963): 1–173.

R. Korona. "Genetic Divergence and Fitness Convergence under Uniform Selection in Experimental Populations of Bacteria." *Genetics,* June 1996: 637–644.

D. D. Kosambi. *Ancient India: A History of Its Culture and Civilization.* New York: Pantheon, 1965.

S. M. Kosslyn. "Seeing and Imaging in the Cerebral Hemispheres." *Psychological Bulletin* 94 (1987): 148–175.

————. *Image and Brain.* Cambridge, Mass.: MIT Press, 1994.

S. M. Kosslyn, C. F. Chabris, C. J. Marsolek, and O. Koenig. "Categorical versus Coordinate Spatial Relations: Computational Analyses and Computer Simulations." *Journal of Experimental Psychology: Human Perception and Performance,* May 1992: 562–577.

S. M. Kosslyn, O. Koenig, A. Barrett, C. B. Cave, T. J. Tang, and J. D. Gabrieli. "Evidence for Two Types of Spatial Representations: Hemispheric Specialization for Categorical and Coordinate Relations." *Journal of Experimental Psychology: Human Perception and Performance,* November 1989: 723–735.

R. H. Kramer. "Patch Cramming: Monitoring Intracellular Messengers in Intact Cells with Membrane Patches Containing Detector Ion Channels." *Neuron,* March 1990: 335–341.

M. Krams, M. F. Rushworth, M. P. Deiber, R. S. Frackowiak, and R. E. Passingham. "The Preparation, Execution and Suppression of Copied Movements in the Human Brain." *Experimental Brain Research,* June 1998: 386–398.

D. W. Krause, Z. Kielan-Jaworowska, and J. F. Bonaparte. "Ferugliotherium Bonaparte, the First Known Multituberculate from South America." *Journal of Vertebrate Paleontology* 12:3 (1992): 351–376.

J. Krause. "Differential Fitness Returns in Relation to Spatial Position in Groups." *Biological Reviews of the Cambridge Philosophical Society,* May 1994: 187–206.

Alan G. Kraut and Sarah Brookhart. "Human Development: The Origins of Behavior." *APS Observer, Special Issue: Basic Research in Psychological Science: A Human Capital Initiative Report.* Washington, D.C.: American Psychological Society, February 1998: 27–28.

Edward Kravitz. "Hormonal Control of Behavior: Amines and the Biasing of Behavioral Output in Lobsters." *Science,* September 30, 1988: 1775–1781.

V. A. Krischik. "Choice of Odd and Conspicuous Prey by Largemouth Bass, *Micropterus salmoides.*" M.S. thesis. College Park: University of Maryland, 1977.

A. L. Kroeber. *The Nature of Culture.* Chicago: University of Chicago Press, 1952.

Julian Krolik. *Active Galactic Nuclei: From Central Black Hole to the Galactic Environment.* Princeton, N.J.: Princeton University Press, 1999.

Thomas Kuhn. *The Structure of Scientific Revolutions.* 2nd ed. Chicago: University of Chicago Press, 1970.

J. A. Kulik, H. I. Mahler, and A. Earnest. "Social Comparison and Affiliation under Threat: Going beyond the Affiliate-Choice Paradigm." *Journal of Personality and Social Psychology,* February 1994: 301–309.

J. J. Kulikowski and T. R. Vidyasagar. "Space and Spatial Frequency: Analysis and Representation in the Macaque Striate Cortex." *Experimental Brain Research* 64:1 (1986): 5–18.

Sudhir Kumar and S. Blair Hedges. "A Molecular Timescale for Vertebrate Evolution." *Nature,* April 30, 1998: 917–920.

H. Kummer. *Weisse Affen am Roten Meer: Das Soziale Leben der Wüstenpaviane.* Munich: R. Piper, 1992.

Hans Kummer. "Tripartite Relations in Hamadryas Baboons." In *Social Communication among Primates,* edited by Stuart A. Altmann. Chicago: University of Chicago Press, 1967: 63–72.

———. "Two Variations in the Social Organization of Baboons." In *Primates: Studies in Adaptation and Variability,* edited by Phyllis C. Jay. New York: Holt, Rinehart and Winston, 1968: 293–312.

———. *Primate Societies: Group Techniques of Ecological Adaptation.* Chicago: Aldine Press, 1971.

Conrad C. Labandeira. Personal communication. February 16, 1999.

Conrad C. Labandeira and Tom L. Phillips. "Trunk Borings and Rhachis Skulls of Tree Ferns from the Late Pennsylvanian (Kasimovian) of Illinois; Implications for the Origin of the Galling Functional Feeding Group and Holometabolous Insects." *Palaeontographica, Part A:* in press.

Robert F. Lachlan, Lucy Crooks, and Kevin N. Laland. "Who Follows Whom? Shoaling Preferences and Social Learning of Foraging Information in Guppies." *Animal Behaviour,* July 1998: 181–190. http://www.susqu.edu/facstaff/b/brakke/complexity/hagey/flock.html. February 1999.

R. Ladouceur, F. Talbot, and M. J. Dugas. "Behavioral Expressions of Intolerance of Uncertainty in Worry: Experimental Findings." *Behavior Modification,* July 1997: 355–371.

Diogenes Laertius. *Lives of Eminent Philosophers.* Translated by R. D. Hicks. 1925. Reprint, Cambridge, Mass.: Harvard University Press, 1972.

———. *The Life of Pythagoras.* In *The Pythagorean Sourcebook and Library,* edited by David Fideler and compiled and translated by Kenneth Sylvan Guthrie. Grand Rapids, Mich.: Phanes Press, 1987: 141–158.

N. Lahav. "Prebiotic Co-evolution of Self-replication and Translation or RNA World?" *Journal of Theoretical Biology,* August 1991: 531–539.

R. Lahoz-Beltra, S. R. Hameroff, and J. E. Dayhoff. "Cytoskeletal Logic: A Model for Molecular Computation via Boolean Operations in Microtubules and Microtubule-Associated Proteins." *Biosystems* 29:1 (1993): 1–23.

J. A. Lake. "Origin of the Eukaryotic Nucleus: Eukaryotes and Eocytes Are Genotypically Related." *Canadian Journal of Microbiology,* January 1989: 109–118.

R. T. Lakoff. *Language and Woman's Place.* New York: Harper and Row, 1975.

Keven N. Laland and Kerry Williams. "Social Transmission of Maladaptive Information in the Guppy." *Behavioral Ecology,* September 1998: 493–499.

Angus I. Lamond and William C. Earnshaw. "Structure and Function in the Nucleus." *Science,* April 24, 1998: 547–553.

J. G. Landels. *Engineering in the Ancient World.* Berkeley: University of California Press, 1978.

H. J. Landy, L. Keith, and D. Keith. "The Vanishing Twin." *Acta Geneticae Medicae et Gemellologiae,* 31:3–4 (1982): 179–194.

H. J. Landy, S. Weiner, S. L. Corson, F. R. Batzer, and R. J. Bolognese. "The 'Vanishing Twin': Ultrasonographic Assessment of Fetal Disappearance in the First Trimester." *American Journal of Obstetrics and Gynecology,* July 1986: 14–19.

C. C. Lange, L. P. Wackett, K. W. Minton, and M. J. Daly. "Engineering a Recombinant *Deinococcus Radiodurans* for Organopollutant Degradation in Radioactive Mixed Waste Environments." *Nature Biotechnology,* October 1998: 929.

Justine H. Lange. "Dominance in Crayfish." *Science.* April 5, 1996: 18.

Ellen Langer. *Mindfulness.* Reading, Mass.: Addison-Wesley, 1989.

J. H. Langlois, L. A. Roggman, and L. S. Vaughn. "Facial Diversity and Infant Preferences for Attractive Faces." *Developmental Psychology* 27 (1991): 79–84.

Judith H. Langlois, Lori A. Roggman, and Lisa Musselman. "What Is Average and What Is Not Average about Attractive Faces." *Psychological Science,* July 1994: 214–220.

Roy Larick and Russell L. Ciochon. "The African Emergence and Early Asian Dispersals of the Genus *Homo.*" *American Scientist,* November–December 1996. http://www.sigmaxi.org/amsci/articles/96articles/Larick3.html. Downloaded June 1999.

Bibb Latane, Andrzej Nowak, and James H. Liu. "Measuring Emergent Social Phenomena: Dynamism, Polarization and Clustering as Order Parameters of Social Systems." *Behavioral Science* 39 (1994): 1–24.

Sing Lau and Sau M. Cheung. "Reading Interests of Chinese Adolescents: Effects of Personal and Social Factors." *International Journal of Psychology* 23:6 (1988): 695–705.

P. Laudedale, P. Smith-Cunnien, J. Parker, and J. Inverarity. "External Threat and the Definition of Deviance." *Journal of Personality and Social Psychology,* May 1984: 1058–1068.

E. O. Laumann. "Friends of Urban Men: An Assessment of Accuracy in Reporting Their Socioeconomic Attributes, Mutual Choice, and Attitude Agreement." *Sociometry* 32 (1969): 54–70.

W. Graeme Laver, Norbert Bischofberger, and Robert G. Webster. "Disarming Flu Viruses." *Scientific American,* January 1999: 78–87.

Richard E. Leakey and Roger Lewin. *People of the Lake: Mankind and Its Beginnings.* New York: Avon Books, 1983.

"Lean Company vs. Job Security: Sides Battle over GM's Future." ABC news.com—Business, June 13, 1998. http://www.abcnews.com/sections/business/DailyNews/gmstrike2_980613/ index.html. November 1998.

M. R. Leary, E. S. Tamber, S. K. Terdal, and D. L. Downs. "Self-esteem as an Interpersonal Monitor: The Sociometer Hypothesis." *Journal of Personality and Social Psychology* 68 (1995): 518–530.

Joseph E. LeDoux. "Emotion, Memory and the Brain." *Scientific American,* June 1994: 50–57.

Richard Borshay Lee. "The Hunters: Scarce Resources in the Kalahari." In *Conformity and Conflict: Readings in Cultural Anthropology,* edited by James P. Spradley and David W. McCurdy. Boston: Little, Brown, 1987: 192–207.

Y. L. Lee, T. Cesario, R. Lee, S. Nothvogel, J. Nassar, N. Farsad, and L. Thrupp. "Colonization by Staphylococcus Species Resistant to Methicillin or Quinolone on Hands of Medical Personnel in a Skilled-Nursing Facility." *American Journal of Infection Control,* December 1994: 346–351.

Herbert M. Lefcourt. *Locus of Control: Current Trends in Theory and Research.* 2nd ed. Hillsdale, N.J.: Lawrence Erlbaum Associates, 1982.

R. Legroux and J. Magrou. "État organisé des colonies bactériennes." *Annales de l'Institut Pasteur* 34 (1920): 417–431.

F. Le Guyader, M. Pommepuy, and M. Cormier. "Implantation of *Escherichia Coli* in Pilot Experiments and the Influence of Competition on the Flora." *Canadian Journal of Microbiology,* February 1991: 116–121.

Christiana M. Leonard, Linda J. Lombardino, Laurie R. Mercado, Samuel R. Browd, Joshua I. Breier, and O. Frank Agee. "Cerebral Asymmetry and Cognitive Development in Children: A Magnetic Resonance Imaging Study." *Psychological Science,* March 1996: 89–95.

Doris May Lessing. *Going Home.* London: Panther, 1968.

A. A. Lev, Y. E. Korchev, T. K. Rostovtseva, C. L. Bashford, D. T. Edmonds, and C. A. Pasternak. "Rapid Switching of Ion Current in Narrow Pores: Implications for Biological Ion Channels." *Proceedings of the Royal Society of London. Series B: Biological Sciences,* June 22, 1993: 187–192.

R. W. Levenson, P. Ekman, and W. Friesen. "Voluntary Facial Action Generates Emotion-Specific Autonomic Nervous System Activity." *Psychophysiology,* July 1997: 363–384.

Daniel S. Levine. "Survival of the Synapses." *The Sciences,* November–December 1988: 46–52.

Ricardo Levi-Setti. *Trilobites: A Photographic Atlas.* Chicago: University of Chicago Press, 1975.

Claude Lévi-Strauss. *The Elementary Structures of Kinship.* Translated by James Harle Bell, John Richard von Sturmer, and Rodney Needham. Boston: Beacon Press, 1969.

Steven Levy. *Artificial Life: A Report from the Frontier Where Computers Meet Biology.* New York: Vintage Books, 1993.

Stuart B. Levy. *The Antibiotic Paradox: How Miracle Drugs Are Destroying the Miracle.* New York: Plenum Press, 1992.

———. "Multidrug Resistance—a Sign of the Times." *New England Journal of Medicine,* May 7, 1998.

Chia-Wei Li, Jun-Yuan Chen, and Tzu-En Hua. "Precambrian Sponges with Cellular Structures." *Science,* February 6, 1998: 879–882.

B. Libet, E. W. Wright Jr., and C. A. Gleason. "Preparation- or Intention-to-Act, in Relation to Pre-event Potentials Recorded at the Vertex." *Electroencephalography and Clinical Neurophysiology* 56, October 1983: 367–372.

Benjamin Libet. "Unconscious Cerebral Initiative and the Role of Conscious Will in Voluntary Action." *Behavioral and Brain Sciences* 8 (1985): 529–566.

Robert Jay Lifton. "Beyond Armageddon: New Patterns of Ultimate Violence." *Modern Psychoanalysis* 22:1 (1997): 17–29.

Edward J. Lincoln. "Japan's Financial Mess." *Foreign Affairs*, May–June 1998: 57–66.

S. E. Lindley, K. J. Lookingland, and K. E. Moore. "Dopaminergic and Beta-adrenergic Receptor Control of Alpha-melanocyte-stimulating Hormone Secretion during Stress." *Neuroendocrinology*, July 1990: 46–51.

Gilbert N. Ling. *A Physical Theory of the Living State: The Association-Induction Hypothesis; with Considerations of the Mechanics Involved in Ionic Specificity.* New York: Blaisdell, 1962.

———. "The Functions of Polarized Water and Membrane Lipids: A Rebuttal." *Physiological Chemistry and Physics* 9:4–5 (1977): 301–311.

———. "Maintenance of Low Sodium and High Potassium Levels in Resting Muscle Cells." *Journal of Physiology* (Cambridge), July 1978: 105–123.

———. "Peer Review and the Progress of Scientific Research." *Physiological Chemistry and Physics* 10:1 (1978): 95–96.

———. "Oxidative Phosphorylation and Mitochondrial Physiology: A Critical Review of Chemiosmotic Theory and Reinterpretation by the Association-Induction Hypothesis." *Physiological Chemistry and Physics and Medical NMR* 13 (1981): 29–96.

———. *In Search of the Physical Basis of Life.* New York: Plenum Press, 1984.

———. "Solute Exclusion by Polymer- and Protein-Dominated Water: Correlation with Results of Nuclear Magnetic Resonance (NMR) and Calorimetric Studies and Their Significance for the Understanding of the Physical State of Water in Living Cells." *Scanning Microscopy,* June 1988: 871–884.

———. "Can We See Living Structure in a Cell?" *Scanning Microscopy,* June 1992: 405–439.

———. *A Revolution in the Physiology of the Living Cell.* Malabar, Fla.: Krieger, 1992.

———. "The New Cell Physiology: An Outline, Presented against Its Full Historical Background, Beginning from the Beginning." *Physiological Chemistry and Physics and Medical NMR* 26:2 (1994): 121–203.

———. "Debunking the Alleged Resurrection of the Sodium Pump Hypothesis." *Physiological Chemistry and Physics and Medical NMR* 29:2 (1997): 123–198.

———. Personal communications. December 1, 1998–July 14, 1999.

G. N. Ling and A. Fisher. "Cooperative Interaction among Cell Surface Sites: Evidence in Support of the Surface Adsorption Theory of Cellular Electrical Potentials." *Physiological Chemistry and Physics and Medical NMR* 15:5 (1983): 369–378.

G. N. Ling and W. Hu. "Studies on the Physical State of Water in Living Cells and Model Systems. X. The Dependence of the Equilibrium Distribution Coefficient of a Solute in Polarized Water on the Molecular Weights of the Solute: Experimental Confirmation of the 'Size Rule' in Model Studies." *Physiological Chemistry and Physics and Medical NMR* 20:4 (1988): 293–307.

G. N. Ling and M. M. Ochsenfeld. "Control of Cooperative Adsorption of Solutes and Water in Living Cells by Hormones, Drugs, and Metabolic Products." *Annals of the New York Academy of Sciences* 204 (March 30, 1973): 325–336.

G. N. Ling, M. M. Ochsenfeld, C. Walton, and T. J. Bersinger. "Experimental Confirmation, from Model Studies, of a Key Prediction of the Polarized Multilayer Theory of Cell Water." *Physiological Chemistry and Physics* 10:1 (1978): 87–88.

David Livingstone. *Missionary Travels and Researches in South Africa.* New York: Harper and Brothers, 1860.

Werner R. Loewenstein. *The Touchstone of Life: Molecular Information, Cell Communication, and the Foundations of Life.* New York: Oxford University Press, 1999.

Elizabeth Loftus. *Memory: Surprising New Insights into How We Remember and Why We Forget.* Reading, Mass.: Addison-Wesley, 1980.

————. "When a Lie Becomes Memory's Truth: Memory Distortion after Exposure to Misinformation." *Current Directions in Psychological Science,* August 1992: 121–123.

Konrad Lorenz. *On Aggression.* New York: Harcourt Brace Jovanovich, 1974.

Irving Lorge. "Prestige, Suggestion, and Attitudes." *Journal of Social Psychology* 7 (1936): 386–402.

Louisiana State University. "Physicist Finds Out Why 'We Are Stardust . . .'" Posted June 25, 1999. http://www.sciencedaily.com/releases/1999/06/990625080416.html. Downloaded August 1999.

Michael J. Lovaglia and Jeffrey A. Houser. "Emotional Reaction and Status in Groups." *American Sociological Review,* October 1996: 867–883.

J. Lucy. *Grammatical Categories and Cognition: A Case Study of the Linguistic Relativity Hypothesis.* Cambridge, U.K.: Cambridge University Press, 1992.

L. O. Lundqvist. "Facial EMG Reactions to Facial Expressions: A Case of Facial Emotional Contagion?" *Scandinavian Journal of Psychology,* June 1995: 130–141.

Aaron Lynch. *Thought Contagion: How Belief Spreads through Society.* New York: Basic Books, 1996.

J. J. Lynch. *The Broken Heart: The Medical Consequences of Loneliness.* New York: Basic Books, 1979.

J. J. Lynch and I. F. McCarthy. "The Effect of Petting on a Classically Conditioned Emotional Response." *Behavioral Research and Therapy* 5 (1967): 55–62.

————. "Social Responding in Dogs: Heart Rate Changes to a Person." *Psychophysiology* 5 (1969): 389–393.

S. A. Lytaev and V. I. Shostak. "The Thalamic Integration of Afferent Flows in Man during Image Recognition." *Zhurnal Vysshei Nervnoi Deiatelnosti Imeni I. P. Pavlova,* January–February 1992: 12–20.

James L. McClelland, David E. Rumelhart, and the PDP Research Group. *Parallel Distributed Processing: Explorations in the Microstructure of Cognition.* Vol. 2, *Psychological and Biological Models.* Cambridge, Mass.: MIT Press, 1986.

N. McConaghy and R. Zamir. "Sissiness, Tomboyism, Sex-Role, Sex Identity and Orientation." *Australian and New Zealand Journal of Psychiatry,* June 1995: 278–283.

John McCrone. *The Ape That Spoke: Language and the Evolution of the Human Mind.* New York: William Morrow, 1990.

William McDougall. *An Introduction to Social Psychology.* Boston: John W. Luce, 1908.

B. S. McEwen. "Corticosteroids and Hippocampal Plasticity." *Brain Corticosteroid Receptors: Studies on the Mechanism, Function, and Neurotoxicity of Corticosteroid Action.* Annals of the New York Academy of Sciences, 746, November 30, 1994: 142–144, 178–179.

H. McGinley, R. Lefevre, and P. McGinley. "The Influence of a Communicator's Body Position on Opinion Change in Others." *Journal of Personality and Social Psychology* 31 (1975): 686–690.

H. A. McGregor, J. D. Lieberman, J. Greenberg, S. Solomon, J. Arndt, L. Simon, and T. Pyszczynski. "Terror Management and Aggression: Evidence that Mortality Salience Motivates Aggression against Worldview-Threatening Others." *Journal of Personality and Social Psychology,* March 1998: 590–605.

Charles Mackay, LL.D. *Extraordinary Popular Delusions and the Madness of Crowds.* 1841. Reprint, New York: Farrar, Straus and Giroux, 1932.

Katelyn Y. A. McKenna and John A. Bargh. "Coming Out in the Age of the Internet: Identity 'Demarginalization' through Virtual Group Participation." *Journal of Personality and Social Psychology,* September 1998: 681–694.

D. McKenney, K. E. Brown, and D. G. Allison. "Influence of *Pseudomonas Aeruginosa* Exoproducts on Virulence Factor Production in *Burkholderia Cepacia:* Evidence of Interspecies Communication." *Journal of Bacteriology,* December 1995: 6989–6992.

Ian MacKenzie. "Philippine Negotiators Agree on Peace Accord." Reuters, August 29, 1996.

Paul MacLean. "The Imitative-Creative Interplay of Our Three Mentalities." In *Astride the Two Cultures: Arthur Koestler at 70*, edited by H. Harris. New York: Random House, 1976: 187–213.

Paul D. MacLean. *A Triune Concept of the Brain and Behaviour.* Toronto: University of Toronto Press, 1973.

David McLellan. *Karl Marx: His Life and Thought.* New York: Harper Colophon, 1973.

C. MacLeod and I. L. Cohen. "Anxiety and the Interpretation of Ambiguity: A Text Comprehension Study." *Journal of Abnormal Psychology,* May 1993: 238–247.

Kenneth J. McNamara. "Dating the Origin of Animals." *Science,* December 20, 1996: 1993f–1997f.

S. J. McNaughton, F. F. Banyikwa, and M. M. McNaughton. "Promotion of the Cycling of Diet-Enhancing Nutrients by African Grazers." *Science,* December 5, 1997: 1788–1800.

William H. McNeill. *Keeping Together in Time: Dance and Drill in Human History.* Cambridge, Mass.: Harvard University Press, 1995.

————. *Plagues and Peoples.* 1976. Reprint, New York: Anchor Books, 1998.

E. F. MacNichol Jr. "Retinal Processing of Visual Data." *Proceedings of the National Academy of Sciences of the United States of America,* June 1966: 1331–1344.

G. Maenhaut-Michel and J. A. Shapiro. "The Roles of Starvation and Selective Substrates in the Emergence of araB-lacZ Fusion Clones." *EMBO Journal,* November 1, 1994: 5229–5239.

A. T. Malcolm and M. P. Janisse. "Imitative Suicide in a Cohesive Organization: Observations from a Case Study." *Perceptual and Motor Skills* 79, December 1994: 1475–1478.

Bronislaw Malinowski. *Argonauts of the Western Pacific.* 1922. Reprint, Prospect Heights, Ill.: Waveland Press, 1984.

J. P. Mallory. "Speculations on the Xinjiang Mummies." *Journal of Indo-European Studies,* fall 1995: 371–398.

William Raymond Manchester. *The Arms of Krupp, 1587–1968.* Boston: Little, Brown, 1968.

————. *The Glory and the Dream: A Narrative History of America, 1932–1972.* New York: Bantam Books, 1974.

————. *The Last Lion, Winston Spencer Churchill.* Vol. 1, *Visions of Glory, 1874–1932.* New York: Dell Publishing, 1984.

"Manila Extends Truce in Southern Philippines." Xinhua News Agency, China, July 10, 1997.

Mike Mansfield. "The U.S. and Japan: Sharing Our Destinies." *Foreign Affairs,* spring 1989: 3–15.

Curtis W. Marean. "A Critique of the Evidence for Scavenging by Neandertals and Early Modern Humans: New Data from Kobeh Cave (Zagros Mountains, Iran) and Die Kelders Cave 1 Layer 10 (South Africa)." *Journal of Human Evolution,* August 1998: 111–136.

Christian Marendaz. "Nature and Dynamics of Reference Frames in Visual Search for Orientation: Implications for Early Visual Processing." *Psychological Science,* January 1998: 27–32.

Lynn Margulis. *Symbiosis in Cell Evolution: Microbial Communities in the Archean and Proterozoic Eons.* 2nd ed. New York: W. H. Freeman, 1993.

————. Personal communication. March 22, 1997.

Lynn Margulis and Dorion Sagan. *Microcosmos: Four Billion Years of Microbial Evolution.* New York: Summit Books, 1986.

M. M. Marinkovic. "Importance of Introversion for Science and Creativity." *Analytische Psychologie* 12:1 (1981): 1–35.

Alexander Marshack. "Evolution of the Human Capacity: The Symbolic Evidence." In *Yearbook of Physical Anthropology.* New York: Wiley-Liss, 1989.

————. "The Tai Plaque and Calendrical Notation in the Upper Palaeolithic." *Cambridge Archaeological Journal,* April 1991: 25.

————. "On 'Close Reading' and Decoration versus Notation." *Current Anthropology,* February 1997: 81.

Charles Richard Marshall. Personal communication. January 31, 1997.

Charles Richard Marshall and J. William Schopf. *Evolution and Molecular Revolution.* Sudbury, Mass.: Jones and Bartlett, 1996.

William Martin and Miklos Müller. "The Hydrogen Hypothesis for the First Eukaryote." *Nature,* March 5, 1998: 37–41.

Martin Marty. *Fundamentalisms Observed.* Chicago: University of Chicago Press, 1991.

Martin E. Marty and R. Scott Appleby. *The Glory and the Power: The Fundamentalist Challenge to the Modern World.* Boston: Beacon Press, 1992.

Martin E. Marty and R. Scott Appleby, eds. *The Fundamentalism Project: A Study Conducted by the American Academy of Arts and Sciences.* Chicago: University of Chicago Press, 1991.

———. *Accounting for Fundamentalisms: The Dynamic Character of Movements.* Chicago: University of Chicago Press, 1994.

Gerald Marwell and Pamela Oliver. *The Critical Mass and Collective Action.* Cambridge, U.K.: Cambridge University Press, 1993.

M. W. Marzke, K. L. Wullstein, and S. F. Viegas. "Evolution of the Power ('Squeeze') Grip and Its Morphological Correlates in Hominids." *American Journal of Physical Anthropology,* November 1992: 283–298.

Mary W. Marzke, N. Toth, and K. N. An. "EMG Study of Hand Muscle Recruitment during Hard Hammer Percussion Manufacture of Oldowan Tools." *American Journal of Physical Anthropology,* March 1998: 315–332.

John M. Marzluff, Bernd Heinrich, and Colleen S. Marzluff. "Raven Roosts Are Mobile Information Centres." *Animal Behaviour* 51 (1996): 89–103.

Abraham Maslow. "Dominance Feeling, Behaviour and Status." In *Dominance, Self-esteem, Self-actualization: Germinal Papers of A. H. Maslow,* edited by R. J. Lowry. Pacific Grove, Calif.: Brooks/Cole, 1973: 49–70.

R. Masuda and K. Tsukamoto. "The Ontogeny of Schooling Behaviour in the Striped Jack." *Journal of Fish Biology,* March 1998: 483–493.

R. J. Mathew, A. A. Swihart, and M. L. Weinman. "Vegetative Symptoms in Anxiety and Depression." *British Journal of Psychiatry,* August 1982: 162–165.

R. J. Mathew, M. L. Weinman, and D. L. Barr. "Personality and Regional Cerebral Blood Flow." *British Journal of Psychiatry,* May 1984: 529–532.

Roy J. Mathew and W. H. Wilson. "Intracranial and Extracranial Blood Flow during Acute Anxiety." *Psychiatry Research,* May 16, 1997: 93–107.

Tetsuro Matsuzawa. "Field Experiments on Use of Stone Tools by Chimpanzees in the Wild." In *Chimpanzee Cultures,* edited by Richard W. Wrangham, W. C. McGrew, Frans B. M. de Waal, and Paul G. Heltne with assistance from Linda A. Marquardt. Cambridge, Mass.: Harvard University Press, 1994: 351–370.

James Mattson and Merrill Simon. *The Pioneers of NMR and Magnetic Resonance in Medicine: The Story of MRI.* Ramat Gan, Israel: Bar-Ilan University Press, 1996.

Humberto R. Maturana. "Reality: The Search for Objectivity or the Quest for a Compelling Argument." *Irish Journal of Psychology* 9:1 (1988): 25–82.

Humberto R. Maturana and Francisco J. Varela. *The Tree of Knowledge: The Biological Roots of Human Understanding.* Boston: Shambhala, 1988.

Marcel Mauss. *Sociology and Psychology: Essays by Marcel Mauss.* Translated by Ben Brewster. London: Routledge and Kegan Paul, 1979.

Gottfried Mayer-Kress and Cathleen Barczys. "The Global Brain as an Emergent Structure from the Worldwide Computing Network, and Its Implications for Modelling." *Information Society* 11:1 (1995): 1–28.

J. Maynard Smith. "Group Selection and Kin Selection." *Nature* 201 (1964): 1145–1147.

Ernst Mayr. *Populations, Species, and Evolution.* Cambridge, Mass.: Harvard University Press, 1970.

Didier Mazel, Broderick Dychinco, Vera A. Webb, and Julian Davies. "A Distinctive Class of Integron in the Vibrio Cholerae Genome." *Science,* April 24, 1998: 605–608.

Margaret Mead. *Sex and Temperament in Three Primitive Societies.* London: Routledge and Kegan Paul, 1977.

H. Meade and C. Ziomek. "Urine as a Substitute for Milk?" *Nature Biotechnology,* January 1998: 21.

Linda Mealey. "The Sociobiology of Sociopathy: An Integrated Evolutionary Model." *Behavioral and Brain Sciences* 18:3 (1995): 523–599.

S. Mecklenbräuker. "Input- and Output-Monitoring in Implicit and Explicit Memory." *Psychological Research* 57 (1995): 179–191.

Meggie. "Can a Christian Hate?" *Christian Identity* (online magazine). http://www4.stormfront.org/posterity/ci/hate.html. November 1998.

A. Mehrabian and N. Williams. "Non-Verbal Concomitants of Perceived and Intended Persuasiveness." *Journal of Personality and Social Psychology* 13 (1969): 37–58.

Albert Mehrabian. *Silent Messages: Implicit Communication of Emotions and Attitudes.* Belmont, Calif.: Wadsworth, 1981.

James Mellaart. *Catal-Huyuk: A Neolithic Town in Anatolia.* New York: McGraw-Hill, 1967.

P. A. Mellars. "The Character of the Middle-Upper Palaeolithic Transition in Southwest France." In *The Explanation of Culture Change: Models in Prehistory,* edited by Colin Renfrew. Pittsburgh: University of Pittsburgh Press, 1973: 255–276.

E. Mellet, L. Petit, B. Mazoyer, M. Denis, and N. Tzourio. "Reopening the Mental Imagery Debate: Lessons from Functional Anatomy." *Neuroimage,* August 1998: 129–139.

M. T. Menard, S. M. Kosslyn, W. L. Thompson, N. M. Alpert, and S. L. Rauch. "Encoding Words and Pictures: A Positron Emission Tomography Study." *Neuropsychologia,* March 1996: 185–194.

Jin Meng and Malcolm C. Mckenna. "Faunal Turnovers of Palaeogene Mammals from the Mongolian Plateau." *Nature,* July 23, 1998: 364–367.

K. Michio. "Japanese Responses to the Defeat in World War II." *International Journal of Social Psychiatry,* autumn 1984: 178–187.

R. E. Michod, M. F. Wojciechowski, and M. A. Hoelzer. "DNA Repair and the Evolution of Transformation in the Bacterium *Bacillus subtilis.*" *Genetics,* January 1988: 31–39.

Edward M. Miller. "Paternal Provisioning versus Mate Seeking in Human Populations." *Personality and Individual Differences,* August 1994: 227–255.

———. "African Exodus: Re-evaluating an Hypothesis." *Mankind Quarterly,* fall–winter 1997: 69–84.

Joan G. Miller. "Culture and the Self: Uncovering the Cultural Grounding of Psychological Theory." In *The Self across Psychology: Self-Recognition, Self-Awareness, and the Self-Concept,* edited by Joan Gay Snodgrass and Robert L. Thompson. New York: New York Academy of Sciences, 1997: 217–232.

W. R. Miller and M. E. Seligman. "Depression and Learned Helplessness in Man." *Journal of Abnormal Psychology,* June 1975: 228–238.

William R. Miller, Robert A. Rosellini, and Martin E. P. Seligman. "Learned Helplessness and Depression." In *Psychopathology: Experimental Models,* edited by Jack D. Maser and Martin E. P. Seligman. San Francisco: W. H. Freeman and Company, 1977: 104–130.

S. Mishra, T. R. Venkatesan, and B. L. K. Somayajulu. "Earliest Acheulian Industry from Peninsular India." *Current Anthropology,* December 1995: 847–851.

Sheila Mishra. "The Age of the Acheulian in India: New Evidence." *Current Anthropology,* June 1992: 325–328.

Steven Mithen. *The Prehistory of the Mind: The Cognitive Origins of Art, Religion and Science.* London: Thames and Hudson, 1996.

Dr. Masao Miyamoto. "Government of the Bureaucrats, by the Bureaucrats, and for the Bureaucrats." *Tokyo Business Today,* February 1996: 6–10.

S. Moens, K. Michiels, V. Keijers, F. Van Leuven, and J. Vanderleyden. "Cloning, Sequencing, and Phenotypic Analysis of laf1, Encoding the Flagellin of the Lateral Flagella of *Azospirillum Brasilense* Sp7." *Journal of Bacteriology,* October 1995: 5419–5426.

Anne Simon Moffat. "Down on the Animal Pharm." *Science,* December 18, 1998: 2177.

————. "Toting Up the Early Harvest of Transgenic Plants." *Science,* December 18, 1998: 2176–2178.

B. Moghaddam, M. L. Bolinao, B. Stein-Behrens, and R. Sapolsky. "Glucocorticoids Mediate the Stress-Induced Extracellular Accumulation of Glutamate." *Brain Research,* August 1994: 251–254.

S. J. Mojzsis, G. Arrhenius, K. D. Mckeegan, T. M. Harrison, A. P. Nutman, and C. R. L. Friend. "Evidence for Life on Earth before 3,800 Million Years Ago." *Nature,* November 7, 1996: 55–59.

Svend Erik Moller, E. L. Mortensen, L. Breum, C. Alling, O. G. Larsen, T. Boge-Rasmussen, C. Jensen, and K. Bennicke. "Aggression and Personality: Association with Amino Acids and Monoamine Metabolites." *Psychological Medicine,* March 1996: 323–331.

J. Money and D. Mathews. "Prenatal Exposure to Virilizing Progestins: An Adult Follow-Up Study of Twelve Women." *Archives of Sexual Behavior,* February 1982: 73–83.

R. L. Montgomery, S. W. Hinkle, and R. F. Enzie. "Arbitrary Norms and Social Change in High- and Low-Authoritarian Societies." *Journal of Personality and Social Psychology,* June 1976: 698–708.

B. Moore. "The Evolution of Imitative Learning." In *Social Learning in Animals: The Roots of Culture,* edited by Cecilia M. Heyes and Bennett G. Galef, Jr. San Diego: Academic Press, 1996: 245–265.

J. M. Moore, trans. *Aristotle and Xenophon on Democracy and Oligarchy.* Berkeley: University of California Press, 1986.

Alan Moorehead. *The Russian Revolution.* New York: Bantam Books, 1959.

Margret I. Moré, L. David Finger, Joel L. Stryker, Clay Fuqua, Anatol Eberhard, and Stephen C. Winans. "Enzymatic Synthesis of a Quorum-Sensing Autoinducer through Use of Defined Substrates." *Science,* June 14, 1996: 1655–1658.

Virginia Morell. "Life's Last Domain." *Science,* August 23, 1996: 1043–1045.

————. "How the Malaria Parasite Manipulates Its Hosts." *Science,* October 10, 1997: 223.

————. "Genes May Link Ancient Eurasians, Native Americans." *Science,* April 24, 1998: 520.

————. "Ecology Returns to Speciation Studies." *Science,* June 25, 1999: 2106–2108.

Michio Morishima. *Why Has Japan Succeeded: Western Technology and the Japanese Ethos.* Cambridge, U.K.: Cambridge University Press, 1982.

Akio Morita. "The Politics of Business: The Business of Politics." *Business in the Contemporary World* 2:1 (1989): 11–13.

————. "Partnering for Competitiveness: The Role of Japanese Business." *Harvard Business Review,* May 1992: 76–83.

————. "The Case for a New World Economic Order." *PHP Intersect,* June 1993: n.p.

Donald R. Morris. *The Washing of the Spears: A History of the Rise of the Zulu Nation under Shaka and Its Fall in the Zulu War of 1879.* New York: Simon and Schuster, 1986.

W. N. Morris, S. Worchel, J. L. Bios, J. A. Pearson, C. A. Rountree, G. M. Samaha, J. Wachtler, and S. L. Wright. "Collective Coping with Stress: Group Reactions to Fear, Anxiety, and Ambiguity." *Journal of Personality and Social Psychology,* June 1976: 674–679.

Douglass H. Morse. *Behavioral Mechanisms in Ecology.* Cambridge, Mass.: Harvard University Press, 1980.

M. J. Morwood, P. B. O'Sullivan, F. Aziz, and A. Raza. "Fission-Track Ages of Stone Tools and Fossils on the East Indonesian Island of Flores." *Nature,* March 12, 1998: 173–176.

S. Moscovici, E. Lage, and M. Naffrechoux. "Influence of a Consistent Minority on the Responses of a Majority in a Color Perception Task." *Sociometry* 32 (1969): 365–380.

S. Moscovici and B. Personnaz. "Studies in Social Influence. Part V: Minority Influence and Conversion Behavior in a Perceptual Task." *Journal of Experimental Social Psychology* 16 (1980): 270–282.

Cynthia Moss. *Elephant Memories: Thirteen Years in the Life of an Elephant Family.* New York: William Morrow, 1988.

Noel Mostert. *Frontiers: The Epic of South Africa's Creation and the Tragedy of the Xhosa People.* New York: Knopf, 1992.

H. L. Movius. "The Lower Paleolithic Cultures of Southern and Eastern Asia." *Transactions of the American Philosophical Society* 38 (1948): 329–420.

E. R. Moxon, P. B. Rainey, M. A. Nowak, and R. E. Lenski. "Adaptive Evolution of Highly Mutable Loci in Pathogenic Bacteria." *Current Biology,* January 1, 1994: 24–33.

E. Richard Moxon and Christopher Wills. "DNA Microsatellites: Agents of Evolution." *Scientific American,* January 1999: 94–99.

M. Mulder and A. Stemerding. "Threat, Attraction to Group, and Need of Strong Leadership." *Human Relations* 16 (1963): 317–334.

J. Muñoz-Dorado and J. M. Arias. "The Social Behavior of Myxobacteria." *Microbiologia,* December 1995: 429–438.

Oswyn Murray. "Life and Society in Classical Greece." In *The Oxford History of the Classical World: Greece and the Hellenistic World,* edited by John Boardman, Jasper Griffin, and Oswyn Murray. New York: Oxford University Press, 1988.

M. Mushinski. "Violence in America's Public Schools." *Statistical Bulletin—Metropolitan Insurance Companies,* April–June 1994: 2–9.

D. G. Myers. *The Pursuit of Happiness.* New York: Avon Books, 1993.

David G. Myers and Ed Diener. "Who Is Happy?" *Psychological Science,* January 1995: 10–19.

Arne Naess. *Ecology, Community, and Lifestyle: Outline of an Ecosophy.* Revised and translated by David Rothenberg. New York: Cambridge University Press, 1989.

T. Nagayama, H. Aonuma, and P. L. Newland. "Convergent Chemical and Electrical Synaptic Inputs from Proprioceptive Afferents onto an Identified Intersegmental Interneuron in the Crayfish." *Journal of Neurophysiology,* May 1997: 2826–2830.

Thomas Nagel. "The Sleep of Reason." *New Republic,* October 12, 1998: 32–38.

C. T. Nagoshi, R. C. Johnson, and G. P. Danko. "Assortative Mating for Cultural Identification as Indicated by Language Use." *Behavior Genetics,* January 1990: 23–31.

Yu Y. Nakamura, A. Pashkin, and J. S. Tsai. "Coherent Control of Macroscopic Quantum States in a Single-Cooper-Pair Box." *Nature,* April 29, 1999: 786–788.

Christine Nalepa. Personal communication. April 6, 1997.

Christine A. Nalepa. "Nourishment and the Origin of Termite Eusociality." In *Nourishment and Evolution in Insect Societies,* edited by James H. Hunt and Christine A. Nalepa. Boulder, Colo.: Westview Press, 1994: 57–96.

National Aeronautics and Space Administration. " 'Old Faithful' Black Hole in Our Galaxy Ejects Mass Equal to an Asteroid at Fantastic Speeds." Posted January 9, 1998. http://www.sciencedaily.com/releases/1998/01/980109074724.html. Downloaded August 1999.

Joseph Needham. *Science and Civilisation in China.* 7 vols. Cambridge, U.K.: Cambridge University Press, 1954–1998.

J. V. Neel and R. H. Ward. "Village and Tribal Genetic Distances among American Indians, and the Possible Implications for Human Evolution." *Proceedings of the National Academy of Sciences of the United States of America* 65 (1970): 323–330.

Martin Nesirky. "FOCUS-CIS Faces International Terror Menace—Putin." Moscow: Reuters, September 15, 1999.

Randolph M. Nesse, M.D., and George C. Williams, Ph.D. *Why We Get Sick: The New Science of Darwinian Medicine.* New York: Random House, 1995.

V. F. Nettles, J. M. Wood, and R. G. Webster. "Wildlife Surveillance Associated with an Outbreak of Lethal H5N2 Avian Influenza in Domestic Poultry." *Avian Diseases,* July–September 1985: 733–741.

J. Netz, U. Ziemann, and V. Hömberg. "Hemispheric Asymmetry of Transcallosal Inhibition in Man." *Experimental Brain Research* 104:3 (1995): 527–533.

T. M. Newcomb. *Personality and Social Change: Attitude Formation in a Student Community.* New York: Dryden, 1943.

————. *The Acquaintance Process.* New York: Holt, Rinehart and Winston, 1961.

————. "Stabilities Underlying Changes in Interpersonal Attraction." *Journal of Abnormal and Social Psychology* 66 (1963): 393–404.

The New English Bible: New Testament. New York: Oxford University Press, 1961.

L. S. Newman, K. J. Duff, and R. F. Baumeister. "A New Look at Defensive Projection: Thought Suppression, Accessibility, and Biased Person Perception." *Journal of Personality and Social Psychology,* May 1997: 980–1001.

H. D. Niall. "The Evolution of Peptide Hormones." *Annual Review of Physiology 1982* 44: 615–624.

G. Nicolis and I. Prigogine. *Self-Organization in Nonequilibrium Systems: From Dissipative Structures to Order through Fluctuations.* New York: John Wiley and Sons, 1977.

R. Nisbett and L. Ross. *Human Inference: Strategies and Shortcomings of Social Judgment.* Englewood Cliffs, N.J.: Prentice-Hall, 1980.

Toshisada Nishida. "Review of Recent Findings on Mahale Chimpanzees: Implications and Future Research Directions." In *Chimpanzee Cultures,* edited by Richard W. Wrangham, W. C. McGrew, Frans B. M. de Waal, and Paul G. Heltne with assistance from Linda A. Marquardt. Cambridge, Mass.: Harvard University Press, 1994: 373–396.

No Compromise: The Militant, Direct Action Magazine of Grassroots Animal Liberationists and Their Supporters. http://www.nocompromise.org/. November 1998.

R. Noe, F. B. de Waal, and J. A. van Hooff. "Types of Dominance in a Chimpanzee Colony." *Folia Primatologica* 34:1–2 (1980): 90–110.

Dennis Normile. "New Views of the Origins of Mammals." *Science,* August 7, 1998: 774–775.

————. "New Ground-Based Arrays to Probe Cosmic Powerhouses." *Science,* April 30, 1999: 734–735.

V. Norris and G. J. Hyland. "Do Bacteria Sing? Sonic Intercellular Communication between Bacteria May Reflect Electromagnetic Intracellular Communication Involving Coherent Collective Vibrational Modes That Could Integrate Enzyme Activities and Gene Expression." *Molecular Microbiology,* May 1997: 879–880.

Gary North. *Backward Christian Soldiers? An Action Manual for Christian Reconstruction.* Tyler, Tex.: Institute for Christian Economics, 1984.

————. *The Dominion Covenant—Genesis.* Tyler, Tex.: Institute for Christian Economics, 1987.

F. Nottebohm, M. E. Nottebohm, and L. Crane. "Developmental and Seasonal Changes in Canary Song and Their Relation to Changes in the Anatomy of Song-Control Nuclei." *Behavioral and Neural Biology,* November 1986: 445–471.

M. J. Novacek. "Evidence for Echolocation in the Oldest Known Bats." *Nature,* May 9–15, 1985: 140–141.

Michael Nunley. "Stress, Anxiety, Power, and Healing: A Hypothesis." Paper presented at the Kroeber Anthropological Society Annual Meeting, University of California at Berkeley, Berkeley, Calif., April 1986.

Peter Nyikos. Personal communication. August 24, 1997.

Robert L. O'Connell. *Of Arms and Men: A History of War, Weapons, and Aggression.* New York: Oxford University Press, 1989.

Office of the United Nations High Commissioner for Refugees. "Background Paper on Refugees and Asylum Seekers from Afghanistan." UNHCR Centre for Documentation and Research, Geneva, June 1997. http://www.unhcr.ch/refworld/country/cdr/cdrafg.html. November 1998.

Yasushi Okada and Nobutaka Hirokawa. "A Processive Single-Headed Motor: Kinesin Superfamily Protein KIF1A." *Science,* February 19, 1999: 1152–1157.

D. R. Omark and M. S. Edelman. "The Development of Attention Structures in Young Children." In *Attention Structures in Primates and Man,* edited by M. R. A. Chance and R. Larson. New York: John Wiley and Sons: 1977.

D. R. Omark, M. Omark, and M. S. Edelman. "Formation of Dominance Hierarchies in Young Children: Action and Perception." In *Psychological Anthropology,* edited by T. Williams. The Hague: Mouton, 1975.

Y. Omura. "Inhibitory Effect of NaCl on Hog Kidney Mitochondrial Membrane-Bound Monoamine Oxidase: pH and Temperature Dependences." *Japanese Journal of Pharmacology,* December 1995: 293–302.

A. Onderdonk, B. Marshall, R. Cisneros, and S. B. Levy. "Competition between Congenic *Escherichia Coli* K-12 Strains In Vivo." *Infection and Immunity,* April 1981: 74–79.

One Nation. *Pauline Hanson's One Nation* (homepage). http://www.gwb.com.au/onenation/. November 1998.

C. J. Ong, M. L. Wong, and J. Smit. "Attachment of the Adhesive Holdfast Organelle to the Cellular Stalk of *Caulobacter Crescentus.*" *Journal of Bacteriology,* March 1990: 1448–1456.

James Opie. "Xinjiang Remains and 'the Tocharian Problem.' " *The Journal of Indo-European Studies,* fall 1995: 431–445.

George Ordish. *The Year of the Ant.* New York: Charles Scribner's Sons, 1978.

George Ostrogorsky. *History of the Byzantine State.* Translated by Joan Hussey. New Brunswick, N.J.: Rutgers University Press, 1969.

Gerald D. Otis and John L. Louks. "Rebelliousness and Psychological Distress in a Sample of Introverted Veterans." *Journal of Psychological Type* 40 (1997): 20–30.

Alondra Yvette Oubré. *Instinct and Revelation: Reflections on the Origins of Numinous Perception.* Amsterdam: Gordon and Breach, 1997.

Ovid. *Metamorphoses.* 15.352. The Perseus Project, edited by Gregory R. Crane. http://hydra.perseus.tufts.edu. December 1998.

Norman R. Pace. "A Molecular View of Microbial Diversity and the Biosphere." *Science,* May 2, 1997: 734–740.

M. Pagel and R. A. Johnstone. "Variation across Species in the Size of the Nuclear Genome Supports the Junk-DNA Explanation for the C-value Paradox." *Proceedings of the Royal Society of London. Series B: Biological Sciences,* August 22, 1992: 119–124.

Roberta L. Paikoff and Ritch C. Savin-Williams. "An Exploratory Study of Dominance Interactions among Adolescent Females at a Summer Camp." *Journal of Youth and Adolescence,* October 1983: 419–433.

Jeffrey D. Palmer. "Organelle Genomes: Going, Going, Gone!" *Science,* February 7, 1997: 790–791.

R. F. Paloutzian and C. W. Ellison. "Loneliness, Spiritual Well-being and the Quality of Life." In *Loneliness: A Sourcebook of Current Theory, Research, and Therapy,* edited by L. A. Peplau and D. Perlman. New York: John Wiley and Sons, 1982: 224–237.

Daniel R. Papaj and Alcinda C. Lewis. *Insect Learning: Ecological and Evolutionary Perspectives.* New York: Chapman and Hall, 1993.

J. C. Parikh and B. Pratap. "An Evolutionary Model of a Neural Network." *Journal of Theoretical Biology,* May 7, 1984: 31–38.

L. L. Parker and B. G. Hall. "A Fourth *Escherichia Coli* Gene System with the Potential to Evolve Beta-glucoside Utilization." *Genetics,* July 1988: 485–490.

A. Pascual-Leone and F. Torres. "Plasticity of the Sensorimotor Cortex Representation of the Reading Finger in Braille Readers." *Brain* 116 (1993): 39–52.

R. Patel, M. S. Rouse, K. E. Piper, F. R. Cockerill III, and J. M. Steckelberg. "In Vitro Activity of LY333328 against Vancomycin-Resistant Enterococci, Methicillin-Resistant *Staphylococcus aureus,* and Penicillin-Resistant *Streptococcus pneumoniae.*" *Diagnostic Microbiology and Infectious Disease,* February 1998: 89–92.

Tony Pawson and John D. Scott. "Signaling through Scaffold, Anchoring, and Adaptor Proteins." *Science*, December 19, 1997: 2075–2080.

Pausanius. *Description of Greece*. The Perseus Project, edited by Gregory R. Crane. http:// hydra.perseus.tufts.edu. August 1998.

C. M. Payne, C. Bernstein, and H. Bernstein. "Apoptosis Overview Emphasizing the Role of Oxidative Stress, DNA Damage and Signal-Transduction Pathways." *Leukemia and Lymphoma*, September 1995: 43–93.

P. J. E. Peebles. "Evolution of the Cosmological Constant." *Nature*, March 4, 1999: 25–26.

Wilder Penfield. "Consciousness, Memory and Man's Conditioned Reflexes." In *On the Biology of Learning*, edited by Karl H. Pribram. New York: Harcourt, Brace and World, 1969.

Elizabeth Pennisi. "Versatile Gene Uptake System Found in Cholera Bacterium." *Science*, April 24, 1998: 521–522.

———. "Direct Descendants from an RNA World." *Science*, May 1, 1998: 673.

———. "Genome Data Shake Tree of Life." *Science*, May 1, 1998: 672–674.

———. "Molecular Evolution: How the Genome Readies Itself for Evolution." *Science*, August 21, 1998: 1131–1134.

L. A. Perez, Z. F. Peynirciolu, and T. A. Blaxton. "Developmental Differences in Implicit and Explicit Memory Performance." *Journal of Experimental Child Psychology*, September 1998: 167–185.

Timothy Perper. Personal communication. June 13, 1997.

Wendy A. Petka, James L. Harden, Kevin P. McGrath, Denis Wirtz, and David A. Tirrell. "Reversible Hydrogels from Self-Assembling Artificial Proteins." *Science*, July 17, 1998: 389–392.

Henry G. Pflaum. *Essai sur le Cursus Publicus sous le Haut-Empire Romain* (Study of the Cursus-Publicus in the High Roman Empire). Paris: Imprimerie Nationale, 1950.

J. A. Phillips, A. C. Alberts, and N. C. Pratt. "Differential Resource Use, Growth, and the Ontogeny of Social Relationships in the Green Iguana." *Physiology and Behavior*, January 1993: 81–88.

A. Pierro, H. K. van Saene, S. C. Donnell, J. Hughes, C. Ewan, A. J. Nunn, and D. A. Lloyd. "Microbial Translocation in Neonates and Infants Receiving Long-term Parenteral Nutrition." *Archives of Surgery*, February 1996: 176–179.

I. M. Piliavin, J. A. Piliavin, and J. Rodin. "Costs, Diffusion, and the Stigmatized Victim." *Journal of Personality and Social Psychology* 32 (1975): 429–438.

Alessandra Piontelli. *From Fetus to Child: An Observational and Psychoanalytic Study*. New York: Tavistock/Routledge, 1992.

A. Piontelli, L. Bocconi, A. Kustermann, B. Tassis, C. Zoppini, and U. Nicolini. "Patterns of Evoked Behaviour in Twin Pregnancies during the First 22 Weeks of Gestation." *Early Human Development*, November 1997: 39–45.

Steven Pinker. *The Language Instinct*. New York: William Morrow, 1994.

Plato. *Apology*. In *Library of the Future*. 4th ed., ver. 5.0. Irvine, Calif.: World Library, 1996. CD-ROM.

———. *Charmides*. In *Library of the Future*. 4th ed., ver. 5.0. Irvine, Calif.: World Library, 1996. CD-ROM.

———. *Euthyphro*. In *Library of the Future*. 4th ed., ver. 5.0. Irvine, Calif.: World Library, 1996. CD-ROM.

———. *Laws*. In *Library of the Future*. 4th ed., ver. 5.0. Irvine, Calif.: World Library, 1996. CD-ROM.

———. *Phaedo*. In *Library of the Future*. 4th ed., ver. 5.0. Irvine, Calif.: World Library, 1996. CD-ROM.

———. *Protagoras*. In *Library of the Future*. 4th ed., ver. 5.0. Irvine, Calif.: World Library, 1996. CD-ROM.

————. *Theaetetus.* In *Library of the Future.* 4th ed., ver. 5.0. Irvine, Calif.: World Library, 1996. CD-ROM.

————. *Timaeus.* In *Library of the Future.* 4th ed., ver. 5.0. Irvine, Calif.: World Library, 1996. CD-ROM.

————. *Republic.* The Perseus Project, edited by Gregory R. Crane. http://hydra.perseus.tufts. edu. August 1998.

Hidde L. Ploegh. "Viral Strategies of Immune Evasion." *Science,* April 10, 1998: 248–253.

Plutarch. *Lycurgus.* In *Aristotle and Xenophon on Democracy and Oligarchy.* Translated by J. M. Moore. Berkeley: University of California Press, 1986.

————. *Agis.* In *Library of the Future.* 4th ed., ver. 5.0. Irvine, Calif.: World Library, 1996. CD-ROM.

————. *Numa Pompilius.* In *Library of the Future.* 4th ed., ver. 5.0. Irvine, Calif.: World Library, 1996. CD-ROM.

————. *Solon.* In *Library of the Future.* 4th ed., ver. 5.0. Irvine, Calif.: World Library, 1996. CD-ROM.

————. *Themistocles.* The Perseus Project, edited by Gregory R. Crane. http://hydra.perseus. tufts.edu. August 1998.

————. *Octavius.* Para 2. The Perseus Project, edited by Gregory R. Crane. http://hydra. perseus.tufts.edu. December 1998.

G. G. Pope and J. E. Cronin. "The Asian *Hominidae.*" *Journal of Human Evolution* 13 (1984): 377–396.

Adam Posen. "Some Background Q&A on Japanese Economic Stagnation. Briefing Memo Submitted to the Trade Subcommittee Committee on Ways and Means, U.S. House of Representatives, July 13, 1998." Institute for International Economics. Washington, D.C. http://www.iie.com/japanwm.html. November 1998.

Richard Potts. "Environmental Hypotheses of Hominid Evolution." *Yearbook of Physical Anthropology.* New York: John Wiley and Sons, 1998.

Sally J. Power. Personal communication. July 28, 1998.

Sally J. Power and Lorman L. Lundsten. "Studies That Compare Type Theory and Left-Brain/Right-Brain Theory." *Journal of Psychological Type* 43 (1997): 22–28.

Clyde Prestowitz. "Japanese vs. Western Economies: Why Each Side Is a Mystery to the Other." *Technology Review,* May–June 1988: 26–37.

Clyde V. Prestowitz Jr. *Trading Places: How We Allowed Japan to Take the Lead.* New York: Basic Books, 1988.

John Price. "Self-esteem." *Lancet,* October 22, 1988: 943–944.

John S. Price and Anthony Stevens. "The Human Male Socialization Strategy Set: Cooperation, Defection, Individualism, and Schizotypy." *Evolution and Human Behavior,* January 1998: 57–70.

Heather Pringle. "Death in Norse Greenland." *Science,* February 14, 1997: 924–926.

————. "Ice Age Communities May Be Earliest Known Net Hunters." *Science,* August 29, 1997: 1203–1204.

————. "The Slow Birth of Agriculture." *Science,* November 20, 1998: 1446.

M. E. Procidano and L. H. Rogler. "Homogamous Assortative Mating among Puerto Rican Families: Intergenerational Processes and the Migration Experience." *Behavior Genetics,* May 1989: 343–354.

Stephen Prudhoe. *A Monograph on Polyclad Turbellaria.* London: British Museum, 1985.

PTA AG, Telekom Austria. "Kommunikation und Informationsaustausch im Wandel der Jahrhunderte—Vor 2000 Jahren: Der Römische 'Cursus Publicus.'" *Geschichte des Post- und Fernmeldewesens.* http://www.pta.at/fr/ag/geschichte/index-fr.html. December 1998.

Archbishop Lazar Puhalo. "The Tragedy of Kosovo: A Knife in the Heart of Serbia—a Warning to the World." New Ostrog Monastery, Dewdney, B.C., Canada. http://www.new.ostrog.org/kosovo.html. November 1998.

Pursuing Excellence. A Study of U.S. Eighth Grade Mathematics and Science Teaching, Learning, Curriculum and Achievement in International Context. U.S. Department of Education. Washington, D.C. http://nces.ed.gov/timss/97198-4.html. November 1998.

A. Pusey, J. Williams, and J. Goodall. "The Influence of Dominance Rank on the Reproductive Success of Female Chimpanzees." *Science,* August 8, 1997: 828–831.

Katrin Pütsep, Carl-Ivar Brändén, Hans G. Boman, and Staffan Normark. "Antibacterial Peptide from *H. Pylori.*" *Nature,* April 22, 1999: 671–672.

"The Pythagorean Pagan—The Delphic Oracle Site." http://www.aros.net/~eriugena/pita.htm. July 1998.

David C. Queller, Joan E. Strassman, and Colin R. Hughes. "Genetic Relatedness in Colonies of Tropical Wasps with Multiple Queens." *Science,* November 1988: 1155–1157.

Anne Querrien. "The Metropolis and the Capital." In *Zone 1 / 2: The Contemporary City.* New York: Zone Books, 1987: 219–221.

Howard Rachlin. "Self and Self-control." In *The Self across Psychology: Self-Recognition, Self-Awareness, and the Self-Concept,* edited by Joan Gay Snodgrass and Robert L. Thompson. New York: New York Academy of Sciences, 1997: 85–98.

Roland Radloff. "Opinion Evaluation and Affiliation." *Journal of Abnormal and Social Psychology* 62 (1961): 578–585.

M. Radman, I. Matic, J. A. Halliday, and F. Taddei. "Editing DNA Replication and Recombination by Mismatch Repair: From Bacterial Genetics to Mechanisms of Predisposition to Cancer in Humans." *Philosophical Transactions of the Royal Society of London. Series B: Biological Sciences,* January 30, 1995: 97–103.

Martin C. Raff, Barbara A. Barres, Julia F. Burne, Harriet S. Coles, Yasuki Ishizaki, and Michael D. Jacobson. "Programmed Cell Death and Control of Cell Survival: Lessons from the Nervous System." *Science,* October 29, 1993: 695–700.

A. Raine, P. Brennam, and S. A. Mednick. "Birth Complications Combined with Early Maternal Rejection at Age One Year Predispose to Violent Crime at Age Eighteen Years." *Archives of General Psychiatry* 51 (1994): 984–988.

Kharlena Maria Ramanan. "Neandertal Architecture." *Neandertals: A Cyber Perspective.* http://thunder.indstate.edu/~ramanank/structures.html. May 1999.

G. G. Rao. "Risk Factors for the Spread of Antibiotic-Resistant Bacteria." *Drugs,* March 1998: 323–330.

S. Chenchal Rao, Gregor Rainer, and Earl K. Miller. "Integration of What and Where in the Primate Prefrontal Cortex." *Science,* May 2, 1997: 821–824.

H. Rasmussen, P. Barrett, J. Smallwood, W. Bollag, and C. Isales. "Calcium Ion as Intracellular Messenger and Cellular Toxin." *Environmental Health Perspectives,* March 1990: 17–25.

J. Ratliff-Crain, L. Temoshok, J. K. Kiecolt-Glaser, and L. Tamarkin. "Issues in Psychoneuroimmunology Research." *Health Psychology* 8:6 (1989): 747–752.

O. Rauprich, M. Matsushita, C. J. Weijer, F. Siegert, S. E. Esipov, and J. A. Shapiro. "Periodic Phenomena in *Proteus Mirabilis* Swarm Colony Development." *Journal of Bacteriology,* November 1996: 6525–6538.

Bertram H. Raven. Personal communication. November 6, 1998.

Bertram H. Raven and Jeffrey Z. Rubin. *Social Psychology.* New York: John Wiley and Sons, 1983.

John Reader. *Man on Earth.* Austin: University of Texas Press, 1988.

L. P. Reagan and B. S. McEwen. "Controversies Surrounding Glucocorticoid-Mediated Cell Death in the Hippocampus." *Journal of Chemical Neuroanatomy,* August 1997: 149–167.

L. A. Real. "Animal Choice Behavior and the Evolution of Cognitive Architecture." *Science,* August 30, 1991: 980–986.

G. Recanzone, C. Schreiner, and M. Merzenich. "Plasticity in the Frequency Representation of Primary Auditory Cortex following Discrimination Training in Adult Owl Monkeys." *Journal of Neuroscience* 13 (1993): 97–103.

Dennis T. Regan. "Effects of a Favor and Liking on Compliance." *Journal of Experimental Social Psychology* 7 (1971): 627–639.

Larry Reibstein with Nadine Joseph. "Mimic Your Way to the Top." *Newsweek,* August 8, 1988: 50.

H. T. Reis. "Physical Attractiveness in Social Interaction." *Journal of Personality and Social Psychology* 38 (1980): 604–617.

Edwin O. Reischauer. *Japan Past and Present.* 3rd ed. Tokyo: Charles E. Tuttle, 1964.

————. *The Japanese.* Cambridge, Mass.: Harvard University Press, 1981.

Mary Renault. *The Nature of Alexander.* London: Allen Lane, 1975.

————. *Fire from Heaven.* New York: Vintage Books, 1977.

Colin Renfrew. *Archaeology and Language: The Puzzle of Indo-European Origins.* New York: Cambridge University Press, 1988.

Richard M. Restak, M.D. *The Mind.* New York: Bantam Books, 1988.

————. *The Modular Brain: How New Discoveries in Neuroscience Are Answering Age-Old Questions about Memory, Free Will, Consciousness, and Personal Identity.* New York: Charles Scribner's Sons, 1994.

S. M. Resnick, I. I. Gottesman, and M. McGue. "Sensation Seeking in Opposite-Sex Twins: An Effect of Prenatal Hormones?" *Behavior Genetics,* July 1993: 323–329.

Reuters. "INTERVIEW—Thai Banks Bad Debts Yet to Peak—Moody's." August 26, 1999.

Vernon Reynolds. "Ethology of Social Change." In *The Explanation of Culture Change: Models in Prehistory,* edited by Colin Renfrew. Pittsburgh: University of Pittsburgh Press, 1973: 467–480.

Gillian Rhodes, Alex Sumich, and Graham Byatt. "Are Average Facial Configurations Attractive Only Because of Their Symmetry?" *Psychological Science,* January 1999: 52–58.

Gillian Rhodes and Tanya Tremewan. "Averageness, Exaggeration, and Facial Attractiveness." *Psychological Science,* March 1996: 105–110.

S. Ribeiro and G. B. Golding. "The Mosaic Nature of the Eukaryotic Nucleus." *Molecular Biology and Evolution,* July 1998: 779–788.

Thomas H. Rich, Patricia Vickers-Rich, Andrew Constantine, Timothy F. Flannery, Lesley Kool, and Nicholas van Klaveren. "A Tribosphenic Mammal from the Mesozoic of Australia." *Science,* November 21, 1997: 1438–1442.

John Paul Ridge, Ephraim J. Fuchs, and Polly Matzinger. "Neonatal Tolerance Revisited: Turning on Newborn T Cells with Dendritic Cells." *Science,* March 22, 1996: 1723–1726.

Cecilia Ridgeway. "Status in Groups: The Importance of Motivation." *American Sociological Review* 47 (1982): 76–88.

James Ridgeway. "The Far Right's Bomb Squad." *Freedom Writer Magazine,* March–April 1997: 7–8.

————. "Identity Politics." *Freedom Writer Magazine,* July–August 1997: 3–5.

N. Rieder. "Early Stages of Toxicyst Development in *Didinium Nasutum.*" *Zeitschrift für Naturforschung B,* April 1968: 569.

N. Rikitomi, M. Akiyama, and K. Matsumoto. "Role of Normal Microflora in the Throat in Inhibition of Adherence of Pathogenic Bacteria to Host Cells: In Vitro Competitive Adherence between *Corynebacterium Pseudodiphtheriticum* and *Branhamella Catarrhalis.*" *Kansenshogaku Zasshi,* February 1989: 118–124.

Michael Ray Rinzler and Alan Rinzler, eds. *The New Paradigm in Business: Emerging Strategies for Leadership and Organizational Change.* New York: J. P. Tarcher/Perigee, 1993.

Suzanne Ripley. "Intertroop Encounters among Ceylon Gray Langurs *(Presbytis Entellus).*" In *Social Communication among Primates,* edited by Stuart A. Altmann. Chicago: University of Chicago Press, 1967: 237–254.

T. A. Rizzo and W. A. Corsaro. "Social Support Processes in Early Childhood Friendship: A Comparative Study of Ecological Congruences in Enacted Support." *American Journal of Community Psychology,* June 1995: 389–417.

Trevor W. Robbins. "Refining the Taxonomy of Memory." *Science,* September 6, 1996: 1353–1354.

John Storm Roberts. *Black Music of Two Worlds: African, Caribbean, Latin, and African-American Traditions.* New York: Schirmer, 1998.

R. G. Roberts, R. Jones, and M. A. Smith. "The Human Colonisation of Australia: Optical Dates of 53,000 and 60,000 Years Bracket Human Arrival at Deaf Adder Gorge, Northern Territory." *Quaternary Science Reviews* 13 (1994): 575–583.

L. C. Robertson and M. R. Lamb. "Neuropsychological Contributions to Theories of Part/Whole Organization." *Cognitive Psychology,* April 1991: 299–330.

Robert S. Robins. "Paranoid Ideation and Charismatic Leadership." *Psychohistory Review,* fall 1986: 15–55.

Eugenio Rodriguez, Nathalie George, Jean-Philippe Lachaux, Jacques Martinerie, Bernard Renault, and Francisco J. Varela. "Perception's Shadow: Long-distance Synchronization of Human Brain Activity." *Nature,* February 4, 1999: 430–433.

Henry L. Roediger. "Memory Illusions." *Journal of Memory and Language,* April 1, 1996: 76–100.

Henry L. Roediger III and Kathleen B. McDermott. "Creating False Memories: Remembering Words Not Presented in Lists." *Journal of Experimental Psychology* 21:4 (July 1995): 803.

R. A. Rosellini and M. E. Seligman. "Frustration and Learned Helplessness." *Journal of Experimental Psychology:Animal Behavior Processes,* April 1975: 149–157.

J. F. Rosenbaum, J. Biederman, E. A. Bolduc-Murphy, S. V. Faraone, J. Chaloff, D. R. Hirshfeld, and J. Kagan. "Behavioral Inhibition in Childhood: A Risk Factor for Anxiety Disorders." *Harvard Review of Psychiatry,* May–June 1993: 2–16.

A. A. Rosenberg and Jerome Kagan. "Physical and Physiological Correlates of Behavioral Inhibition." *Developmental Psychobiology* 22:8 (December 1989).

M. Rosenberg. *Myxobacteria: Development and Cell Interactions.* New York: Spring, 1984.

Mark R. Rosenzweig. "Environmental Complexity, Cerebral Change, and Behavior." *American Psychologist* 21 (1966): 321–342.

Joël de Rosnay. *Le Cerveau Planétaire.* Paris: Olivier Orban, 1986.

————. *L'Homme symbiotique: Regards sur le troisième millénaire.* Paris: Seuil, 1996.

Morris Rossabi. *Khubilai Khan: His Life and Times.* Los Angeles: University of California Press, 1988.

H. J. Rothgerber. "External Intergroup Threat as an Antecedent to Perceptions of In-group and Out-group Homogeneity." *Journal of Personality and Social Psychology,* December 1997: 1206–1212.

J. Rotton, T. Barry, J. Frey, and E. Soler. "Air Pollution and Interpersonal Attraction." *Journal of Applied Social Psychology* 8 (1978): 57–71.

Linda Rowan. "Stellar Birth and Death." *Science,* May 30, 1997: 1315.

T. Rowe. "Definition, Diagnosis, and Origin of Mammalia." *Journal of Vertebrate Paleontology* 8:3 (1988): 241–264.

A. L. Rowse. *The Expansion of Elizabethan England.* London: MacMillan, 1955.

K. S. Rózsa. "The Pharmacology of Molluscan Neurons." *Progress in Neurobiology* 23:1–2 (1984): 79–150.

J. Z. Rubin, F. J. Provenzano, and Z. Luria. "The Eye of the Beholder: Parents' Views on Sex of Newborns." *American Journal of Orthopsychiatry* 44 (1974): 512–519.

Earl Rubington. "Deviant Subcultures." In *Sociology of Deviance,* edited by M. Michael Rosenberg, Robert A. Stebbins, and Allan Turowitz. New York: St. Martin's, 1982: 42–70.

R. Rudner, O. Martsinkevich, W. Leung, and E. D. Jarvis. "Classification and Genetic Characterization of Pattern-Forming Bacilli." *Molecular Microbiology,* February 1998: 687–703.

Erkki Ruoslahti. "Stretching Is Good for a Cell." *Science,* May 30, 1997: 1345–1346.

R. Rupprecht, N. Wodarz, J. Kornhuber, B. Schmitz, K. Wild, H. U. Braner, O. A. Müller, and P. Riederer. "In Vivo and In Vitro Effects of Glucocorticoids on Lymphocyte Proliferation in Man: Relationship to Glucocorticoid Receptors." *Neuropsychobiology* 24:2 (1990–1991): 61–66.

Br. Abu Ruqaiyah. "The Islamic Legitimacy of the 'Martyrdom Operations.'" Translated by Br. Hussein El-Chamy. *Nida'ul Islam Magazine (The Call of Islam: A Comprehensive Intellectual Magazine),* December–January 1996–1997. http://www.islam.org.au/articles/16/martyrdom.html. August 1998.

J. Philippe Rushton, Robin J. Russell, and Pamela A. Wells. "Genetic Similarity Theory: Beyond Kin Selection." *Behavior Genetics,* May 1984: 179–193.

R. C. Russ, J. A. Gold, and W. F. Stone. "Attraction to a Dissimilar Stranger as a Function of Level of Effectance Arousal." *Journal of Experimental Social Psychology* 15 (1979): 459–466.

C. L. Russell, K. A. Bard, and L. B. Adamson. "Social Referencing by Young Chimpanzees *(Pan troglodytes)." Journal of Comparative Psychology,* June 1997: 185–193.

Peter Russell. *The Global Brain.* Los Angeles: Jeremy P. Tarcher, 1983.

———. *The Global Brain Awakens: Our Next Evolutionary Leap.* Palo Alto, Calif.: Global Brain, 1995.

J. Rydell and J. R. Speakman. "Evolution of Nocturnality in Bats: Potential Competitors and Predators during Their Early History." *Food Science and Technology,* May 1995: 183–191.

F. Sachs and F. Qin. "Gated, Ion-Selective Channels Observed with Patch Pipettes in the Absence of Membranes: Novel Properties of a Gigaseal." *Biophysical Journal,* September 1993: 1101–1107.

Marshall D. Sahlins. "Poor Man, Rich Man, Big-Man, Chief." In *Conformity and Conflict: Readings in Cultural Anthropology,* edited by James P. Spradley and David W. McCurdy. Boston: Little, Brown, 1987: 304–319.

Eisuke Sakakibara. *Beyond Capitalism: The Japanese Model of Market Economics.* Lanham, Md.: University Press of America, 1993.

G. P. Salmond, B. W. Bycroft, G. S. Stewart, and P. Williams. "The Bacterial 'Enigma': Cracking the Code of Cell-Cell Communication." *Molecular Microbiology,* May 1995: 615–624.

U. Sandler and A. Wyler. "Non-Coding DNA Can Regulate Gene Transcription by Its Base Pair's Distribution." *Journal of Theoretical Biology,* July 7, 1998: 85–90.

Kathleen Sandman and John N. Reeve. "Origin of the Eukaryotic Nucleus." *Science,* April 24, 1998: 499d.

Robert Sapolsky. "Lessons of the Serengeti." *The Sciences,* May–June 1988: 38–42.

Robert M. Sapolsky. "Stress in the Wild." *Scientific American,* January 1990: 116–123.

———. "Stress, Social Status, and Reproductive Physiology in Free-Living Baboons." In *Psychobiology of Reproductive Behavior: An Evolutionary Perspective,* edited by David Crews. Englewood Cliffs, N.J.: Prentice-Hall, 1987.

———. *Why Zebras Don't Get Ulcers: An Updated Guide to Stress, Stress-Related Diseases, and Coping.* New York: W. H. Freeman, 1998.

I. G. Sarason, B. R. Sarason, and G. R. Pierce. "Social Support, Personality, and Health." In *Topics in Health Psychology,* edited by S. Maes, C. D. Spielberger, P. B. Defares, and I. G. Sarason. New York: John Wiley and Sons, 1988: 245–256.

H. B. Sarnat and M. G. Netsky. "The Brain of the Planarian as the Ancestor of the Human Brain." *Canadian Journal of Neurological Sciences,* November 1985: 296–302.

I. Sarnoff and P. G. Zimbardo. "Anxiety, Fear, and Social Affiliation." *Journal of Abnormal and Social Psychology* 61 (1962): 356–363.

S. Scarr. "Theoretical Issues in Investigating Intellectual Plasticity." In *Plasticity of Development,* edited by S. E. Brauth, W. S. Hall, and R. J. Dooling. Cambridge, Mass.: MIT Press, 1991: 57–71.

John Scarry. "World Prehistory: The Middle Paleolithic." Chapel Hill: Research Laboratories of Archaeology, University of North Carolina at Chapel Hill. http://www.unc.edu/courses/anth100/midpaleo.html. Downloaded June 1999.

Stanley Schachter. *The Psychology of Affiliation.* Stanford, Calif.: Stanford University Press, 1959.

J. M. Schaeffer and A. J. Hsueh. "Alpha-Bungarotoxin-Luciferin as a Bioluminescent Probe for Characterization of Acetylcholine Receptors in the Central Nervous System." *Journal of Biological Chemistry,* February 1984: 2055–2058.

J. R. Schäfer, Y. Kawaoka, W. J. Bean, J. Süss, D. Senne, and R. G. Webster. "Origin of the Pandemic 1957 H2 Influenza A Virus and the Persistence of Its Possible Progenitors in the Avian Reservoir." *Virology,* June 1993: 781–788.

Richard Louis Schanck. "A Study of a Community and Its Groups and Institutions Conceived of as Behaviors of Individuals." *Psychological Monographs* 43:2 (1933): 1–133.

Steven J. Schapiro, Pramod N. Nehete, Jaine E. Perlman, Mollie A. Bloomsmith, and Jagannadha K. Sastry. "Effects of Dominance Status and Environmental Enrichment on Cell-Mediated Immunity in Rhesus Macaques." *Applied Animal Behaviour Science,* March 1998: 319–332.

J. H. Scharf. "Turning Points in Cytology." *Acta Histochemica: Supplementband* 39 (1990): 11–47.

Bruce Schechter. "How the Brain Gets Rhythm." *Science,* October 18, 1996: 339–340.

Robert I. Scheinman, Patricia C. Cogswell, Alan K. Lofquist, and Albert S. Baldwin Jr. "Role of Transcriptional Activation of I B in Mediation of Immunosuppression by Glucocorticoids." *Science,* October 13, 1995: 283–286.

Klaus R. Scherer and Paul Ekman, eds. *Approaches to Emotion.* Hillsdale, N.J.: Lawrence Erlbaum Associates, 1984.

D. Schluter. "Experimental Evidence That Competition Promotes Divergences in Adaptive Radiation." *Science,* November 4, 1994: 798.

C. P. Schmidt and J. W. McCutcheon. "Reexamination of Relations between the Myers-Briggs Type Indicator and Field Dependence-Independence." *Perceptual and Motor Skills,* December 1988: 691–695.

J. Schmidt and B. K. Ahring. "Granular Sludge Formation in Upflow Anaerobic Sludge Blanket (UASB) Reactors." *Biotechnology and Bioengineering* 49 (1996): 229–246.

L. A. Schmidt and N. A. Fox. "Fear-Potentiated Startle Responses in Temperamentally Different Human Infants." *Developmental Psychobiology,* March 1998: 113–120.

W. Schmidt-Lorenz. "The Fine Structure of the Swarm Cells of Myxobacteria." *Journal of Applied Bacteriology,* March 1969: 22–23.

W. Schmidt-Lorenz and H. Kühlwein. "Contributions to the Knowledge of the Myxobacteria-Cell. 2. Surface Structures of the Swarm Cells." *Archives of Microbiology* 68:4 (1969): 405–426.

Mark A. Schneegurt. "Cyanosite: A Webserver for Cyanobacterial Research." http://www.cyanosite. bio.purdue.edu/images/images.html. Downloaded September 1999.

M. Schneider, D. W. Ow, and S. H. Howell. "The In Vivo Pattern of Firefly Luciferase Expression in Transgenic Plants." *Plant Molecular Biology,* June 1990: 935–947.

Michael Schneider. "Understanding the Brain—a New Look at Seeing: The Visual Cortex and Self-Organization." In *Projects in Scientific Computing, 1995.* Pittsburgh: Pittsburgh Supercomputing Center, 1995. http://www.psc.edu/science/Miikkulainen/Miikkul-vis.html. May 1999.

C. Scholtissek. "Source for Influenza Pandemics." *European Journal of Epidemiology,* August 1994: 455–458.

Renee Schoof. "China Cult Could Be Formidable Foe." Associated Press, April 27, 1999.

J. W. Schopf. "Are the Oldest 'Fossils', Fossils?" *Origins of Life and Evolution of the Biosphere,* January 1976: 19–36.

J. William Schopf. "Microfossils of the Early Archean Apex Chert: New Evidence of the Antiquity of Life." *Science,* April 30, 1993: 640–646.

———. *Cradle of Life: The Discovery of Earth's Earliest Fossils.* Berkeley: University of California Press, 1999.

J. William Schopf and Cornelius Klein, eds. *Proterozoic Biosphere: A Multidisciplinary Study.* New York: Cambridge University Press, 1992.

Ethan J. Schreier, Alessandro Marconi, and Chris Packham. "Evidence for a 20 Parsec Disk at the Nucleus of Centaurus A." *The Astrophysical Journal,* June 1, 1998: 2.

R. Schropp. "Children's Use of Objects—Competitive or Interactive?" Paper presented at the Nineteenth International Ethological Conference, Toulouse, France, 1985.

Andrea Schulte-Peevers. "The Brothers Grimm and the Evolution of the Fairy Tale." *German Life,* March 31, 1996: 18–26.

G. J. Schütz, W. Trabesinger, and T. Schmidt. "Direct Observation of Ligand Colocalization on Individual Receptor Molecules." *Biophysical Journal,* May 1998: 2223–2226.

"Search for Control of the Hamas." *Seattle Times,* March 5, 1996. http://www.seattletimes.com/ extra/browse/html/hama_030596.html. November 1998.

Ole Seehausen, Jacques J. M. van Alphen, and Frans Witte. "Cichlid Fish Diversity Threatened by Eutrophication That Curbs Sexual Selection." *Science,* September 19, 1997: 1808–1810.

Thomas D. Seeley. *Honeybee Ecology: A Study of Adaptation in Social Life.* Princeton, N.J.: Princeton University Press, 1985.

————. *The Wisdom of the Hive: The Social Physiology of Honey Bee Colonies.* Cambridge, Mass.: Harvard University Press, 1995.

Thomas D. Seeley and Royce A. Levien. "A Colony of Mind: The Beehive as Thinking Machine." *The Sciences,* July–August, 1987: 38–42.

M. H. Segall, D. T. Campbell, and M. J. Herskovitz. *The Influence of Culture on Visual Perception.* Indianapolis: Bobbs-Merrill, 1966.

Adolf Seilacher, Pradip K. Bose, and Friedrich Pflüger. "Triploblastic Animals More Than One Billion Years Ago: Trace Fossil Evidence from India." *Science,* October 2, 1998: 80–83.

M. E. Seligman. "Learned Helplessness." *Annual Review of Medicine* 23 (1972): 407–412.

M. E. Seligman and G. Beagley. "Learned Helplessness in the Rat." *Journal of Comparative Psychology,* February 1975: 534–541.

Martin E. P. Seligman. *Learned Optimism.* New York: Alfred A. Knopf, 1990.

S. Semaw, P. Renne, J. W. K. Harris, C. S. Feibel, R. L. Bernor, N. Fesse-ha, and K. Mowbray. "2.5-Million-Year-Old Stone Tools from Gona, Ethiopia." *Nature,* January 23, 1997: 333.

D. A. Senne, B. Panigrahy, Y. Kawaoka, J. E. Pearson, J. Süss, M. Lipkind, H. Kida, and R. G. Webster. "Survey of the Hemagglutinin (HA) Cleavage Site Sequence of H5 and H7 Avian Influenza Viruses: Amino Acid Sequence at the HA Cleavage Site as a Marker of Pathogenicity Potential." *Avian Diseases,* April–June 1996: 425–437.

Richard Sennett. *The Corrosion of Character: The Personal Consequences of Work in the New Capitalism.* New York: W. W. Norton, 1998.

Serbiainfo, www.serbia-info.com, October 1998.

Robert F. Service. "Microbiologists Explore Life's Rich, Hidden Kingdoms." *Science,* March 21, 1997: 1740–1742.

————. "Quantum Computing Makes Solid Progress." *Science,* April 30, 1999: 722–723.

————. "Organic Molecule Rewires Chip Design." *Science,* July 16, 1999: 313–315.

P. Servos and M. Peters. "A Clear Left Hemisphere Advantage for Visuo-Spatially Based Verbal Categorization." *Neuropsychologia* 28:12 (1990): 1251–1260.

"Sex, Lies, and Grand Schemes of Thought in Closed Groups." *Cultic Studies Journal* 14:1 (1997): 58–84.

Robert M. Seyfarth and Dorothy L. Cheney. "Some General Features of Vocal Development in Nonhuman Primates." In *Social Influences on Vocal Development,* edited by Charles T. Snowdon and Martine Hausberger. Cambridge, U.K.: Cambridge University Press, 1996: 249–273.

J. A. Shapiro. "Bacteria as Multicellular Organisms." *Scientific American,* June 1988: 82–89.

————. "Natural Genetic Engineering in Evolution." *Genetica* 86:1–3 (1992): 99–111.

————. "Pattern and Control in Bacterial Colony Development." *Science Progress* 76: 301–302, pt. 3–4 (1992): 399–424.

————. "Natural Genetic Engineering of the Bacterial Genome." *Current Opinion in Genetics and Development,* December 1993: 845–848.

————. "The Significances of Bacterial Colony Patterns." *Bioessays,* July 17, 1995: 597–607.

————. "Genome Organization, Natural Genetic Engineering and Adaptive Mutation." *Trends in Genetics,* March 13, 1997: 98–104.

————. "Thinking about Bacterial Populations as Multicellular Organisms." *Annual Review of Microbiology.* Palo Alto, Calif.: Annual Reviews, 1998: 81–104.

————. Personal communications. February 9–September 24, 1999.

James A. Shapiro, ed. *Mobile Genetic Elements.* New York: Academic Press, 1983.

Lucy Shapiro and Richard Losick. "Protein Localization and Cell Fate in Bacteria." *Science,* May 2, 1997: 712–718.

Robert Shapiro. "Prebiotic Cytosine Synthesis: A Critical Analysis and Implications for the Origin of Life." *Proceedings of the National Academy of Sciences of the United States of America,* April 1999: 4396–4401.

———. *Planetary Dreams: The Quest to Discover Life beyond Earth.* New York: John Wiley and Sons, 1999.

G. B. Sharp, Y. Kawaoka, D. J. Jones, W. J. Bean, S. P. Pryor, V. Hinshaw, and R. G. Webster. "Coinfection of Wild Ducks by Influenza A Viruses: Distribution Patterns and Biological Significance." *Journal of Virology,* August 1997: 6128–6135.

N. Shashar and T. W. Cronin. "Polarization Contrast Vision in Octopus." *Journal of Experimental Biology* 199 (April 1996): 999–1004.

Y. Shavit. "Endogenous Opioids May Mediate the Effects of Stress on Tumor Growth and Immune Function." *Proceedings of the Western Pharmacology Society* 26 (1983).

Barry Shelby. "Japan and Africa." *World Press Review,* January 1989: 44.

Muzafer Sherif and Hadley Cantril. *The Psychology of Ego-Involvements: Social Attitudes and Identifications.* New York: John Wiley and Sons, 1947.

Muzafer Sherif, O. H. Harvey, B. Jack White, William R. Hood, and Carolyn W. Sherif. *The Robbers Cave Experiment: Intergroup Conflict and Cooperation.* Middletown, Conn.: Wesleyan University Press, 1988.

D. F. Sherry and B. G. Galef. "Social Learning without Imitation: More about Milk Bottle Opening by Birds." *Animal Behavior* 40 (1990): 987–989.

Shi-xu. "Cultural Perceptions: Exploiting the Unexpected of the Other." *Culture and Psychology* 1 (1995): 315–342.

L. J. Shimkets. "Control of Morphogenesis in Myxobacteria." *Critical Reviews in Microbiology* 14:3 (1987): 195–227.

———. "The Role of the Cell Surface in Social and Adventurous Behaviour of Myxobacteria." *Molecular Microbiology,* September 1989: 1295–1299.

———. "Social and Developmental Biology of the Myxobacteria." *Microbiology Reviews,* December 1990: 473–501.

T. D. Shou and Y. F. Zhou. "Orientation and Direction Sensitivity of Cells in Subcortical Structures of the Visual System." *Sheng Li Hsueh Pao,* April 1996: 105–112.

T. J. Shors, T. B. Seib, S. Levine, and R. F. Thompson. "Inescapable versus Escapable Shock Modulates Long-Term Potentiation in the Rat Hippocampus." *Science,* April 14, 1989: 224–226.

L. P. Shu, G. B. Sharp, Y. P. Lin, E. C. Claas, S. L. Krauss, K. F. Shortridge, and R. G. Webster. "Genetic Reassortment in Pandemic and Interpandemic Influenza Viruses: A Study of 122 Viruses Infecting Humans." *European Journal of Epidemiology,* February 1996: 63–70.

C. G. Sibley and J. E. Ahlquist. "The Phylogeny of the Hominoid Primates, as Indicated by DNA-DNA Hybridization." *Journal of Molecular Evolution* 20:1 (1984): 2–15.

———. "DNA Hybridization Evidence of Hominoid Phylogeny: Results from an Expanded Data Set." *Journal of Molecular Evolution* 26:1–2 (1987): 99–121.

C. G. Sibley, J. A. Comstock, and J. E. Ahlquist. "DNA Hybridization Evidence of Hominoid Phylogeny: A Reanalysis of the Data." *Journal of Molecular Evolution,* March 1990: 202–236.

R. E. Sicard. "Hormones, Neurosecretions, and Growth Factors as Signal Molecules for Intercellular Communication." *Developmental and Comparative Immunology,* spring 1986: 269–272.

H. Sigall and R. Helmreich. "Opinion Change as a Function of Stress and Communicator Credibility." *Journal of Experimental and Social Psychology* 5 (1969): 70–78.

P. V. Simonov. "The 'Bright Spot of Consciousness.' " *Zhurnal Vysshei Nervnoi Deiatelnosti Imeni I. P. Pavlova,* November–December 1990: 1040–1044.

Dean Keith Simonton. *Greatness: Who Makes History and Why.* New York: Guilford Press, 1994.

Margaret Thaler Singer and Janja Lalich. *Cults in Our Midst.* San Francisco: Jossey-Bass Publishers, 1995.

Wolf Singer. "Neurobiology: Striving for Coherence." *Nature,* February 4, 1999: 391–393.

J. Singh and R. T. Knight. "Prefrontal Lobe Contribution to Voluntary Movements in Humans." *Brain Research* 531 (1990): 45–54.

———. "Effects of Posterior Association Cortex Lesions on Brain Potentials Preceding Self-Initiated Movements." *Journal of Neuroscience,* May 1993: 1820–1829.

Purushottam Singh. *Neolithic Cultures of Western Asia.* New York: Seminar Press, 1974.

John R. Skoyles. "Alphabet and the Western Mind." *Nature* 309 (1984): 409–410.

———. "Did Ancient People Read with Their Right Hemispheres? A Study in Neuropalaeo-graphology." *New Ideas in Psychology* 3 (1985): 243–252.

———. "The Origin of Classical Greek Culture: The Transparent Chain Theory of Literacy/Society Interaction." *Journal of Social and Biological Structures* 13 (1990): 321–353.

———. Personal communications. November 1997–April 2000.

———. "Motor Perception and Anatomical Realism in Classical Greek Art." *Medical Hypothesis.* 51 (1998): 69–70.

———. "Complete Online Text of Selected Papers and Two Books (*Odyssey* and *Leviathan*)." http://www.users.globalnet.co.uk/~skoyles/. May 1999.

David Smillie. "Human Nature and Evolution: Language, Culture, and Race." Paper given at the Biennial Meeting of the International Society of Human Ethology, Amsterdam, August 1992.

———. "Darwin's Tangled Bank: The Role of Social Environments." *Perspectives in Ethology.* Vol. 10, *Behavior and Evolution,* edited by P. P. G. Bateson et al. New York: Plenum Press, 1993: 119–141.

———. "Darwin's Two Paradigms: An 'Opportunistic' Approach to Group Selection Theory." *Journal of Social and Evolutionary Systems* 18:3 (1995): 231–255.

———. "Group Processes and Human Evolution: Sex and Culture as Adaptive Strategies." Paper presented at the Nineteenth Annual Meeting of the European Sociobiological Society, Alfred, N.Y., July 25, 1996.

S. M. Smirnakis, M. J. Berry, D. K. Warland, W. Bialek, and M. Meister. "Adaptation of Retinal Processing to Image Contrast and Spatial Scale." *Nature,* March 6, 1997: 69–73.

J. Smit and N. Agabian. "Cell Surface Patterning and Morphogenesis: Biogenesis of a Periodic Surface Array during Caulobacter Development." *Journal of Cell Biology,* October 1982: 41–49.

Adam Smith. *An Inquiry into the Nature and Causes of the Wealth of Nations.* Dublin: Whitestone, 1776.

D. R. Smith and M. Dworkin. "Territorial Interactions between Two Myxococcus Species." *Journal of Bacteriology,* February 1994: 1201–1205.

Frederick Winston Furneaux Smith, Earl of Birkenhead. *Rudyard Kipling.* New York: Random House, 1978.

M. D. Smith, C. I. Masters, E. Lennon, L. B. McNeil, and K. W. Minton. "Gene Expression in *Deinococcus Radiodurans.*" *Gene,* February 1991: 45–52.

John Maynard Smith. "When Learning Guides Evolution." *Nature* 329 (1987): 761–762.

Patrick Smith. "You Have a Job, but How about a Life?" *Business Week,* November 16, 1998. http://www.businessweek.com/1998/46/b3604087.html. Downloaded February 2000.

A. G. Smithers and D. M. Lobley. "Dogmatism, Social Attitudes and Personality." *British Journal of Social and Clinical Psychology,* June 1978: 135–142.

Elliott Sober and David Sloan Wilson. *Unto Others: The Evolution and Psychology of Unselfish Behavior.* Cambridge, Mass.: Harvard University Press, 1988.

Olga Soffer. *The Upper Paleolithic of the Central Russian Plain.* New York: Academic Press, 1985.

———. Personal communication. December 7, 1997.

Alan Sokal and Jean Bricmont. *Fashionable Nonsense: Postmodern Intellectuals' Abuse of Science.* New York: Picador, 1998.

K. Sokolowski and H. D. Schmalt. "Emotional and Motivational Influences in an Affiliation Conflict Situation." *Zeitschrift für Experimentelle Psychologie* 43:3 (1996): 461–482.

J. M. Solomon and A. D. Grossman. "Who's Competent and When: Regulation of Natural Genetic Competence in Bacteria." *Trends in Genetics* 12 (1996): 150–155.

Joseph Soltis, Robert Boyd, and Peter J. Richerson. "Can Group Functional Behaviors Evolve by Cultural Group Selection? An Empirical Test." *Current Anthropology,* June 1995: 473–494.

Paul Somerson. "I've Got a Server in My Pants." *ZD Equip Magazine.* http://www.zdnet.co.uk/athome/feature/serverinpants/welcome.html. Downloaded August 1999.

J. M. Sommer and A. Newton. "Turning Off Flagellum Rotation Requires the Pleiotropic Gene pleD: pleA, pleC, and pleD Define Two Morphogenic Pathways in *Caulobacter Crescentus.*" *Journal of Bacteriology,* January 1989: 392–401.

H. Sompolinsky and R. Shapley. "New Perspectives on the Mechanisms for Orientation Selectivity." *Current Opinion in Neurobiology,* August 1997: 514–522.

P. Y. Sondaar. "Pleistocene Man and Extinctions of Island Endemics." *Société Géologique de France: Memoires* 150 (1987): 159–165.

Sorin Sonea. "The Global Organism: A New View of Bacteria." *The Sciences,* July–August 1988: 38–45.

Sorin Sonea and Maurice Panisset. *A New Bacteriology.* Boston: James and Bartlett, 1983.

Charles H. Southwick. "Genetic and Environmental Variables Influencing Animal Aggression." In *Animal Aggression: Selected Readings,* edited by Charles H. Southwick. New York: Van Nostrand Reinhold, 1970.

Space Telescope Science Institute. "Hubble Provides Multiple Views of How to Feed a Black Hole." Press release no. STScl-PR98-14. Posted May 15, 1998. http://oposite.stsci.edu/pubinfo/pr/1998/14/. Downloaded August 1998.

———. "Hubble Completes Eight-Year Effort to Measure Expanding Universe." Press release no. STScl-PR99-19. Baltimore: Space Telescope Science Institute, May 15, 1999. http://oposite.stsci.edu/pubinfo/pr/1999/19/pr.html. Downloaded June 1999.

Albert Speer. *Inside the Third Reich—Memoirs.* Translated by Richard Winston and Clara Winston. New York: Collier Books, 1970.

Stephan E. Speyer. "Comparative Taphonomy and Paleoecology of Trilobite Lagerstatten." *Alcheringa* 11 (1987): 205–232.

Rene A. Spitz. *Hospitalism: An Inquiry into the Genesis of Psychiatric Conditions in Early Childhood.* The Psychoanalytic Study of the Child, vol. 1. New York: International Universities Press, 1945.

Rene A. Spitz, M.D., with Katherine M. Wolf, Ph.D. *Anaclitic Depression: An Inquiry into the Genesis of Psychiatric Conditions in Early Childhood, II.* The Psychoanalytic Study of the Child, vol. 2. New York: International Universities Press, 1946.

Nicholas C. Spitzer and Terrence J. Sejnowski. "Biological Information Processing: Bits of Progress." *Science,* August 22, 1997: 1060–1061.

H. Philip Spratt. "The Marine Steam-Engine." In *A History of Technology.* Vol. 5, *The Late Nineteenth Century, c. 1850 to c. 1900,* edited by Charles Singer, E. J. Holmyard, A. R. Hall, and Trevor I. Williams. Oxford: Oxford University Press, 1958.

V. Spruiell. "Crowd Psychology and Ideology: A Psychoanalytic View of the Reciprocal Effects of Folk Philosophies and Personal Actions." *International Journal of Psychoanalysis* 69:2 (1988): 171–178.

S. Stack. "The Effect of the Media on Suicide: Evidence from Japan, 1955–1985." *Suicide and Life-threatening Behavior,* summer 1996: 132–142.

John Stackhouse. "Will Karachi Be the Next Kabul?" *Toronto Globe and Mail,* reprinted in *World Press Review,* October 1998: 17.

B. M. Staw, L. E. Sandelands, and J. E. Dutton. "Threat-Rigidity Effects in Organizational Behavior: A Multilevel Analysis." *Administrative Science Quarterly* 26 (1981): 501–524.

Lyle Steadman. Personal communication. December 7, 1997.

Peter N. Stearns. "The Rise of Sibling Jealousy in the Twentieth Century." In *Emotion and Social Change: Toward a New Psychohistory,* edited by Carol Z. Stearns and Peter N. Stearns. New York: Holmes and Meier, 1988: 69–85.

G. L. Stebbins Jr. *Variation and Evolution in Plants.* New York: Columbia University Press, 1950.

Volker Stefanski. "Social Stress in Loser Rats: Opposite Immunological Effects in Submissive and Subdominant Males." *Physiology and Behavior,* April 1998: 605–613.

Robert M. Stelmack. "The Psychophysics and Psychophysiology of Extraversion and Arousal." *The Scientific Study of Human Nature: Tribute to Hans J. Eysenck at Eighty,* edited by Helmuth Nyborg. New York: Elsevier Science, 1997: 388–403.

Christopher Stephens. "Modelling Reciprocal Altruism." *British Journal for the Philosophy of Science,* December 1996: 533–551.

Mitchell Stephens. *A History of News: From the Drum to the Satellite.* New York: Viking Press, 1988.

Richard S. Stephens, Sue Kalman, Claudia Lammel, Jun Fan, Rekha Marathe, L. Aravind, Wayne Mitchell, Lynn Olinger, Roman L. Tatusov, Qixun Zhao, Eugene V. Koonin, and Ronald W. Davis. "Genome Sequence of an Obligate Intracellular Pathogen of Humans: *Chlamydiatrachomatis.*" *Science,* October 23, 1998: 754–759.

K. Stern and M. K. McClintock. "Regulation of Ovulation by Human Pheromones." *Nature,* March 12, 1998.

Kenneth S. Stern. "Militias and the Religious Right." *Freedom Writer,* October 1996. http://apocalypse.berkshire.net/~ifas/fw/9610/militias.html. November 1998.

Anthony Stevens and John Price. *Evolutionary Psychiatry: A New Beginning.* London: Routledge, 1996.

———. "The Group Splitting Hypothesis of Schizophrenia." In *Evolutionary Psychiatry: A New Beginning.* London: Routledge, 1996: prepublication draft.

O. C. Stine, G. J. Dover, D. Zhu, and K. D. Smith. "The Evolution of Two West African Populations." *Journal of Molecular Evolution,* April 1992: 336–344.

Gregory Stock. *Metaman: The Merging of Humans and Machines into a Global Superorganism.* New York: Simon and Schuster, 1993.

A. Stolba. "Entscheidungsfindung in Verbanden von Papio Hamadryas." Ph.D. diss. Zurich: Universitat Zurich, 1979.

Ross M. Stolzenberg, Mary Blair-Loy, and Linda J. Waite. "Religious Participation in Early Adulthood: Age and Family Life Cycle Effects on Church Membership." *American Sociological Review,* February 1995: 84–103.

Valentina Stosor, Gary A. Noskin, and Lance R. Peterson. "The Management and Prevention of Vancomycin-Resistant Enterococci." *Infections in Medicine* 13:6 (1996): 487–488, 493–498.

Evelyn Strauss. "Mob Action: Peer Pressure in the Bacterial World." *Science News,* August 23, 1997: 124–125.

———. "Society for Developmental Biology Meeting: How Embryos Shape Up." *Science,* July 10, 1998: 166–167.

F. L. Strodtbeck, R. J. James, and C. Hawkins. "Social Status in Jury Deliberations." *American Sociological Review* 22 (1957): 713–719.

R. Stroufe, A. Chaikin, R. Cook, and V. Freeman. "The Effects of Physical Attractiveness on Honesty: A Socially Desirable Response." *Personality and Social Psychology Bulletin* 3 (1977): 59–62.

Charles B. Strozier. "Christian Fundamentalism, Nazism, and the Millennium." *Psychohistory Review,* winter 1990: 207–217.

Shirley C. Strum. *Almost Human: A Journey into the World of Baboons.* New York: Random House, 1987.

S. A. Sturdza. "Expansion Phenomena of Proteus Cultures. I. The Swarming Expansion." *Zentralblatt für Bakteriologie,* December 1975: 505–530.

C. Sturmbauer and A. Meyer. "Genetic Divergence, Speciation and Morphological Stasis in a Lineage of African Cichlid Fishes." *Nature,* August 13, 1992: 578.

Deyan Sudjic. "Tokyo's 'Spectacular' Stores." *Times* (London), reprinted in *World Press Review,* October 1989: 72.

Suetonius. *C. Suetonius Tranquillus. Divi Augusti Vita.* Edited by Michael Adams. London: Macmillan, 1939.

Yukimaru Sugiyama. "Social Characteristics and Socialization of Wild Chimpanzees." In *Primate Socialization,* edited by Frank E. Poirier. New York: Random House, 1972: 145–164.

C. H. Summers and N. Greenberg. "Activation of Central Biogenic Amines following Aggressive Interaction in Male Lizards, *Anolis Carolinensis.*" *Brain Behavior and Evolution* 45:6 (1995): 339–349.

Cliff H. Summers and Neil Greenberg. "Somatic Correlates of Adrenergic Activity during Aggression in the Lizard, *Anolis carolinensis.*" *Hormones and Behavior* 28 (1994): 29–40.

Curt Suplee. *Physics in the 20th Century.* New York: Harry N. Abrams, 1999.

S. J. Suomi, H. F. Harlow, and J. K. Lewis. "Effect of Bilateral Frontal Lobectomy on Social Preferences of Rhesus Monkeys." *Journal of Comparative Physiology,* March 1970: 448–453.

J. Süss, J. Schäfer, H. Sinnecker, and R. G. Webster. "Influenza Virus Subtypes in Aquatic Birds of Eastern Germany." *Archives of Virology* 135:1–2 (1994): 101–114.

Tad Szulc. *Fidel: A Critical Portrait.* New York: William Morrow, 1986.

Richard J. A. Talbert, trans. *Plutarch on Sparta.* New York: Penguin Books, 1988.

Robert H. Tamarin. *Principles of Genetics.* New York: McGraw-Hill, 1999.

A. Tanabe-Tochikura, M. T. Ang Singh, H. Tsuchie, J. Zhang, F. J. Paladin, and T. Kurimura. "A Newly Developed Immunofluorescence Assay for Simultaneous Detection of Antibodies to Human Immunodeficiency Virus Type 1 and Type 2." *Journal of Virological Methods,* April 1995: 239–246.

Sheridan M. Tatsuno. *Created in Japan: From Imitators to World-Class Innovators.* New York: Harper and Row, 1989.

Carol Tavris. *Anger: The Misunderstood Emotion.* New York: Simon and Schuster, 1984.

S. E. Taylor and M. Lobel. "Social Comparison Activity under Threat: Downward Evaluation and Upward Contacts." *Psychological Review,* October 1989: 569–575.

"Teetering toward Fascism." *Economist,* reprinted in *World Press Review,* October 1998: 8–9.

L. Temoshok. "Psychoimmunology and AIDS." *Advances in Biochemical Psychopharmacology* 44 (1988): 187–197.

Lydia Temoshok, Craig Van Dyke, and Leonard S. Zegans. *Emotions in Health and Illness: Theoretical and Research Foundations.* Orlando: Grune and Stratton, 1983.

A. R. Templeton. "Out of Africa? What Do Genes Tell Us?" *Current Opinion in Genetics and Development* 7, 1997: 841–847.

G. W. Terman, Y. Shavit, J. W. Lewis, J. T. Cannon, and J. C. Liebeskind. "Intrinsic Mechanisms of Pain Inhibition: Activation by Stress." *Science,* December 14, 1984: 1270–1277.

B. Terrana and A. Newton. "Requirement of a Cell Division Step for Stalk Formation in *Caulobacter Crescentus.*" *Journal of Bacteriology,* October 1976: 456–462.

Romila Thapar. *A History of India.* Vol. 1. New York: Penguin Books, 1985.

Robert E. Thayer. *The Biopsychology of Mood and Arousal.* New York: Oxford University Press, 1989.

J. G. M. Thewissen, editor. "The Emergence of Whales: Evolutionary Patterns in the Origin of Cetacea." New York: Plenum, 1998.

J. W. Thibaut and H. W. Riecken. "Some Determinants and Consequences of the Perception of Social Causality." *Journal of Personality* 24 (1955): 113–133.

Paul Thibodeau. *Christian Reconstruction: God's Glorious Millennium?—Anthology of Quotations.* http://www.serve.com/thibodep/cr/worldcnq.html. November 1998.

———. "The New Canaan." *Christian Reconstruction: God's Glorious Millennium?—Anthology of Quotations.* http://www.serve.com/thibodep/cr/canaan.html. November 1998.

Hartmut Thieme. "Lower Paleolithic Hunting Spears from Germany." *Nature,* February 27, 1997: 807–810.

Earl Thomas and Louise DeWald. "Experimental Neurosis: Neuropsychological Analysis." In *Psychopathology: Experimental Models,* edited by Jack D. Maser and Martin E. P. Seligman. San Francisco: W. H. Freeman and Company, 1977: 214–231.

Lewis Thomas. *Lives of a Cell: Notes of a Biology Watcher.* New York: Bantam Books, 1975.

J. M. Thompson. *Napoleon Bonaparte.* Oxford: Basil Blackwell, 1988.

John N. Thompson. "The Evolution of Species Interactions." *Science,* June 25, 1999: 2116–2118.

Alan Thorne, Rainer Grün, Graham Mortimer, Nigel A. Spooner, John J. Simpson, Malcolm McCulloch, Lois Taylor, and Darren Curnoe. "Australia's Oldest Human Remains: Age of the Lake Mungo 3 Skeleton." *Journal of Human Evolution,* June 1999: 591–612.

Thucydides. *History of the Peloponnesian War.* Translated by Richard Crawley. In *Library of the Future.* 4th ed., ver. 5.0. Irvine, Calif.: World Library, 1996. CD-ROM.

Lionel Tiger and Robin Fox. *The Imperial Animal.* New York: Holt, Rinehart and Winston, 1971.

Niko Tinbergen. *The Herring Gull's World: A Study of the Social Behavior of Birds.* New York: Basic Books, 1961.

S. A. Tishkoff, E. Dietzsch, W. Speed, et al. "Global Patterns of Linkage Disequilibrium at the CD4 Locus and Modern Human Origins." *Science,* March 8, 1996: 1380–1387.

Phillip Tobias. "The Brain of *Homo Habilis:* A New Level of Organisation in Cerebral Evolution." *Journal of Human Evolution* 16 (1987): 741–761.

Today's Japan. Tokyo: NHK-TV, January 22, 1990, June 22, 1990, and July 2, 1990. Television news broadcasts.

A. J. Tomarken and R. J. Davidson. "Frontal Brain Activation in Repressers and Nonrepressers." *Journal of Abnormal Psychology* 103 (1994): 339–349.

A. J. Tomarken, R. J. Davidson, R. Wheeler, and R. Doss. "Individual Differences in Anterior Brain Asymmetry and Fundamental Dimensions of Emotions." *Journal of Personality and Social Psychology* 62 (1992): 676–687.

Michael Tomasello. "The Question of Chimpanzee Culture." In *Chimpanzee Cultures,* edited by Richard W. Wrangham, W. C. McGrew, Frans B. M. de Waal, and Paul G. Heltne with assistance from Linda A. Marquardt. Cambridge, Mass.: Harvard University Press, 1994: 301–317.

David Tomei and Frederick O. Cope, eds. *Apoptosis: The Molecular Basis of Cell Death.* Plainview, N.Y.: Cold Spring Harbor Laboratory Press, 1991.

E. P. Torrance. "Some Consequences of Power Differences on Decision Making in Permanent and Temporary Three-Man Groups." *Research Studies* (Washington State College) 22 (1954): 130–140.

Maurizio Tosi. "Early Urban Evolution and Settlement Patterns in the Indo-Iranian Borderland." In *The Explanation of Culture Change: Models in Prehistory,* edited by Colin Renfrew. Pittsburgh: University of Pittsburgh Press, 1973: 429–446.

N. Toth, K. D. Schick, and E. S. Savage-Rumbaugh. "Pan the Tool-Maker: Investigations into the Stone Tool-Making and Tool-Using Capabilities of a Bonobo *(Pan paniscus).*" *Journal of Archaeological Science,* January 1993: 81–91.

Nicholas Toth, Desmond Clark, and Giancarlo Ligabue. "The Last Stone Ax Makers." *Scientific American,* July 1992: 88–93.

E. G. Trams. "On the Evolution of Neurochemical Transmission." *Differentiation* 19:3 (1981): 125–133.

J. Travis. "Novel Bacteria Have a Taste for Aluminum." *Science News,* May 30, 1998: 341.

John Travis. "Outbound Traffic." *Science News,* November 15, 1997: 316–317.

Tom Tregenza and Roger K. Butlin. "Speciation without Isolation." *Nature,* July 22, 1999: 311–312.

Albert A. Trever. *History of Ancient Civilization.* Vol. 2, *The Roman World.* New York: Harcourt, Brace, 1939.

Harry C. Triandis. "Collectivism and Individualism as Cultural Syndromes." *Cross-Cultural Research,* August–November 1993: 155–180.

Robert L. Trivers. "The Evolution of Reciprocal Altruism." *Quarterly Review of Biology* 46:4 (1971): 35–57.

K. R. Truett, L. J. Eaves, J. M. Meyer, A. C. Heath, and M. G. Martin. "Religion and Education as Mediators of Attitudes: A Multivariate Analysis." *Behavior Genetics,* January 1992: 43–62.

L. Tsimiring, H. Levine, I. Aranson, E. Ben-Jacob, I. Cohen, O. Shochet, and W. N. Reynolds. "Aggregation Patterns in Stressed Bacteria." *Physical Review Letters* 75:9 (1995): 1859–1862.

Tsu-Ming Han and Bruce Runnegar. "Megascopic Eukaryotic Algae from the 2.1-Billion-Year-Old Negaunee Iron-Formation, Michigan." *Science,* July 10, 1992: 232–235.

T. Tully. "Drosophila Learning: Behavior and Biochemistry." *Behavioral Genetics,* September 1984: 527–557.

————. "Discovery of Genes Involved with Learning and Memory: An Experimental Synthesis of Hirschian and Benzerian Perspectives." *Proceedings of the National Academy of Sciences of the United States of America,* November 26, 1996: 13460–13467.

————. "Regulation of Gene Expression and Its Role in Long-term Memory and Synaptic Plasticity." *Proceedings of the National Academy of Sciences of the United States of America,* April 29, 1997: 4239–4241.

T. Tully, G. Bolwig, J. Christensen, J. Connolly, M. DelVecchio, J. DeZazzo, J. Dubnau, C. Jones, S. Pinto, M. Regulski, B. Svedberg, and K. Velinzon. "A Return to Genetic Dissection of Memory in Drosophila." *Cold Spring Harbor Symposia on Quantitative Biology* 61 (1996): 207–218.

T. Tully, G. Bolwig, J. Christensen, J. Connolly, J. DeZazzo, J. Dubnau, C. Jones, S. Pinto, M. Regulski, F. Svedberg, and K. Velinzon. "Genetic Dissection of Memory in Drosophila." *Journal of Physiology* (Paris) 90:5–6 (1996): 383.

Peter Turney, Darrell Whitley, and Russell Anderson, eds. "Evolution, Learning, and Instinct: One Hundred Years of the Baldwin Effect." *Evolutionary Computation* (Special Issue on the Baldwin Effect) 4:3 (1996): iv–viii. http://ai.iit.nrc.ca/baldwin/editorial.html. Downloaded February 2000.

Bert N. Uchino, Darcy Uno, and Julianne Holt-Lunstad. "Social Support, Physiological Processes, and Health." *Current Directions in Psychological Science,* October 1999: 145–148.

United Nations Office for the Coordination of Humanitarian Affairs. "Humanitarian Report 1997: Gender Issues in Afghanistan." http://www.reliefweb.int/dha_ol/pub/humrep97/gender.html. November 1998.

U.S. Department of Energy. DOE Microbial Genome Program. http://www.er.doe.gov/production/ober/EPR/mig_top.html. January 1999.

University of Illinois at Urbana-Champaign. "Simulation Reveals Very First Stars That Formed in the Universe." Posted March 25, 1999. http://www.sciencedaily.com/releases/1999/03/990325054316.html. Downloaded August 1999.

University of Michigan. "Possible Human Ancestors Lived in Spain 780,000 Years Ago." Posted October 28, 1998. http://www.umich.edu/~newsinfo/Releases/1998/Oct98/r102698c.html. Downloaded August 1999.

Rod Usher and Yuri Zarakhovich. "The New Russian Nazis." *Time,* June 9, 1997. http://cgi.pathfinder.com/time/November 1998.

David Ussishkin. "Notes on the Fortifications of the Middle Bronze II Period at Jericho and Shechem." *Bulletin of the American Schools of Oriental Research,* November 1989: n.p.

S. J. van Enk, J. I. Cirac, and P. Zoller. "Photonic Channels for Quantum Communication." *Science,* January 9, 1998: 205–208.

Paul van Geert. "Green, Red and Happiness: Towards a Framework for Understanding Emotion Universals." *Culture and Psychology,* June 1995: 259–268.

Jan A. R. A. M. Van Hooff. "Facial Expressions in Higher Primates." *Symposium of the Zoological Society of London* 8 (1962): 97–125.

————. "Facial Expressions in Higher Primates." *Symposium of the Zoological Society of London* 10 (1963): 103–104.

————. "Understanding Chimpanzee Understanding." In *Chimpanzee Cultures,* ed. Richard W. Wrangham, W. C. McGrew, Frans B. M. de Waal, and Paul G. Heltne with assistance from Linda A. Marquardt. Cambridge, Mass.: Harvard University Press, 1994: 267–284.

R. H. Van Zelst. "Sociometrically Selected Work Teams Increase Production." *Personnel Psychology* 5 (1952): 175–185.

Francisco J. Varela. "Autonomy and Autopoiesis." In *Self-organizing Systems: An Interdisciplinary Approach,* edited by Gerhard Roth and Helmut Schwegler. New York: Campus Verlag, 1981: 14–23.

Harold Varmus and Neal Nathanson. "Science and the Control of AIDS." *Science,* June 19, 1998: 1815a.

Thorstein Veblen. *The Theory of the Leisure Class; an Economic Study of Institutions.* New York: Modern Library, 1934.

L. Vedeler. "Dramatic Play: A Format for 'Literate' Language?" *British Journal of Educational Psychology* 67:2 (June 1997): 153–167.

R. Veitch, R. DeWood, and K. Bosko. "Radio News Broadcasts: Their Effects on Interpersonal Helping." *Sociometry* 40 (1977): 383–386.

J. C. Venter, U. di Porzio, D. A. Robinson, S. M. Shreeve, J. Lai, A. R. Kerlavage, S. P. Fracek Jr., K. U. Lentes, and C. M. Fraser. "Evolution of Neurotransmitter Receptor Systems." *Progress in Neurobiology* 30:2–3 (1988): 105–169.

Geerat J. Vermeij. "Animal Origins." *Science,* October 25, 1996: 525–526.

Glaucia N. R. Vespa, Linda A. Lewis, Katherine R. Kozak, Miriana Moran, Julie T. Nguyen, Linda G. Baum, and M. Carrie Miceli. "Galectin-1 Specifically Modulates TCR Signals to Enhance TCR Apoptosis but Inhibit IL-2 Production and Proliferation." *Journal of Immunology,* January 15, 1999: 799–806.

Paul Veyne, ed. *A History of Private Life.* Vol. 1, *From Pagan Rome to Byzantium.* Cambridge, Mass.: Harvard University Press, 1987.

P. K. Visscher. "Collective Decisions and Cognition in Bees." *Nature,* February 4, 1999: 400.

Ezra Vogel. "Pax Nipponica?" *Foreign Affairs,* spring 1986: 752–767.

Gretchen Vogel. "Scientists Probe Feelings behind Decision-Making." *Science,* February 28, 1997: 1269.

———. "Cell Biology: Parasites Shed Light on Cellular Evolution." *Science,* March 7, 1997: 1422–1430.

———. "Searching for Living Relics of the Cell's Early Days." *Science,* September 12, 1997: 1604.

———. "Did the First Complex Cell Eat Hydrogen?" *Science,* March 13, 1998: 1633–1634.

———. "Chimps in the Wild Show Stirrings of Culture." *Science,* June 25, 1999: 2070–2073.

K. S. Vogel. "Development of Trophic Interactions in the Vertebrate Peripheral Nervous System." *Molecular Neurobiology,* fall–winter 1993: 363–382.

Vladimir Voinovich. "An Exile's Dilemma." *Wilson Quarterly,* August 1990: 114–120.

M. V. Vol'kenshten. "Molecular Drive." *Molekuliarnaia Biologiia,* September–October 1990: 1181–1199.

Ernst von Glasersfeld. "An Introduction to Radical Constructivism." Amherst, Mass.: Scientific Reasoning Research Institute, University of Amherst. http://www.umass.edu/srri/vonGlasersfeld/onlinePapers.html. Downloaded June 1999.

P. D. Wadhwa, C. Dunkel-Schetter, A. Chicz-DeMet, M. Porto, and C. A. Sandman. "Prenatal Psychosocial Factors and the Neuroendocrine Axis in Human Pregnancy." *Psychosomatic Medicine,* September–October 1996: 432–446.

George Wald. "The Cosmology of Life and Mind." In *New Metaphysical Foundations of Modern Science,* edited by Willis Harman with Jane Clark. Sausalito, Calif.: Institute of Noetic Sciences, 1994: 123–132.

M. J. C. Waller. "Darwinism and the Enemy Within." *Journal of Social and Evolutionary Systems* 18:3 (1995): 217–229.

———. "Organization Theory and the Origins of Consciousness." *Journal of Social and Evolutionary Systems* 19:1 (1996): 17–30.

L. L. Wallrath, Q. Lu, H. Granok, and S. C. Elgin. "Architectural Variations of Inducible Eukaryotic Promoters: Preset and Remodeling Chromatin Structures." *Bioessays,* March 1994: 165–170.

James Walsh. "Shoko Asahara: The Making of a Messiah." *Time,* April 3, 1995. http://cgi.pathfinder.com/time/. September 1998.

Yvonne Walsh and Robert Bor. "Psychological Consequences of Involvement in a New Religious Movement or Cult." *Counselling Psychology Quarterly* 9:1 (1996): 47–60.

E. Walster, V. Aronson, D. Abrahams, and L. Rottmann. "Importance of Physical Attractiveness in Dating Behavior." *Journal of Personality and Social Psychology* 4 (1966): 508–516.

Ning Wang, James P. Butler, and Donald E. Ingber. "Mechanotransduction across the Cell Surface and through the Cytoskeleton." *Science,* May 21, 1993: 1124–1127.

Geoffrey C. Ward. *A First-Class Temperament: The Emergence of Franklin Roosevelt.* New York: Harper and Row, 1989.

M. J. Ward and D. R. Zusman. "Regulation of Directed Motility in *Myxococcus Xanthus.*" *Molecular Microbiology,* June 1997: 885–893.

P. Ward and A. Zahavi. "The Importance of Certain Assemblages of Birds as 'Information-centres' for Food Finding." *Ibis* 115:4 (1973): 517–534.

S. L. Washburn and D. A. Hamburg. "Aggressive Behavior in Old World Monkeys and Apes." In *Primates: Studies in Adaptation and Variability,* edited by Phyllis C. Jay. New York: Holt, Rinehart and Winston, 1968: 458–478.

Andrew Watson. "Inflation Confronts an Open Universe." *Science,* March 6, 1998: 1455.

Paul H. Weaver. *The Suicidal Corporation.* New York: Simon and Schuster, 1988.

B. Weber, E. W. Fall, A. Berger, and H. W. Doerr. "Reduction of Diagnostic Window by New Fourth-Generation Human Immunodeficiency Virus Screening Assays." *Journal of Clinical Microbiology,* August 1998: 2235–2239.

Max Weber. *Charisma and Institution Building.* Chicago: University of Chicago Press, 1968.

R. G. Webster. "Predictions for Future Human Influenza Pandemics." *Journal of Infectious Diseases* (Supplement 1), August 1997: S14–S19.

———. "Influenza Virus: Transmission between Species and Relevance to Emergence of the Next Human Pandemic." *Archives of Virology Supplementum* 13 (1997): 105–113.

Daniel W. Weedman. "Making Sense of Active Galaxies." *Science,* October 16, 1998: 423–424.

Steve Weiner, Qinqi Xu, Paul Goldberg, Jinyi Liu, and Ofer Bar-Yosef. "Evidence for the Use of Fire at Zhoukoudian, China." *Science,* July 10, 1998: 251–253.

R. M. Weinshilboum. "Catecholamine Biochemical Genetics in Human Populations." In *Neurogenetics,* edited by X. O. Breakefield. New York: Elsevier, 1979: 257–282.

J. M. Weiss. "Influence of Psychological Variables on Stress-Induced Pathology." *Ciba Foundation Symposium* 8 (1972): 253–265.

Jay M. Weiss. "Effects of Coping Behavior on Development of Gastrointestinal Lesions in Rats." *Proceedings of the Annual Convention of the American Psychological Association* 2 (1967): 135–136.

———. "Effects of Coping Responses on Stress." *Journal of Comparative and Physiological Psychology* 65:2 (1968): 251–260.

———. "Effects of Predictable and Unpredictable Shock on Development of Gastrointestinal Lesions in Rats." *Proceedings of the Annual Convention of the American Psychological Association* 3 (1968): 263–264.

———. "Effects of Coping Behavior in Different Warning Signal Conditions on Stress Pathology in Rats." *Journal of Comparative and Physiological Psychology,* October 1971: 1–13.

J. Welkowitz, L. Kaufman, and S. Sadd. "Attribution of Psychological Characteristics from Masked and Unmasked Conversations." *Journal of Communication Disorders,* September 1981: 387–397.

E. L. Wells and G. Marwell. *Self-Esteem: Its Conceptualization and Measurement.* Beverly Hills, Calif.: Sage Publications, 1976.

Thomas Welte, David Leitenberg, Bonnie N. Dittel, Basel K. al-Ramadi, Bing Xie, Yue E. Chin, Charles A. Janeway Jr., Alfred L. M. Bothwell, Kim Bottomly, and Xin-Yuan Fu. "STAT5 Interaction with the T Cell Receptor Complex and Stimulation of T Cell Proliferation." *Science,* January 8, 1999: 222–225.

Brant Wenegrat, Eleanor Castillo-Yee, and Lisa Abrams. "Social Norm Compliance as a Signaling System. II. Studies of Fitness-Related Attributions Subsequent on a Group Norm Violation." *Ethology and Sociobiology* 17 (1996): 417–429.

K. R. Wentzel and S. R. Asher. "The Academic Lives of Neglected, Rejected, Popular, and Controversial Children." *Child Development,* June 1995: 754–763.

C. Wehrhahn and D. Rapf. "ON- and OFF-Pathways Form Separate Neural Substrates for Motion Perception: Psychophysical Evidence." *Journal of Neuroscience,* June 1992: 2247–2250.

J. F. Werker and J. E. Pegg. "Infant Speech Perception and Phonological Acquisition." In *Phonological Development: Research, Models and Implications,* edited by C. E. Ferguson, L. Menn, and C. Stoel-Gammon. Parkton, Md.: York Press, 1992.

Janet F. Werker. "Becoming a Native Listener." *American Scientist,* January–February 1989: 54–59.

————. "Exploring Developmental Changes in Cross-Language Speech Perception." In *Language,* edited by L. Gleitman and M. Liberman. *An Invitation to Cognitive Science,* edited by D. Osherson, vol. 1. Cambridge, Mass.: MIT Press, 1995: 87–106.

Janet F. Werker and Renee N. Desjardins. "Listening to Speech in the First Year of Life: Experiential Influences on Phoneme Perception." *Current Directions in Psychological Science,* June 1995: 76–81.

Janet F. Werker and Richard C. Tees. "The Organization and Reorganization of Human Speech Perception." *Annual Review of Neuroscience* 15 (1992): 377–402.

Martin West. "Early Greek Philosophy." In *The Oxford History of the Classical World: Greece and the Hellenistic World,* edited by John Boardman, Jasper Griffin, and Oswyn Murray. New York: Oxford University Press, 1988: 107–117.

M. J. West-Eberhard. "Sexual Selection, Social Competition, and Speciation." *Quarterly Review of Biology* 58 (1983): 155–183.

R. E. Wheeler, R. J. Davidson, and A. J. Tomarken. "Frontal Brain Asymmetry and Emotional Reactivity: A Biological Substrate of Affective Style." *Psychophysiology,* January 1993: 82–89.

William Morton Wheeler. "The Ant Colony as an Organism." *Journal of Morphology* 22 (1911): 307–325.

Randall White. "Beyond Art: Toward an Understanding of the Origins of Material Representation in Europe." *Annual Review of Anthropology* 21 (1992): 537–564.

————. "Technological and Social Dimensions of 'Aurignacian-Age' Body Ornaments across Europe." In *Before Lascaux: The Complex Record of the Early Upper Paleolithic,* edited by H. Knecht, A. Pike-Tay, and R. White. Boca Raton, Fla.: CRC Press, 1993: 247–299.

————. "Substantial Acts: From Materials to Meaning in Upper Paleolithic Representation." In *Beyond Art: Upper Paleolithic Symbolism,* edited by D. Stratmann, M. Conkey, and O. Soffer. San Francisco: California Academy of Sciences, 1996.

A. Whiten, J. Goodall, W. C. McGrew, T. Nishida, V. Reynolds, Y. Sugiyama, C. E. G. Tutin, R. W. Wrangham, and C. Boesch. "Cultures in Chimpanzees." *Nature,* June 17, 1999: 682–685.

Robert Whiting. *Ya Gotta Have Wa: When Two Cultures Collide on the Baseball Diamond.* New York: MacMillan, 1989.

William B. Whitman, David C. Coleman, and William J. Wiebe. "Prokaryotes: The Unseen Majority." *Proceedings of the National Academy of Sciences of the United States of America,* June 9, 1998: 6578–6583.

F. W. Wicker and E. L. Young. "Instance-Based Clusters of Fear and Anxiety: Is Ambiguity an Essential Dimension?" *Perceptual and Motor Skills,* February 1990: 115–121.

S. L. Wiesenfeld. "Sickle-cell Trait in Human Biological and Cultural Evolution: Development of Agriculture Causing Increased Malaria Is Bound to Gene-Pool Changes Causing Malaria Reduction." *Science,* September 8, 1967: 1134–1140.

G. S. Wilkinson. "Reciprocal Food Sharing in the Vampire Bat." *Nature* 308 (1984): 181–184.

————. "The Social Organization of the Common Vampire Bat." *Behavioral Ecology and Sociobiology* 17 (1985): 111–121.

B. J. Willett, K. Adema, N. Heveker, A. Brelot, L. Picard, M. Alizon, J. D. Turner, J. A. Hoxie, S. Peiper, J. C. Neil, and M. J. Hosie. "The Second Extracellular Loop of CXCR4 Determines Its Function as a Receptor for Feline Immunodeficiency Virus." *Journal of Virology,* August 1998: 6475–6481.

Terry Williams. *The Cocaine Kids.* Reading, Mass.: Addison-Wesley, 1989.

C. Wilson and J. W. Szostak. "In Vitro Evolution of a Self-Alkylating Ribozyme." *Nature,* April 27, 1995: 777–782.

D. S. Wilson and E. Sober. "Reintroducing Group Selection to the Human Behavioral Sciences." *Behavioral and Brain Sciences,* December 1994: 585–654.

David Sloan Wilson. "Adaptive Genetic Variation and Human Evolutionary Psychology." *Ethology and Sociobiology* 15 (1994): 219–235.

————. "Reply to Joseph Soltis, Robert Boyd, and Peter J. Richerson: Can Group Functional Behaviors Evolve by Cultural Group Selection? An Empirical Test." *Current Anthropology,* June 1995: 473–494.

————. "Incorporating Group Selection into the Adaptationist Program: A Case Study Involving Human Decision Making." In *Evolutionary Social Psychology,* edited by J. Simpson and D. Kendricks. Mahwah, N.J.: Lawrence Erlbaum, 1997: 345–386.

————. "Human Groups as Units of Selection." *Science,* June 30, 1997: 1816–1817.

E. O. Wilson. "Epigenesis and the Evolution of Social Systems." *Journal of Heredity,* March–April 1981: 70–77.

Edward O. Wilson. *The Insect Societies.* Cambridge, Mass.: Harvard University Press, 1971.

————. *Sociobiology: The Abridged Edition.* Cambridge, Mass.: Harvard University Press, 1980.

G. W. Wilson. "Making an Imprint on Blood-Glucose Monitoring." *Nature Biotechnology,* April 14, 1997: 322.

Margo Wilson and Martin Daly. "The Man Who Mistook His Wife for a Chattel." In *The Adapted Mind: Evolutionary Psychology and the Generation of Culture,* edited by Jerome H. Barkow, Leda Cosmides, and John Tooby. New York: Oxford University Press, 1992: 289–322.

Paul R. Wilson. "Perceptual Distortion of Height as a Function of Ascribed Academic Status." *Journal of Social Psychology,* February 1968: 97–102.

William Van Dusen Wishard. "The Twenty-first Century Economy." In *The 1990s and Beyond,* edited by Edward Cornish. Bethesda, Md.: World Future Society, 1990: 131–136.

Lawrence M. Witmer. "Flying Feathers." *Science,* May 23, 1997: 1209–1210.

Nina Witoszek and Andrew Brennan, eds. *Philosophical Dialogues: Arne Naess and the Progress of Ecophilosophy.* Lanham, Md.: Rowman and Littlefield, 1998.

M. Wobst. "Locational Relationships in Paleolithic Society." *Journal of Human Evolution* 5 (1976): 49–58.

A. J. Wolfe and H. C. Berg. "Migration of Bacteria in Semisolid Agar." *Proceedings of the National Academy of Sciences of the United States of America,* September 1989: 6973–6977.

Jeremy M. Wolfe. "What Can One Million Trials Tell Us about Visual Search?" *Psychological Science,* January 1998: 33–39.

H. Wolfgang, R. Miltner, Christoph Braun, Matthias Arnold, Herbert Witte, and Edward Taub. "Coherence of Gamma-Band EEG Activity as a Basis for Associative Learning." *Nature,* February 4, 1999: 434–436.

J. J. Wolken. "*Euglena:* The Photoreceptor System for Phototaxis." *Journal of Protozoology,* November 1977: 518–522.

D. J. Woodbury. "Pure Lipid Vesicles Can Induce Channel-like Conductances in Planar Bilayers." *Journal of Membrane Biology,* July 1989: 145–150.

World Health Organization. "Division of Emerging and Other Communicable Diseases Surveillance and Control: Partnerships to Meet the Challenge." http://www.who.int/emc/ aboutemc.html. December 1998.

Richard W. Wrangham. "Overview—Ecology, Diversity, and Culture." In *Chimpanzee Cultures,* edited by Richard W. Wrangham, W. C. McGrew, Frans B. M. de Waal, and Paul G. Heltne with assistance from Linda A. Marquardt. Cambridge, Mass.: Harvard University Press, 1994: 21–24.

Richard W. Wrangham, Frans B. M. de Waal, and W. C. McGrew. "The Challenge of Behavioral Diversity." In *Chimpanzee Cultures,* edited by Richard W. Wrangham, W. C. McGrew, Frans B. M. de Waal, and Paul G. Heltne with assistance from Linda A. Marquardt. Cambridge, Mass.: Harvard University Press, 1994: 1–18.

Richard Wrangham and Dale Peterson. *Demonic Males: Apes and the Origins of Human Violence.* Boston: Houghton Mifflin, 1997.

Gregory A. Wray, Jeffrey S. Levinton, and Leo H. Shapiro. "Molecular Evidence for Deep Precambrian Divergences among Metazoan Phyla." *Science,* October 25, 1996: 568–573.

Robert Wright. *Three Scientists and Their Gods: Looking for Meaning in an Age of Information.* New York: Times Books, 1988.

————. *The Moral Animal: Why We Are the Way We Are: The New Science of Evolutionary Psychology.* New York: Vintage Books, 1994.

W. G. Wright. "Evolution of Nonassociative Learning: Behavioral Analysis of a Phylogenetic Lesion." *Neurobiology of Learning and Memory,* May 1998: 326–337.

Bernice Wuethrich. "How Reptiles Took Wing." *Science,* March 7, 1997: 1419–1420.

————. "Will Fossil from Down Under Upend Mammal Evolution?" *Science,* November 21, 1997: 1401–1402.

————. "Geological Analysis Damps Ancient Chinese Fires." *Science,* July 10, 1998: 165–166.

V. C. Wynne-Edwards. *Animal Dispersion in Relation to Social Behaviour.* New York: Hafner, 1962.

————. *Evolution through Group Selection.* Oxford: Blackwell Scientific, 1986.

W. Wyrwicka. "Imitative Behavior: A Theoretical View." *Pavlovian Journal of Biological Sciences,* July–September, 1988: 125–131.

Xenophon. *Spartan Society.* In *Plutarch on Sparta.* Translated by Richard J. A. Talbert. New York: Penguin Books, 1988.

————. *Memorabilia.* The Perseus Project, edited by Gregory R. Crane. http://hydra.perseus. tufts.edu. August 1998.

Shuhai Xiao, Yun Zhang, and Andrew H. Knoll. "Three-dimensional Preservation of Algae and Animal Embryos in a Neoproterozoic Phosphorite." *Nature,* February 4, 1998: 553–558.

K. Yamamoto, A. Soliman, J. Parsons, and O. L. Davies Jr. "Voices in Unison: Stressful Events in the Lives of Children in Six Countries." *Journal of Child Psychology and Psychiatry and Allied Disciplines,* November 1987: 855–864.

Kaoru Yamamoto. "Children's Ratings of the Stressfulness of Experiences." *Developmental Psychology,* September 1979: 580–581.

L. Yamasaki, L. P. Kanda, and P. R. E. Lanford. "Identification of Four Nuclear Transport Signal-Binding Proteins That Interact with Diverse Transport Signals." *Molecular and Cellular Biology,* July 1989: 3028–3036.

K. Yamatani, T. Ono, H. Nishijo, and A. Takaku. "Activity and Distribution of Learning-Related Neurons in Monkey (*Macaca Fuscata*) Prefrontal Cortex." *Behavioral Neuroscience,* August 1990: 503–531.

Michael B. Yarmolinsky. "Programmed Cell Death in Bacterial Populations." *Science,* February 10, 1995: 836–837.

Barbara M. Yarnold. "Steroid Use among Miami's Public School Students, 1992: Alternative Subcultures: Religion and Music versus Peers and the 'Body Cult.'" *Psychological Reports,* February 1998: 19–24.

M. R. Yarrow and C. Zahn-Waxler. "The Emergence and Functions of Prosocial Behaviors in Young

Children." In *Readings in Child Development and Relationships,* edited by R. Smart and M. Smart. New York: MacMillan, 1977.

S. Yasuoka. "Introduction." In *Family Business in the Era of Industrial Growth: Its Ownership and Management,* edited by S. Yasuoka. Tokyo: University of Tokyo Press, 1984.

Shih-Rung Yeh, Russell Fricke, and Donald Edwards. "The Effect of Social Experience on Serotonergic Modulation of the Escape Circuit of Crayfish." *Science,* January 19, 1996: 366–369.

Shih-Rung Yeh, Barbara E. Musolf, and Donald H. Edwards. "Neuronal Adaptations to Changes in the Social Dominance Status of Crayfish." *Journal of Neuroscience,* January 1997: 697–708.

J. C. Yin, M. DelVecchio, H. Zhou, and T. Tully. "CREB as a Memory Modulator: Induced Expression of a DCREB2 Activator Isoform Enhances Long-term Memory in Drosophila." *Cell,* April 7, 1995: 107–115.

J. C. Yin and T. Tully. "CREB and the Formation of Long-Term Memory." *Current Opinion in Neurobiology,* April 6, 1996: 264–268.

S. Stanley Young, Andrew Rusinko, and Dimitri Zaykin. "The Discovery of Multiple, Interacting Genes." Paper delivered at North Carolina State University, October 17, 1997. http://www.stat.ncsu.edu/~dietz/seminars/young.html. Downloaded June 21, 1999.

Thomas J. Young. "Judged Political Extroversion-Introversion and Perceived Competence of U.S. Presidents." *Perceptual and Motor Skills,* October 1996: 578.

Boguslaw Zajac. " 'Death to the Jews and Masons.' " *Rzeczpospolita,* reprinted in *World Press Review,* October 1998: 11–12.

R. Zazzo. "The Twin Condition and the Couple Effects on Personality Development." *Acta Geneticae Medicae et Gemellologiae* 25 (1976): 343–352.

———. "Genesis and Peculiarities of the Personality of Twins." *Progress in Clinical and Biological Research* 24A (1978): 1–11.

Q. K. Zhao. "Intergroup Interactions in Tibetan Macaques at Mt. Emei, China." *American Journal of Physical Anthropology,* December 1997: 459–470.

Richang Zheng and Yijun Qiu. "A Study of the Personality Trend of the Players in the Chinese First-rate Women's Volleyball Teams." *Information on Psychological Sciences* 3 (1984): 22–27.

Hassan Ziady. "An African View of Debt." *Jeune Afrique Economie,* reprinted in *World Press Review,* August 1989: 50.

Marvin Zuckerman. "Good and Bad Humors: Biochemical Bases of Personality and Its Disorders." *Psychological Science,* November 1995: 325–332.

S. Zuckerman. *The Social Life of Monkeys and Apes.* Reprint, with a postscript, London: Routledge and Kegan Paul, 1981.

Charles S. Zuker and Rama Ranganathan. "The Path to Specificity." *Science,* January 29, 1999: 650–651.

H. M. Zullow and M. Seligman. "Pessimistic Rumination Predicts Defeat of Presidential Candidates: 1900–1984." *Psychological Inquiry* I (1990): 52–61.

ACKNOWLEDGMENTS

For providing me with prepublication copies of scholarly papers, unpublished research findings, or simply diving for wayward facts: Lynn Margulis, Eshel Ben-Jacob, Elizabeth Loftus, Amotz Zahavi, Robert M. Seyfarth, James A. Shapiro, Dr. Christopher Boehm, Nancy Segal, David Sloan Wilson, John Marzluff, Bernd Heinrich, Anne B. Clark, William S. Condon, Gilbert Ling, Inon Cohen, John F. Eisenberg, Dorion Sagan, Neil Greenberg, Gordon Burghardt, David Smillie, Don Beck, Robert Huber, Don Edwards, William F. Herrnkind, Charles D. Derby, Stephen T. Hasiotis, Janet Werker, David Berreby, Dr. Charles Marshall, James Vaughn Kohl, Dr. Charles F. Delwiche, Elisabet Sahtouris, Robert Burton, Michael Waller, Bradley Fisk, Anabela Pinto Jensen, Daniel Fessler, A. J. Figueredo, Peter Corning, Herbert Lefcourt, Margaret Carruthers, Kevin Brett, Ben Waggoner, Ron Richardson, Bob Guccione Jr., the late and much-missed Dr. Kerry B. Clark, Nancy Moran, James L. Gould, Ted Galen, Christine Nalepa, Dr. Kenneth Parks, Dr. Peter Frost, William Tillier, Alan F. Dixson, Irving Wolfson, Timothy Perper, Bill Benzon, Koen DePryck, Carol Shelley, Sally P. Power, Steve E. Hartman, Patricia Hausman, Henry Levi, Lorraine Rice, Caleb Rosado, Frank Forman, Louis R. Andrews, Alisdair Daws, Chris McCulloch, Ben Edlund, Kenny Laguna, Karol Kamin, Richard Metzger, Alex Burns, Peter Giblin, Dr. Alexander Elder, Emily Loose, Richard Curtis, Scott Antes, Stephen Power, Gerard Helferich, and Diane Aronson. Despite the gallant efforts of these experts and friends to yank me from missteps, all errors are my own.

INDEX

Abdullah (Saudi crown prince), 199
aborigines, Australian, 87, 110
absolutism. *See* authoritarianism
Academy, Plato's, 175
Accademia del Cimento, Italy, 183
accounting systems, 180
acetylcholine, 145
Achaeans, 123–24
Achilles, 124, 169–70
active galactic nucleus, 223n
adenine, 30
adenosine monophosphate, 22–23
adenosine triphosphate, 15, 186n
Adler, Felix, 79
adolescents
    as conformity enforcers, 85
    in Spartan society, 139–40
Aegean Sea, 165
Afghanistan, 175, 198–99
Africa, 58, 59, 126, 181
afterimages, 74
Agamemnon, 124
Age of Reason, 115, 177
aggression
    group unity in face of, 197
    introvert vs. extrovert, 156
    selective breeding and, 99
    See also violence
agoge (Lycurgan military discipline), 138
agriculture, 117, 121, 209
    in ancient Greece, 125
    drug-producing crops, 212
    genetic engineering and, 211
    landed-aristocracy and development of mass,
        116n
    productivity advances in, 102–3
    and spread of civilization, 105
AIDS/HIV, 115, 190, 212, 213, 216
aircraft, 142, 221–22
Ajivika sect, 158
Albania, 209
Alcibiades, 160, 174
Alcman, 150
Alexander, Lamar, 173
Alexander the Great, 169, 176, 180
algae, 26, 109
Allehadi, Atiqur Rehman, Sultan, 199–200
alleles, 24n
Allen, Woody, 64
Alpha-1-antitrypsin, 212
alphabets, 125, 180
alpha rays, 91
Altmann, Stuart, 53
altruism, 5, 8–10
    reciprocal, 5n, 111n

amoeba, 22, 25, 169
AMP. *See* adenosine monophosphate
amygdala, 62, 146
anaclitic shock, 6
Anatolia. *See* Catal Hüyük
Anaxagoras, 144, 166
Anaximander, 158
Andromache, 124
angular gyrus, 63
animals
    domestication of, 103, 105, 114, 121, 209
    emotions and, 8, 54n
    genetic engineering of, 212
    group hunting by, 60, 132
    hunting of, 60, 102, 103
    inner-judges in, 148
    likeness as attractor in, 154, 196
    motherhood in, 55
    personality variation in, 146
    reciprocity among, 109–10
    as religious imagery, 103–4
    See also humans; mammals; primates; *specific*
        animals
Anomalocaris canadensis, 27
anterior cingulate, 69
antiabortion violence, 202, 203
antibodies, 9, 43, 212
antigay crimes, 202
anti-Semitism, 199, 201
antithrombin III, 212
ants, 94, 205
    collaboration among, 37–38
    conquest and, 118–19
    enemies of, 93
    stored attention and, 169
apes, 6, 52, 59, 61, 87, 98, 167, 170
apocalypticism, 197–98
Apollo, 136
apoptosis, 7–8, 9, 74n, 144, 187n
appearance. *See* physical appearance
Aquino, Corazon, 173
Arabia, 158, 179, 180. *See also* Saudi Arabia
Archelaus, 144
archetypal forms, Platonic, 175
Archimedes, 192
Archytas, 162
Argentina, 50, 172
Argos, 179
aristocracy, landed, 116n
Aristophanes, 182n
Aristotle, 162, 182n
    on dialectic, 152
    heritage of, 177, 221
    as philosophical bridge-builder, 176–77
    on Platonic archetypal forms, 175

Aristotle *(continued)*
  on Spartan system, 140
  on Thales, 126, 128
Army of the Pure, 199–200
Arnold, Matthew, 178
art and sculpture, 103–4, 122, 123
Artavanis-Tsakonas, Spyridon, 213
Artemis, 123n
Aryan Nations movement, 202
Aryan Republic Army, 202
Asahara, Shoko, 198n
asceticism, 158
Asch, Solomon, 73
Asian flu, 214
Aspasia, 166
Assyria, 118, 179, 182
astronomy, 127–28, 158, 182–83
Athens, 178, 220
  Alcibiades and, 160
  cultural perspective of, 2, 135, 141–44,
    149–50, 152–53, 164, 165, 166, 205
  as democracy, 138, 141, 165, 175, 177
  Golden Age of, 164, 165–66, 174
  Pericles and, 160, 165–66, 174
  Plato's philosophical influence in, 174–75
  Pythagoras's philosophical influence in, 162
  Sparta compared with, 136–37, 141, 150, 176,
    205
  Sparta's wars with, 166–67, 174, 177
  Thirty Tyrants' coup in, 174
atoms, 15, 185, 223
ATP. *See* adenosine triphosphate
attention, stored, 169, 170
attention deficit disorder, 8
attention structures, 168–74
attractants, 47
attraction
  as a force, 14, 15, 109, 219
  between organisms, 32, 40, 101, 120, 153–54
  signals of, 17, 38, 41, 47, 109, 148, 169
Augustine, Saint, 160
Aum Shinrikyo, 198
Austen, Jane, 141
Australian aborigines, 87, 110
authoritarianism
  Aristotle's rejection of, 177
  fundamentalist, 192, 197–206, 220
  Platonic, 175–76, 203
  as real threat to democracy, 203
  small-town mind-set of, 205
  Spartan exemplification of, 141, 177, 203,
    205–6
  See also fascism
autonomic nervous system, 62
Autonomous Region of Muslim Mindanao, 198
aviation, 221–22
AZT (drug), 213
Aztecs, 112

baboons, 5–6, 130–34
  hierarchy among, 167, 173n
  hostility toward deformity in, 82

information pooling by, 52–54, 60
male alliances among, 109–10
social isolation impact, 7
Babylonians, 118
Bacillus subtilus, 44
bacteria, 1–2, 25, 34, 120, 190, 194
  chemical gradient detection, 23
  chromosomes, 24, 210
  diversity generators and, 92, 94–95, 96
  genetic adaptation by, 18–19, 29, 44–46
  genetic engineering and, 210–12
  global brain and, 207–9, 210, 216
  nanotechnology and, 222–23
  origins of, 15–22, 219
  as plague cause, 181
  receptors, 167
  reciprocity among, 109
  as social learning machines, 46–48
bacteriocins, 211
Bahrain, 105
Bakker, Robert, 39, 48
Baldwin Effect, 111
Balkans, 191, 196, 202
Baltimore, David, 213
Bantus, 88, 110
Barash, David, 83
Barkashov, Alexander, 201
barter. *See* trade and commerce
Bartlett, John G., 213
bats, 55
B. catarrhalis, 48
Beattie, Melody, 151
beauty, physical, 83–84
bees, 5, 34–37, 119, 169
behavioral memes, 51, 57–59, 61–63
Belgium, 39
Ben-Jacob, Eshel, 16, 17, 18, 20, 26, 44–47, 194,
  222
berdaches, 151
Berreby, David, 151
beta rays, 91
Bible Belt, 204
bickering. *See* creative bickering
biotechnology, 211
birds
  communal information processing by, 12,
    40–41
  flocking behavior, 39–40
  group selection and, 11–12
  imitative learning in, 40–41, 58
  interspecies global mind and, 207
  as social learning machines, 48, 57–58
  species distinguishing in, 78–79
  as virus transporters, 214
Birdsell, J. B., 94
Bischof-Köhler, D., 75
Black Death, 115
Blattabacterium cuenoti, 208
Boccaccio, Giovanni, 116
Boehm, Christopher, 53
Bonaparte, Jose (paleontologist), 50
Bosnia, 196

Bougainville (Solomon Islands), 110
Boxer Rebellion, 181, 182
Brady, Joseph, 10
brain, 32
    Broca's area, 61, 63
    cells in newborn, 130
    cerebellum, 62
    cerebral cortex, 55, 71–72, 157n
    humor and, 87
    introvert functioning of, 157
    neural networks, 9–10, 148, 174, 204, 220
    physiology of, 74
    split-brain patients, 67–68
    synapses, 108
    See also global brain; limbic system; mass mind;
        nervous system
Bray, Dennis, 167
Brett, Kevin, 27
British Royal Society, 183
broadband technology, 2
Broca's area, 61, 63
brontotheres, 50
Brown, Judy E., 212
Bruner, Jerome, 155
Bryson, Jeff B., 156, 157
Buddha, 158
Buddhism, 120
bulls, 103, 104
    Bos taurus, 103, 114
bullying, 84, 85–86
Burkholderia cepacia, 210
Burma, 172
Burnham, Daniel, 129
Burstein, Daniel, 172
Byzantine Empire, 83, 181

Caesar, Julius, 126, 169
Cambrian era, 30–32, 34
Cambrian explosion, 20, 30
cAMP (cyclic AMP), 22–23, 29–30, 169
Cannon, Walter, 4
capitalist development, 174, 192. See also trade
    and commerce
Caporael, Linnda, 74–75
Capp, Al, 87
carbon films, 26
Carchesium, 26n
cardinal adsorbents, 186
Carlyle, Thomas, 170, 185
Carthaginians, 110
Casey, Bob, 188
caste system, 135
Castro, Fidel, 85–86
Catal Hüyük, 113, 121, 122, 141, 143, 178
    social structure of, 103–8
Cayley, George, 221–22
cells
    apoptosis in, 7–8, 9, 74n, 144, 187n
    attraction between, 153–54
    communication among, 22, 27, 28
    development ability by, 22
    division of, 24–25

division of labor by, 15
first primitive (prokaryotes), 15, 16, 21, 23–26
meiosis, 24
in newborn's brain, 130
nucleus, 21, 28, 29
physiological study of, 182, 183–90
proteins, 21, 23, 185, 186, 210, 211
receptors, 167., See also neurotransmitters
T type, 9, 10
See also DNA; genes; multicellularity
Celts. See Gauls
censorship
    by Afghani Taliban, 199
    Platonist, 175
Centers for Disease Control, U.S., 216
Central Asian republics, 200
central nervous system, 31, 32
centrosomes, 186
cerebellum, 62
cerebral cortex, 55, 71–72, 157n
Chance, Michael, 168
character displacement, 93
chariots, 118, 122
Charmides, 175
Chechnya, 198, 200
Cheney, Dorothy, 55
Chesterfield, Lord, 91n
Chibcha, 112
children
    Christian fundamentalist restraints on learning
        by, 203
    conformity's importance to, 75, 76, 81–82,
        84–85, 88–89
    empathy in, 75
    gift giving between, 111
    humiliation and, 88–89
    likeness as attractor in, 155
    mothering of, 55, 75
    personality variation in, 146–49
    sibling rivalry in, 79
    social organizers' effects on, 116–17
    in Spartan society, 138–40
    stigmatization of introverted, 156
    See also adolescents; infants
Child Study Association of America, 79
chimpanzees, 52, 110
    behavioral memes, 58–61
    collective intelligence in, 52, 54
    grooming as medium of exchange, 110
    hierarchy among, 167, 170, 173
    hunting among, 60
    likeness as attractor in, 154, 196
    perceptual linkage among, 75
    social exclusion by, 8, 87
    tool use by, 58, 59, 60
China, 156–57, 170, 173
    ancient empire, 118, 180
    Boxer Rebellion, 181, 182
    conformity's importance in, 86
    contemporary cults in, 194–95
    Indo-European invasions of, 123
    modern technologies of, 195

China (continued)
  Mongol conquest of, 119
  prehistoric, 59, 60
  religious fundamentalists in, 200
  trade and commerce by, 174, 179, 180, 181
chlamydia, 210
chloroplasts, 21
chordates, 23
Chou, 118
Christian fundamentalism, 200, 202–3, 205
Christian Gallery, 203
Christian Reconstructionists, 202
chromosomes, 23–25, 65–66, 91–92, 210
Chukchee, 78, 110
Churchill, Winston, 115n
Cicero, 182n
cilia, 21n, 23, 26
cities
  development of, 101–2, 104–8, 114, 116,
    117, 135, 220
  subcultures in, 143, 152
civility, 204
clams, 26–27, 29, 30
clans, 101, 126
Clark, Kerry B., 27, 31, 34
class, inner-judges and, 148
cockroaches, 34, 208
coined money, 127, 138
Cold Spring Harbor Laboratory, N.Y., 30
collagen, 185
collective intelligence. See intelligence, collective
collective learning, 42, 43, 46–47, 54
collective memory, 38, 208
collective mind, 204–5, 209. See also mass mind
color, perception of, 78
commerce. See trade and commerce
communication
  alphabets and, 125, 180
  in ancient societies, 179–80
  among bacteria, 17, 18–19
  among cells, 22, 28
  diversity generators and, 97–98
  electronic, 179, 196
  global data connection, 1, 3
  information exchange as, 118–20, 218
  mammals and, 51, 55
  writing systems as, 125, 135n
  See also information pooling; language
comparator mechanism, 47
complementary gene sets, 112n
complex adaptive systems, 9–10, 12
  bacterial colonies as, 46–48
  biological role in, 184
  cells as, 186–87
  elements of, 42–44, 56, 83, 91, 144, 165
  hypotheses and, 129, 130, 132, 134, 135, 140,
    141, 218–19
  inner-judges and, 43, 144, 148
  as nested hierarchies, 129
  resource shifters and, 106, 107–8
computers, 1, 195, 196
  neural networks and, 9–10

Condon, William, 75–76
conflict and conquest
  in ancient Greece, 122–24, 135
  as diversity generator, 92–94
  information exchange and, 118–20
  See also violence; warfare
conformity enforcers, 81–91, 153, 165
  for birds, 57
  cellular, 186
  among children, 75, 76, 81–82, 84–85,
    88–89
  cities and, 101–2, 105
  collective learning and, 42, 46
  creative bickering and, 220–21
  cruelty as, 81–83, 84–85, 87, 90
  dangers of, 198–204
  humor as, 87
  in mammals, 51–52, 56, 58
  politicians as, 115
  reality and, 72–77, 193
  social exclusion and, 86–90
  in Spartan society, 137
  threatened groups and, 194
Confucius, 155, 179
Confuciusornis, 40n, 48n
conquest. See conflict and conquest; warfare
consciousness, 69
conspiracy theorists, 203
Constantine, emperor of Rome, 179–80
control, 6, 11, 86, 98, 139, 168, 171, 178, 193,
  201, 203, 204, 205
Coolidge, 181
cooperative states, molecular, 185
Copernicus, 163
corporations, conformity enforcement by, 86
corpus callosum, 67n
Covenant, Sword and Arm of the Lord Christian
  Identity, 202
C. pseudodiphtherticum, 48
crayfish, 32, 33
creative bickering, 93–94, 95, 97, 99, 100, 122,
  184, 220–21
creative webs, 17, 45–46
creativity, 149
Crete, 105, 121, 122, 141, 142
Critias, 138, 174–75
Croesus of Lydia, 127
Cromwell, Oliver, 85, 86
Croton, Italy, 159, 161–62
cruelty, 81–83, 84–85, 87, 90
crustaceans, 30–31, 32–33
Culianu, Ioan Petru, 123n
cults, 194–95, 198
culture
  communication systems as disseminators of,
    180
  emotional expressiveness and, 155
  inner-judges and, 148, 149
  self-control linked with horizons of, 204–5
  social norms and, 115, 194
  See also subcultures; specific aspects and cultures
curiosity, 53

cursus publicus (Roman postal service), 179
cyanobacteria, 15–16, 21, 22, 207, 222
Cyclades, 121, 122
cyclic adenosine monophosphate. *See* cAMP
Cyclops, 124
Cynics, 153
cystic fibrosis, 212
cytomegalovirus, 210
cytosine, 30
cytoskeletons, 22, 186, 187
Czechoslovakia, 97

Dagestan, 200
Damadian, Raymond, 189
Dark Ages, 177
Darnell, Alfred, 203
Darwin, Charles, 8–9, 42, 49, 81, 87, 98, 221n
data connection, global, 1, 3
Davies, Paul, 91
Dawkins, Richard, 30, 32, 62, 219
dCREB2 (gene), 30
death wish. *See* self-destruct mechanism
Decameron (Boccaccio), 116
decision making, 6, 10, 23, 32, 46, 53, 69, 83,
    131, 136, 138, 165, 167, 218
Decline and Fall of the Roman Empire, The (Gibbon),
    179–80
deformity, physical, 82, 83, 87
Deinococcus radiodurans, 212
Delano, Warren, 181
DeLillo, Don, 71
Delos, 159
Delphic oracle, 136
democracy
    Athenian, 138, 141, 165, 175, 177
    fundamentalist authoritarian threats to, 203
    as Pythagorean adversary, 162
democratic socialists, 192
deoxyribonucleic acid. *See* DNA
depression, 6, 7–8
development, cellular, 22
dialectic, 152
Dialogs (Plato), 143
Diamond, Jared, 78–79
dinosaurs, 39, 50, 51
Diodorus, 162, 175
Dionysius II, 175
disease, 190
    AIDS/HIV, 115, 190, 212, 213, 216
    influenza strains, 115, 213–16
    microbial sources of, 209–11
    new drug development and, 211, 212–14
    plagues and epidemics, 115–16, 181, 213–16
    resistance to, 114, 145, 208, 214
    See also immune system
disfigurement, 83
diversity generators, 91–99, 153, 165, 220–21
    in Athenian society, 142–43, 150
    birds and, 57–58
    cellular, 186
    cities as, 102, 104, 105
    collective learning and, 42, 43, 46–47

creative bickering as, 93–94, 95, 97, 99, 100,
    122
fissioning of groups by, 94–96, 99, 100–101,
    108
group differences and, 97–99
human evolution and, 60
innovation and, 96–97, 99
mammals and, 53, 56, 58
See also subcultures
division of labor
    by bees, 35
    cellular, 15
    cities and, 102
DNA (deoxyribonucleic acid)
    bacterial, 18, 210–11
    cell formation and, 15
    maintenance of bodily functions by, 71
    multicellular organisms and, 29, 207–8, 220
    sexuality and, 24
Dolomedes triton, 122n
domestication of animals, 103, 105, 114, 121,
    209
dominance hierarchy, 33, 120, 167–68
dopamine, 145
Dorians, 124, 125, 135, 137, 140, 142
Drews, Robert, 122
Driver, Michael, 156, 157
drugs, development of, 211, 212–14
Dudinov, Andrei, 201
Dugatkin, Lee, 32
Dunbar, Robin, 55
Durant, Will, 160
Durkheim, Emile, 93
Dymanes (Dorian tribe), 137

eagle emblem, 209
Earle, Timothy, 86
Earth Liberation Frontists, 196
echinoderms, 30, 34
E. coli. See Eschericia coli
Economist (British journal), 201–2
economy
    capitalist, 174, 192
    global, 195–96
    Japanese leadership in, 171–74
    See also trade and commerce
ecoterrorists, 196
eco-"warriors," 202
Edelmann, Ludwig, 188
Eder, Donna, 76
Ediacarans, 27
Edser, John, 164
EEA. *See* Environment of Evolutionary
    Adaptedness
Egypt, 105, 158
    contemporary Islamic fundamentalism in, 200
    as mathematical influence on Thales, 127
    Nile-centered empire of, 179
    as Roman colony, 126
Eibl-Eibesfeldt, Irenaus, 81–82
Eisenberg, John F., 50, 51
Ekman, Paul, 75

electronic communication, 179
  via Internet, 1, 196
electrons, 14–15, 91, 185–86
elephants, 54–55
Eliot, T. S., 149
Elm Hollow, N.Y., 192–93, 205
Elohim City, Okla., 202
emblems, 209
embryo, 25
  twin competitiveness in womb, 146–47
emotions
  animals and, 8, 54n
  fear as, 197–98
  introverts vs. extroverts and, 157
  like-mindedness and, 154, 155
  as reality factor, 75
  southern vs. northern males' control of, 204
  See also specific emotions
empathy, 75
endorphins, 144
Energy Department, U.S., 212
England, 60, 116n, 172, 179, 181, 195, 209
Enke, Janet Lynne, 76
enteric nervous system, 68
entrepreneurs, 157
"Environment of Evolutionary Adaptedness"
  (phase), 113
enzymes, 23, 210, 211
Epaminondas, 162
epidemics, 115–16, 181, 213–16
Epstein, Ann, 88–89
Erasmus, 181
Erasmus University (Holland), 215
Erikson, Erik, 98
Eschericia coli (E. coli), 17n, 44, 45, 48, 210, 211
Eskimos, 86, 99
Ethiopia, 105
ethnic cleansing, 196, 202
ethnic liberationists, 192
Euclid, 127, 162
eugenics
  Platonic, 175
  racist, 196, 200, 202
Euglena, 23
eukaryotes, 21–27, 29
Eupatrids, 125
Euripides, 166
Eurypterids, 30–31
Eurytus, 175
evil, 88
evolution, 1–8, 217
  of bacteria, 15–21, 44–45
  Baldwin Effect, 111
  creative bickering's role in, 94, 99, 100
  current theory of, 219
  of eukaryotes, 21–27, 29
  group selection and, 4, 6, 8–13
  individualism and, 34
  individual selection and, 3–8, 12–13
  natural selection and, 4, 94, 112, 209
  neo-Darwinist, 4, 44
exclusion, social, 86–90, 155–56

explicit memes, 49
extinction, overhunting and, 60
extroverts, 145, 147, 156–57, 160
eyeglasses, 182
eyes, 66–67. See also sight
eyespot, cellular, 23
Eysenck, Hans, 156

facial responses, infant, 74–75, 83
family, 7
  conflict, 94
  parent-offspring linkage, 50–51
  sibling rivalry, 79
  See also children; infants; motherhood
Faraday, Michael, 179
farming. See agriculture
fascism, 192, 201–2, 203
fashion cues, 32
Faustian introverts, 198n
  characteristics of, 156
  Pythagoras as, 157–62
  societal contribution of, 205–6
fear, 197–98
  fight-or-flight syndrome, 4–5
Federation of American Societies for Experimental Biology, 187–88
feminism, origin and development of term,
  77–78
Fernald, Russ, 33
fertility, 133–34
fight-or-flight syndrome, 4–5
fire, 59
fish, 37
  conflict among, 93–94
  imitative learning in, 32–33, 170, 220
  interspecies global mind in, 207
  likeness as attractor in, 154
  schooling behavior of, 32
fishing, 96, 122
Fisk, Robert, 199
flagella, 21n, 23
flockbound introverts, 156, 159, 160
flocking behavior, 39–40
Flores (Indonesian island), 60
flying machines, 142
Fonar Corporation, 189
food processing, 211
Forbes (magazine), 172
Forbes family, 181
Ford Motor Company, 86
Forrest, W. G., 136, 139, 140
Fossey, Dian, 167
France, 97, 126, 172, 195
Franklin, Benjamin, 179, 221
Franzoi, Stephen L., 158
Freda, 78
free-market capitalists, 192
Freud, Sigmund, 6, 93, 204
friendship, 7
frontal lobes, 157n
fruit flies, 29, 213
Fujimori, Alberto, 173

fundamentalism, 175, 192, 196, 197–206, 220
fungi, 109

Gabbra, 218
Galen, 80
Galileo, 163, 182–83
Gamaa al-Islamiya (Islamic Group), 200
gamma rays, 91
ganglia, 27, 41
Gates, Bill, 149
Gauls, 94, 117
Gazzaniga, Michael, 68, 69
"geeks," 155–56
gender
    berdaches and, 151
    stereotypes, 77–78
    See also men; sexuality; women
General Motors, 195
genes, 4–8
    altruism and, 5
    bacteria and, 18–19, 44–46, 210–12
    diversity generators and, 91–92, 94, 98
    evolutionary restructuring of, 113–14, 116
    horticultural advances and, 102–3
    memory and, 29–30
    molecular grouping and, 23–24
    mutation by, 44, 45
    as personality variation factor, 146–47
    reciprocity and, 111, 112–13, 122
    reproductive isolation, 122n
    selfish gene concept, 4, 34, 94, 219
    viruses and, 207–8, 210
    See also DNA (deoxyribonucleic acid)
genetic engineering, 18, 210–13
Genghis Khan, 117
gennetai (Athenian group), 142
genocide, 198, 199
genome, 24
Gerhart, John, 187
Germany, 60, 195, 197, 209
    Nazi takeover of, 203
germ cells, 25n
Gerousia (Spartan governmental body), 138
Gibbon, Edward, 179–80
gift giving, 111, 112
Gilligan, Carol, 85
Gilman, A., 97
Gilmour, J. S. L., 64
global brain, 3, 179, 190, 217–23
    interspecies, 207–9, 210, 216
    microbial, 18–19, 20
Global Brain Study Group, 3
global economy, 195–96
glucocorticoids, 144–45
Glynn, I. M., 189
goddesses, 104, 123n
Goethe, Johann Wolfgang von, 207
golden mean, 177, 221
Gorgias, 144, 153
gorillas, 87, 167
gossip, 76–77
Goths, 180

Grant, Michael, 159, 160
gravity, 15, 219
Greece, 77, 105, 152, 182n, 220
    origins of, 122–26
    philosophers of, 174–77
    seafaring traders of, 113, 122, 135, 179
    wars with Persia, 127, 158, 165
    war with Persia, 165
    See also Athens; Crete; Sparta
groups, 121, 221n
    in cities, 108
    fear as unifier of, 197–98
    fissioning of, 94–96, 99, 100–101, 108
    like-mindedness in, 153, 154
    threat response by, 194
    See also intergroup tournaments
group selection, 4, 6, 8–13
Group Selection Squad, 6
Grypania, 21
guanosine, 30
Guatemala, 113
guilt, sense of, 89
guns and artillery, 181
guppies, 32
gymnasium (Athenian club), 142

Hall, Edward T., 76
Hall, K. R. L., 52, 53, 54
Hamblin, Dora Jane, 102
Hamilton, William, 5, 6, 12
Hannibal, 94
Harappan civilization, 105, 115
Harlow, Harry, 6, 7, 82, 154
Harrer, Heinrich, 87
Hattusas, 105
Hawking, Stephen, 3
Hazara Shiites, 199
heart, 26
heat, as violence factor, 95–96
Hecate, 123n
Heilman, Carole, 213
Heinrich, Bernd, 41
Helen of Troy, 135
helium, 14
Hellenism, 177
helots (Spartan social class), 135, 142
helplessness, learned, 6, 10–12
hepatitis B, 212
Hercules, 142
herd instinct, 35, 75–77
Herodotus, 162
herpes virus, 210
Herskovits, Melville, 113
Hesiod, 123–24
Heylighen, Francis, 3, 179
He Zexiu, 195
H5N1 (virus), 215
hierarchy
    collective intelligence and, 35, 36–37
    dominance, 33, 120, 167–68, 173n
    nested, 129
    social, 32–33

Hippias, 144
hippocampus, 63, 146
Hippocrates, 152n
Hitler, Adolf, 115n, 169, 201, 203
Hittites, 118
HIV (human immunodeficiency virus). *See* AIDS/HIV
Hobbes, Thomas, 87
Holland, John, 18n, 42, 167
Homer, 123, 135, 169–70
Homo erectus, 59, 60, 94, 96
Homo habilis, 59, 61, 63
homophobia, 202
Homo sapiens, 60, 96, 110, 126
Hong Kong, 172, 195, 215–16
Hong Kong flu, 214
Hooke, Robert, 183
Hopi, 114n
hormones, 33–34, 156
  prenatal, 160
  stress-related, 43, 144, 146, 147
horses, 123
Horsley, Neal, 191
horticulture, 102–3, 209
human immunodeficiency virus. *See* AIDS/HIV
human rights, 192, 199
humans
  attention's importance to, 168–74
  conformity enforcers in, 72
  hunting by, 60, 96, 102, 103–4, 117, 209
  interconnectivity of, 1, 2, 152, 179, 180, 219
  interspecies global brain and, 208–9
  likeness as attractor in, 153, 154–55, 196
  personality variation in, 145–49
  reality for, 64–67, 69–74, 170–71, 219
  reciprocity among, 110–13, 120, 122
  tools and toolmaking and, 59–61, 96, 104, 106, 138, 179
humiliation, children's feelings of, 88–89
humor, as conformity enforcer, 87
Hungary, 60, 173
Huns, 117
hunting
  by animal groups, 60, 132
  humans and, 60, 96, 102, 103–4, 117, 209
Hyde, Samuel C., 204
Hyksos, 118
Hylleis (Dorian tribe), 137
hypothalamic-pituitary-adrenal axis, 68
hypothalamus, 67, 87
hypotheses, 2, 196, 221n
  complex adaptive systems and, 129, 130, 132, 134, 135, 140, 141, 218–19
  information center theory, 41
  mass mind and, 190, 206, 220
  scientific, 190
  subcultures as, 164–65

Iamblichus, 158–59, 161
Ibn Battuta, 117
Ibn Khaldun, 205

Ice Age, end of, 100, 102–3
ideas, mass mind and, 121
identity, 42, 126, 152
Ienaga, Saburo, 179
ijime (bullying), 85
Iliad (Homer), 123, 169
imitation
  collective intelligence and, 35
  as conformity enforcer, 52
  of higher-ups, 170
imitative learning, 31–32, 35, 40–41, 58, 170, 220
immigration, 195
immune system, 8, 9, 10, 210
  disease resistance and, 114, 145, 208, 214
  inner-judges' impact on, 43, 144, 148
  self-destruct mechanism and, 11, 43
Imperial Canal (China), 180
implicit memes, 49, 54–55, 56
Incas, 180
India, 123, 158, 171, 172, 195
  caste system in, 135
  expansive conquest by, 119–20
  trade activity by, 179, 180, 181
individualism, evolution and, 34
individual selection theory, 3–8, 12–13
Indo-Europeans, 77, 118, 122–24
Indonesia, 195
infants, 99
  brain cell formation in, 130
  facial responses by, 74–75, 83
  gender stereotypes and, 77
  personality variation in, 145–49
  in Spartan society, 138–39
  synchrony in, 76
infectability, molecule receptors and, 167
influence, resource shifters and, 167, 220
influenza, 115, 213–16
information center hypothesis, 41
information exchange, 218
  war and conquest as means of, 118–20
information pooling, 6
  by birds, 12, 40–41
  collective intelligence and, 35, 37, 46
  mammals and, 53–54
inner-judges, 7, 150, 152, 165, 221
  cellular, 186, 187
  collective learning and, 42, 43, 47
  psychobiology and, 144–49
innovation, 96–97, 99, 162, 168
insects, 34, 94, 205
  altruism theory and, 5
  collaboration among, 37–38, 119
  collective intelligence and, 34–37, 169
  conquest and, 118–19
  enemies of, 93
  genetics of, 213
  interspecies global brain and, 208
  memory in, 29
  reproductive isolation and, 122n
  stored attention and, 169
  See also specific insects

instincts, 62, 111, 113
intelligence (IQ), 2, 16
intelligence, collective, 57, 83, 220
    effects of violence on, 206
    information pooling and, 35, 37, 46
    inner-judges and, 148, 150
    intergroup tournaments and, 44
intelligence, networked, 1, 12–13, 209
    bacteria and, 18–19, 45–46
    mammals and, 52, 59
intelligence, superorganismic, 1, 3, 30
interconnectivity, human, 1, 2, 152, 179, 180,
    219
intergroup tournaments, 57, 166, 181, 221
    attention's importance to, 171
    cellular, 187
    cities and, 101, 108
    collective learning and, 42, 44, 47–48
    mass mind shifts and, 165
    science and, 184
    Spartan example of, 176
    warrior-leaders of, 115
internationalists, 141, 205
International Monetary Fund, 195
Internet, 1, 196
intolerance, 83, 192
    Platonist, 175, 177
introverts, 145, 147, 149, 155–61, 220
    Faustian, 156–62, 198n, 205
    flockbound, 156, 159, 160
    social exclusion of, 155–56
invention. See innovation
Ionia, 124, 125, 128
IQ, 2, 16
Iran, 123, 173–74
    fundamentalist morality, 175, 198
iron, 124, 223
Iroquois, 101
Islam, 110–11, 171
    fundamentalists, 175, 196, 198–200
Islamic Group (Gamaa al-Islamiya), 200
Islamic traders, 110–11
isolation, social, 6–8
isolationism, 141
Israel, 60, 105
Italy, 126, 159, 161–62. See also Roman
    Empire

Jainism, 158
Japan, 39, 118
    Aum Shinrikyo cult, 198
    bullying (Ijime), 85
    economic implosion in, 195
    economic leadership by, 171–74
    importance of social acceptance in, 89
Jericho, 101, 102, 105, 114–15, 129, 141
Jesus of Nazareth, 9, 43–44, 81, 130
Jews. See anti-Semitism
Johnson, Allen, 86
Jung, Carl, 145
Justinian, emperor of Byzantine Empire, 181

Kafka, Franz, 149
Kagan, Jerome, 84, 145–46, 147, 149
Karlish, S. J. D., 189
Kauffman, Stuart, 204
Kautilya, 119–20
kidneys, 22
kin selection, 5n
Kirschner, Marc, 187
Kissane, Kelly K., 122n
Ki-Zerbo, Joseph, 172
Knight, Chris, 103
Knossos, Crete, 105, 142
Koch, Robert, 209
Konsler, Reed, 151
Kosovo, 191, 196
Kotzker rebbe, 217
Krupp family, 115
Kummer, Hans, 5–6, 7, 131, 173n
!Kung bushmen, 217–18
Kwakiutl Indians, 83

labor. See division of labor
Lacedaemonia, 135
Laden, Osama Bin. See Osama Bin Laden
LA gene, 114
Lamberg-Karlovsky, C. C., 102
language, 220
    alphabets, 125, 180
    in ancient cultures, 180
    animal, 55
    Indo-European contributions, 118, 122, 123
    as reality factor, 63, 77–80
    syntactic speech, 49, 61
    writing systems, 125, 135n
Lao-tzu, 155, 179
Laurent, Francois, 221n
Laws, The (Plato), 174–75, 203
learned helplessness, 6, 10–12
learning, 30, 83
    collective, 42, 43, 46–47, 54
    imitative, 31–32, 35, 40–41, 58, 170, 220
    secular schooling and, 203–4
    social learning machines, 9, 41–48, 144, 145,
        148
Leibniz, Gottfried Wilhelm, 163
lens development, 182–83
Leonardo da Vinci, 221
Leontius, 67
leptons, 14–15n, 185, 219
Levi-Montalcini, Rita, 149
liberals, 192
libertarianism, 141, 192
life, origins of, 15–16
Lifton, Robert Jay, 198
Li Hongzhi, 194
like-mindedness. See subcultures
likeness, as attractor, 153–55, 196
Lillienthal, Otto, 222
limbic system, 62, 72, 147, 159, 160
Ling, Gilbert, 178–79, 181, 182, 183–90, 220
Lippershey, Hans, 182

Lister, Joseph, 209
Livia, 180
Livingstone, David, 4–5
LLaden, Osama Bin. *See* Osama Bin Laden
locus coeruleus, 67
Loftus, Elizabeth, 72–73
logical positivism, 64
Luria, Z., 77
Lycurgus, 127, 129, 135–40, 150
Lydians, 127, 128
lymphocytes, 145, 148
Lysis, 162

Macedon, 177
MacKay, Charles, 81
magnifying lens, 182
Malaysia, 195
Malphighi, Marcello, 183
mammals, 49–56, 58, 59. *See also* humans
Mardonius, 165
Margulis, Lynn, 21–22, 24, 25, 27
Mario (Roman politician), 126
Marvell, Andrew, 217
Marx, Karl, 121
Marzluff, John and Colleen, 41
masculinity models, 98–99
Maskariputra, Goshala, 158
mass mind, 207, 220, 221
    hypotheses and, 190, 206, 220
    ideas and options of, 121, 206
    intergroup tournaments and, 165
    resource shifters and, 165, 166–74
    See also collective mind
mastery, personal sense of, 145
materials science, 44, 124
Matlin, Mary, 191
matriarchy, 55, 131, 132–33
Maupassant, Guy de, 71
Maurice of Nassau, Prince, 182
Mauryan Empire, 120
Mayer-Kress, Gottfried, 3, 179
McDougall, William, 4
McVeigh, Timothy, 202
Mead, Margaret, 42, 100
measles, 114
Medes, 128
Mein Kampf (Hitler), 201
meiosis, 24
memes, 30–32, 35
    behavioral, 51, 57–59, 61–63
    explicit, 49
    implicit, 49, 54–55, 56
    verbal, 62–63
memory, 62, 63
    collective, 38, 208
    genes and, 29–30
    reality and, 72–73, 146
men
    masculinity models, 98–99
    regional violence differentials in, 204
Menelaus, 135
mental modules. *See* instincts

merchants, 115
Mesoamerica, 112, 113
Mesopotamia, 105, 122
Messenian War, 150
Mexico, 39, 173, 181
microbes, 2, 19, 207–16, 220
    See also bacteria; *other specific types*
Microbial Genome Program, 212
micro-coordination, 75
microscopy, 183
microtubules, 26, 186–87
Middle East. *See specific civilizations and countries*
Miletus (city), 125–27, 128, 142, 178
militia groups, 202–3
Mill, J. S., 81
Milosevic, Slobodan, 202
Milton, John, 86, 163
mind
    collective, 204–5, 209
    conformity enforcer impact on, 203–4
    decision making, 69
    See also global brain; mass mind
Minotaur, 142
Misuari, Nur, 198
MIT Experimental Study Group, 44
mitochondria, 21, 22
Moguls, 171
moieties, 100–101, 107
molecular biology, 211
molecules, 21, 23, 167, 185, 219
mollusks, 26–27, 29, 30, 31
Mongols, 117, 119
monkeys
    cruelty among, 82
    decision making by, 10
    female promiscuity in, 92
    intergroup tournaments and, 171
    language and, 55
    likeness as attractor in, 154, 196
    social isolation effects on, 7
    social rejection effects on, 88
monoclonal antibodies, 212
Montaigne, Michel de, 163
Montgolfier brothers, 221
Moore, J. M., 140
Moore, Richard D., 213
morality, authoritarian, 175, 198–99
Morita, Akio, 172
Moro Islamic Liberation Front, 198
Moss, Cynthia, 54n
motherhood, 50–51, 55, 75
motor strip, 69
multicellularity, 2, 19, 25, 26–28, 208, 219–20
multiple gene effects, 112n
Mundugumor, 75
murder. *See* violence
Murrah Federal Building (Oklahoma City),
    bombing of, 202
Muslims. *See* Islam
mutation, 44, 45
Mycenaeans, 124–25
mysticism, 158

mythology, 97–98, 123n, 124, 128, 142
myxobacteria, 16n, 17, 25, 47–48

nanotechnology, 222–23
Napoleon, 169
National Institutes of Health, 188, 216
Native Americans, 101, 112, 114, 181
natural selection, 4, 94, 112, 209
Navajo, 114n
Nazism, 201, 203, 205
neo-Darwinism, 4, 44
Neolithic
    cities, 101, 102, 121
    religion, 103, 104, 107, 108
    trade, 105, 121
Nero, emperor of Rome, 182n
nervous system, 22, 23, 26, 27, 31, 32
    autonomic, 62
    central, 31, 32
    enteric, 68
networked intelligence. *See* intelligence,
    networked
networking, cellular, 27
networking, computer, 1, 9–10
neural networks, 9–10, 148, 174, 204, 220
neurobiology, 121
neurons, 23, 41, 66–67, 74, 130, 154
neurotransmitters, 33, 145, 146, 147, 168
neutrons, 14, 91
New Caledonia, 110
New Conservatives, 86
New Guinea, 78–79
New York Times, 170
NIH. *See* National Institutes of Health
Nile River, 179
nitrogen, 208, 209
NMR imagers. *See* nuclear magnetic resonance
    imagers
nodes
    complex adaptive systems, 129, 130, 150
    neural network, 2, 9–10, 148, 174, 204
nomads, 117
norepinephrine, 145, 146, 147
normalizers, 84
norms, social, 115, 194
North American Militia, 202
nuclear magnetic resonance imagers, 184–85,
    189
nuclear weapons, 198, 200
nucleoside analogs, 213
nucleotides, 21, 23n, 91
nucleus, cell, 21, 28, 29
Numa Pompilius, 162

obsidian, 104, 105, 106, 107, 121, 122
occipital cortex, 67
octopamine, 33
octopus, 31, 37
Odysseus, 124
Oedipus, 142
Oklahoma City bombing, 202
Olmec civilization, 113

onychophorans, 30
open-mindedness, 205–6
opium trade, 181–82
optic nerve, 66
optics, 182–83
Orwell, George, 81
Osama Bin Laden, 199, 200
ostracism, 89
Ovid, 182n
oxygen, 22
oystercatchers, 57–58

P. aeruginosa, 210
painting. *See* art and sculpture
Pakistan, 105, 199–200
Paleozoic era, 2
Pamphyloi (Dorian tribe), 137
parallel distributed processing systems, 165
parallelism, 165
paranoia, 203
parent-offspring linkage, 50–51
Parmenides, 144
Pasteur, Louis, 209
Patanowä-teri, 101n
Paul, Saint, 179
Pausanias, 162
PCR (polymerase chain reaction), 51
Peloponnesian War, 166–67, 174, 177
Penfield, Wilder, 65
perception, 64–70
    of color, 78
    mass, 174. *See also* mass mind
    reality and, 71–80
    visual, 66–67
Pericles, 160, 165–66, 174
perioeci (Spartan social class), 135, 136
Persia, 118, 176
    wars with Greece, 127, 158, 165
    See also Iran
personal computers, 1
personality, variation in, 145–49
Peru, 173, 181
Peugeot, 172
phages, 18
pheromones, 37, 38, 208
Phidias, 166
Philippines, 172, 173, 181, 195, 198
Philolaus, 162, 175
philosophy, 174–77
    Aristotelian, 176–77
    psychobiological aspects of, 144
    Pythagorean, 159–60
    Pythagorean influences on Platonic, 174–75
    Thales as "father" of, 128, 158
    Zeno and basic domains of, 158
    See also specific philosophers
Phineas Priesthood, 202
Phlious, 162
Phoenicians, 179
photosynthesis, 15, 21
phototaxis, 23
physical appearance, 82–84, 87

Physiological Chemistry and Physics and Medical NMR (journal), 190
Piontelli, Alessandra, 146–47
PKA (enzyme), 23n
plagues, 115–16, 181
plasmids, 18, 211
plasmodial slime molds, 25
plasmodium, 25
Plato, 67, 125
    background of, 174–75
    on Miletus, 126
    political philosophy of, 175–76, 203
    Pythagoras as influence on, 155, 175
    Socrates and, 143, 144
    Sophists and, 152n
Pleistocene era, 98
Pliny the Elder, 182n
pluralism
    Aristotelian, 177
    Athenian, 141–44, 149–50, 164, 165, 166
    open-mindedness and, 205–6
Plutarch, 124, 127, 137, 138, 140, 162, 182n
Poland, 209
Polarized Multilayer Theory of Cell Water, 185–86
political elitism, 158
political philosophy, 174–76
    Platonic/Spartan ideal, 175–76
    See also authoritarianism; democracy; fascism
politicians, 115
polygenes, 12n
polymerase chain reaction, 211
popularity, 145
Porifera, 30
Portugal, 181
postal service, 179, 181
postmodern fundamentalism, 197–98
potassium, 183
pottery, 106, 141
Precambrian era, 14–19
prefrontal cortex, 69
prestige, 170
Priestly, Joseph, 179
priests, 107, 115, 116
Prigogine, Ilya, 28
primates, 6, 52
    hierarchy among, 167–68
    hostile behaviors, 82, 156n
    reciprocity among, 109–10
    See also specific types
problem solving, 10–12, 12, 47
prokaryotes, 15, 16, 21, 23–26
Promise Ministries, 202
Prorus of Cyrene, 162
Protagoras, 144, 152n, 153
protease inhibitors, 213
proteins, 21, 23, 185, 186, 210, 211
protons, 14–15, 91
protozoans, 22, 23, 26, 208
Provenzano, F. J., 77
pseudospeciation, 98
psychobiology, 143–45

psychoneuroimmunology, 6
pulley, invention of, 162
Pumphouse Gang (baboon group), 52, 53–54, 131–34
Puritans, 89
Putin, Vladimir, 200
Puyallup-Nisqually, 110
Pythagoras, 157–64, 166, 168
    downfall of, 161–62
    as Faustian introvert, 157
    as influence on Plato, 155, 175
    philosophy of, 159–60
Pythagoreanism, 153, 159–64
    Archytas's treatises on, 162
    as scientific influence on Aristotle, 177

quarks, 14–15n, 185, 219

Rabi, Isidor, 184
racism, 196, 200–202
Rad (baboon), 131–32
radical constructionism, 64–66, 69, 70
reality, 71–80
    conformity enforcers and, 72–77, 193
    connectivity and, 219
    emotions and, 75
    language's role in, 63, 77–80
    memory's role in, 72–73, 146
    perception and, 64–67, 69–70, 71–80
    as shared hallucination, 2, 71–80, 170–71, 193
receptors, T cell, 9
reciprocal altruism, 5n, 111n
reciprocity, 109–13, 120, 122
recombinant protein therapeutics, 212
reincarnation, 158, 161
Reischauer, Edwin, 89
religion, 103–4, 105, 107, 112, 155, 209
    fundamentalist, 175, 192, 196, 197–206, 220
Renaissance, 115, 163, 177
reproduction, cellular, 25
Republic, The (Plato), 175
resource shifters, 115–16, 184, 221
    cellular, 187
    cities as, 106, 107–8
    collective learning and, 42, 43–44, 47
    mass mind and, 165, 166–74
    mass perception and, 174
retina, 66, 67
retraction (force), 109
retroviruses, 212
Rhodesia, 116n
rhythm, 75–76
Rickover, Hyman, 178
Rockefeller, John D., 149
Roman Empire, 77, 126, 163, 177, 179–80, 182n
Roosevelt, Franklin Delano, 115n, 181
Roslin Institute, 212
Rosnay, Joel de, 3, 179
Rozier, Pilatre de, 221n
Rubin, J. Z., 77
Rudolph, Eric, 202
Ruqaiyah, Abu, 191

Russell, Peter, 3, 179
Russia, 125, 126, 172, 173
    current disintegration of, 194
    Islamic fundamentalists and, 200
    neofascist Russian Unity practices in, 201–2
    Stalinist morality and, 175
Russian National Unity, 201
Rwanda, 191

Sagan, Dorion, 27
salmonella, 44, 48
Salmonella abortusovis, 17n
Salt, Mrs., 193, 220
Samos, 158, 159
San, 130
Santa Fe Institute, 42
Sapolsky, Robert, 7
Sappho, 64
sarin gas, 198
Saudi Arabia, 198, 199, 200
Sauer, Helmut, 26
Savin-Williams, Ritch, 85
Schachter, Stanley, 154
Schanck, Richard, 192–93
Schleiden, Matthias, 183
schooling, secular, 203–4
schooling behavior, 32
Schropp, Reinhard, 111
Schwann, Theodor, 183
science, 86, 182–86, 220
    Aristotelian influence on, 177
    cell physiology, 182, 183–90
    subculture of, 184, 189, 190
    See also specific fields
Science (journal), 212
screw, invention of, 162
sculpture. *See* art and sculpture
seafaring, 60, 122, 124, 125, 135, 179
seaweeds, 26
Seeley, Thomas, 34–35
selective breeding, 98–99
self-control, 204–5
self-destruct mechanism, 6–8, 11, 12, 43
self-interest, 4, 5–6
selfish gene concept, 4, 34, 94, 219
selfishness, 4, 8–9
selflessness. *See* altruism
Seligman, Martin, 6–7
sense-data, 64
senses, 30, 31, 38, 66–67, 78. *See also* perception
Serbia, 196, 202
serial processing systems, 165
serotonin, 33, 145, 168
sewage treatment, 211
sex, diversity generators and, 91–92, 98–99
sexuality, 24, 25, 27, 50
    artistic imagery and, 103–4, 122, 123
    Spartan approach to, 137–38
    war and conquest and, 123, 124
Seyfarth, Robert, 55
Shakespeare, William, 45, 163
shamans, 107, 151

Shapiro, James A., 16, 23, 44, 47
Sheridan, Greg, 200
Sherkat, Darren E., 203
Shihadeh, Edward, 204
shunning, 83, 152
shyness, 146, 147, 149, 161. *See also* introverts
sibling rivalry, 79, 94
    prenatal competition between twins, 146–47
sickle-cell anemia, 114
siege mentality, 197
sight, 23, 30, 31, 66–67, 78
silent trade, 110–11
Singapore, 173, 195
Sioux, 114n
slavery, 125
    in ancient Greece, 135, 136, 138, 165
    in Roman Empire, 180
smallpox, 114, 181
small-town certainties, 205
smell, 30, 31
Smillie, David, 98
Smith, Adam, 168
social class, inner-judges and, 148
social cues, 74–75
social exclusion, 86–90, 155–56
social hierarchy, 32–33
social integrators, 115n, 116, 117
social isolation, 6–8
sociality, 1, 14, 207
    of bacteria, 16–20
    in the big bang, 14
    Ice Age end and, 100
    insects and, 34
    mammalian, 49–50
    multicellularity and, 27
social learning machines, 9, 41–48, 144, 145, 148
social norms, 115, 194
social organizers, 116–17
sociobiology, 34
Socrates, 152, 179
    death of, 175
    influence on Aristotle by, 177
    as magnet for extroverts, 160
    Pericles and, 166, 174
    Plato's *Dialogs* and, 143–44
Socratic method, 143, 152, 164
sodium, cellular, 183, 187–89
sogo shosha (Japanese trading companies), 171
Solon, 127, 129, 135
somatic cells, 25n
Sophists, 152–53
sound, 30
Southeastern Louisiana University, 204
South Korea, 172, 195
Soviet Union (former). *See* Russia; *names of other*
    former republics
Spain, 60, 125, 126, 179, 180, 181
Spanish flu, 213–14
Sparta, 159, 165, 178, 220
    Alcibiades and, 160
    contemporary fundamentalist thought and,
        191, 203, 205–6

Sparta *(continued)*
 cultural perspective of, 2, 135–42, 150, 164, 205
  as Platonist influence, 175–76, 177
  war with Athens, 166–67, 177, 174
spectacles (eyeglasses), 182
speech. *See* language
Speer, Albert, 115n
spiders, aquatic, 122n
spiny lobsters, 2, 32–33, 111, 170, 220
spirochetes, 21, 22, 24, 25
Spitz, Rene, 6, 7
split-brain patients, 67–68
sponges, 30, 153, 196
springtails, 34
Stackhouse, John, 199
Stalin, Joseph, 175
Statesman (Plato), 175
steamships, 181
stereoscopic vision, 23
stereotypes, gender, 77–78
Stokes, Henry Scott, 172
Stone Age, 2, 121, 220
stone tools, 59–61, 96, 104, 179
  Acheulean, 60, 96, 127
  oldowan, 60
stored attention, 169, 170
"stored labor" (innovation), 168
stress hormones, 43, 144, 146, 147. *See also* glucocorticoids
striatum, 62, 157
stromatolites, 15, 16, 17, 46–47, 222
Strum, Shirley, 52, 53–54, 131, 133, 134
subcultures, 178, 191–93, 220
  Athenian, 152
  Athenian vs. Spartan view of, 150
  attention's importance in, 171
  basic rules of, 192–93
  competition among, 108, 121
  emotional expression and, 155
  fundamentalism and, 205
  as hypotheses, 164–65
  infant personality type linked with, 145–49
  Internet communication and, 196
  introverts and, 147, 149, 160, 220
  psychobiology of, 143–45
  Pythagorean pioneering of, 163
  scientific, 184, 189, 190
  See also diversity generators
Suberde (Turkish village), 104
Sudan, 198
suicide, 83, 89
Sui Wendi, emperor of China, 180
Sulla, 126
Sully, James, 84
Sumerians, 135n
Sun-tzu, 127, 213
superior colliculus, 67
superior parietal region, 69
supernormal stimuli, 115n
superorganismic intelligence, 1, 3, 30
supramarginal gyrus, 63

Suzuki, 173
symposion (Athenian drinking club), 142
synapses, 23, 108
synchrony, 76, 169, 208
  emotional, 115n
syntactic speech, 49, 61
Syracuse, 175
Szent-Gyorgi, Albert, 184

Taiwan, 172, 195
Tajikistan, 200
Taliban, 175, 198–99
Tasmanians, 79
T cells, 9, 10
telescope, 182–83
Tellegen, Auke, 95
Tell Mureybit (Syria), 102
Temoshok, Lydia, 6
temporal lobes, 63
Tepe Yahya (Iran), 105
terrorism, 196, 202
Terry, Randall, 191
Teutonic knights, 116n
Thailand, 172, 195
thalamus, 66, 67, 87
Thales, 129, 134, 135, 162n, 220
  background and accomplishments of, 126–28
  as "father" of philosophy, 128, 158
Thayer, Robert E., 4
Thebes, 141–42, 162
Thermus aquaticus, 211
Theseus, 142
Thirty Tyrants coup (Athens), 174
Thomas, Bruno, 172
Thomas, Lewis, 108
Thompson, John N., 113–14
Thrasybulus, 126–27
Thucydides, 136, 158
thymine, 30
Tibet, 87
Timon of Athens, 161
Tinbergen, Niko, 82, 115n
Tlingit, 101
Tokai University, 204
Tomlin, Lily, 71
tools and toolmaking, 58–61, 96, 104, 106, 138, 179
totalitarianism. *See* authoritarianism
totems, 101
touch, 31
Touch the Future Movement, 81
toxicysts, 23
trade and commerce, 117, 129, 179–81, 220
  by Athens, 141, 142, 165
  by Catal Hüyük, 104–6
  by China, 179, 180, 181
  by Japan, 171–74
  restoration by Renaissance, 177
  by Roman Empire, 180
  seafaring and, 124, 125–26
  among tribal peoples, 110–11, 113
transnationality, 155

transportation systems, in ancient cultures, 179–80
transposons, 18
Traube, Moritz, 183
tribes, 126, 136–37, 140, 151, 155. *See also* *specific peoples*
trilobites, 27, 34
Trivers, Robert, 111n
Trobriand Islands, 113
Troeltsch, Ernst, 197
Trojan War, 135
Troshin, A. S., 184
tuberculosis, 115, 181
tubulin, 186
Turchin, Valentin, 3
Turing, Alan, 149
Turkey. *See* Catal Hüyük
Turkmenistan, 200
twins, prenatal competitiveness of, 146–47
tyranny, 165
Tyrtaeus, 135, 150

Ubaidian culture, 105
Uganda, 170
ultranationalists, 192, 201–2, 203
ultraviolet light, 91, 92
United States, 173, 195–96, 209
    China and, 181, 182
    Christian fundamentalist violence in, 202–3, 205
    colonial-era ostracism in, 89–90
    immigration to, 195
    Islamic fundamentalist opposition to, 199–200
    New Conservatives in, 86
    regional variation in male violence in, 204
universe, 14–15, 109
University of Michigan, 204
Upanishads, 158
Ur, 122
urbanization. *See* cities
Utku Eskimos, 86
utopian communities, 136
    Platonic concept, 175
Uzbekistan, 200

vacuum energy density, 223n
vagus nerve, 68
validation, 155
Vardhamana, 158
vectors, 18
vegetarianism, 158, 161
Venezuela, 196
verbal memes, 62–63
violence, 206
    fundamentalist-inspired, 198–203
    heat as factor in, 95–96
    male propensity in southern vs. northern U.S. regions, 204
    See also conflict and conquest; warfare
viruses (microorganisms), 190, 207–8, 210, 213–16
vision. *See* eyes; sight

visual cortex, 66
Vogel, Ezra, 171
volvox, 25.

Waal, Frans de, 111
waggle dance, of bees, 36, 37, 169
Wald, George, 14–15n
Walker, Clive, 59
Waller, Michael, 47
warfare, 2, 99, 101, 114–15, 116, 220
    Boxer Rebellion, 181, 182
    ethnic cleansing and, 196
    Greco-Persian, 127, 158, 165
    among Greek city-states, 166, 166–67, 174, 177
    gun and artillery advances, 181
    Indo-Europeans and, 122–24
    as information exchange, 118–20
    among insects, 37, 118–19
    iron weaponry's impact on, 124
    nuclear weapons and, 198, 200
    Peloponnesian War, 166
    Trojan War, 135
    weapons of mass destruction and, 197
    women as spoils of, 123, 124
    World War I, 197, 204
warrior-leaders, 115, 116
wasps, 93
water. *See* Polarized Multilayer Theory
weapons, 106
    guns and artillery, 181
    for hunting, 104
    iron's introduction, 124
    of mass destruction, 198
    nuclear, 198, 200
Weaver, Paul, 86
weaving, 60–61
Weber, Max, 86
Webster, Robert, 214, 215, 216
weights and measures, 180
Weiss, Jay, 10
Whitehead, Alfred North, 149
Whyte, Lancelot Law, 184
Wilson, David Sloan, 6, 12
Wilson, Edward O., 34, 93, 94, 119, 178
Winthrop, John, 89
witchcraft, 88
Wodehouse, P. G., 164
wolf emblem, 209
wolves, 51–52
women
    adolescent cruelty among, 85
    Afghani suppression of, 198
    artistic portrayals of, 103–4, 122, 123
    gender stereotypes and, 77–78
    Islamic fundamentalist view of, 198
    masculinity models and, 98–99
    as matriarchs, 55, 131, 132–34
    as mothers, 50–51, 55, 75
    pheromones and, 208
    as Pythagoras's followers, 161
    as spoils of war, 123, 124

women *(continued)*
  status in ancient Greece and Rome, 77
  status in Sparta, 137–38
Wordsworth, William, 81
World Health Organization, 215, 216
World War I, 197, 204
World Wide Web, 1
Wright brothers, 222
Wright, Robert, 4
writing systems, 125, 135n
Wynne-Edwards, V. C., 11–12

Xenophon, 135, 137, 138, 139–40, 150
Xerxes, emperor of Persia, 165

Yamamoto, Kaoru, 88
Yanomamo, 94, 98–99, 151
Yaqatlenlis, 170

Zahavi, Amotz, 12, 41
Zeno, 152
Zoothamnium, 26n
Zoroaster, 155

ABOUT THE AUTHOR

Howard Bloom is the founder of the International Paleopsychology Project, executive editor of the New Paradigm book series, and a board member of the Epic of Evolution Society. He's also a member of the New York Academy of Sciences, the National Association for the Advancement of Science, the American Psychological Society, the Human Behavior and Evolution Society, the European Sociobiological Society, and the Academy of Political Science. He has been featured in every edition of *Who's Who in Science and Engineering* since the publication's inception. He is a visiting scholar at New York University.

In 1995 Bloom founded an informal academic circle called "The Group Selection Squad," whose efforts precipitated radical reevaluations of neo-Darwinist dogma within the scientific community. In 1997, he founded a new discipline, paleopsychology, whose participants have included physicists, psychologists, microbiologists, paleontologists, entomologists, neuroscientists, paleoneurologists, invertebrate zoologists, and systems theorists.

The introduction to Bloom's theories, *The Lucifer Principle: A Scientific Expedition into the Forces of History* (Atlantic Monthly Press, 1995), has undergone eleven printings and been the focus of considerable attention in the scientific community. Evolutionary biologist David Sloan Wilson (coauthor of *Unto Others: The Evolution and Psychology of Unselfish Behavior*) said Bloom "raced ahead of the timid scientific herd" with a "grand vision" that "we do strive as individuals, but we are also part of something larger than ourselves, with a complex physiology and mental life that we carry out but only dimly understand." Added Elizabeth F. Loftus, past president of the American Psychological Society: Bloom has achieved "a revolutionary vision of the relationship between psychology and history. *The Lucifer Principle* will have a profound impact on our concepts of human nature. It is astonishing that a book of this importance could be such a pleasure to read."

Howard Bloom lives and works in New York City.